*Abstract Algebra:
A First Course*

TO NORB

*even though
mathematics applied to electrical engineering
is more his style*

Abstract Algebra: A First Course

Violet Hachmeister Larney
STATE UNIVERSITY OF NEW YORK AT ALBANY

PRINDLE, WEBER & SCHMIDT, INCORPORATED
BOSTON, MASSACHUSETTS

© Copyright 1975 by Prindle, Weber & Schmidt, Incorporated
20 Newbury Street, Boston, Massachusetts, 02116

All rights reserved. No part of this book may be reproduced or transmitted in any form or by any means, electronic or mechanical, including photocopying, recording, or any information storage and retrieval system, without permission, in writing, from the publisher.

ISBN 87150–209–7

Printed in the United States of America

Library of Congress Cataloging in Publication Data

Larney, Violet Hachmeister.
  Abstract algebra.

  Includes index.
  1. Algebra, Abstract.   I. Title.
QA162.L37   512'.02   75–28495
ISBN 0–87150–209–7

Preface

This book is designed to be the text for an introductory course in modern algebra and was written with these purposes in mind:

(1) To introduce the student, at an early stage, to topics and techniques that will be needed in other mathematics courses.

(2) To develop mathematical maturity in the student whose most likely experience with college mathematics has been a standard calculus course.

(3) To provide a solid algebraic background that will enable a student to handle confidently either a more advanced undergraduate course, independent study, or a first-year graduate course in algebra.

(4) To give an adequate and interesting survey of basic properties of groups, rings and fields to the student who may not be planning to take additional algebra courses.

Because the text is geared to the needs of the beginning student, it includes an unusually large number (225) of completely worked-out examples. Almost every new definition or theorem is illustrated at the time it is introduced. The problem set at the end of each section gives the reader abundant opportunity to handle "concrete" situations, as well as to prove theorems. Furthermore, the student can get immediate feedback concerning progress, since the back of the book contains answers to almost 400 of the more than 700 problems. [A separate *Answer Book* provides answers to the other 300 problems.] In summary, my aim has been to write a leisurely-paced mathematics book with enough explanations and examples so that a student can easily read and understand it, and so that the instructor can be more than a translator.

The manner of organization of the six chapters makes it possible for the instructor to offer a unified course of varying lengths, depending upon the number of class periods available and the mathematical background of the class. Excluding the Epilogue [which can be read by any class], there are 50 sections. Although many variations are possible, the instructor may find one of the following three alternatives useful as a starting point in planning a course. The topics mentioned under each plan provide a quick survey of the course content, but are by no means a comprehensive listing of items covered in the sections.

Plan I—32 Sections

1.1–1.7 Relations, maps, equivalence classes, factor sets, binary operations.

2.1–2.10 An overview of the algebraic systems that will be studied in more detail in later chapters; elementary properties of groups, rings, integral domains, fields; concept of morphisms and isomorphic algebras.

3.1–3.6 Beginning number theory: divisibility, unique factorization, mathematical induction, ring of integers modulo n.

4.1–4.6 Polynomial rings; factorization of (and roots of) polynomials over the complex, real, and rational fields.

5.1–5.3 Cyclic groups, cosets, Lagrange theorem.

Plan II—44 Sections

Sections in Plan I less time devoted to Chapter 1) plus:

4.7 Algebraic number fields.

5.4–5.8 Normal subgroups, factor groups, direct products, representations of groups, fundamental morphism theorem.

6.1–6.6 Ideals, factor rings, ring morphisms, characteristic, quotient fields, algebraic extension fields.

Plan III—50 Sections

Sections in Plan II (less time spent on Chapters 1, 2) plus:

4.8–4.10 Quadratic fields, quadratic Euclidean domains and quadratic unique factorization domains.

5.9 Conjugacy classes, Sylow p-subgroups.

6.7–6.8 Splitting fields, Galois groups, applications of Galois theory.

A word is in order about the use to be made of the five appendices.

Appendix A. Mathematical Logic, Its Use in Proving Theorems. This brief introduction to symbolic logic can be taken up at the beginning of the course (under Plans I, II, or III), or it can be referred to whenever a question arises as to the rationale behind some method of proof being used.

Appendix B. Permutations: Cycles, Transpositions, Parity. One of the many proofs concerning the parity of a permutation is given. This is optional material for Plan II.

Appendix C. From Peano's Postulates to the Rational Integers. The properties of the natural numbers are developed from the Peano postulates, and the set of natural numbers is then extended to the ordered integral domain of rational integers. This is an optional topic for Plan III.

Appendix D. Reference List of Groups of Order n, for $n \leq 12$. The operation table is given for each noncyclic group of order 12 or less; these tables provide a readily accessible source of examples for class discussion and for use in solving problems.

Appendix E. Thumbnail Sketches of 22 Mathematicians. This material has been included for the purpose of enriching the course, adding human interest, and making the mathematics come more alive. Here we find historical notes on 22 mathematicians whose names appear somewhere in the text in conjunction with a definition or theorem.

If space permitted, I would mention by name each student who has studied algebra from a mimeographed version of this book, and whose suggestions, interest, and words of encouragement kept me going. I am grateful to the editor, John Martindale, for his unfailing optimism and enthusiasm throughout the project, and to the mathematicians who read and reviewed the manuscript. Especially helpful were the comments of Robert L. Wilson, Jr., of Washington and Lee University. I am deeply indebted to my colleague Hugh Gordon for the hours he spent in reading the final version most carefully. With his penchant for detail, he spotted many of the little "bugs" that are easily missed by the casual reader; his assistance is greatly appreciated.

Violet H. Larney
Albany, New York

Contents

Preface   *vii*

1   Relations and Maps

1.1	Prologue	1
1.2	Sets	4
1.3	Relations	11
1.4	Equivalence Classes and Factor Sets	17
1.5	Maps	22
1.6	Composite Maps	28
1.7	Binary Operations	33

2   Abstract Algebras: An Overview

2.1	Key Definitions	38
2.2	Properties Common to All Abstract Algebras	42
2.3	Abstract Algebras with One Operation	46
2.4	Some Basic Properties of Groups	51
2.5	Groups of Permutations	55
2.6	Abstract Algebras with Two Operations	66
2.7	Some Basic Properties of Rings	73
2.8	Ordered Integral Domains	78
2.9	Factorization in Integral Domains	83
2.10	Morphisms and Isomorphic Algebras	87

3   The Rational Integers

| 3.1 | Divisors and the Division Theorem | 95 |
| 3.2 | Greatest Common Divisor, Least Common Multiple | 100 |

3.3	Mathematical Induction	108
3.4	The Fundamental Theorem of Arithmetic	114
3.5	Congruence Modulo n	117
3.6	Factor Ring of Integers Modulo n	123

4 Polynomials

4.1	Polynomial Rings	129
4.2	Division in Polynomial Domains	135
4.3	Unique Factorization of Polynomials Over a Field	140
4.4	Roots of Polynomials	144
4.5	Polynomials Over the Real and Complex Fields	150
4.6	Polynomials Over the Rational Field	153
4.7	Minimal Polynomials and Algebraic Number Fields	158
4.8	Quadratic Fields and Quadratic Domains	168
4.9	Units and Primes in Quadratic Domains	172
4.10	Quadratic Unique Factorization Domains	176

5 Groups

5.1	Integral Powers of Group Elements	183
5.2	Cyclic Groups	188
5.3	Cosets and Lagrange's Theorem	195
5.4	Normal Subgroups and Factor Groups	202
5.5	Group Morphisms	209
5.6	Direct Products	217
5.7	Group Representations	221
5.8	The Kernel and The Fundamental Morphism Theorem	228
5.9	A Sylow Theorem	234

6 Rings and Field Theory

6.1	Subrings and Ideals	240
6.2	Factor Rings	246
6.3	Ring Morphisms	251
6.4	The Characteristic of a Ring	256
6.5	Quotient Fields and Other Ring Extensions	261
6.6	Algebraic Extension Fields	268
6.7	Splitting Field and Galois Group of a Polynomial	276
6.8	Two Applications of Galois Theory	282
6.9	Epilogue	291

Appendices

A		Mathematical Logic, Its Use in Proving Theorems	293
B		Permutations: Cycles, Transpositions, Parity	300
C		From Peano's Postulates to the Rational Integers	305
D		Reference List of Groups of Order n, for $n \leq 12$	312
E		Thumbnail Sketches of 22 Mathematicians	316
	1	Euclid (4th Century B.C.)	316
	2	Diophantus (3d Century A.D.)	316
	3	Cardan (1501–1576)	317
	4	Fermat (1601–1665)	317
	5	Euler (1707–1783)	318
	6	Lagrange (1736–1813)	319
	7	Wilson (1741–1793)	320
	8	Gauss (1777–1855)	320
	9	Cauchy (1789–1857)	321
	10	Abel (1802–1829)	322
	11	Hamilton (1805–1865)	323
	12	Galois (1811–1832)	323
	13	Boole (1815–1864)	325
	14	Cayley (1821–1895)	325
	15	Eisenstein (1823–1852)	326
	16	Kronecker (1823–1891)	326
	17	Dedekind (1831–1916)	327
	18	Klein (1849–1925)	328
	19	Peano (1858–1932)	328
	20	Hardy (1877–1947)	329
	21	Noether (1882–1935)	330
	22	Stark, Harold (1939–)	332

A Short Reading List — 334

Answers — 336

List of Symbols — 358

Index — 361

*Abstract Algebra:
A First Course*

1
Relations and Maps

Mathematicians are like lovers. ... Grant a mathematician the least principle, and he will draw from it a consequence which you must also grant him, and from this consequence another.
—Bernard de Fontenelle (1657–1757)

1.1 Prologue

You are about to enter the fascinating world of abstract algebra. Some mathematicians might prefer to call this book *Modern Algebra*, although several topics are drawn from the realm of *classical algebra*. Can it be both modern and classical? According to one dictionary, *modern* is defined to be "characteristic of recent times or the present." The algebra that you will be learning is certainly modern, in that it is the kind of mathematics that a college student typically studies these days. If by *modern* one means that the algebra was discovered (or invented) just last year, or during the past two or three decades, then a large part of it is *not* modern. Much of the mathematics in this book was known a century or two ago. If we accept the definition of *classical* as "having lasting significance or recognized worth," then we must agree that most topics have withstood the test of time, and so have become a part of classical algebra.

Instead of worrying about the semantics of the situation, let us say that a beginning abstract algebra course usually deals with subject matter that ranges from the classical to the modern, but it gives a modern flavor to the topics drawn from classical algebra. In this book we discuss ideas that were known in the fourth century B.C. as well as results that were achieved as recently as 1966. It is safe to say that more research work in algebra has been published in the past half century than in all the centuries before that. However, you need to have a knowledge of the ideas put forth by the earlier algebraists before you can understand much about present-day developments. In that sense this book is a prelude to algebra.

We couldn't begin to list the original source of each definition and theorem. However, certain postulates, theorems and algebras have become associated, through the years, either with the mathematicians who discovered them, or with those who first published the results. For example, there is the Italian gambler Cardan of the 16th century, who stole and published another man's work; there is the French genius Galois of the early 1800's, who made some great contributions to algebra while still a teenager, and who was killed at the age of twenty in a duel over a woman; there is the brilliant algebraist, Emmy Noether, who was forced by Hitler to flee her native Germany in 1933. In Appendix E you will find a collection of historical sketches of many of the individuals whose names are mentioned in the text. We hope that these brief sketches of the mathematicians behind the mathematics will be of interest, and will give you a somewhat greater appreciation of the mathematics itself.

In previous courses you have had some experience in proving theorems in a "logical manner." In studying abstract algebra you will be using a *two-valued logic*, as usual. In a two-valued logic, a statement p is either true or false, but not both. If p is true, then "not p," the negation of p, is false, and if p is false, then "not p" is true. The truth or falsity of p depends, of course, on the definitions and postulates of the system within which we're working.

Often you will be asked to verify that a statement p *logically implies* a statement q. To do this, you must show that q is true whenever p is true. Our mathematical shorthand for a *logical implication* is $p \Rightarrow q$. This may be expressed in words in any one of the following ways:

(1) p implies q.
(2) If p, then q.
(3) p only if q.
(4) p is sufficient for q.
(5) q is necessary for p.

Statement p is called the *premise*, or *hypothesis*, and q is called the *conclusion*.

We can write the following, for instance:

$$3a - 5 = 0 \Rightarrow 3a = 5.$$

This says that *if* $3a - 5 = 0$, *then* $3a = 5$. Of course, if a is 2, then both the premise and the conclusion are false. However, the *implication* is still true.

To say that two statements, p and q, are *logically equivalent* means that p implies q, and q implies p. Thus, p and q are either both true or both false. Our mathematical shorthand for logical equivalence is $p \Leftrightarrow q$. Other ways of stating this are:

(1) p is equivalent to q.
(2) p if and only if q.
(3) p is necessary and sufficient for q.
(4) p iff q.
(5) $p \Rightarrow q, \quad q \Rightarrow p$.

In the previous example we said that $3a - 5 = 0 \Rightarrow 3a = 5$. It is also true that $3a = 5 \Rightarrow 3a - 5 = 0$. So we can write

$$3a - 5 = 0 \Leftrightarrow 3a = 5.$$

For a more detailed discussion of elementary mathematical logic and its application to proving theorems, we refer you to Appendix A.

A struggling young mathematics student once asked the great mathematician Felix Klein the secret of mathematical discovery. Klein replied: "You must have a problem. Choose one definite objective and drive ahead toward it. You may never reach your goal, but you will find something of interest on the way." And that is how many significant advances in mathematics have been made—as interesting by-products of another goal that a mathematician had set for himself. Sometimes the by-product turns out to be of greater consequence than the original goal.

Perhaps your objective at the moment is simply to get an A in this course. Whether you reach that (rather inconsequential) goal or not, it is hoped that you will discover some interesting, and even exciting, mathematics along the way!

1.1 Problems

1 You are given a statement p and a statement q. Give the most appropriate of these answers:

$$p \Rightarrow q, \qquad q \Rightarrow p, \qquad p \Leftrightarrow q, \qquad \text{none}.$$

(a) p: Polygon $ABCD$ is a rectangle.
q: Polygon $ABCD$ is a quadrilateral.
(b) p: $r^2 \geq 0$
q: r is a real number.
(c) p: Tomorrow is Sunday.
q: Today is Saturday.
(d) p: Triangle RST has three sides of equal length.
q: Triangle RST has three angles of equal size.
(e) p: $a^2 = 4$
q: $a = 2$
(f) p: $a = 3$
q: $a + b = 7$
(g) p: $a + b < c + d$
q: $a < c, \quad b < d$

2 Write the *converse* and the *contrapositive* of each implication. Look these words up in Appendix A if they are not familiar to you. Then indicate the truth or falsity of the converse, and of the contrapositive.

(a) If $x = 3$, then $x^2 = 9$.
(b) $a + b$ is an integer only if a is an integer and b is an integer.
(c) $7 < 2 \Rightarrow 8 < 3$.
(d) $x = y$ is a necessary condition for $2x + y = 12$.
(e) "x is an even integer" is a sufficient condition for "xy is an even integer." (Assume that x and y are integers.)

3 Based upon your *present* understanding of mathematics, explain why each of the following statements is true. (Note: Keep your answers to Prob. 3, 4, and 5, and re-read them after you complete the course.)

(a) $3 \cdot 0 = 0$
(b) $(-3)(-7) = 21$
(c) $\frac{5}{8} \cdot \frac{3}{7} = \frac{15}{56}$
(d) $-(5/8) = (-5)/8 = 5/(-8)$

4 (a) Explain the difference in the two uses of the symbol $-$ in these expressions: $a - b, \quad -b$.
(b) Is it ever possible that $2 + 2 \neq 4$?
(c) In elementary school you may have talked about the "four fundamental operations of arithmetic." Which two operations are really more fundamental, and why?

(d) Is it always true that $(ab)^2 = a^2b^2$?

(e) Comment on the truth or falsity of the statement:
$$(a + b)^2 = a^2 + b^2.$$

(f) Determine which of the following polynomials can be factored.
$$x^2 - 4, \quad x^2 - 3, \quad x^2 + 3$$

Can you criticize the question itself?

5   Mathematicians have been accused of taking common words and giving them uncommon meanings. Each of the nouns in the following list means something special to a person who has studied abstract algebra. Do you recognize any? Take each word that you think you may have used before (mathematically speaking), and describe briefly what it means to you. (You are not really expected to score very high on this question.)

Class	Group	Ideal
Map	Kernel	Field
Operation	Ring	Root
Identity	Unit	Extension

6   Pretend that you are a newspaper reporter, and write a *nonmathematical* cohesive paragraph in which you use each of the twelve words given in the vocabulary list of Problem 5. Underline those 12 words in your paragraph.

7   Study Appendix A, and make a list of the tautologies mentioned that you think you have used, either in previous mathematics courses or in everyday discourse.

1.2 Sets

Since we must start somewhere, we rely on your intuition to understand the meaning of the undefined terms *element* and *set*. Notationally, $a \in S$ means that element a is in set S. A set is also called a *class* or a *collection*. The elements, or objects, in a set can be most anything; for example, they could be real numbers, points, triangles, or dishes. The notation $b \notin S$ means that element b is not in set S. If we write $a, b \in S$, we mean that $a \in S$ and $b \in S$; this statement can also be written $S = \{a, b, \dots\}$.

If $A = \{a_1, a_2, \dots, a_n\}$, then A shall denote a set consisting of n distinct elements. Set A is called a *finite set of order n*. The statement that "the order of set A is n" is abbreviated $|A| = n$. If $|\emptyset| = 0$, then $\emptyset$ is a set containing no elements, and is called the *empty set*, or the *null set*. If $B = \{b_1, b_2, b_3, \dots\}$, then B is called an *infinite set*. The notation means that if k is any positive integer, then $b_k \in B$, and if k and m are two distinct positive integers, then b_k and b_m are two distinct elements of set B.

1.2 Sets

Let $p(x)$ be a statement involving element x. The set U from which x is to be taken must be specified; this set U is the *universal set* under consideration. If $S = \{x : p(x)\}$, then S, by definition, contains each element x that makes $p(x)$ a true statement.

EXAMPLE 1.1

We illustrate the preceding definitions, letting

$$U = \{0, 1, -1, 2, -2, \ldots\}.$$

Examples of sets whose elements are in U are the following:

$A = \{2, -5, 2, 0, 0\} = \{2, -5, 0\} = \{-5, 2, 0\}$. Note that repetitions of an element are eliminated, and that the elements can be listed in any order. We see that $|A| = 3$.

$B = \{2, 4, 6, \ldots\} = \{2, 4, \ldots, 2n, \ldots\} = \{2n : n = 1, 2, 3, \ldots\}$. So B is an infinite set consisting of the positive even integers.

$C = \{x : x \text{ is a divisor of } 6\} = \{1, 2, 3, 6, -1, -2, -3, -6\}$, and $|C| = 8$.

$D = \{x : x^2 \geq 0\} = U$.

$E = \{x : x^2 = -1\} = \emptyset$. □

Note that the symbol □ appears at the end of Example 1.1. This symbol will be found at the end of each numbered example and at the end of the proof of each theorem. So before you begin reading, you can determine the length of an example or of a proof.

We have been using the symbol $=$ of *equality*. Are we clear as to its meaning? By definition, the statement $a = b$ means that a and b are symbols for the same object, whether that object is an element of a set or a set itself. For instance, we can write that $\frac{3}{5} = 0.6$, because $\frac{3}{5}$ and 0.6 are merely symbols for the same rational number. Likewise, if $S = \{3, 4, 7\}$, $T = \{7, 3, 4\}$, and $V = \{3, 7, 4, 3\}$, then $S = T = V$, since each set contains precisely the same three elements: 3, 4, and 7.

With this definition of equality, we can see that it makes sense to postulate the *Substitution Principle*, which says, roughly, that an object may always be substituted for its equal in any statement or expression.

EXAMPLE 1.2

By the Substitution Principle, given the statement $3a - b = b^2$ and the information that $b = 4$, we can conclude that $3a - 4 = b^2$. Observe that it is not necessary to substitute 4 for b in every occurrence of b. Also, it is this Principle that allows one to say that the expression $(3 + 2)^7$ equals the expression $(5)^7$. □

If each element in a set S is also in a set T, we say that S is a *subset* of T. Our shorthand for "S is a subset of T" is $S \subseteq T$, or $T \supseteq S$. Thus, the empty set is a subset of every set. If S is a subset of T, but T contains at least one element that is not in S, then S is called a *proper subset of T*, and we use the notation $S \subset T$, or $T \supset S$. Hence the statement $S \subseteq T$ implies that either $S = T$ or $S \subset T$. Also, $S = T$ if and only if $S \subseteq T$ and $T \subseteq S$. We usually prove that two sets S and T are equal by verifying that $x \in S \Rightarrow x \in T$, and $y \in T \Rightarrow y \in S$.

The collection consisting of all subsets of a given set S is called the *power set* of S, and is denoted as $P(S)$.

EXAMPLE 1.3

If $S = \{1, 2, 3\}$, then S has eight subsets, and

$$P(S) = \{\varnothing, S, \{1, 2\}, \{1, 3\}, \{2, 3\}, \{1\}, \{2\}, \{3\}\}.$$

Note that $\{1\} \neq \{\{1\}\}$, since $1 \neq \{1\}$. Observe that

$$1 \in S, \quad \{1\} \subset S, \quad \{1\} \in P(S), \quad \{\{1\}\} \subset P(S), \quad \{1, 2\} \subset S,$$
$$\{1, 2\} \in P(S), \quad \{\{1, 2\}\} \subset P(S), \quad S \subseteq S, \quad S \in P(S), \quad \varnothing \subset S,$$
$$\varnothing \in P(S), \quad \varnothing \subset P(S). \quad \square$$

The *union*, $A \cup B$, and the *intersection*, $A \cap B$, of sets A and B are defined as follows:

$$A \cup B = \{x : x \in A \text{ or } x \in B\},$$
$$A \cap B = \{x : x \in A \text{ and } x \in B\}.$$

If $A \cap B = \varnothing$, then A and B are said to be *disjoint sets*.

A so-called *Venn diagram*, named after the English logician John Venn, is often used to represent subsets of a universal set. The universal set U is frequently represented as the set of points inside a rectangle. A subset A of U is represented as the set of points inside a circle, or inside some other simple curve that is drawn inside the rectangle.

EXAMPLE 1.4

In the Venn diagram shown in Figure 1–1 we represent subsets A, B, $A \cup B$, and $A \cap B$ of the universal set U.

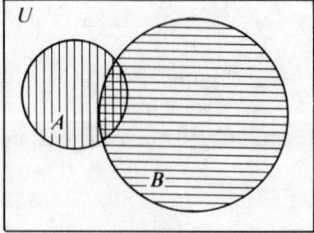

Figure 1–1

The shaded region represents $A \cup B$, while the cross-hatched region represents $A \cap B$. It is apparent that $A \cap B \subseteq A \cup B$. $\square$

The *relative complement*, $A - B$, of B with respect to A, and the *symmetric difference*, $A + B$, of A and B are defined as follows:

$$A - B = \{x : x \in A \text{ and } x \notin B\},$$
$$A + B = (A \cup B) - (A \cap B).$$

EXAMPLE 1.5

Venn diagrams for $A - B$ and $A + B$ are shown in Figure 1-2.

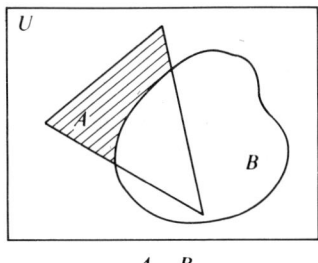
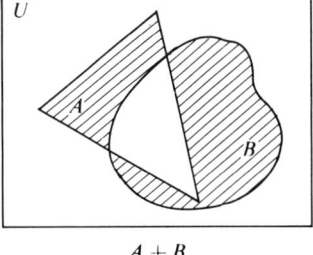

Figure 1-2

In each diagram, A is the region bounded by a triangle, and B is the region bounded by a simple closed curve of irregular shape. □

EXAMPLE 1.6

If $A = \{1, 2, 3, 4, 5\}$ and $B = \{4, 5, 6\}$, then

$$A \cup B = \{1, 2, 3, 4, 5, 6\}, \quad A \cap B = \{4, 5\},$$
$$A - B = \{1, 2, 3\}, \quad B - A = \{6\},$$
$$A + B = B + A = \{1, 2, 3, 6\}. \quad \square$$

EXAMPLE 1.7

Venn diagrams can be drawn to illustrate the fact that

$$A - (B \cup C) = (A - B) \cap (A - C).$$

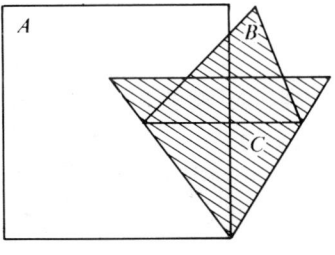
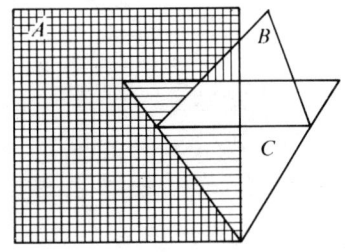

$B \cup C$: ⧗
$A - (B \cup C)$: Unshaded region.

$A - B$: ≡, $A - C$: ||||||
$(A - B) \cap (A - C)$: ▦

Figure 1-3 Unshaded region in left diagram = Cross-hatched region in right diagram.

A picture does not constitute a proof, of course. Using the definitions of this section, and some logic, it is not difficult to verify that

$$x \in A - (B \cup C) \Rightarrow x \in (A - B) \cap (A - C),$$

and $\quad y \in (A - B) \cap (A - C) \Rightarrow y \in A - (B \cup C).$

Therefore, $A - (B \cup C) = (A - B) \cap (A - C)$. $\square$

We can talk about the union and intersection of any number of sets. Suppose there are sets $A_1, A_2, \ldots, A_n$. Letting $I = \{1, 2, \ldots, n\}$, we define and write the *union* and *intersection* of the A_i as follows:

$$A_1 \cup A_2 \cup \cdots \cup A_n = \{x : x \in A_i \text{ for } at \text{ } least \text{ } one \text{ } i \in I\}$$
$$= \bigcup_{i=1}^{n} A_i = \bigcup_{i \in I} A_i.$$
$$A_1 \cap A_2 \cap \cdots \cap A_n = \{x : x \in A_i \text{ for } each \text{ } i \in I\}$$
$$= \bigcap_{i=1}^{n} A_i = \bigcap_{i \in I} A_i.$$

If $I = \{1, 2, 3, \ldots\}$, then we write:

$$\bigcup_{i \in I} A_i = \bigcup_{i=1}^{\infty} A_i = A_1 \cup A_2 \cup \cdots = \{x : x \in A_i \text{ for } at \text{ } least \text{ } one \text{ } i \in I\},$$
$$\bigcap_{i \in I} A_i = \bigcap_{i=1}^{\infty} A_i = A_1 \cap A_2 \cap \cdots = \{x : x \in A_i \text{ for } each \text{ } i \in I\}.$$

Set I is called the *indexing set*.

You have probably worked with subsets of the real numbers known as *intervals*. Recall that an *open interval* is a subset (a, b), and a *closed* interval is a subset $[a, b]$, where $a < b$ and

$$(a, b) = \{x : a < x < b\}, \quad [a, b] = \{x : a \leq x \leq b\}.$$

The interval $[a, b)$ is neither open nor closed, where

$$[a, b) = \{x : a \leq x < b\}.$$

EXAMPLE 1.8

(a) If $A_1 = \{1, 3, 4, 7\}$, $A_2 = \{2, 3, 4\}$, and $A_3 = \{1, 3, 4, 7, 8\}$, then

$$\bigcup_{i=1}^{3} A_i = \{1, 2, 3, 4, 7, 8\} \quad \text{and} \quad \bigcap_{i=1}^{3} A_i = \{3, 4\}.$$

(b) Let $I = \{1, 2, 3, \ldots\}$, and define $B_i = (i - 1, i]$, where $i \in I$. Then

$$\bigcup_{i \in I} B_i = (0, 1] \cup (1, 2] \cup (2, 3] \cup \cdots = \{x : x > 0\} \quad \text{and} \quad \bigcap_{i \in I} B_i = \varnothing. \quad \square$$

1.2 Sets

Certain sets will be used frequently enough to warrant giving them special names. Throughout the text we shall use the following notation:

$\mathbf{Z}^+ = \{1, 2, 3, \ldots\}$ = the set of *natural numbers*.
$\mathbf{Z} = \{0, 1, -1, 2, -2, \ldots\}$ = the set of *rational integers*.
$\mathbf{Q} = \{a/b : a, b \in \mathbf{Z}, b \neq 0\}$ = the set of *rational numbers*.
$\mathbf{Q}^+$ = the set of *positive rational numbers*.
$\mathbf{Q}^* = \mathbf{Q} - \{0\}$ = the set of *nonzero rational numbers*.
$\mathbf{R}$ = the set of *real numbers*.
$\mathbf{R}^+$ = the set of *positive real numbers*.
$\mathbf{R}^* = \mathbf{R} - \{0\}$ = the set of *nonzero real numbers*.
$\mathbf{C} = \{a + bi : a, b \in \mathbf{R}, i^2 = -1\}$ = the set of *complex numbers*.
$\mathbf{C}^* = \mathbf{C} - \{0\}$ = the set of *nonzero complex numbers*.

1.2 Problems

1. Let the universal set be $\mathbf{R}$. Find $A \cup B$, $A \cap B$, $A - B$, $B - A$, and $A + B$, if A and B are the following subsets of $\mathbf{R}$.

$$A = \{x : 0 \leq x < 2\} = [0, 2), \quad B = \{x : x \geq 1\} = [1, \infty).$$

2. Let A and B be the following sets of points in the Cartesian plane:

$$A = \{(x, y) : x^2 + y^2 < 1\}, \quad B = \{(x, y) : y > x^2\}.$$

Sketch the regions of the plane that correspond to:

$$A, \quad B, \quad A \cup B, \quad A \cap B, \quad A - B, \quad B - A, \quad A + B.$$

3. (a) Let $S = \{x, \{x\}, \{y\}\}$. Determine which of the following statements are true.
   - (1) $y \in S$
   - (2) $y \subset S$
   - (3) $\{y\} \in S$
   - (4) $\{y\} \subset S$
   - (5) $\{\{y\}\} \in S$
   - (6) $\{\{y\}\} \subset S$
   - (7) $\{x, y\} \subset S$

   (b) If $S = \{x, y, \{y\}\}$, determine which of statements (1)–(7) of part (a) are true.

4. Given that $\emptyset$ is the empty set, simplify each of the following.
   - (a) $\{\emptyset\} \cap \{\emptyset\}$
   - (b) $\emptyset \cap \{\emptyset\}$
   - (c) $\{\emptyset, \{\emptyset\}\} - \emptyset$
   - (d) $\{\emptyset, \{\emptyset\}\} - \{\emptyset\}$
   - (e) $\{\emptyset, \{\emptyset\}\} - \{\{\emptyset\}\}$
   - (f) $\{\emptyset, \{\emptyset\}\} + \{\{\emptyset\}\}$
   - (g) $\{\emptyset\} \cup \{\{\emptyset\}\}$

5. Give an example of nonempty sets A, B, S, T, and V which satisfy all four of the following statements at the same time.

$$A \subset B, \quad B \in S, \quad S \subset T, \quad \{T\} = V$$

6 In the manner of Example 1.7, draw Venn diagrams to illustrate each of these statements.
(a) $A - (B \cap C) = (A - B) \cup (A - C)$
(b) $A \cap (B \cup C) = (A \cap B) \cup (A \cap C)$
(c) $A \cap (B + C) = (A \cap B) + (A \cap C)$

7 Draw a Venn diagram for subsets A, B, and C of a universal set U in such a way that U is divided into eight nonoverlapping regions. Label each region in terms of sets A, B, C, and U.

8 (a) If $A - B = B - A$, what conclusion can you reach concerning A and B?
(b) What does $A \cup B = A + B$ imply about A and B?

9 Determine which of the following statements are true. Convert each false statement into a true one by changing the statement in some way.
(a) $A \cap B \subseteq A$
(b) $A \subset A \cup B$
(c) $A - B \subset A + B$
(d) $A + B \subseteq A \cup B$
(e) $A + B = A \cup B \Leftrightarrow A \cap B = \emptyset$
(f) $\emptyset \cup A = A$
(g) $\emptyset \cap A = \emptyset$
(h) $\emptyset + A = A$
(i) $\emptyset - A = A$
(j) $A \cup A = A$
(k) $A \cap A = A$
(l) $A + A = A$
(m) $A - A = \emptyset$

10 Prove: If $A \cap B = A$, then $A \cup B = B$, and conversely.

11 Prove: $A \subseteq B \Leftrightarrow A \cup B = B \Leftrightarrow A \cap B = A$

12 (a) If $A = \{1, 2, 3, 4\}$, list all subsets of A.
(b) If $|A| = n$, verify that $|P(A)| = 2^n$.

13 Prove: $\{\{a\}, \{a, b\}\} = \{\{c\}, \{c, d\}\} \Leftrightarrow a = c$ and $b = d$.

14 Let the universal set be $\mathbf{R}$, and define
$$A_i = \{x: -i \leq x \leq i, \text{ with } i \in \mathbf{Z}^+\} = [-i, +i].$$
Determine each of these sets:
(a) A_1, A_2, A_3
(b) $A_3 - A_1$, $A_1 - A_3$, $A_1 + A_3$
(c) $\bigcup_{i=1}^{5} A_i$, $\bigcap_{i=1}^{5} A_i$, $\bigcup_{i=1}^{\infty} A_i$, $\bigcap_{i=1}^{\infty} A_i$

15 Let the universal set be $\mathbf{R}$, and the indexing set be $\mathbf{Q}^+$.
(a) If $A_i = [0, i]$, find: A_3, $A_{3/2}$, $A_{4/5}$, $\bigcup_{i \in \mathbf{Q}^+} A_i$, $\bigcap_{i \in \mathbf{Q}^+} A_i$
(b) If $B_i = (0, i]$, find: $\bigcup_{i \in \mathbf{Q}^+} B_i$, $\bigcap_{i \in \mathbf{Q}^+} B_i$.
(Note: The meaning of $\bigcup_{i \in I} A_i$ and $\bigcap_{i \in I} A_i$ should be apparent, where the indexing set I is any subset of $\mathbf{R}$.)

1.3 Relations

Suppose we are given two sets, S and T. If $s \in S$ and $t \in T$, then an *ordered pair* is an object (s, t) with the property that $(s, t) = (u, v)$ if and only if $s = u$ and $t = v$. The *Cartesian product*, $S \times T$, is defined to be the set of all ordered pairs (s, t), with $s \in S$ and $t \in T$. That is,

$$S \times T = \{(s, t) : s \in S, t \in T\}.$$

The Cartesian product is also called the *direct product*.

EXAMPLE 1.9

If $S = \{1, 2, 3\}$ and $T = \{3, 4\}$, then

$$S \times T = \{(1, 3), (1, 4), (2, 3), (2, 4), (3, 3), (3, 4)\},$$
$$T \times T = \{(3, 3), (4, 4), (3, 4), (4, 3)\}. \quad \square$$

EXAMPLE 1.10

Corresponding to each point in the Cartesian plane is an ordered pair of real numbers, (x, y). Hence the plane is the geometrical representation of the Cartesian product, $\mathbf{R} \times \mathbf{R}$. $\quad \square$

If ρ is any subset of a Cartesian product, $S \times T$, then ρ is called a *relation from S to T*. (Note that we shall be using lower case Greek letters to denote relations.) If ρ is a subset of $S \times S$, then ρ is called a *relation on S*.

EXAMPLE 1.11

Let $S \times T$ be the Cartesian product of Example 1.9. The following are relations from S to T:

$$\rho_1 = \{(1, 3), (3, 4), (2, 3)\}, \qquad \rho_2 = \{(2, 4)\}, \qquad \rho_3 = S \times T.$$

How many relations are there from S to T? By Problem 1.2–12(b), a set containing n elements has 2^n subsets, including the null set as one of these subsets. The order of $S \times T$ is 6, and so there are 2^6, or 64, relations from S to T. $\quad \square$

EXAMPLE 1.12

The following are relations on $\mathbf{R}$:

$$\rho_1 = \{(x, y) : x = y^2\},$$
$$\rho_2 = \{(x, y) : x^2 + y^2 \leq 4\}.$$

The *graph* of a relation on $\mathbf{R}$ is the set of points in the plane that correspond to the ordered pairs in the relation. The graph of ρ_1 is a parabola, and the graph of ρ_2 is a circle and its inside. $\quad \square$

There is a special kind of relation on a set that we shall find to be extremely useful in later chapters. An *equivalence relation* on a set S is a relation ρ on S with these three properties:

(1) *Reflexive property*: $(a, a) \in \rho$, for *all* $a \in S$.
(2) *Symmetric property*: If a and b are any elements in S such that $(a, b) \in \rho$, then $(b, a) \in \rho$.
(3) *Transitive property*: If a, b, and c are any elements in S such that $(a, b) \in \rho$ and $(b, c) \in \rho$, then $(a, c) \in \rho$.

EXAMPLE 1.13

Let us determine all possible equivalence relations on S, if $S = \{1, 2, 3\}$. Since $S \times S$ contains 9 pairs, we know that there are 2^9, or 512 relations on S. Of the 512 relations, only these five are equivalence relations:

$$\rho_1 = \{(1, 1), (2, 2), (3, 3)\},$$
$$\rho_2 = \{(1, 1), (2, 2), (3, 3), (1, 2), (2, 1)\},$$
$$\rho_3 = \{(1, 1), (2, 2), (3, 3), (1, 3), (3, 1)\},$$
$$\rho_4 = \{(1, 1), (2, 2), (3, 3), (2, 3), (3, 2)\},$$
$$\rho_5 = S \times S.$$

Observe that, because of the reflexive property, ρ_1 is a subset of every equivalence relation on S. □

EXAMPLE 1.14

If $\rho = \{(x, y): x, y \in \mathbf{R}, \text{ and } y = x\} = \{(r, r): r \in \mathbf{R}\}$, then ρ is an equivalence relation on $\mathbf{R}$. The graph of ρ is the line that bisects the first and third quadrants. This line must be part of the graph of every equivalence relation on $\mathbf{R}$. □

As an alternative to writing $(a, b) \in \rho$ to express the fact that the pair (a, b) is in the relation ρ, we introduce a somewhat simpler notation. By definition,

$$a \, \rho \, b \Leftrightarrow (a, b) \in \rho.$$

We read $a \, \rho \, b$ as "a relation b." Instead of using the Greek letter ρ to represent an equivalence relation, we often use the tilde, $\sim$. We can read $a \sim b$ as "a is equivalent to b." Using this notation, the definition of an equivalence relation becomes:

An equivalence relation on a set S is a relation $\sim$ on S such that for all $a, b, c \in S$, these properties hold:

(1) $a \sim a$
(2) $a \sim b \Rightarrow b \sim a$
(3) $a \sim b, \; b \sim c \Rightarrow a \sim c$

EXAMPLE 1.15

The following are examples of relations, each of which possesses one or more of the properties of an equivalence relation:

(a) Let S be the set of all triangles, and let $a \sim b$ mean that triangle a is similar to triangle b. It is readily verified that similarity of triangles is an equivalence relation on the set of all triangles.

(b) Let $\leq$ be the relation on **R**. Now if a is any real number, it is true that $a \leq a$, so the reflexive property holds. However, $a \leq b$ does not imply that $b \leq a$, so the symmetric property does not hold. The relation is transitive, for if $a \leq b$ and $b \leq c$, then $a \leq c$. Note that in ordered pair notation, $\leq \, = \{(x, y) : x \leq y\}$.

(c) As a rather nonmathematical example, we see that the relation, "has the same first name as" is an equivalence relation on the set of citizens of the United States.

(d) If the relation on U.S. citizens is: "$a \rho b$" if and only if "a and b have a parent in common," then ρ is reflexive and symmetric, but not transitive. □

We shall have more than one occasion to refer to the equivalence relation properties of ordinary equality, $=$. That $=$ is an equivalence relation on any set follows immediately from the definition given in Section 1.2, namely that $a = b$ if and only if a and b are symbols for the same object. For future reference, we state this result as our first theorem.

THEOREM 1.1

If S is any set, the relation $=$ of equality is an equivalence relation on S.

EXAMPLE 1.16

Every high school mathematics student has used Theorem 1.1 in working algebra and geometry problems. Let's write the three properties as a high school student might:

Statement	*Reason*
1. $a = a$.	1. Identity.
2. If $a = b$, then $b = a$.	2. Two sides of an equation may be interchanged without destroying the equality.
3. If $a = b$ and $b = c$, then $a = c$.	3. Things equal to the same thing are equal to each other. □

A relation ρ on a set S is said to be *antisymmetric* if, for all $a, b \in S$,

$$a \rho b, \quad b \rho a \Rightarrow a = b.$$

A relation ρ on a set S is called a *partial order relation on S* if ρ is

(1) Reflexive.   (2) Antisymmetric.   (3) Transitive.

Set S is called a *partially ordered set* with respect to ρ.

EXAMPLE 1.17

We give two well-known examples of partial order relations.

(a) In Example 1.15(b), we saw that $\leq$ is reflexive and transitive, but not symmetric. However, if $a \leq b$ and $b \leq a$, then $a = b$, so $\leq$ is antisymmetric. Therefore $\leq$ is a partial order relation on the set of real numbers.

(b) Let S be any set, and consider the relation of set inclusion, $\subseteq$, on a collection T of subsets of S. If $A, B, C \in T$, then:

(1) $A \subseteq A$.
(2) $A \subseteq B, B \subseteq A \Rightarrow A = B$.
(3) $A \subseteq B, B \subseteq C \Rightarrow A \subseteq C$.

Therefore, T is a partially ordered set with respect to $\subseteq$. □

A finite set S that is partially ordered by a relation ρ may be represented by an *inclusion diagram*. If a, b, c, d, and e are elements of S, the diagram in Figure 1–4 represents a partial order relation on S.

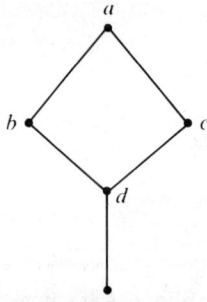

Figure 1–4

The fact that b is below a and is joined to a by a line means that $b \rho a$, and that there is no s, $s \neq a$, $s \neq b$, such that $b \rho s$ and $s \rho a$. That is, there is no element included between a and b. From the diagram sketched, we see that $e \rho d$, $d \rho b$, $d \rho c$, $b \rho a$, and $c \rho a$. From the transitive property, we also conclude that $e \rho b$, $e \rho c$, $e \rho a$, $d \rho a$.

EXAMPLE 1.18

We discuss three partially ordered sets and their inclusion diagrams.

(a) The inclusion diagram of set $\{1, 2, 7\}$ is very simple if $\leq$ is the partial order relation on the set, since $1 \leq 2$, and $2 \leq 7$.

Figure 1–5

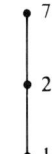

(b) Let $S = \{1, 2, 3, 4, 6, 12\}$. Then "is a divisor of" is a partial order relation on S. The inclusion diagram is shown in Figure 1–6.

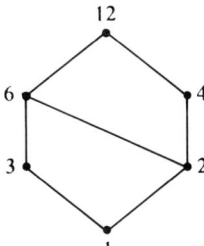

Figure 1–6

The diagram indicates that 1 is a divisor of each number, and that 2 is a divisor of 4, 6, and 12.

(c) Set inclusion, $\subseteq$, is a partial order relation on $P(S)$, where $P(S)$ is the power set of $\{a, b, c\}$.

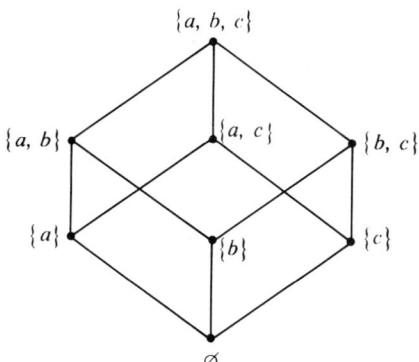

Figure 1–7

In this section we have defined and illustrated two kinds of relations on a set that are of great significance to mathematicians: equivalence relations, and partial order relations. In the next section we discuss the important effect that an equivalence relation has on a set.

1.3 Problems

1. Let $A = \{1, 2, 3, 4\}$, and $B = \{5, 6\}$.
   (a) How many relations exist from A to B?
   (b) List all relations on B. Which of these are equivalence relations on B?
   (c) How many equivalence relations are there on A?

2   A relation may satisfy all, some, or none of the three properties: (1) reflexive, (2) symmetric, (3) transitive. In each case, determine which of these properties are satisfied by the relation ρ, making sure that you justify your answers. (Hint: All eight possible cases are represented, and so no two answers will be the same.)
   (a) $S = \mathbf{Z}$. $x \rho y \Leftrightarrow x + y$ is divisible by 5.
   (b) $S = \mathbf{Z}$. $x \rho y \Leftrightarrow x + y$ is an even integer.
   (c) $S = \mathbf{Z} \cup \{\sqrt{3}\}$. $x \rho y \Leftrightarrow x + y \in \mathbf{Z}$.
   (d) $S = \mathbf{R}$. $x \rho y \Leftrightarrow x < y$.
   (e) $S = \mathbf{Z}$. $x \rho y \Leftrightarrow x$ is a multiple of y, or y is a multiple of x.
   (f) $S = \mathbf{Z}$. $x \rho y \Leftrightarrow x - y < 1$.
   (g) $S = \{x : x \in \mathbf{R}, x > 1\}$. $x \rho y \Leftrightarrow x^2 > y$.
   (h) $S = \mathbf{R}$. $x \rho y \Leftrightarrow x^2 = y$.

3   (a) The graph of any equivalence relation on $\mathbf{R}$ always contains the line $y = x$. Why?
   (b) Verify that ρ is an equivalence relation on $\mathbf{R}$, if $\rho = \{(x, y) : xy = 12, \text{ or } x = y\}$. Graph this relation.

4   Prove: $(A \cap B) \times (C \cap D) = (A \times C) \cap (B \times D)$, where A, B, C, and D are any subsets of a given set S.

5   Find four sets, A, B, C, and D for which
$$(A \cup B) \times (C \cup D) \neq (A \times C) \cup (B \times D).$$

6   Prove: $A \times (B \cup C) = (A \times B) \cup (A \times C)$.

7   An alternate definition of an ordered pair (s, t) is the following:
$$(s, t) = \{\{s\}, \{s, t\}\}.$$
Verify that this definition is equivalent to the definition given in this section.

8   Let $S = \{1, 2, 3\} \cup \{4, 5\} \cup \{6\}$. Referring to these three subsets of S, define relation ρ such that $(a, b) \in \rho$ iff a and b are in the same subset of S. (Thus, $(1, 2) \in \rho$, $(5, 4) \in \rho$, and $(1, 4) \notin \rho$.)
   (a) List all the elements of ρ.
   (b) Verify that ρ is an equivalence relation on S.

9   Let $S = \{1\} \cup \{2, 3, 5\} \cup \{4, 5\}$. Referring to these three subsets of S, define ρ to be a relation such that $(a, b) \in \rho \Leftrightarrow a$ and b are in the same subset of S.
   (a) Give a reason why ρ is *not* an equivalence relation on S.
   (b) List all elements of ρ.

10 If $\subseteq$ is the partial order relation, construct inclusion diagrams of:
   (a) The power set of the set $\{1, 2\}$.
   (b) The power set of the set $\{1, 2, 3, 4\}$.
   (c) The set $\{\{1, 2, 3, 4\}, \{1, 2, 3\}, \{1, 2, 4\}, \{1, 3\}, \{2, 3\}, \{3\}, \{4\}, \emptyset\}$.

1.4 Equivalence Classes and Factor Sets

A collection $\{A_i\}$ of nonempty sets is called a *disjoint collection* if $A_j \cap A_k = \emptyset$, for every two distinct sets, A_j and A_k, in the collection. We also say that the sets are *pairwise disjoint*. A *partition* of a set S is a collection $P = \{A_i\}$ of pairwise disjoint subsets of S such that $S = \cup A_i$, where the union is over all A_i in P. In other words, if S can be expressed as the union of subsets of S, no two of which have any element in common, then that collection of subsets is a partition of S. Note that a partition P is a set of sets.

EXAMPLE 1.19

Let $S = \{a, b, c, d, e, f, g\}$. Then S can also be written:

$$S = \{a, b, c\} \cup \{d, e\} \cup \{f\} \cup \{g\}.$$

The subsets in the union are pairwise disjoint. So if

$$P = \{\{a, b, c\}, \{d, e\}, \{f\}, \{g\}\},$$

then P is a partition of S.

Another partition of S is $\{\{a, c, e, g\}, \{b, d, f\}\}$. At the two extremes, we have the partition $\{S\}$ and the partition

$$\{\{a\}, \{b\}, \{c\}, \{d\}, \{e\}, \{f\}, \{g\}\}. \quad \square$$

EXAMPLE 1.20

We give two partitions of the set of positive real numbers.

(a) $\mathbf{R}^+ = (0, 1] \cup (1, 2] \cup (2, 3] \cup \cdots$. So if $A_i = (i - 1, i]$, then $\{A_1, A_2, A_3, \ldots\}$ is a partition of $\mathbf{R}^+$.
(b) If $B_i =$ the open interval $(i - 1, i)$, then $\{\mathbf{Z}^+, B_1, B_2, B_3, \ldots\}$ is a partition of $\mathbf{R}^+$. $\quad \square$

There is a very interesting and important connection between *partitions* of a set and *equivalence relations* on the set, which we shall now explore in detail. Our plan is to show that each partition P of a set S determines an equivalence relation $\sim$ on S, and conversely, that each equivalence relation $\sim$ on S determines a partition P of S. The first part of our plan is carried out in Theorem 1.2.

THEOREM 1.2

Let $P = \{A_i\}$ be a partition of a set S, and define a relation $\sim$ on S as follows:

$$a \sim b \Leftrightarrow a, b \in A_k, \quad \text{for some } A_k \in P.$$

Then $\sim$ is an equivalence relation on S.

Proof

We shall verify that $\sim$ satisfies the three properties of an equivalence relation.

(1) $a \in S \Rightarrow a \in A_k$ for some k (since P is a partition of S)
$\Rightarrow a, a \in A_k \Rightarrow a \sim a$.
(2) $a \sim b \Rightarrow a, b \in A_j \Rightarrow b, a \in A_j \Rightarrow b \sim a$.
(3) $a \sim b, b \sim c \Rightarrow a, b \in A_k$ for some k and $b, c \in A_t$ for some t
$\Rightarrow A_k = A_t$ (since distinct sets are pairwise disjoint)
$\Rightarrow a, c \in A_k \Rightarrow a \sim c$. $\square$

If you look back at Problem 1.3–8, you will see that you were dealing there with a special case of Theorem 1.2.

EXAMPLE 1.21

Let $S = \{a, b, c, d\}$. We give some partitions of S, and the corresponding equivalence relations.

(a) If $P = \{\{a, b\}, \{c, d\}\}$, then P determines an equivalence relation such that $a \sim b, b \sim a, c \sim d, d \sim c$, and of course, $a \sim a, b \sim b, c \sim c$, and $d \sim d$.

(b) If $P = \{S\}$, then $x \sim y$ for all $x, y \in S$. In terms of ordered pairs, $(x, y) \in \sim$, for all $x, y \in S$, and so the equivalence relation is the Cartesian product, $S \times S$.

(c) If $P = \{\{a\}, \{b\}, \{c\}, \{d\}\}$, then each element is equivalent only to itself, and the equivalence relation is that of ordinary equality, $=$.

An interesting combinatorial problem would be to determine *all* possible partitions of S. Recall that

$$\binom{n}{k} = \frac{n!}{k!(n-k)!},$$

where $\binom{n}{k}$ is the symbol for the number of *combinations* of n objects, taken k at a time. That is, $\binom{n}{k}$ is the number of distinct subsets of order k of a set of order n.

If $\{a, b, c, d\}$ is to be partitioned into three subsets, one subset thus being of order 2, then the number of such partitions is $\binom{4}{2}$, which is 6. The results can be summarized as illustrated in table on opposite page.

According to Theorem 1.2, these 15 distinct partitions of S determine 15 distinct equivalence relations on S.

1.4 Equivalence Classes and Factor Sets

No. of Elements in the Subsets of the Partition	Number of Partitions of this Type
1, 1, 1, 1	1
1, 1, 2	$6 = \binom{4}{2}$
1, 3	$4 = \binom{4}{3}$
2, 2	$3 = \frac{1}{2}\binom{4}{2}$
4	$1 = \binom{4}{4}$
	$\overline{15}$ □

EXAMPLE 1.22

Let S be the set of all triangles in the Euclidean plane. Define $T_{\psi\theta\lambda}$ to be the set of triangles, each of which has positive angles ψ, θ, λ, where $\psi + \theta + \lambda = 180°$, and define

$$P = \{T_{\psi\theta\lambda} : \psi + \theta + \lambda = 180°\}.$$

This partition P of S gives us the equivalence relation: $x \sim y \Leftrightarrow$ triangle x has the same size angles as triangle y. In the geometric sense, this means that x and y are similar triangles. So the partition $\{T_{\psi\theta\lambda}\}$ determines the well-known equivalence relation of similarity of triangles. □

We now turn our attention to the question of how an equivalence relation on a set determines a partition of the set. If $\sim$ is an equivalence relation on a set S, and if $a \in S$, then *equivalence class* $\bar{a}$ is defined to be the following subset of S:

$$\bar{a} = \{x : x \in S \text{ and } x \sim a\}.$$

We must show that the collection $\{\bar{a}\}$ of distinct equivalence classes is a partition of S. To this end we verify the statements in the next theorem.

THEOREM 1.3

Let $\bar{a}, \bar{b}, \ldots,$ be the equivalence classes of a set S with respect to an equivalence relation, $\sim$. Then:

(a) $a \in S \Rightarrow a \in \bar{a}$
(b) $a \in \bar{b} \Leftrightarrow a \sim b \Leftrightarrow \bar{a} = \bar{b}$
(c) Either $\bar{a} \cap \bar{b} = \varnothing$, or $\bar{a} = \bar{b}$

Proof

(a) $a \sim a$, and so it follows from the definition of $\bar{a}$ that $a \in \bar{a}$.
(b) By the definition of an equivalence class, $a \in \bar{b} \Leftrightarrow a \sim b$.

To prove that $a \sim b \Rightarrow \bar{a} = \bar{b}$, we first verify that $\bar{a} \subseteq \bar{b}$. Let x be any element in $\bar{a}$. Then $x \sim a$. But $a \sim b$, by hypothesis. So $x \sim b$, and hence $x \in \bar{b}$. Therefore, $\bar{a} \subseteq \bar{b}$. We leave it to you to show, by a similar argument, that $a \sim b \Rightarrow \bar{b} \subseteq \bar{a}$. Therefore, we conclude that $a \sim b \Rightarrow \bar{a} = \bar{b}$.

Using the fact that $a \in \bar{a}$, we have $\bar{a} = \bar{b} \Rightarrow a \in \bar{b} \Rightarrow a \sim b$.

(c) We must show that any two classes are either disjoint or identical. Suppose that $\bar{a} \cap \bar{b} \neq \varnothing$. Let $c \in \bar{a} \cap \bar{b}$. We then get this string of implications:

$$c \in \bar{a} \cap \bar{b} \Rightarrow c \in \bar{a}, c \in \bar{b} \Rightarrow c \sim a, c \sim b \Rightarrow$$
$$a \sim c, c \sim b \Rightarrow a \sim b \Rightarrow \bar{a} = \bar{b}. \quad \square$$

We have established the fact that any two distinct equivalence classes are disjoint. The next theorem is essentially a corollary to Theorem 1.3.

THEOREM 1.4

If $\bar{a}, \bar{b}, \ldots,$ are the equivalence classes of a set S with respect to an equivalence relation $\sim$, then the collection $P = \{\bar{a}\}$ of all distinct equivalence classes of S is a partition of S.

Proof

Let $\cup \bar{a}$ be the union of all classes of the collection P. Obviously, $\cup \bar{a} \subseteq S$. Also, $a \in S \Rightarrow a \in \bar{a}$, for an $\bar{a} \in P \Rightarrow a \in \cup \bar{a}$. So $S \subseteq \cup \bar{a}$, and therefore, $S = \cup \bar{a}$. Now the elements of P are pairwise disjoint subsets of S. Hence, P is a partition of S. $\quad \square$

Theorems 1.2 and 1.4 together form what is called the *Fundamental Partition Theorem*. In addition to being called a *partition* of S, the set $\{\bar{a}\}$ of equivalence classes with respect to equivalence relation $\sim$ on S is also called the *factor set of S with respect to $\sim$*.

EXAMPLE 1.23

Let $S = \{a, b, c, d, e, f\}$. One factor set, or partition, of S is $P = \{\{a, b, c\}, \{d, e\}, \{f\}\}$. The elements of P are these three equivalence classes:

$$\{a, b, c\} = \bar{a} = \bar{b} = \bar{c}, \quad \{d, e\} = \bar{d} = \bar{e}, \quad \{f\} = \bar{f}.$$

In writing these last equations, we are using Theorem 1.3, which states that $\bar{x} = \bar{y}$ iff $x \sim y$. Thus, each equivalence class has as many names as there are elements in the class. $\quad \square$

EXAMPLE 1.24

We define a relation $\sim$ on $\mathbf{Z}$ as follows:

$$a \sim b \Leftrightarrow a - b = 2k, \text{ for some integer } k.$$

So two integers are equivalent iff their difference is an even integer. That $\sim$ is an equivalence relation is verified as follows:

(1) $a \sim a$, since $a - a = 0 = 2 \cdot 0$.
(2) $a \sim b \Rightarrow a - b = 2k \Rightarrow b - a = 2(-k) \Rightarrow b \sim a$.
(3) $a \sim b, b \sim c \Rightarrow a - b = 2m, b - c = 2n \Rightarrow$
$$a - c = 2(m + n) \Rightarrow a \sim c.$$

We now determine the factor set of $\mathbf{Z}$ with respect to $\sim$.

$$\begin{aligned}\bar{1} = \{x : x \sim 1\} &= \{x : x - 1 = 2k, k \in \mathbf{Z}\} \\ &= \{x : x = 2k + 1, k \in \mathbf{Z}\} \\ &= \{2k + 1 : k \in \mathbf{Z}\} = \text{the set of odd integers.} \\ \bar{2} = \{x : x \sim 2\} &= \{x : x - 2 = 2k, k \in \mathbf{Z}\} \\ &= \{x : x = 2(k + 1), k \in \mathbf{Z}\} \\ &= \{x : x = 2m, m \in \mathbf{Z}\} = \text{the set of even integers.}\end{aligned}$$

Therefore, $\{\bar{1}, \bar{2}\}$ is the factor set of $\mathbf{Z}$ with respect to $\sim$. □

1.4 Problems

1 Let $S = \{a, b, c\}$.
(a) List all possible factor sets of S.
(b) List all possible equivalence relations on S.
(c) Match each factor set of part (a) with the equivalence relation of part (b) that it determines.

2 If you like combinatorial problems, calculate (without actually finding them) the total number of distinct factor sets of S, if $S = \{a, b, c, d, e\}$.

3 Describe four distinct partitions of all college students in the United States.

4 Find the error in the following reasoning: "If a relation $\sim$ on S is symmetric and transitive, then it is automatically reflexive, for $a \sim b \Rightarrow b \sim a$, and hence $a \sim a$."

5 For each set S and each equivalence relation $\sim$, describe the partition, equivalence classes, and give the order of the factor set of S.
(a) S is the set of all polygons. $a \sim b \Leftrightarrow a$ has the same number of sides as b.
(b) $S = \mathbf{Z}^+$. $a \sim b \Leftrightarrow a$ has the same remainder as b when a and b are each divided by 5. (Note: The remainder will be one of these integers: 0, 1, 2, 3, 4.)
(c) $S = \mathbf{Z}$. $\sim = \{(a, b) : a - b \text{ is divisible by } 4\}$.
(d) S is the set of all people in the world. $a \sim b \Leftrightarrow a$ has the same birthday as b.
(e) S is the set of points in the plane that lie inside a circle of radius 1 with center at a point c. $a \sim b \Leftrightarrow |a - c| = |b - c|$, where $|s - t|$ denotes the undirected distance between points s and t.

6 Make a list of the words that have been defined thus far in this chapter, and be sure that you know the meaning of each term defined. Start this vocabulary list on a separate sheet of paper, so that you can add to it as you study each section.

1.5 Maps

The concept of a *map* is one of the most useful in mathematics. Earlier we defined a *relation* from a set S to a set T to be any subset of $S \times T$. A map is a special kind of relation. By definition, a *map from S to T* is a subset μ of $S \times T$ such that, for *each* $s \in S$ there is *exactly one* $t \in T$ for which $(s, t) \in \mu$. Set S is called the *domain* of the map, and set T is called the *codomain*. If $(s, t) \in \mu$, then t is said to be the *image* of s, and s is an *antecedent* of t. A map is also called a *mapping*, a *function*, or a *transformation*. When you studied calculus you probably talked about "functions." An algebraist is inclined to talk about "maps." The two words can be used interchangeably. A map from S to S is called a *map on S*, or a *function on S*. In a calculus course, a map on **R** is called a *function*.

A discussion of the notation currently used to represent maps from S to T is in order. Such a map is denoted as follows:

$$\mu: \quad S \to T$$

If t is the image of s under map μ, we say that *s maps to t under μ*. Each of these four statements says this in symbols:

(1) $(s, t) \in \mu$ (2) $\mu: s \to t$

(3) $t = \mu(s)$ (4) $t = s\mu$.

In (3) we have used *left-hand function notation* (where μ is to the left of s), and in (4) we have used *right-hand function notation* (where μ is to the right of s). From calculus, you are familiar with the left-hand notation. In this course it is convenient to use the right-hand notation.

The map μ from S to T can be represented geometrically as a kind of Venn diagram, as shown in Figure 1-8.

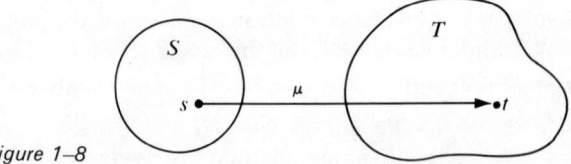

Figure 1-8

EXAMPLE 1.25

Let μ and σ be these two relations on **R**:

$$\mu = \{(x, y): y = x^2\} = \{(x, x^2): x \in \mathbf{R}\},$$
$$\sigma = \{(x, y): x = y^2\} = \{(y^2, y): y \in \mathbf{R}\}.$$

1.5 Maps

Under the map μ, each x in **R** has a *unique* image, x^2, while σ contains, for instance, the ordered pairs $(9, 3)$ and $(9, -3)$. Although the graphs of these relations are both parabolas, μ is a function, while σ is not. ☐

EXAMPLE 1.26

Let $S = \{1, 2, 3, 4\}$, $T = \{a, b, c, d, e\}$. We shall define a map, $\mu: S \to T$, and exhibit, for future reference, five commonly accepted ways that a mathematician might use to represent μ. Each way has the same purpose: to give the image of each element of S.

(1) $\mu = \{(1, a), (2, c), (3, c), (4, e)\}$ (2) $\mu = \begin{pmatrix} 1 & 2 & 3 & 4 \\ a & c & c & e \end{pmatrix}$

(3) $\mu:\ \begin{aligned} 1 &\to a \\ 2 &\to c \\ 3 &\to c \\ 4 &\to e \end{aligned}$ (4) $\begin{aligned} \mu(1) &= a \\ \mu(2) &= c \\ \mu(3) &= c \\ \mu(4) &= e \end{aligned}$ (5) $\begin{aligned} 1\mu &= a \\ 2\mu &= c \\ 3\mu &= c \\ 4\mu &= e \end{aligned}$ ☐

In defining a map from the set S to the set T of Example 1.26, we could have chosen any one of five elements to be the image of 1, any one of five elements to be the image of 2, etc. Thus, there are 5^4, or 625, different maps from a set of order 4 to a set of order 5.

The *range* of a map, $\mu: S \to T$, is a subset $S\mu$ of T that consists of the images of all elements of S under map μ. That is,

$$S\mu = \{t: t \in T, (s, t) \in \mu \text{ for some } s \in S\}.$$

In Example 1.25, $\mathbf{R}\mu$ is the set of nonnegative real numbers. In Example 1.26, $S\mu = \{a, c, e\}$, the elements b and d having no antecedents.

It is apparent that if $\mu: S \to T$ is any map, then either $S\mu \subset T$ or $S\mu = T$. If $S\mu = T$, then μ is said to be a *surjective* map, or an *onto* map, and we say that μ maps S *onto* T. Quite simply, if μ is an onto map, then *every* element of T has an antecedent.

EXAMPLE 1.27

Let $\mu: \mathbf{R} \to \mathbf{R}$ be the map such that $x\mu = 3x + 5$, for all $x \in \mathbf{R}$. Is μ surjective? Let y be any element in **R**. We must find an x such that $x\mu = y$. Solving the equation, $3x + 5 = y$, for x, we get that $x = (y - 5)/3$.

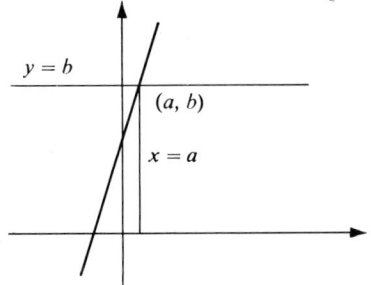

Figure 1–9

This means that the antecedent of y is $(y - 5)/3$, and so μ maps **R** *onto* **R**. That the map is onto can also be seen from its graph, which is a straight line with slope 3. We see this as follows: a horizontal line through any y-intercept b will intersect the graph. The x-coordinate a of this point of intersection is the antecedent of b. □

In a map $\mu: S \to T$, an element t of $S\mu$ may have more than one antecedent. Such is the case in Example 1.26, since 2 and 3 are both antecedents of c. Figure 1–10 represents such a situation.

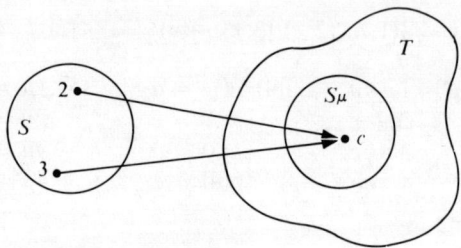

Figure 1–10

If each element of $S\mu$ has *exactly* one antecedent, then the map μ is called an *injective* map, or a *one-to-one* map, abbreviated as 1–1. Hence μ is injective if each element of T has *at most* one antecedent. To show that a map μ is 1–1, we need to verify that if $a, b \in S$ and $a\mu = b\mu$, then $a = b$.

EXAMPLE 1.28

Let $\mu: \mathbf{Z}^+ \to \mathbf{Z}^+$ be such that $n\mu = 2n$. That is,

$$\mu = \begin{pmatrix} 1 & 2 & 3 & 4 & 5 & \cdots & n & \cdots \\ 2 & 4 & 6 & 8 & 10 & & 2n & \end{pmatrix}.$$

Now $a\mu = b\mu \Rightarrow 2a = 2b \Rightarrow a = b$. So μ is 1–1. Each odd positive integer has no antecedent, so μ is not onto. □

EXAMPLE 1.29

Let $\mu: \mathbf{R} \to \mathbf{R}$ be the map defined by: $x\mu = 3 - 2x$. Then $y = 3 - 2x \Rightarrow x = (3 - y)/2$, so μ is surjective. Also,

$$a\mu = b\mu \Rightarrow 3 - 2a = 3 - 2b \Rightarrow -2a = -2b \Rightarrow a = b.$$

Hence, μ is injective. □

A map that is both 1–1 and onto is called a *bijective* map. If $\mu: S \to T$ is bijective, we say that μ is a *one-to-one correspondence* between S and T.

EXAMPLE 1.30

We exhibit three types of maps, given the sets

$$A = \{1, 2, 3\}, B = \{n, p, q, r\}, C = \{x, y, z\}, D = \{4, 5, 6\}.$$

Injective	Surjective	Bijective
$v: A \to B$	$\rho: B \to C$	$\sigma: C \to D$
$1 \to n$	$n \to y$	$x \to 6$
$2 \to q$	$p \to x$	$y \to 4$
$3 \to r$	$q \to x$	$z \to 5$
	$r \to z$	

The definitions of maps can be summarized in this fashion:

$\mu: S \to T$ is a map. $\Leftrightarrow$ Each $s \in S$ has *exactly one* image in T.
Map μ is surjective. $\Leftrightarrow$ Each $t \in T$ has *at least one* antecedent.
Map μ is injective. $\Leftrightarrow$ Each $t \in T$ has *at most one* antecedent.
Map μ is bijective. $\Leftrightarrow$ Each $t \in T$ has *exactly one* antecedent.

EXAMPLE 1.31

Let $\mu: \mathbf{R} \times \mathbf{R} \to \mathbf{R} \times \mathbf{R}$ be defined as follows:

$$(x, y) \to (x + y, x - y) = (x, y)\mu.$$

For example, $(2, 5) \to (7, -3)$ and $(4, -1) \to (3, 5)$.

To determine if μ is onto, we let (a, b) be any given element in the codomain, and then attempt to find at least one element (x, y) in the domain such that $(x, y) \to (a, b)$. We do this as follows:

$$(x, y)\mu = (a, b) \Rightarrow (x + y, x - y) = (a, b)$$
$$\Rightarrow x + y = a, \quad x - y = b$$
$$\Rightarrow x = (a + b)/2, \quad y = (a - b)/2.$$

Since $\left(\dfrac{a + b}{2}, \dfrac{a - b}{2}\right)$ is the *unique* solution of the pair of linear equations, we conclude that (a, b) has *exactly one* antecedent, for each (a, b) in $\mathbf{R} \times \mathbf{R}$. Therefore, μ is both onto and 1–1, and so is bijective. □

EXAMPLE 1.32

Another map on $\mathbf{R} \times \mathbf{R}$ is:

$$v: (x, y) \to (x - y, 3x - 3y).$$

Suppose that (a, b) is in the range. Then

$$(x, y) \to (a, b) \Rightarrow (x - y, 3x - 3y) = (a, b)$$
$$\Rightarrow x - y = a, 3x - 3y = b.$$

When we solve this last pair of equations for x and y, we get that $b = 3a$. Pair (x, y) is an antecedent of $(a, 3a)$ iff $x - y = a$. The range is $\{(a, 3a) : a \in \mathbf{R}\}$, or $\{(x, y) : y = 3x\}$. Geometrically speaking, the entire xy-plane is mapped onto the line whose equation is $y = 3x$. Therefore, map v is certainly not surjective.

From the above we also conclude that v is not 1–1. We note that (x, y) and (s, t) have the same image if and only if $x - y = s - t$. For instance, $(7, 3)v = (10, 6)v = (4, 12)$. □

Given maps $\mu: S \to T$ and $v: S \to T$, our definition of $=$ tells us that $\mu = v$ if and only if $s\mu = sv$ for each $s \in S$. Notice that not only must μ and v consist of the same set of ordered pairs, but also μ and v must have the same codomain.

EXAMPLE 1.33

Let $B = \{1, -1, i, -i\}$, where B is a subset of the set $\mathbf{C}$ of complex numbers. Consider these maps:

$$\mu: \quad B \to \mathbf{C} \qquad v: \quad B \to \mathbf{C} \qquad \rho: \quad B \to B$$
$$x \to x^4 = x\mu \qquad x \to 1 = xv \qquad x \to 1 = x\rho$$

Then $x\mu = xv = x\rho$, for each $x \in B$. Therefore, $\mu = v$. However, $v \ne \rho$, since v and ρ do not have the same codomain. □

A bijective map on a set S is also called a *permutation of S*. The map of Example 1.31 is a permutation of the Cartesian plane, $\mathbf{R} \times \mathbf{R}$. The subject of permutations will be of great interest in later sections.

EXAMPLE 1.34

Let $\lambda: \mathbf{Z} \to \mathbf{Z}$ be the map such that $n\lambda = n + 3$. An antecedent of any integer m is $m - 3$, so λ is onto. Further,

$$n\lambda = k\lambda \Rightarrow n + 3 = k + 3 \Rightarrow n = k,$$

so λ is 1–1. Therefore, λ is a permutation of $\mathbf{Z}$. Permutation λ can be written as follows:

$$\lambda = \begin{pmatrix} 0 & 1 & -1 & 2 & -2 & 3 & -3 & 4 & -4 & & n & \\ 3 & 4 & 2 & 5 & 1 & 6 & 0 & 7 & -1 & \cdots & n+3 & \cdots \end{pmatrix}. \quad □$$

By definition, the *identity map* on a set S is the map ε such that $s\varepsilon = s$, for all s in S. Obviously, ε is bijective, and so it is a permutation of S. We call ε the *identity permutation* of S.

1.5 Maps

EXAMPLE 1.35

Let $S = \{1, 2, 3, 4\}$. Each permutation λ of S can be expressed in the form,

$$\lambda = \begin{pmatrix} 1 & 2 & 3 & 4 \\ 1\lambda & 2\lambda & 3\lambda & 4\lambda \end{pmatrix},$$

where 1λ can be any one of the four elements: 1, 2, 3, or 4. After 1λ has been chosen, one has three choices for the value of 2λ, then two choices for 3λ, and only one choice for 4λ. So there are 4!, or 24, different permutations of S. Three of these permutations are:

$$\mu = \begin{pmatrix} 1 & 2 & 3 & 4 \\ 3 & 4 & 2 & 1 \end{pmatrix}, \quad \eta = \begin{pmatrix} 1 & 2 & 3 & 4 \\ 4 & 1 & 3 & 2 \end{pmatrix}, \quad \varepsilon = \begin{pmatrix} 1 & 2 & 3 & 4 \\ 1 & 2 & 3 & 4 \end{pmatrix},$$

where ε is the identity permutation. □

1.5 Problems

1. Let $A = \{1, 2, 3\}$, and $B = \{4, 5\}$.
   (a) List all maps from A to B, and indicate which of these maps are surjective.
   (b) Find all injective maps from A to B and from B to A.

2. Let $S = \{a, b, c\}$.
   (a) Calculate the number of relations on S, and the number of maps.
   (b) List all permutations of S.

3. (a) If $|S| = n$, calculate the number of: (1) relations on S, (2) maps on S, and (3) permutations of S.
   (b) Give the number of relations, maps, and bijective maps on S if $|S| = 2, 3,$ or 4.

4. If $|S| = n$, can a map $\mu: S \to S$ be onto and not 1–1, or 1–1 and not onto? Explain.

5. Give an example of a map on $\mathbf{Z}$ that is:
   (a) 1–1 but not onto.   (b) Onto but not 1–1.
   (c) Neither 1–1 nor onto.   (d) Bijective.

6. Give a brief discussion which clarifies these statements.
   (a) An ordinary road map of the United States is the codomain of a bijective map.
   (b) An example of a real function is f, if $f(x) = \sin x$. However, "$\sin x$" is *not* the function.

7. (a) Verify that μ is a permutation of $\mathbf{R}$, if $a, b \in \mathbf{R}$, $a \neq 0$, and

   $$\mu = \{(x, y): y = ax + b, x \in \mathbf{R}\}.$$

   (b) What is the graph of the identity permutation of $\mathbf{R}$?

8   In each case μ is a real function. Determine whether μ is injective and/or surjective. Give the range, $\mathbf{R}\mu$.
   (a) $x\mu = 3x + 4$
   (b) $x\mu = x^2 + 1$
   (c) $x\mu = 2^x$
   (d) $x\mu = \begin{cases} x + 1 & \text{if } x \text{ is rational.} \\ 3x & \text{if } x \text{ is irrational.} \end{cases}$
   (e) $x\mu = \begin{cases} 1/x & \text{if } x \neq 0 \\ 0 & \text{if } x = 0 \end{cases}$
   (f) $x\mu = 4\sin x$
   (g) $x\mu = \begin{cases} \tan x & \text{if } x \neq n\pi + \pi/2 \\ 0 & \text{if } x = n\pi + \pi/2, \text{ where } n \in \mathbf{Z}. \end{cases}$
   (h) $x\mu = x^3 - 1$

9   Let $\mu: \mathbf{R} \times \mathbf{R} \to \mathbf{R} \times \mathbf{R}$ be a map as defined below. Describe the range in terms of a set of points in the Cartesian plane. In which cases is μ 1–1?
   (a) $(x, y) \to (y, x)$
   (b) $(x, y) \to (y, x^3)$
   (c) $(x, y) \to (x, 1)$
   (d) $(x, y) \to (x^2, 1 - y^2)$
   (e) $(x, y) \to (x + y, 2x + y)$
   (f) $(x, y) \to (x + y + 1, 2x + y)$
   (g) $(x, y) \to (x + y + 1, 2x + 2y)$
   (h) $(x, y) \to (x^2 - y^2, 2xy)$

10 (a) If $\mu: \mathbf{R} \to \mathbf{R}$ is defined by $x\mu = x^2 + 1$, give a surjective map v such that $x\mu = xv$, for all $x \in \mathbf{R}$. Can $\mu = v$?
   (b) Suppose that $x\mu = x^2 + 2$, and $xv = 3$. Find a domain S and a codomain T such that $\mu = v$.

1.6 Composite Maps

Suppose that we have two maps:

$$\lambda: S \to T \qquad \mu: T \to V$$
$$s \to s\lambda = t \qquad t \to t\mu = v$$

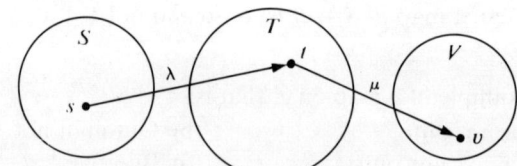

Figure 1–11

Figure 1–11 is a geometric representation of these two maps. Since element s maps to t, and t maps to v, the net effect is that s maps to v. By definition, the *composite* of maps $\lambda: S \to T$ and $\mu: T \to V$ is the map $\lambda \circ \mu: S \to V$, such that $s(\lambda \circ \mu) = (s\lambda)\mu$, for all $s \in S$. This definition can be pictured by means of a diagram as in Figure 1–12.

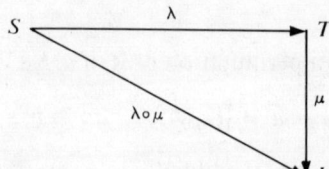

Figure 1–12

1.6 Composite Maps

A *commutative diagram* is a diagram (such as Figure 1–12) that shows that a map is equal to the composite of several other maps, or that the composites of different sequences of maps are equal.

EXAMPLE 1.36

Let $A = \{1, 2, 3,\}$, $B = \{n, p, q, r\}$, and $C = \{x, y, z\}$. Define maps λ and μ as follows:

$$\lambda: \ A \to B \qquad \mu: \ B \to C$$
$$1 \to p \qquad\qquad n \to x$$
$$2 \to n \qquad\qquad p \to y$$
$$3 \to q \qquad\qquad q \to z$$
$$\qquad\qquad\qquad\quad r \to z$$

The composite map, $\lambda \circ \mu$, is as follows:

$$\lambda \circ \mu: \ A \to C$$
$$1 \to 1(\lambda \circ \mu) = (1\lambda)\mu = p\mu = y$$
$$2 \to 2(\lambda \circ \mu) = (2\lambda)\mu = n\mu = x$$
$$3 \to 3(\lambda \circ \mu) = (3\lambda)\mu = q\mu = z$$

If you look back at Example 1.30, and compute the map, $\nu \circ \rho: A \to C$, you will find that $\nu \circ \rho = \lambda \circ \mu$. This gives us the commutative diagram in Figure 1–13.

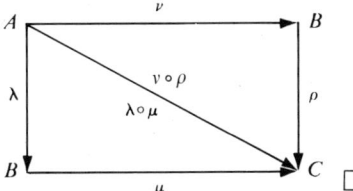

Figure 1–13

EXAMPLE 1.37

In this example we compare the right-hand function notation with the left-hand function notation. Let λ and μ be these functions on **R**:

$$\lambda: \ \mathbf{R} \to \mathbf{R} \qquad \mu: \ \mathbf{R} \to \mathbf{R}$$
$$x \to \sin x \qquad\quad x \to 3x + 2$$

Left: $(\mu \circ \lambda)(x) = \mu[\lambda(x)] = \mu(\sin x) = 3 \sin x + 2$.
Right: $x(\lambda \circ \mu) = (x\lambda)\mu = (\sin x)\mu = 3 \sin x + 2$.

Thus we have obtained the composite map,

$$\lambda \circ \mu: \quad x \to 3 \sin x + 2$$

Left: $(\lambda \circ \mu)(x) = \lambda[\mu(x)] = \lambda(3x + 2) = \sin(3x + 2)$.
Right: $x(\mu \circ \lambda) = (x\mu)\lambda = (3x + 2)\lambda = \sin(3x + 2)$.

$$\mu \circ \lambda: \quad x \to \sin(3x + 2)$$

In either notation, one always first maps x by the map whose "name" is adjacent to x. In the right-hand notation, however, one works naturally, as one reads, from left to right. □

EXAMPLE 1.38

Suppose we are given maps λ, μ and η on **R**, where

$$x\lambda = \sin x, \quad x\mu = 3x + 2, \quad x\eta = x^4.$$

Note that λ and μ are the maps of Example 1.37. Then

$$x[(\lambda \circ \mu) \circ \eta] = [x(\lambda \circ \mu)]\eta = (3 \sin x + 2)\eta = (3 \sin x + 2)^4.$$

Now $\quad x(\mu \circ \eta) = (x\mu)\eta = (3x + 2)\eta = (3x + 2)^4$,

and so $\quad x[\lambda \circ (\mu \circ \eta)] = (x\lambda)(\mu \circ \eta) = (\sin x)(\mu \circ \eta) = (3 \sin x + 2)^4.$

Therefore, in this case we have that $(\lambda \circ \mu) \circ \eta = \lambda \circ (\mu \circ \eta)$. □

THEOREM 1.5

If $\lambda: A \to B$, $\mu: B \to C$, and $\eta: C \to D$ are any three maps, then $(\lambda \circ \mu) \circ \eta = \lambda \circ (\mu \circ \eta)$.

Proof

Let a be any element in A. Then, by using repeatedly the definition of a composite map, we have that

$$a[(\lambda \circ \mu) \circ \eta] = [a(\lambda \circ \mu)]\eta = [(a\lambda)\mu]\eta = (a\lambda)(\mu \circ \eta) = a[\lambda \circ (\mu \circ \eta)].$$

Therefore, by the definition of equal maps, $(\lambda \circ \mu) \circ \eta = \lambda \circ (\mu \circ \eta)$. □

Because of Theorem 1.5, we can eliminate the parentheses, and write the composite of three maps simply as: $\lambda \circ \mu \circ \eta$.

The next theorem takes note of the fact that if two maps are injective, then so is their composite, and if the two maps are surjective, then so is their composite.

THEOREM 1.6

Let maps, $\lambda: A \to B$ and $\mu: B \to C$, be given.

(a) If λ and μ are each 1–1, then so is $\lambda \circ \mu$.
(b) If λ and μ are each onto, then so is $\lambda \circ \mu$.
(c) If λ and μ are each bijective, then so is $\lambda \circ \mu$.
(d) Let $A = B = C$. If λ and μ are each permutations of A, then so is $\lambda \circ \mu$.

Proof

(a) Let x and y be any elements of A. Then

$$x(\lambda \circ \mu) = y(\lambda \circ \mu) \Rightarrow (x\lambda)\mu = (y\lambda)\mu \quad \text{(By definition)}$$
$$\Rightarrow x\lambda = y\lambda \quad \text{(Given, } \mu \text{ is 1-1)}$$
$$\Rightarrow x = y. \quad \text{(Given, } \lambda \text{ is 1-1)}$$

Therefore, $\lambda \circ \mu$ is 1-1.

(b) We must show that any element c in C has an antecedent in A under the map $\lambda \circ \mu$. Now μ is onto, and so there is a b in B such that $b\mu = c$. Also, λ is onto, and so there is an a in A such that $a\lambda = b$. Hence,

$$a(\lambda \circ \mu) = (a\lambda)\mu = b\mu = c.$$

Therefore, a is an antecedent of c.

(c) This is an immediate consequence of parts (a) and (b).

(d) This follows from (c) and the definition of a permutation. ☐

EXAMPLE 1.39

Let $S = \{1, 2, 3, 4, 5\}$. Three permutations of S are the following:

$$\alpha = \begin{pmatrix} 1 & 2 & 3 & 4 & 5 \\ 2 & 3 & 4 & 5 & 1 \end{pmatrix}, \quad \beta = \begin{pmatrix} 1 & 2 & 3 & 4 & 5 \\ 2 & 5 & 1 & 3 & 4 \end{pmatrix},$$

$$\theta = \begin{pmatrix} 1 & 2 & 3 & 4 & 5 \\ 3 & 1 & 4 & 5 & 2 \end{pmatrix}.$$

The order in which the elements of the domain are listed is of no consequence. For instance, $2\beta = 5$, no matter in which position 2 appears in the first line of the representation of β. Let us write β so that the first line of β looks exactly like the second line of α. Then

$$\alpha \circ \beta = \begin{pmatrix} 1 & 2 & 3 & 4 & 5 \\ 2 & 3 & 4 & 5 & 1 \end{pmatrix} \circ \begin{pmatrix} 2 & 3 & 4 & 5 & 1 \\ 5 & 1 & 3 & 4 & 2 \end{pmatrix} = \begin{pmatrix} 1 & 2 & 3 & 4 & 5 \\ 5 & 1 & 3 & 4 & 2 \end{pmatrix}.$$

Also

$$\beta \circ \alpha = \begin{pmatrix} 1 & 2 & 3 & 4 & 5 \\ 2 & 5 & 1 & 3 & 4 \end{pmatrix} \circ \begin{pmatrix} 2 & 5 & 1 & 3 & 4 \\ 3 & 1 & 2 & 4 & 5 \end{pmatrix} = \begin{pmatrix} 1 & 2 & 3 & 4 & 5 \\ 3 & 1 & 2 & 4 & 5 \end{pmatrix}.$$

Thus, $\alpha \circ \beta \neq \beta \circ \alpha$. It can be verified similarly that $\theta \circ \beta = \beta \circ \theta = \varepsilon$, where ε is the identity permutation, and that $\alpha \circ \varepsilon = \varepsilon \circ \alpha, = \alpha$, $\beta \circ \varepsilon = \varepsilon \circ \beta = \beta$. ☐

THEOREM 1.7

If μ is any permutation of a set S, and ε is the identity permutation of S, then $\varepsilon \circ \mu = \mu \circ \varepsilon = \mu$.

Proof

Let a be any element of S. Then $a\mu = b$, for some b in S. So $a(\varepsilon \circ \mu) = (a\varepsilon)\mu = a\mu = b$, and $a(\mu \circ \varepsilon) = (a\mu)\varepsilon = b\varepsilon = b$. Since the image of a is b under each of the maps: μ, $\varepsilon \circ \mu$, and $\mu \circ \varepsilon$, we conclude that the three maps are equal. □

1.6 Problems

1. Using the maps defined in Example 1.30, find the following maps.
   (a) $\nu \circ \rho$ (b) $(\nu \circ \rho) \circ \sigma$ (c) $\rho \circ \sigma$ (d) $\nu \circ (\rho \circ \sigma)$

2. Let $S = \{1, 2, 3, 4\}$ and let λ and μ be these permutations of S.

   $$\lambda = \begin{pmatrix} 1 & 2 & 3 & 4 \\ 3 & 4 & 1 & 2 \end{pmatrix}, \quad \mu = \begin{pmatrix} 1 & 2 & 3 & 4 \\ 4 & 3 & 1 & 2 \end{pmatrix}$$

   (a) Find $\lambda \circ \mu$ and $\mu \circ \lambda$.
   (b) Find η and ν such that $\lambda \circ \eta = \varepsilon$ and $\nu \circ \mu = \varepsilon$ where ε is the identity map.

3. Let λ and μ be permutations of $\mathbf{Z}$, with $n\lambda = n + 5$, and $n\mu = n - 9$. Answer the questions asked in Problem 2.

4. Let λ, μ, and η be maps on $\mathbf{R}$ such that

   $$x\lambda = x^2 + 1, \quad x\mu = \sin(x - 1), \quad x\eta = 4 - x.$$

   Find the image of x under each of the following maps, and the range of each map.
   (a) $\lambda \circ \mu, \mu \circ \lambda$ (b) $\lambda \circ \eta, \eta \circ \lambda$
   (c) $\mu \circ \eta, \eta \circ \mu$ (d) $\mu \circ \mu, \eta \circ \eta$.

5. Let η and μ be maps on $\mathbf{R}$ such that $x\eta = (x - 1)/2$, and $x\mu = 2x + 1$. Find $\eta \circ \mu$ and $\mu \circ \eta$, and draw a commutative diagram that relates η, μ, and ε.

6. Draw a commutative diagram to illustrate Theorem 1.5.

7. (a) Given these maps:

   $$\lambda: A \to B, \quad \mu: B \to C, \quad \lambda \circ \mu: A \to C;$$

   prove: $\lambda \circ \mu$ is 1–1 $\Rightarrow \lambda$ is 1–1.

   (b) Given these maps:

   $$\lambda: A \to B, \quad \mu: B \to C, \quad \lambda \circ \mu: A \to C;$$

   prove: μ is 1–1, $\lambda \circ \mu$ is onto $\Rightarrow \lambda$ is onto.

1.7 Binary Operations

If you were asked what you are doing when you calculate that $9 + 7 = 16$, you would most likely answer that you are adding, or that you are performing the operation of addition. But what does it mean to "perform an operation"? We can give a precise answer in terms of maps.

By definition, a *binary operation* λ *on a set* A is a map,

$$\lambda: A \times A \to A.$$

Thus, a binary operation is determined whenever we give a rule, method, or scheme for associating with an ordered pair of elements of A a unique element of A.

EXAMPLE 1.40

We give examples of three different binary operations on $\mathbf{Z}$:

$$\lambda: (a, b) \to (a, b)\lambda = a + b.$$
$$\eta: (a, b) \to \text{the larger of } a \text{ and } b.$$
$$\theta: (a, b) \to 2a - b.$$

The images of a few pairs under each of these maps are:

$$\lambda: (1, 3) \to 4 \quad \eta: (1, 3) \to 3 \quad \theta: (1, 3) \to -1$$
$$(3, 1) \to 4 \quad\quad (3, 1) \to 3 \quad\quad (3, 1) \to 5$$
$$(2, -2) \to 0 \quad (2, -2) \to 2 \quad (2, -2) \to 6$$
$$(-1, 0) \to -1 \quad (-1, 0) \to 0 \quad (-1, 0) \to -2 \quad \square$$

In this text an *operation* on A will mean a binary operation on A, for binary operations are the only ones with which we shall be concerned.

In Example 1.40 we saw that $(1, 3)\theta = -1$. Another way to express the fact that $(1, 3)$ maps to -1 under θ is $1 \theta 3 = -1$. In general, if θ is an operation on A, we define $a \theta b = (a, b)\theta$. Thus,

$$\theta: (a, b) \to c = (a, b)\theta = a \theta b.$$

We read $a \theta b = c$ as: "a operation b equals c".

Up to this point we have been using small Greek letters to denote maps. If the map is an operation, we shall often represent the operation by a symbol that is not a letter of the alphabet. Actually, any symbol would do, but we shall usually use one of these:

$$\otimes, \star, \circ, +, \cdot, \oplus, \odot,$$

EXAMPLE 1.41

(a) Suppose we want to define an operation $\star$ on set A, where $A = \{r, s, t\}$. We must determine an image of each ordered pair (x, y). So there are nine images to be defined. We make this definition:

$x \neq y \Rightarrow x ☆ y = z$, where z is the third element of A,

$x ☆ x = x$, for all $x \in A$.

Then $r ☆ s = t$ and $r ☆ r = r$. It is convenient to list the nine images in a table, known as an *operation table*, as follows:

☆	r	s	t
r	r	t	s
s	t	s	r
t	s	r	t

(b) By means of the following table, a second operation, $\otimes$, is defined on A.

$\otimes$	r	s	t
r	s	t	t
s	r	s	s
t	s	t	r

□

In any operation table, the "x" of the ordered pair (x, y) is found in the left-hand column, and the "y" is found in the top row. The element at the intersection of the row preceded by x and the column headed by y is the image of (x, y) under the map $\otimes$, if $\otimes$ is the operation (as shown in the following diagram).

$\otimes$	$\cdots$	y
.		.
.		.
.		.
x	$\cdots$	$x \otimes y$

In attempting to define an operation $\otimes$ on a set A, one must check to be sure that the definition assigns to each ordered pair *exactly one* element of A. Otherwise, $\otimes$ is not an operation on A.

EXAMPLE 1.42

(a) If $a, b, c \in \mathbf{R}$, define $a \otimes b = c$ if and only if $c^2 = a^2 + b^2$. But then $4 \otimes 3 = 5$ and $4 \otimes 3 = -5$. So $\otimes$ is not an operation on $\mathbf{R}$. Observe, however, that $\otimes$ *is* an operation on $\mathbf{R}^+$.

(b) As another example, note that ordinary addition, $+$, is an operation on $\mathbf{Z}$ and on E, if E is the set of even integers. But $+$ is not an operation on D, the set of odd integers, since the sum of two odd integers is an even integer. □

Suppose that $\otimes$ is an operation on a set S, and T is a subset of S. If $a \otimes b \in T$ for all $a, b \in T$, then we say that T is *closed* under $\otimes$. If $a \otimes b \notin T$ for some $a, b \in T$, then T is not closed under $\otimes$, and hence $\otimes$ is not a binary operation on T. In Example 1.42(b) we saw that E is closed under $+$, while D is not.

1.7 Binary Operations

Let $a_i \in A$, $i = 1, 2, \ldots, n$, and let $\otimes$ be an operation on A. By $a_1 \otimes a_2 \otimes a_3$ is meant $(a_1 \otimes a_2) \otimes a_3$. In general, we define

$$a_1 \otimes a_2 \otimes \cdots \otimes a_{n-1} \otimes a_n = (a_1 \otimes a_2 \otimes \cdots \otimes a_{n-1}) \otimes a_n,$$

for each integer $n \geq 3$. (A definition like this is called a *recursive* definition.) For instance,

$$a \otimes b \otimes c \otimes d = [a \otimes b \otimes c] \otimes d = [(a \otimes b) \otimes c] \otimes d.$$

Quite simply, we're working from left to right in performing the operation.

EXAMPLE 1.43

Let ☆ be the operation that was defined in Example 1.41(a). Then

$$s \star r \star s \star t = [(s \star r) \star s] \star t = [t \star s] \star t = r \star t = s.$$

If we associate the four elements in some other way, we may or may not get the same result. For instance:

$$s \star [(r \star s) \star t] = s \star [t \star t] = s \star t = r,$$
$$s \star [r \star (s \star t)] = s \star [r \star r] = s \star r = t,$$
$$(s \star r) \star (s \star t) = t \star r = s. \qquad \square$$

Let $a \in S$, and let $\otimes$ be a binary operation on S. If n is any positive integer, the *nth power of a*, a^n, is defined recursively as follows:

$$a^1 = a,$$
$$a^k = a^{k-1} \otimes a, \text{ if } k > 1.$$

These results are obtained directly from the definition:

$$a^2 = a^1 \otimes a = a \otimes a,$$
$$a^3 = a^2 \otimes a = (a \otimes a) \otimes a = a \otimes a \otimes a,$$
$$a^4 = a^3 \otimes a = (a \otimes a \otimes a) \otimes a = a \otimes a \otimes a \otimes a.$$

Continuing this process, we see intuitively that

$$a^n = a \otimes a \otimes a \otimes \cdots \otimes a,$$

where a appears n times to the right of the equals sign.

A few words to summarize what we have done in this chapter are in order. We have been describing various *subsets of the Cartesian product*, $S \times T$. These subsets we called *relations* from S to T. Under certain conditions, these relations were called *maps*. If $S = T \times T$, then the maps from S to T were called *binary operations* on T. If the relations were from S to S, some of the relations obtained were *equivalence relations*, some were *partial order relations*, and still others were *maps*. The bijective maps on S were also called *permutations* of S. Although it may seem as though we have introduced you to a great variety of topics in this chapter,

1.7 Problems

1 In each of the following, a set S and a definition of $\otimes$ are given. Determine whether or not $\otimes$ is a binary operation on S.
(a) $S = \{-1, -2, -3, \ldots\}$. $a \otimes b = 3a - b^2$.
(b) $S = \mathbf{Z}^+$. $a \otimes b = 20a - b$.
(c) $S = \mathbf{Z}$. $a \otimes b = c$, where $a < c \leq b$.
(d) $S = \mathbf{Z}^+$. $a \otimes b = a + b - 2$.
(e) $S = \mathbf{Q}$. $a \otimes b = c$, where c is the smallest rational number that is larger than $a + b$.
(f) $S = \{1, 3, 5, \ldots\}$. $a \otimes b = a^b$.
(g) $S = \mathbf{C}$. $a \otimes b = c$, where $c = |a| + |b|i$.
(Note: $|x + yi| = \sqrt{x^2 + y^2}$.)

2 (a) Let $S = \{a, b\}$. Construct operation tables for five distinct operations on S.

(b) Compute the total number of distinct operations on a set S, if the order of S is 2, 3, or n.

3 Set A is a subset of the set $\mathbf{C}$ of complex numbers, where $A = \{1, i, -1, -i\}$. If the operation on A is multiplication of complex numbers, construct the operation table of A.

4 Let ☆ and $\otimes$ be the operations that were defined on $\{r, s, t\}$ in Example 1.41. Compute:
(a) $r \star s \star t$, $r \star (s \star t)$.
(b) $r \star r \star s \star t$, $(r \star r) \star (s \star t)$, $r \star [r \star (s \star t)]$, $r \star [r \star s \star t]$.
(c) $t \otimes r \otimes t \otimes s \otimes t$, $(t \otimes r) \otimes (t \otimes s \otimes t)$, $(t \otimes r \otimes t) \otimes (s \otimes t)$, $[t \otimes (r \otimes t)] \otimes (s \otimes t)$.

5 Using the operation table of Example 1.41(b), compute r^2, r^3, r^4, and then give the value of r^n, for any $n \in \mathbf{Z}^+$. Do the same for s and t.

6 Let S be a set of order n.
(a) If M is the set of functions on S, and if $\lambda \circ \eta$ is the composite of functions λ and η, explain why $\circ$ is a binary operation on M.
(b) If P is the set of permutations of S, explain why $\circ$ is a binary operation on P.

7 Given that P is the set of six permutations of $\{1, 2, 3\}$:

$$\varepsilon = \begin{pmatrix} 1 & 2 & 3 \\ 1 & 2 & 3 \end{pmatrix}, \quad \alpha = \begin{pmatrix} 1 & 2 & 3 \\ 2 & 3 & 1 \end{pmatrix}, \quad \beta = \begin{pmatrix} 1 & 2 & 3 \\ 3 & 1 & 2 \end{pmatrix}$$

$$\eta = \begin{pmatrix} 1 & 2 & 3 \\ 1 & 3 & 2 \end{pmatrix}, \quad \rho = \begin{pmatrix} 1 & 2 & 3 \\ 3 & 2 & 1 \end{pmatrix}, \quad \nu = \begin{pmatrix} 1 & 2 & 3 \\ 2 & 1 & 3 \end{pmatrix}$$

and the operation on P is the composition of maps, $\circ$; complete the following operation table.

$\circ$	ε	α	β	η	ρ	ν
ε	ε	α				
α			ε	ρ		
β						
η						
ρ						β
ν					ε	

8 For each permutation θ of Problem 7, find the smallest positive integer n such that $\theta^n = \varepsilon$.

9 Given a set S and an operation $\otimes$, compute c^2, c^3, and c^4, for each c.
(a) $S = \{-1, -2, -3, \ldots\}$, $a \otimes b = 3a - b^2$; $c = -1, -2$.
(b) $S = \{1, 3, 5, \ldots\}$, $a \otimes b = a^b$; $c = 1, 2$.
(c) $S = \{x + yi : x, y \in \mathbf{R}\}$, $(x + yi) \otimes (u + vi) = \sqrt{x^2 + y^2} + (\sqrt{u^2 + v^2})i$; $c = 3 - 4i, 2i$.

10 A *ternary operation* on a set S is a map: $S \times S \times S \to S$. Give an example of a ternary operation on a set of your choice. Also, describe some ternary operation on the set of points in a plane.

2
Abstract Algebras

> *The longer mathematics lives the more abstract—and therefore, possibly, also the more practical—it becomes (1937)—E. T. Bell*

2.1 Key Definitions

This chapter will introduce you to the algebraic structures that you will study in greater depth in later chapters. Hopefully, this orientation will enable you to see where and how each structure fits into the general scheme of things.

In an abstract algebra course one works with various algebraic structures. Each of these structures will be called an *abstract algebra*. By definition, an *abstract algebra* consists of a nonempty set, together with one or more binary operations defined on that set. In this course, when we refer to an *algebra*, we shall mean an abstract algebra. If the algebra consists of a set S and one operation, $\otimes$, the algebra will be denoted by the symbol $(S, \otimes)$. If the algebra consists of a set S and two operations, $\otimes$ and $\star$, the algebra will be denoted by the symbol $(S, \otimes, \star)$. We shall concentrate on algebras that have either one operation or two operations.

EXAMPLE 2.1

(a) Examples of three distinct algebras are the following:

$$(\mathbf{Z}, +), \quad (\mathbf{Z}, \cdot), \quad (\mathbf{Z}, +, \cdot),$$

where the operations are the usual operations of addition and multiplication of integers. The same set can appear in many different algebras.

(b) If $S = \{r, s, t\}$, then $S \times S$ is of order 9. So 3^9 different binary operations can be defined on S (there being 3^9 distinct maps from $S \times S$ to S). If $\otimes$ is one of these operations, then $(S, \otimes)$ is an algebra. Of the 19,683 such algebras, 2 are given in Example 1.41. □

The whole subject of abstract algebra would be trivial if all we did was to put an operation on a set and call it an algebra. But we do much more than that!

2.1 Key Definitions

For one thing, we're interested in whether or not an algebra possesses certain basic properties, among which are the following:

(1) An operation $\otimes$ is *associative* on a set S if $(a \otimes b) \otimes c = a \otimes (b \otimes c)$, for all $a, b, c \in S$.

(2) Set S has an *identity element* under $\otimes$ if there is an $e \in S$ such that $a \otimes e = e \otimes a = a$, for all $a \in S$.

(3) An element $a \in S$ has an *inverse* under $\otimes$ if there is an element $a' \in S$ such that $a \otimes a' = a' \otimes a = e$, where e is an identity element of S.

(4) An operation $\otimes$ is *commutative* on S if $a \otimes b = b \otimes a$, for all $a, b \in S$.

An algebra might possess none, some, or all of these properties.

EXAMPLE 2.2

(a) Consider the algebra $(\mathbf{Z}, +)$. We know, from past experience, that addition is associative and commutative. The identity element of addition is 0, since $a + 0 = 0 + a = a$, for all $a \in \mathbf{Z}$. Also, $a + (-a) = 0$, so $-a$ is the additive inverse of a, for each integer a. Therefore, $(\mathbf{Z}, +)$ possesses all four of the properties we just defined.

(b) Now consider $(\mathbf{Z}, \cdot)$. Multiplication is associative and commutative. The identity element is 1. The inverse of -1 is -1, and the inverse of 1 is 1. But if a is any other integer, there is no integer x such that $a \cdot x = 1$. Thus, only two integers have inverses under multiplication. Structurally, $(\mathbf{Z}, +)$ and $(\mathbf{Z}, \cdot)$ are distinct algebras. □

EXAMPLE 2.3

Let $(S, \otimes)$ be the algebra with the following operation table.

$\otimes$	a	b	c	d
a	a	b	c	d
b	b	d	a	a
c	c	a	b	c
d	d	c	d	b

We see that $c \otimes d = c$, but $d \otimes c = d$, so $\otimes$ is not commutative. Also, $(b \otimes c) \otimes d = a \otimes d = d$, and $b \otimes (c \otimes d) = b \otimes c = a$. Hence $\otimes$ is not associative. (One so-called *counter example* is enough to verify that a property does *not* hold.) An examination of the table reveals that if x is any element, then $a \otimes x = x \otimes a = x$, which means that a is an identity. As for inverses, $b \otimes c = c \otimes b = a$, so b is an inverse of c, and c is an inverse of b. Identity a is its own inverse, and d has no inverse. □

EXAMPLE 2.4

Let $(\mathbf{R}, \otimes)$ be the algebra for which $\otimes$ is defined as follows:

$$a \otimes b = a + b - ab.$$

Then, $3 \otimes 4 = -5$, $\frac{1}{2} \otimes 4 = \frac{5}{2}$, and $2 \otimes 2 = 0$.

The operation is commutative, since

$$a \otimes b = a + b - ab = b + a - ba = b \otimes a.$$

Is $\otimes$ associative? If a, b, c are any elements of **R**, then

$$(a \otimes b) \otimes c = (a + b - ab) \otimes c = a + b - ab + c - (a + b - ab)c$$
$$= a + b - ab + c - ac - bc + abc,$$

and

$$a \otimes (b \otimes c) = a \otimes (b + c - bc) = a + b + c - bc - a(b + c - bc)$$
$$= a + b + c - bc - ab - ac + abc.$$

Therefore, $(a \otimes b) \otimes c = a \otimes (b \otimes c)$, and so $\otimes$ is associative.

We search for an identity. We must find a real number e such that $a \otimes e = e \otimes a = a$ for all a. Now

$$a \otimes e = a \Leftrightarrow a + e - ae = a \Leftrightarrow e(1 - a) = 0 \Leftrightarrow e = 0.$$

So the identity is 0. To find an inverse of a, we look for a number a' such that $a \otimes a' = 0$.

$$a \otimes a' = 0 \Leftrightarrow a + a' - aa' = 0 \Leftrightarrow (a - 1)a' = a \Leftrightarrow a' = \frac{a}{a - 1},$$

if $a \neq 1$. Therefore, every real number except 1 has an inverse under $\otimes$. □

2.1 Problems

1. Which of the following are abstract algebras?
   (a) $(D, \cdot)$, where D is the set of odd integers, and $\cdot$ is ordinary multiplication.
   (b) $(D, +)$, where $+$ is ordinary addition.
   (c) $(\mathbf{Z}^+, \otimes)$, where $a \otimes b = a + b - 2$.
   (d) $(\mathbf{Z}^+, \star)$, where $a \star b = a^b$.

2. Algebra $(S, \otimes)$ is given, where $S = \{a, b, c, d, e\}$, and $\otimes$ is defined by the table:

$\otimes$	e	a	b	c	d
e	e	a	b	c	d
a	a	d	c	b	e
b	b	c	d	e	a
c	c	e	a	d	b
d	d	b	e	a	c

(a) Compute: $a \otimes b, b \otimes a, c \otimes d, d \otimes c$. Is $\otimes$ commutative?

(b) Compute: $(e \otimes a) \otimes b, e \otimes (a \otimes b), (b \otimes c) \otimes d, b \otimes (c \otimes d)$. Is $\otimes$ associative?

(c) Which element is an identity? Does any element have an inverse?

3 (a) Let $S = \{a, b\}$. By constructing operation tables, list all possible algebras $(S, \otimes)$ for which $\otimes$ is commutative.

(b) How many commutative operations can be defined on a set S if $|S| = 3$? Generalize this to the case of $|S| = n$, where n is any positive integer.

4 In each exercise you are given an algebra $(S, \otimes)$, and are to answer the following: (1) Is $\otimes$ commutative? (2) Is $\otimes$ associative? (3) Does S have an identity? (4) If S has an identity, determine which elements, if any, have inverses. (Assume that you know something about addition, subtraction, multiplication, and division of real numbers.)

(a) $(\mathbf{Q}, \otimes), a \otimes b = ab + a + b$.

(b) $(\mathbf{Z}, \otimes), a \otimes b = ab + a + b$.

(c) $(\mathbf{R}, \otimes), a \otimes b = 2ab$.

(d) $(\mathbf{Z}, \otimes), a \otimes b = a - 2b$.

(e) $(\mathbf{Q}, \otimes), a \otimes b = (a + b)/2$.

(f) $(\mathbf{Z}, \otimes), a \otimes b = $ the larger of a and b.

(g) $(\mathbf{Z}, \otimes), a \otimes b = a + b - 2$.

5 Given the operation table of an algebra $(S, \otimes)$; explain how, by inspecting the table, you can determine easily if $\otimes$ is commutative, and if there is an identity element.

6 The following operation table of $S = \{r, s, t, u\}$ has been constructed in such a way that the associative property holds. Fill in the four missing elements so that $(S, \otimes)$ is associative.

$\otimes$	r	s	t	u
r	r	r	—	r
s	r	s	—	u
t	t	t	—	t
u	t	u	—	s

7 Let S_n be the set of $n!$ permutations of $\{1, 2, \ldots, n\}$, and let $\circ$ be composition of maps. Tell why $(S_n, \circ)$ is an algebra, and indicate which properties discussed in this section are possessed by the algebra $(S_n, \circ)$.

8 Let $S = \{a, b\}$, and let T be the set of all functions on S.

(a) List the elements of T.

(b) If $\circ$ is the operation of composition of functions, give the operation table of $(T, \circ)$.

(c) Which properties does algebra $(T, \circ)$ possess?

2.2 Properties Common to all Abstract Algebras

In the last section we spoke of an algebra $(S, \otimes)$ having *an* identity. Can it have more than one identity?

THEOREM 2.1

If a set S has an identity under a binary operation $\otimes$, then the identity is unique.

Proof

Assume that e and f are identity elements of S under $\otimes$. Then

$$e \otimes x = x, \text{ for all } x \in S \Rightarrow e \otimes f = f,$$
$$x \otimes f = x, \text{ for all } x \in S \Rightarrow e \otimes f = e.$$

By the symmetric and transitive properties of $=$, we conclude that $e = f$. □

Let $h, k \in S$. Element h is called a *left identity* under $\otimes$ if $h \otimes x = x$, for all $x \in S$, and k is called a *right identity* under $\otimes$ if $x \otimes k = x$, for all $x \in S$. It is possible for an algebra to have a left identity and not a right identity, or a right and not a left identity.

EXAMPLE 2.5

Suppose that an algebra has this operation table:

$\otimes$	r	s	t
r	r	s	t
s	t	r	s
t	r	s	t

We see that r is a left identity, and t is also a left identity. There is no right identity. □

THEOREM 2.2

If h is a left identity, and k is a right identity of a set S under an operation $\otimes$, then $h = k$, and so h is the identity of $(S, \otimes)$.

Proof

If h is a left identity, then $h \otimes k = k$, and if k is a right identity, then $h \otimes k = h$. Therefore, $h = k$, and h satisfies the definition of an identity. □

Although the identity of an algebra is unique, inverses need not be, as the next example illustrates.

EXAMPLE 2.6

An algebra has the following operation table:

$\otimes$	r	s	t	v
r	r	s	t	v
s	s	r	r	t
t	t	r	s	s
v	v	t	s	v

Note that r is the identity. Element v has no inverse, t has the unique inverse s, while s has two inverses, s and t. □

If we wish to guarantee that we have *unique* inverses, we evidently need to place some additional restriction on the algebra.

THEOREM 2.3

If set S has an identity e under an operation $\otimes$, and if $\otimes$ is associative, then each element of S has at most one inverse under $\otimes$.

Proof

Suppose that b and c are inverses of an element a. We verify that $b = c$ as follows:

$$b = b \otimes e = b \otimes (a \otimes c) = (b \otimes a) \otimes c = e \otimes c = c.$$

(Be sure to supply a reason for each of the above equalities.) □

Assume that an algebra $(S, \otimes)$ has either a left identity e or a right identity e. If $a \in S$, and if there are elements b and c in S such that $b \otimes a = e$ and $a \otimes c = e$, then b is called a *left inverse* of a, and c is called a *right inverse* of a. Is it possible that $b \neq c$?

EXAMPLE 2.7

In the following algebra, a left inverse can differ from a right inverse.

	a	b	c	d
a	a	b	c	d
b	b	d	a	b
c	c	c	d	a
d	d	a	b	c

Element	Left Inverse	Right Inverse
a	a	a
b	d	c
c	b	d
d	c	b

No element except the identity a has an inverse. □

According to Theorem 2.3, if an operation is associative, then an inverse, if it exists, is unique. Associativity plays an important role in the next theorem, also.

THEOREM 2.4

Let $\otimes$ be an operation on a set S such that: (1) $\otimes$ is associative, (2) S has a left identity e under $\otimes$, and (3) each element has a left inverse under $\otimes$. Then:

(a) A left inverse of an element is also its right inverse, and hence is its inverse.

(b) Left identity e is also the right identity, and hence is the identity.

Proof

(a) Let a be any element of S, and let b be a left inverse of a. Then $b \otimes a = e$. Now b also has a left inverse, so $c \otimes b = e$, for some $c \in S$. Then

$$a \otimes b = e \otimes (a \otimes b) = (c \otimes b) \otimes (a \otimes b)$$
$$= [(c \otimes b) \otimes a] \otimes b = [c \otimes (b \otimes a)] \otimes b$$
$$= (c \otimes e) \otimes b = c \otimes (e \otimes b) = c \otimes b = e.$$

Therefore, b is also a right inverse of a.

(b) Let x be any element of S, and let x' be the inverse of x. (By part (a), we know that x has an inverse.) Then

$$x \otimes e = x \otimes (x' \otimes x) = (x \otimes x') \otimes x = e \otimes x = x,$$

and so left identity e is also a right identity. □

If the words "left" and "right" are interchanged in Theorem 2.4, the resulting statements are also true. We leave the proof of this as an exercise.

We say that an operation $\otimes$ on a set S has the *left cancellation property* if, for all $a, b, c \in S$, the equation $a \otimes b = a \otimes c$ implies that $b = c$; similarly, $\otimes$ has the *right cancellation property* if, for all $a, b, c \in S$, the equation $b \otimes a = c \otimes a$ implies that $b = c$.

These two properties are often referred to as the *cancellation laws*.

EXAMPLE 2.8

From past experience with $(\mathbf{Z}, +)$, we know that if $a + b = a + c$, then $b = c$. So the cancellation laws hold in $(\mathbf{Z}, +)$.

In Example 2.3, we see that $b \otimes c = b \otimes d$, but $c \neq d$ (since we assumed that S contained the four distinct elements: a, b, c, and d). Hence, the left cancellation law does not hold in that algebra. However, an examination of the operation table reveals that the right cancellation law does hold in that algebra. □

2.2 Problems

1. Given the set $S = \{e, a\}$, construct an operation table such that: $\otimes$ is associative, e is a left identity but not a right identity, e and a have right inverses, but a has no left inverse.

2   Prove the statements obtained from Theorem 2.4 by interchanging the words, "left" and "right."

3   If $S = \{1, 2\}$, then the power set, $P(S)$, contains these elements: $\emptyset, S, A$, and B, where $A = \{1\}, B = \{2\}$. Give the operation table for each of these three algebras, and examine each algebra with respect to the commutative and associative properties, existence of an identity and of inverses.
(a) $(P(S), \cap)$, the operation being *intersection* of sets.
(b) $(P(S), \cup)$, the operation being *union* of sets.
(c) $(P(S), +)$, the operation being *symmetric difference* of sets.

4   (a) Determine if either of the cancellation laws holds in the algebras of Examples 2.5, 2.6, and 2.7.
(b) If the left cancellation law does not hold in an algebra, describe the situation that exists in the operation table of the algebra.
(c) Replace "left" by "right" in part (b), and answer the question.

5   Given that $(S, \otimes)$ is an algebra with an identity, and that at least one of the cancellation laws holds, prove that each element in S has at most one inverse.

6   Assume that $(S, \otimes)$ is an algebra with an identity, and $\otimes$ is associative. Verify:
(a) Each element has a left inverse $\Rightarrow$ left cancellation law holds.
(b) Each element has a right inverse $\Rightarrow$ the right cancellation law holds.

7   Let S be a finite set, and let $(S, \otimes)$ have an identity.
(a) Explain why it is true that if the left cancellation law holds, then each element in S has exactly one right inverse, and if the right cancellation law holds, then each element has exactly one left inverse.
(b) Determine if the following statement is true or false: If both cancellation laws hold in $(S, \otimes)$, then each element has an inverse.

8   Let $S = \{a, b, c, d\}$. Construct an operation table of an algebra $(S, \otimes)$ with the following properties.
(a) S has a right identity, but not a left identity. The cancellation laws hold, but $\otimes$ is not associative.
(b) S has an identity. One element has two inverses, and each other element has exactly one inverse.
(c) S has an identity. The left cancellation law holds, but the right cancellation law does not hold. Three elements have inverses, but one does not.
(d) S has an identity. Cancellation laws hold, $\otimes$ is commutative, and each element has an inverse.
(e) S has an identity, and each element has exactly one inverse. Neither cancellation law holds, and $\otimes$ is not associative.

2.3 Abstract Algebras With One Operation

An abstract algebra $(S, \otimes)$ is called a *groupoid*. You have worked with many examples of groupoids in the previous sections. For convenience, we give names to special groupoids. The groupoid of greatest significance to us is called a *group*.

By definition, an abstract algebra $(G, \otimes)$ is a *group* if:

(1) $\otimes$ is associative,

(2) G contains an identity, and

(3) each element has an inverse.

Properties (1), (2), and (3) are sometimes called the *group postulates*. The word *group* was first used in a mathematical sense in 1830 by the French mathematician Galois.

An *abelian group*, so named in honor of the mathematician Abel, is a group $(G, \otimes)$ with the additional property:

(4) $\otimes$ is commutative.

An abelian group is also called a *commutative group*. If G is a finite set, then group $(G, \otimes)$ is called a *finite group*. If G is an infinite set, then group $(G, \otimes)$ is called an *infinite group*.

EXAMPLE 2.9

(a) $(\mathbf{Z}, +)$ is an infinite abelian group.

(b) If $S = \{1, -1, i, -i\}$, and the operation is multiplication of complex numbers, then $(S, \cdot)$ is a finite abelian group. (See Problem 1.7-3.)

(c) If S_3 is the set of $3!$ permutations of $\{1, 2, 3\}$, and $\circ$ is composition of maps, then $(S_3, \circ)$ is a finite nonabelian group. (See Problem 1.7-7.)

(d) If $S_\mathbf{Z}$ is the set of all permutations of $\mathbf{Z}$, then $(S_\mathbf{Z}, \circ)$ is an infinite nonabelian group. □

Even though $\otimes$ can be any binary operation, mathematicians often speak of the *product* of a and b when referring to $a \otimes b$, and they call a and b the *factors* of the product. They also refer to the operation table as the *multiplication table*. Thus, they are using the nomenclature that would be appropriate if $\otimes$ were ordinary multiplication of real numbers, for instance. There is no harm in doing this, provided that one interprets the operation correctly. The group operation table is also called a *Cayley table*, in honor of the English mathematician Cayley.

Although G is the set of elements and $(G, \otimes)$ is the group, a mathematician will often refer to the "group G." You may use this abbreviation after you have made it clear what the group operation is, and if there is no chance of ambiguity. Another often-used shorthand form is the expression ab as an abbreviation of $a \otimes b$. This is the same kind of abbreviation that you use when you are expressing the product $a \cdot b$ of two real numbers as ab.

If $(G, \otimes)$ is a group, and if H is a subset of G such that $(H, \otimes)$ is a group, then H is called a *subgroup* of G. Of course, $(G, \otimes)$ is a subgroup of $(G, \otimes)$, and

$(E, \otimes)$ is a subgroup of G, where $E = \{e\}$, e being the identity of G. If H is a subgroup of G such that $E \subset H \subset G$, then H is called a *proper subgroup* of G.

EXAMPLE 2.10

(a) $(\mathbf{Z}, +)$ is a group. If $H = \{0, \pm 3, \pm 6, \ldots\}$, then $(H, +)$ satisfies the group postulates, and so H is a subgroup of $\mathbf{Z}$.

(b) $(\mathbf{C}^*, \cdot)$ is an infinite group. If $H = \{1, -1\}$, $K = \{1, -1, i, -i\}$, and $L = \{a + bi : a, b \in \mathbf{Q}, a^2 + b^2 \neq 0\}$, then H, K, and L are subgroups of $\mathbf{C}^*$. Since $H \subset K \subset L \subset \mathbf{C}^*$, we have that H is a subgroup of K and a subgroup of L, and K is a subgroup of L. We have constructed a *chain* of subgroups. □

EXAMPLE 2.11

It can be verified that $(G, \otimes)$ is a group, if the Cayley table of G is the following:

$\otimes$	e	a	b	c
e	e	a	b	c
a	a	b	c	e
b	b	c	e	a
c	c	e	a	b

If $H = \{e, b\}$, then the Cayley table of H is:

$\otimes$	e	b
e	e	b
b	b	e

So H is closed under the operation $\otimes$, and H is a subgroup of G. This is the only *proper* subgroup of G. □

In addition to groups, there are other groupoids that are of interest to algebraists. We now define four of these one-operation algebras, although we shall not concentrate upon them in this course.

A groupoid $(S, \otimes)$ is called a *semigroup*, a *monoid*, a *quasigroup*, or a *loop*, if $(S, \otimes)$ has these properties:

(1) *Semigroup*: $\otimes$ is associative.

(2) *Monoid*: A semigroup that has an identity element.

(3) *Quasigroup*: For all $a, b \in S$, each of the equations $a \otimes x = b$ and $y \otimes a = b$ has a unique solution.

(4) *Loop*: A quasigroup that has an identity element.

EXAMPLE 2.12

From the definitions, we see that every monoid is a semigroup, and every loop is a quasigroup. In the next section we show that a group satisfies the quasigroup postulate. Hence, every group is a loop, a quasigroup, a monoid, and a semigroup. In Figure 2–1 we have the inclusion diagram for the classes of groupoids that we have defined. In the diagram we list, in brackets, the property that distinguishes a class of algebras from a class immediately above it, and of which it is a subclass.

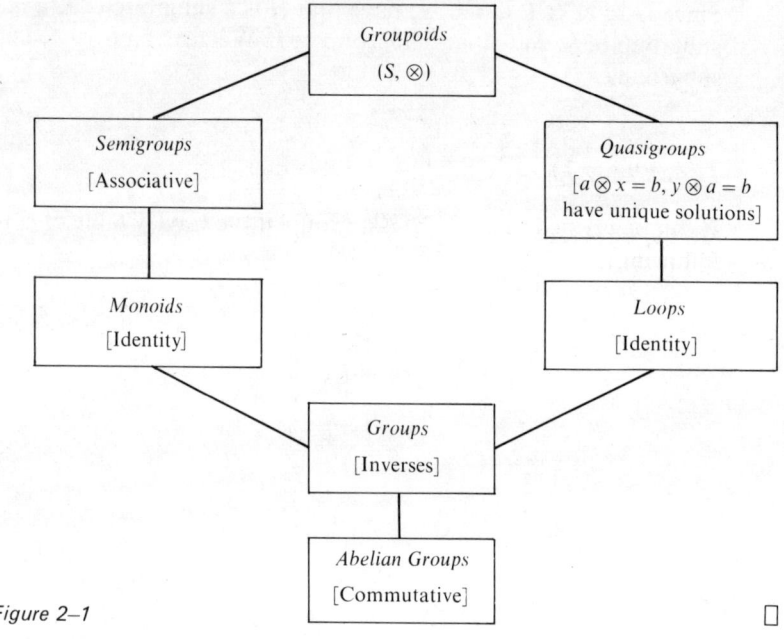

Figure 2–1

EXAMPLE 2.13

(a) A simple example of a semigroup is $(E^+, +)$, where E^+ is the set of positive even integers.

(b) $(\mathbf{Z}, \cdot)$ is a monoid.

(c) Let $(S, \otimes)$ be the algebra with the following operation table:

$\otimes$	a	b	c
a	b	a	c
b	c	b	a
c	a	c	b

Then $(S, \otimes)$ is a quasigroup. A table in which each element appears exactly once in each row and each column is called a *Latin square*. The conditions for a finite quasigroup are satisfied if and only if the operation table is a Latin square. (Why?)

(d) Let $S = \{e, a, b, r, s, t\}$, and let $\otimes$ be defined by the table:

$\otimes$	e	a	b	r	s	t
e	e	a	b	r	s	t
a	a	b	e	s	t	r
b	b	e	a	t	r	s
r	r	t	s	a	e	b
s	s	r	t	b	a	e
t	t	s	r	e	b	a

The identity is e, and the table is a Latin square. Therefore, $(S, \otimes)$ is a loop. Note that r, s, and t do not have inverses. Also, $(rs)t \neq r(st)$. ☐

2.3 Problems

1. Name each of the following algebras, giving the algebra the lowest possible name found on the inclusion diagram of Figure 2–1.
   (a) $(\mathbf{Z}^+, +)$ (b) $(\mathbf{Z}^+, \cdot)$
   (c) $(\mathbf{Z}, +)$ (d) $(\mathbf{Z}, -)$, where $-$ is subtraction.
   (e) $(\mathbf{Z}^+, *), a * b = a + 2b$.
   (f) $(S, \otimes)$, with the following operation table:

$\otimes$	e	a	b	c	d
e	e	a	b	c	d
a	a	e	c	d	b
b	b	d	e	a	c
c	c	b	d	e	a
d	d	c	a	b	e

   (g) $(S, \circ)$, if S is the set of maps on $\mathbf{Z}$, and $\circ$ is composition of maps.
   (h) $(S, \circ)$, if S is the set of bijective maps on $\mathbf{R}$.

2. Using the terms defined in this section, name the algebras found in Examples 2.2–2.7.

3. Give three subgroups of $(\mathbf{R}, +)$, and three subgroups of $(\mathbf{R}^*, \cdot)$.

4. Let G be the group with the following Cayley table:

	e	a	b	r	s	t
e	e	a	b	r	s	t
a	a	b	e	s	t	r
b	b	e	a	t	r	s
r	r	t	s	e	b	a
s	s	r	t	a	e	b
t	t	s	r	b	a	e

   Find four proper subgroups of G.

5 Construct the operation table of the following group:
$G = \{r_0, r_1, r_2, r_3\}$, where r_j is the position in the xy-plane of the positive x-axis after it has been rotated about the origin through j right angles. Operation $\otimes$ is defined by: $r_i \otimes r_j = r_{i+j}$. (For instance, $r_2 \otimes r_3 = r_5 = r_1$.)

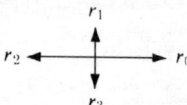

6 Determine if $(S, \otimes)$ is a group. If S is not a group, give it an appropriate name.
(a) $S = \mathbf{Z}.\ a \otimes b = a + b + 3$.
(b) $S = \mathbf{Q} - \{-1\}.\ a \otimes b = a + b + ab$.
(c) $S = \mathbf{Z}.\ a \otimes b = a - 2b$.

7 Verify that $(G, \otimes)$ is a group, and that $\otimes$ is not commutative, if $G = \{(x, y): x, y \in \mathbf{R}, x \neq 0\}$, and $(a, b) \otimes (c, d) = (ac, bc + d)$.

8 Construct the operation table of $(S, \otimes)$, if $S = \{1, 2, 3, 4\}$, and $a \otimes b = c$, where c is the remainder obtained if the product ab is divided by 5. Is S a group? (For instance, $4 \otimes 3 = 2$.)

9 In the groupoid $(\mathbf{Z}, -)$, where $-$ is subtraction of integers, show that there is a right identity but no left identity.

10 Let r, s, t, u, and v be elements in a semigroup $(S, \odot)$.
(a) Using the associative property, verify that

$$rstuv = (rs)(tuv).$$

(b) There are three other ways that the five factors can be associated. Prove that each product equals $rstuv$.

11 (a) If $a \in L$, where $(L, \odot)$ is a loop, does a have a right inverse? Does a have an inverse?
(b) Verify that an associative loop is a group.

12 Show that group $(\mathbf{Z}, +)$ has infinitely many infinite subgroups, and no finite proper subgroups.

13 Let $(\mathbf{Z}, \otimes)$ be the algebra such that $a \otimes b = a + b - 2$, for all $a, b \in \mathbf{Z}$. (See Problem 2.1-4(g).)
(a) Verify that $(\mathbf{Z}, \otimes)$ is an abelian group.
(b) Find an infinite proper subgroup of $(\mathbf{Z}, \otimes)$.
(c) Prove that $(\mathbf{Z}, \otimes)$ contains no finite proper subgroup.

14 Prove: If the order of a loop is less than 5, then the loop is a group.

15 (a) Galois might have said: "Never trust anyone over 30." Why? (See Appendix E–12.)
(b) Abel seemed to have at least three strikes against him. What were they? (See Appendix E–10.)

2.4 Some Basic Properties of Groups

In this section we state and verify some simple, but useful, facts about groups.

THEOREM 2.5

If $(G, \otimes)$ is a group, then (a) the identity element is unique, (b) the inverse of each element is unique, (c) the cancellation laws hold, and (d) the equations, $a \otimes x = b$ and $y \otimes a = b$, have unique solutions, for all $a, b \in G$.

Proof

(a) The identity e is unique, by Theorem 2.1.

(b) The inverse of each element is unique, by Theorem 2.3.

(c) Given that $a \otimes b = a \otimes c$, and that a' is the inverse of a; then

$$a \otimes b = a \otimes c \Rightarrow a' \otimes (a \otimes b) = a' \otimes (a \otimes c)$$
$$\Rightarrow (a' \otimes a) \otimes b = (a' \otimes a) \otimes c$$
$$\Rightarrow e \otimes b = e \otimes c \Rightarrow b = c.$$

The right cancellation law is proved similarly.

(d) $a' \otimes b$ is a solution of $a \otimes x = b$, since

$$a \otimes (a' \otimes b) = (a \otimes a') \otimes b = e \otimes b = b.$$

Now suppose that s and t are solutions of $a \otimes x = b$. Then $a \otimes s = b$ and $a \otimes t = b$, which implies that $a \otimes s = a \otimes t$. Therefore, $s = t$, by part (c), and so $a' \otimes b$ is the unique solution. Proof of the other part is similar. ☐

THEOREM 2.6

If a and b are any elements of group $(G, \otimes)$, then

(a) $(a')' = a$, and

(b) $(a \otimes b)' = b' \otimes a'$, where x' denotes the inverse of x.

Proof

(a) We are asked to verify that a is the inverse of a'. This is obviously true, since $a \otimes a' = a' \otimes a = e$.

(b) We first prove that $b' \otimes a'$ is the left inverse of $a \otimes b$, as follows:

$$(b' \otimes a') \otimes (a \otimes b) = [(b' \otimes a') \otimes a] \otimes b$$
$$= [b' \otimes (a' \otimes a)] \otimes b$$
$$= (b' \otimes e) \otimes b = b' \otimes b = e.$$

Therefore, by Theorem 2.4(a), $b' \otimes a'$ is the inverse of $a \otimes b$. ☐

Note that Theorem 2.6 states that the inverse of the inverse of an element is the element itself, and that the inverse of a product of two elements is the product of their inverses, in reverse order. Of course, if G is an abelian group, we could write that $(a \otimes b)' = a' \otimes b'$, but we must be careful not to write this if G is nonabelian.

Observe that we cut down on our work in proving Theorem 2.6(b) by using Theorem 2.4. Because of Theorem 2.4, we can prove that an element of a group is the identity by simply verifying that it is the *left* identity, and we can prove that s is the inverse of t by simply showing that s is the *left* inverse of t. The next theorem gives a simpler set of conditions which insure that $(G, \otimes)$ is a group.

THEOREM 2.7

If $(G, \otimes)$ is a semigroup such that (1) G contains a left identity, and (2) each element in G has a left inverse, then $(G, \otimes)$ is a group.

Proof

By definition of semigroup, $\otimes$ is associative. By Theorem 2.4 the other group postulates are satisfied. □

We now state the conditions under which a subset H of group G is a subgroup of G.

THEOREM 2.8

Let $(G, \otimes)$ be a group, and let H be a subset of G such that (1) $e \in H$, (2) $a, b \in H \Rightarrow a \otimes b \in H$, (3) $a \in H \Rightarrow a' \in H$. Then $(H, \otimes)$ is a subgroup of $(G, \otimes)$.

Proof

H is nonempty, by condition (1). By (2), $\otimes$ is a binary operation on H. Further,

$$a, b, c \in H \Rightarrow a, b, c \in G \Rightarrow (a \otimes b) \otimes c = a \otimes (b \otimes c).$$

Therefore, all the properties of a group are satisfied by the elements of H, and so $(H, \otimes)$ is a group. □

The shorthand for "H is a subgroup of G" is $H \leq G$. If H is a subgroup of G, and $H \neq G$, we write $H < G$.

EXAMPLE 2.14

We illustrate the technique for verifying that a subset is actually a subgroup. Let $(G, \otimes)$ be a group with identity e, and let a be a fixed element in G. Define N_a to be the set of all elements of G that commute with a. That is,

$$N_a = \{x : x \in G, x \otimes a = a \otimes x\}.$$

(1) $e \otimes a = a \otimes e$, so $e \in N_a$.

(2) Let x and y be any two elements in N_a. Then $x \otimes a = a \otimes x$, and $y \otimes a = a \otimes y$. We prove that $(x \otimes y)$ commutes with a as follows:

$$(x \otimes y) \otimes a = x \otimes (y \otimes a) = x \otimes (a \otimes y)$$
$$= (x \otimes a) \otimes y = (a \otimes x) \otimes y = a \otimes (x \otimes y).$$

Therefore, $x, y \in N_a \Rightarrow x \otimes y \in N_a$.

(3) $x \in N_a \Rightarrow x' \otimes a = (x' \otimes a) \otimes e = (x' \otimes a) \otimes (x \otimes x')$
$$= x' \otimes (a \otimes x) \otimes x' = x' \otimes (x \otimes a) \otimes x'$$
$$= (x' \otimes x) \otimes (a \otimes x') = e \otimes (a \otimes x') = a \otimes x' \Rightarrow x' \in N_a.$$

Subgroup N_a, called the *normalizer of a in G*, is a useful one in developing certain results in group theory. □

If $(G, \otimes)$ is a group and if C is a subset of G such that

$$C = \{x : x \otimes g = g \otimes x, \text{ for all } g \in G\},$$

then C is called the *center* of G. Thus, an element is in C if and only if it commutes with *every* element of G. It can be proved that $C \leq G$. We shall refer to the center of a group upon several occasions.

A study of its subgroups can often tell us quite a bit about the group.
If G is a finite group, we can construct an inclusion diagram of the subgroups of G.

EXAMPLE 2.15

(a) The group G of Example 2.11 has only these subgroups:
$G = \{e, a, b, c\}$, $H = \{e, b\}$, $E = \{e\}$. The inclusion diagram of subgroups is shown in Figure 2–2.

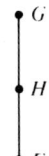

Figure 2–2

(b) The subgroups of group G of Problem 2.3–4 are as follows:
$G = \{e, a, b, r, s, t\}$, $H = \{e, a, b\}$, $K = \{e, r\}$, $L = \{e, s\}$, $M = \{e, t\}$, $E = \{e\}$. Figure 2–3 is the inclusion diagram of these subgroups.

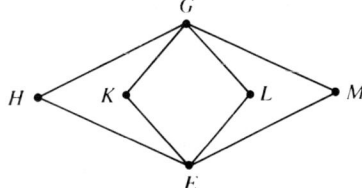

Figure 2–3

2.4 Problems

1. Given the group $(\mathbf{R}^*, \cdot)$, determine which of the following subsets of $\mathbf{R}^*$ are subgroups of $\mathbf{R}^*$.
   (a) $\mathbf{R}^+$
   (b) $\mathbf{Q}^*$
   (c) $\mathbf{Z}^*$
   (d) $\{1, -1\}$
   (e) $\{2^n : n \in \mathbf{Z}\}$
   (f) $\{x : x \text{ is irrational}\}$
   (g) $\{\pi^{2n} : n \in \mathbf{Z}, \pi = 3.1416\ldots\}$

2. Let $(G, \otimes)$ be a group. Show that all three group postulates are used to prove each of the following statements:
   (a) The right cancellation law holds.
   (b) If $a, b \in G$, the equation, $y \otimes a = b$, has a unique solution.

3. Is Theorem 2.7 true if "left" is replaced by "right"? Explain.

4. The *quaternion group*, which we denote as $\mathbf{Q}_8$, is the following: $\mathbf{Q}_8 = \{1, -1, i, -i, j, -j, k, -k\}$. The group operation is defined by these equations: $ij = k, ji = -k, jk = i, kj = -i, ki = j, ik = -j$. The center is $C = \{1, -1\}$. $x^2 = -1$ if $x \notin C$, and $x^2 = 1$ if $x \in C$. Also, $(1)x = x$, $(-1)x = -x$ for all $x \in \mathbf{Q}_8$.
   (a) Construct the group table.
   (b) Find all subgroups of $\mathbf{Q}_8$, and draw the inclusion diagram of subgroups.

5. Let $(H, \otimes)$ and $(K, \otimes)$ be subgroups of a group $(G, \otimes)$, and define $S = H \cap K$. Verify that S is a subgroup of G.

6. Let C be the center of a group G.
   (a) How do you know that C is nonempty?
   (b) Is it possible that $C = G$?
   (c) Prove that C is a subgroup of G.

7. Verify: If e is the identity of group G, and f is the identity of subgroup H, then $e = f$.

8. Given a group $(G, \otimes)$; prove: $(a \otimes b)^2 = a^2 \otimes b^2$ for all $a, b \in G$ if and only if G is abelian.

9. Prove: If $x^2 = e$, for all x in group G, then G is abelian.

10. Use the result of Problem 9 to prove that the algebra of Problem 2.3–1(f) is *not* associative. (Do not exhibit a case in which $(xy)v \neq x(yv)$.)

11. Given that $S = \{e, a, b, c\}$, and that $x^2 = e$, for all $x \in S$. Show that there is exactly one such group that satisfies these conditions. Construct the Cayley table. Also, list the subgroups, and draw the inclusion diagram of subgroups. (This group is known as the *Klein four-group*, and we shall henceforth denote it as K_4.)

12. Prove: If (1) $\odot$ is an associative operation on a set S, and (2) for any $a, b \in S$ there exist $x, y \in S$ such that $a \odot x = b$ and $y \odot a = b$, then $(S, \odot)$ is a group.

(Note: These are the group postulates given in 1906 by a mathematician named E. V. Huntington.)

13 Verify: If $(G, \otimes)$ is a group, and if H is a nonempty subset of G such that: (1) $a, b \in H \Rightarrow a \otimes b \in H$, and (2) $a \in H \Rightarrow a' \in H$, then $H \leq G$.

14 Verify: If $(G, \odot)$ is a group, and if H is a nonempty subset of G such that $a, b \in H \Rightarrow a \otimes b' \in H$, then $H \leq G$.

15 Prove: If there exist two elements, a and b, in a group G such that either $ab = b$ or $ba = b$, then a is the identity of G.

2.5 Groups of Permutations

Our work on bijective maps in Chapter 1, together with the definition of a group, leads readily to the conclusion that the set of all permutations of a set is a group under the operation ∘ of composition of maps.

THEOREM 2.9

If S_A is the set of all permutations of a set A, and ∘ is the operation of composition of maps, then $(S_A, \circ)$ is a group.

Proof

By Theorem 1.6(d), ∘ is an operation on S_A. The operation is associative, by Theorem 1.5. The identity permutation, ε, is the identity element of S_A, by Theorem 1.7. We need prove only that if μ is any permutation of A, then there is an inverse permutation, λ.

Define λ as follows: For all $x, y \in A$, $y\lambda = x$ iff $x\mu = y$. Suppose that a and b are any elements in S, and $a\mu = c$, $b\mu = d$. Then

$$c\lambda = d\lambda \Leftrightarrow a = b \Leftrightarrow a\mu = b\mu \Leftrightarrow c = d,$$

so λ is a map and is 1-1. Further, if s is any element of S, then $s\mu = t$ for some t in S. Hence $t\lambda = s$ and t is the antecedent of s under λ. Therefore, λ is onto. Permutation λ is a left inverse of μ, since

$$y(\lambda \circ \mu) = (y\lambda)\mu = x\mu = y = y\varepsilon.$$

But $\lambda \circ \mu = \varepsilon$ implies that $\mu \circ \lambda = \varepsilon$, by Theorem 2.4(a). Thus λ is the inverse of μ. □

EXAMPLE 2.16

Let $A = \{a, b, c, d, e\}$. Then

$$\mu = \begin{pmatrix} a & b & c & d & e \\ e & b & d & a & c \end{pmatrix} \Rightarrow \mu' = \begin{pmatrix} e & b & d & a & c \\ a & b & c & d & e \end{pmatrix},$$

where μ' is obtained from μ by inverting the rows of μ. Note that $\varepsilon' = \varepsilon$, so the identity permutation is its own inverse. However, the identity is not the only permutation that is its own inverse. For instance, $\sigma = \sigma'$ if

$$\sigma = \begin{pmatrix} a & b & c & d & e \\ b & a & e & d & c \end{pmatrix}. \quad \square$$

If A is of order n, then there are $n!$ permutations of A. This permutation group is known as the *symmetric group of degree n*, and is designated as S_n. In Problem 1.7–7 you were asked to construct the operation table of S_3.

There are instances in which it is important to be able to classify the permutations of S_n as being either even or odd. To this end we now introduce another notation for writing a permutation. Let A be a set of order n. A permutation λ of A is called a *cycle of length k* if there exist k distinct elements of $A: a_1, a_2, \ldots, a_k$, such that

$$a_1\lambda = a_2, \quad a_2\lambda = a_3, \ldots, a_{k-1}\lambda = a_k, \quad a_k\lambda = a_1,$$

and $b\lambda = b$ if $b \in A$ and $b \neq a_i$, for $i = 1, 2, \ldots, k$. The cycle λ may be written as follows:

$$\lambda = (a_1, a_2, a_3, \ldots, a_{k-1}, a_k).$$

The word "cycle" is an apt one for this kind of permutation, as we can depict what is happening by letting the a_i be points on a circle (see Figure 2–4.)

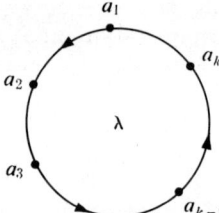

Figure 2–4

As we traverse the circle in the counter-clockwise direction, starting at any point, we cycle back to our original point. Cycle λ is the map:

$$a_1 \to a_2, \quad a_2 \to a_3, \ldots, a_{k-1} \to a_k, \quad a_k \to a_1.$$

EXAMPLE 2.17

Let $A = \{1, 2, 3, 4, 5, 6\}$. Suppose that

$$\mu = \begin{pmatrix} 1 & 2 & 3 & 4 & 5 & 6 \\ 1 & 6 & 4 & 2 & 5 & 3 \end{pmatrix}.$$

We see that $2\mu = 6$, $6\mu = 3$, $3\mu = 4$, $4\mu = 2$. Further, $1\mu = 1$, $5\mu = 5$. So μ is the cycle, $(2, 6, 3, 4)$. If there is no chance of confusion, we omit the commas in writing the cycle. In this case we could write:

$$\mu = (2634) = (6342) = (3426) = (4263).$$

2.5 Groups of Permutations

In writing a cycle, we can start with any element in that cycle. The cycle notation simply tells us that if s is followed by t, then the image of s is t under the mapping.

While μ is a cycle of length 4, we see that λ is a cycle of length 3, if $\lambda = (521)$. In the two-rowed notation:

$$\lambda = \begin{pmatrix} 1 & 2 & 3 & 4 & 5 & 6 \\ 5 & 1 & 3 & 4 & 2 & 6 \end{pmatrix}.$$

Since a cycle is a map, we find the product of two or more cycles in the same way that we have been finding the composite of maps. If $\mu = (2634)$, and $\eta = (51264)$, then $\mu \circ \eta = (246351)$. (For instance, $2\mu = 6, 6\eta = 4 \Rightarrow 2(\mu \circ \eta) = 4$. Hence, 2 is followed by 4 in the cycle notation.) Also, $\eta \circ \mu = (516234)$, and so $\mu \circ \eta \neq \eta \circ \mu$. □

Two cycles, $(a_1, a_2, \ldots, a_k)$ and $(b_1, b_2, \ldots, b_t)$ are said to be *disjoint* if $a_i \neq b_j$, for any i and j, where $i = 1, 2, \ldots, k$, and $j = 1, 2, \ldots, t$. A set of two or more cycles is said to be a *disjoint set* if each pair of cycles in the set is disjoint.

EXAMPLE 2.18

Let $A = \{1, 2, 3, 4, 5, 6\}$. If $\mu = (2634)$ and $\sigma = (6231)$, then μ and σ are not disjoint, but we can write their product as a product of disjoint cycles: $(2634) \circ (6231) = (16) \circ (34)$, and $(6231) \circ (2634) = (13) \circ (24)$. Note that μ and σ do not commute. However, disjoint cycles *do* commute. For instance,

$$(136) \circ (25) = (25) \circ (136) = \begin{pmatrix} 1 & 2 & 3 & 4 & 5 & 6 \\ 3 & 5 & 6 & 4 & 2 & 1 \end{pmatrix}. \quad \square$$

THEOREM 2.10

If $n \geq 3$, then group S_n is not abelian. However, if λ and η are any two disjoint cycles of S_n, then $\lambda \circ \eta = \eta \circ \lambda$.

Proof

S_n is the set of all permutations of n elements. Let a, b, and c be three of those elements. Then

$$(ab) \circ (ac) = (abc), \quad (ac) \circ (ab) = (acb).$$

Therefore, S_n is not abelian.

The proof that disjoint cycles commute will be found in Appendix B. The proof is simple, but the notation looks messy, and we don't want notation to distract you from concentrating on the main ideas of this section. □

A cycle, (a_1, a_2), of length two is called a *transposition*. Obviously, a transposition is its own inverse, since

$$(ab) \circ (ab) = \varepsilon.$$

We now show how we can write a cycle of *any* length as a product of transpositions.

EXAMPLE 2.19

(a) Suppose we have transpositions (ab), (ac), and (ad), where a, b, c, and d are distinct elements. Then one can verify that

$$(ab)(ac)(ad) = (abcd).$$

(b) Using the scheme illustrated in part (a), we can express a cycle as a product of transpositions. For instance,

$$(34781) = (34)(37)(38)(31) = (78134) = (78)(71)(73)(74).$$

Since $(ab)(ab) = \varepsilon$, even the identity permutation can be written as a product of transpositions.

(c) Generalizing, we prove in Appendix B that a cycle of length n can be expressed as a product of $n - 1$ transpositions, as follows:

$$(a_1, a_2, \ldots, a_{n-1}, a_n) = (a_1, a_2)(a_1, a_3) \ldots (a_1, a_{n-1})(a_1, a_n). \quad \square$$

Although a permutation λ can be expressed as a product of *disjoint* cycles in just one way, apart from the order of the cycles, λ can be written as a product of transpositions in many ways.

EXAMPLE 2.20

If $\lambda = \begin{pmatrix} 1 & 2 & 3 & 4 & 5 & 6 & 7 & 8 \\ 6 & 2 & 7 & 1 & 3 & 8 & 5 & 4 \end{pmatrix}$, then

$$\lambda = (1684)(375) = (16)(18)(14)(37)(35),$$

and so λ is a product of 5 transpositions. Using the fact that $(ab)(ab) = \varepsilon$, and $(ab) = (ac)(bc)(ac)$, we can write λ in these ways:

$$\lambda = (16)(18)(14)(17)(13)(17)(13)(15)(13),$$
$$\lambda = (24)(24)(16)(18)(37)(14)(35),$$
$$\lambda = (27)(47)(27)(24)(16)(18)(14)(37)(32)(52)(32).$$

So λ has been written as a product of 5, 7, 9, and 11 transpositions. (Each representation of λ was obtained from the preceding one. We leave it to you to justify these representations.) Observe that, in each case, λ is expressed as a product of an *odd* number of transpositions. $\quad \square$

2.5 Groups of Permutations

Two integers, s and t, are said to be of the *same parity* if they are either both even or both odd. If one of them is even and the other is odd, then s and t are said to be of *opposite parity*. It so happens that if a permutation η can be written as a product of s transpositions, and also as a product of t transpositions, then s and t are of the same parity. The statements that we have been making, without proof, are summarized in Theorem 2.11, the proof of which can be found in Appendix B.

THEOREM 2.11

Let λ be any permutation of S_n.

(a) λ can be factored into a product of a finite number of disjoint cycles. Apart from the order of the factors, this factorization is unique.

(b) A cycle of length k can be written as a product of $(k - 1)$ transpositions. Therefore, λ can be written as a product of a finite number of transpositions.

(c) If λ can be expressed as a product of s transpositions and also as a product of t transpositions, then s and t are of the same parity.

A permutation λ is called an *even permutation* if λ can be expressed as a product of an even number of transpositions, and is called an *odd permutation* if it can be expressed as a product of an odd number of transpositions. We needed the result given in Theorem 2.11 for this definition to make sense.

EXAMPLE 2.21

If

$$\eta = \begin{pmatrix} a & b & c & d & e \\ c & b & a & e & d \end{pmatrix}, \text{ and } \lambda = \begin{pmatrix} a & b & c & d & e \\ b & e & d & c & a \end{pmatrix},$$

then $\eta = (ac)(de)$, and $\lambda = (abe)(cd) = (ab)(ae)(cd)$. So η is an even permutation, and λ is an odd permutation. □

THEOREM 2.12

If T is any set, then the set B of all even permutations of T is a group.

Proof

We first show that T is closed under $\circ$. If $\lambda, \eta \in T$, then

$$\lambda = \alpha_1 \alpha_2 \ldots \alpha_{2k}, \quad \eta = \beta_1 \beta_2 \ldots \beta_{2m},$$

where the α_i and β_j are transpositions. Then $\lambda \circ \eta$ is a product of $2k + 2m$ transpositions, and so is even. Hence, B is closed.

Since the product of any maps is associative, the product of even permutations is associative. B has an identity, since ε is an even permutation. Since a transposition is its own inverse, we see that

$\lambda' = \alpha_{2k}\alpha_{2k-1}\ldots\alpha_2\alpha_1$, and so the inverse of λ is also an even permutation. Therefore, B is a group. □

If T is a set of order n, then the group of even permutations of T is called the *alternating group of degree n*, and is designated as A_n. We now determine the order of A_n.

THEOREM 2.13

If T is any set of order n, then half the permutations of T are even, and half are odd. Therefore, $|A_n| = n!/2$.

Proof

Let A_n be the set of all even permutations of T, and O_n be the set of all odd permutations of T, where

$$A_n = \{\lambda_1, \lambda_2, \ldots, \lambda_s\}, \quad O_n = \{\eta_1, \eta_2, \ldots, \eta_t\}.$$

We must show that $s = t$. Let B and C be the following sets:

$$B = \{\lambda_1\eta_1, \lambda_2\eta_1, \ldots, \lambda_s\eta_1\},$$
$$C = \{\eta_1\eta_1, \eta_2\eta_1, \ldots, \eta_t\eta_1\}.$$

Each element of B is an odd permutation, since it is the product of an even and an odd permutation. So $B \subseteq O_n$, which implies that $s \leq t$. Each element of C is an even permutation, since it is the product of two odd permutations. So $C \subseteq A_n$, which implies that $t \leq s$. Therefore, $s = t$, and hence there are $n!/2$ even permutations. □

EXAMPLE 2.22

If $T = \{1, 2, 3\}$, then the elements of S_3 can be written as follows:

$$e = (1), \quad a = (123), \quad b = (132), \quad r = (23), \quad s = (13), \quad t = (12).$$

Then the alternating group is: $A_3 = \{e, a, b\}$. The Cayley table of S_3 is given in Problem 2.3-4. □

EXAMPLE 2.23

Let $T = \{1, 2, 3, 4\}$. Then $|S_4| = 24$, and $|A_4| = 12$. We shall list each element in A_4 as a product of disjoint cycles. Recall that $(abc) = (ab)(ac)$, and so a three-cycle is an even permutation.

$$A_4 = \{(1), (12)(34), (13)(24), (14)(23), (123), (132), (124), (142), (134), (143),$$
$$(234), (243)\}. \quad \square$$

We conclude this section by looking at an interesting geometrical procedure for obtaining some other subgroups of the symmetric group S_n.

2.5 Groups of Permutations

If P is the set of points of a geometrical figure, then a *symmetry* of the figure is a permutation of P which preserves distance. A permutation *preserves distance* if the distance from p' to q' equals the distance from p to q, where p' is the image of p, and q' is the image of q, and p and q are any two points of P.

It turns out that a *symmetry of a polygon* in the plane is a rotation of the polygon about a center of symmetry (if one exists), or a reflection about a line of symmetry (if one exists), so that after the rotation or reflection has taken place, the polygon occupies the same region of the plane as it did initially.

EXAMPLE 2.24

The pentagon given in Figure 2–5 has just one line of symmetry: the median from the vertex labelled "4". The positions of the vertices after

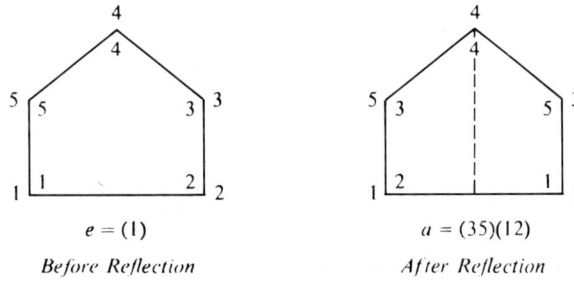

Figure 2–5

reflection through 180° about that median are given by means of a permutation. The initial position of the vertices is represented by the identity permutation e. The reflection a followed by the reflection a brings the pentagon back to its initial position. That is, $a^2 = e$. The set $\{e, a\}$ of symmetries is a group, with this operation table:

	e	a
e	e	a
a	a	e	□

EXAMPLE 2.25

Let's find the symmetries of a rectangle that is not a square, and show that this set of symmetries is a group. You'll understand the procedure better if you do this physically as well as mentally. Draw a rectangle on a piece of paper, and number the vertices as in Figure 2–6:

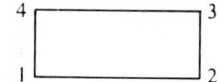

Figure 2–6

Then cut out a rectangle of the same size, and number the vertices, placing the numbers inside the rectangle, and writing them on both sides of the paper. When you place the cut-out rectangle on top of your rectangle above, with numbers matching, you will have Figure 2–7.

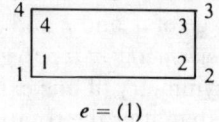

Figure 2–7 $e = (1)$

This is the identity motion, and is represented by the identity permutation, e. There are three other symmetries, shown in Figure 2–8.

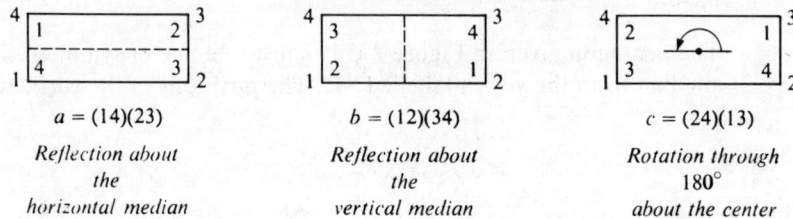

$a = (14)(23)$ $b = (12)(34)$ $c = (24)(13)$

Reflection about the horizontal median   Reflection about the vertical median   Rotation through 180° about the center

Figure 2–8

The product, ab, is that position of the vertices that one obtains by starting from the identity position, reflecting about the horizontal median, *followed by* a reflection about the vertical median. The result is the same as though one had simply rotated the rectangle through 180°, since $ab = c$. Note that this result can be obtained either by physically performing these motions and observing where the vertices fall, or by multiplying the permutations that represent the positions. This group of order 4 introduced in Problem 2.4–11) is named for the mathematician Felix Klein.

Klein four-group, K_4

	e	a	b	c
e	e	a	b	c
a	a	e	c	b
b	b	c	e	a
c	c	b	a	e

A *regular n-gon*, or *regular polygon of n sides*, is a polygon in the plane with n sides of equal length. A regular 3-gon is an equilateral triangle, and a regular 4-gon is a square. The group of symmetries of a regular n-gon is called the *dihedral group of degree n*, and is denoted by D_n. Since the vertices of an n-gon can be labelled $1, 2, 3, \ldots, n$, the symmetries of the n-gon are represented by permutations of these n integers, and hence the group D_n is a subgroup of the symmetric group, S_n.

EXAMPLE 2.26

To construct the dihedral group D_4 we find the symmetries of a square. Four symmetries are obtained by rotating the square about its center

2.5 Groups of Permutations

through 0, 1, 2, or 3 right angles. Let r_i be a rotation through i right angles. Figure 2–9 shows the four rotations.

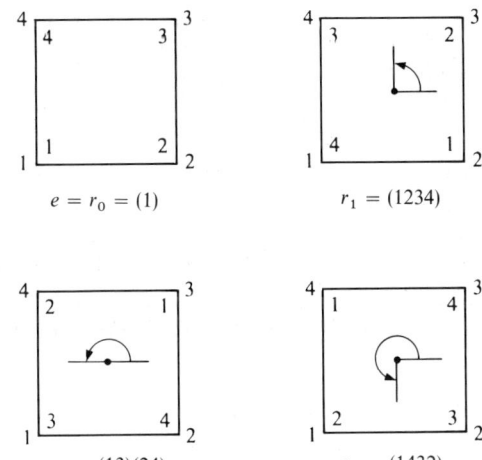

Figure 2–9

There are also four reflections, shown in Figure 2–10.

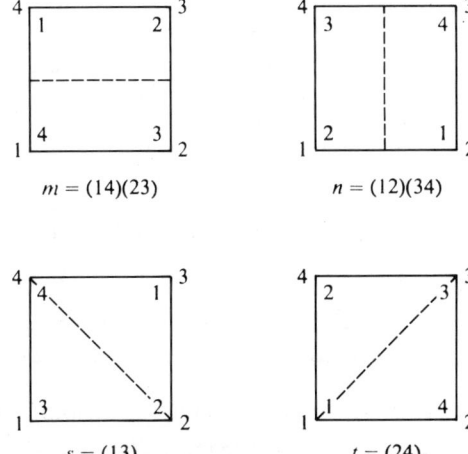

Figure 2–10

Group D_4, also called the *octic group*, consists of these eight permutations. Its multiplication table is given in Figure 2–11.

Figure 2–11 Dihedral group, D_4.

$\circ$	e	r_1	r_2	r_3	m	n	s	t
e	e	r_1	r_2	r_3	m	n	s	t
r_1	r_1	r_2	r_3	e	s	t	n	m
r_2	r_2	r_3	e	r_1	n	m	t	s
r_3	r_3	e	r_1	r_2	t	s	m	n
m	m	t	n	s	e	r_2	r_3	r_1
n	n	s	m	t	r_2	e	r_1	r_3
s	s	m	t	n	r_1	r_3	e	r_2
t	t	n	s	m	r_3	r_1	r_2	e

□

2.5 Problems

1 (a) If $\lambda = \begin{pmatrix} 1 & 2 & 3 & 4 \\ 3 & 4 & 2 & 1 \end{pmatrix}$, find: $\lambda^2, \lambda^3, \lambda^4, \lambda', (\lambda')^2, (\lambda^2)'$.

(b) If $\lambda = \begin{pmatrix} 1 & 2 & 3 & 4 \\ 2 & 1 & 4 & 3 \end{pmatrix}$, find the permutations asked for in part (a).

2 Let $\mu = \begin{pmatrix} a & b & c & d & e \\ d & c & b & a & e \end{pmatrix}, \lambda = \begin{pmatrix} a & b & c & d & e \\ c & a & b & d & e \end{pmatrix}, \eta = \begin{pmatrix} a & b & c & d & e \\ e & c & a & d & b \end{pmatrix}$.

(a) Find a positive integer k such that $\mu^k = \varepsilon$, the identity map. Do the same for λ and η.

(b) Use the results of (a) to write μ', λ', and η' as powers of μ, λ, and η, respectively.

3 Write each permutation μ as a product of disjoint cycles, and then as a product of transpositions.

(a) $\mu: 1 \to 2, 2 \to 5, 3 \to 6, 4 \to 3, 5 \to 1, 6 \to 4$.

(b) $\mu = \begin{pmatrix} 1 & 2 & 3 & 4 & 5 & 6 \\ 5 & 1 & 4 & 6 & 3 & 2 \end{pmatrix}$.

(c) $\mu = \{(1, 1), (2, 4), (3, 5), (4, 2), (5, 3), (6, 6)\}$.

(d) $1\mu = 2, 2\mu = 5, 3\mu = 4, 4\mu = 3, 5\mu = 1, 6\mu = 6$.

4 The product of two odd permutations is ___ (odd or even).
The product of two even permutations is ___.
The product of an odd and an even permutation is ___.

5 Write each odd permutation of $\{1, 2, 3, 4\}$ as a product of disjoint cycles.

6 Write the inverse of each of the following permutations.
(a) (35)   (b) (35)(14)   (c) (35)(14)(34)
(d) (1234) (e) (246)(135) (f) (1234)(23456)

7 Write each permutation as a product of disjoint cycles.
(a) (135)(132)   (b) (345)(2456)
(c) (345)(126)(4612)   (d) (23456)(65432)
(e) (246)(3614)(26153) (f) (24)(25)(26)(13)(14)(45)
(g) (45)(52)(51)(53)

8 There are 24 permutations of the set $\{1, 2, 3, 4\}$, each of which can be written as a product of disjoint cycles of one of the following five types: $(abcd), (abc), (ab), (a), (ab)(cd)$. Calculate the number of permutations there are of each of these types.

9 List the types of permutations of S, in terms of product of disjoint cycles, if $S = \{1, 2, 3, 4, 5\}$, and determine the number of permutations of each type. (See Problem 8.)

10 Explain why a cycle of an odd length is an even permutation, and a cycle of even length is an odd permutation.

11 (a) If $\mu = (12345)$, then $\mu^2 = $ ___.
If $\lambda = (123456)$, then $\lambda^2 = $ ___.
(b) Let $\eta = (123\ldots n)$. If n is odd, then $\eta^2 = $ ___, and if n is even, then $\eta^2 = $ ___.
(c) If η is a cycle, then η^3 is a product of how many disjoint cycles?

12 For each cycle μ, find the smallest positive integer k such that $\mu^k = (1)$, if:
(a) $\mu = (123)$. (b) $\mu = (1234)$. (c) $\mu = (1, 2, \ldots, 100)$.

13 Prove:
(a) If σ and η are two distinct transpositions, then $\sigma\eta$ can be expressed as a cycle of length 3 if σ and η are not disjoint, and as a product of two cycles, each of length 3, if σ and η are disjoint.
(b) Every even permutation can be written as a product of cycles of length 3.

14 The 12 elements of A_4 may be written as follows (from Example 2.23):

$$e = (1) \quad\quad g = (123) \quad\quad r = (132)$$
$$a = (12)(34) \quad\quad h = (134) \quad\quad s = (234)$$
$$b = (13)(24) \quad\quad i = (243) \quad\quad t = (124)$$
$$c = (14)(23) \quad\quad j = (142) \quad\quad u = (143)$$

Construct the operation table of A_4. (In your table heading, list the elements in this order: $e, a, b, c, g, h, i, j, r, s, t, u$.)

15 (a) Represent, by triangles and by the corresponding permutations, the elements of the dihedral group D_3.
(b) Construct the operation table of D_3 by actually performing the rotations and reflections you described in part (a).
(c) What is the relationship between D_3 and S_3?
(d) Interpret the elements of A_3 geometrically, in terms of the symmetries of a triangle.

16 Find all subgroups of D_4, and draw the inclusion diagram of these subgroups. (See Example 2.26.)

17 (a) Express each element of D_5 as a product of disjoint cycles.
(b) List the elements of D_6.

18 Give the order of D_n, and explain briefly how you arrived at your answer.

19 If $S_\mathbf{Z}$ is the group of all permutations of $\mathbf{Z}$, is the symmetric group S_n a subgroup of $S_\mathbf{Z}$? Explain.

20 (a) Prove: If $(G, *)$ is an abelian group with identity e, and if $H = \{x : x \in G, x^2 = e\}$, then H is a subgroup of G.
(b) Use D_4 as a counter example to show that the statement obtained in (a) by omitting the word "abelian" is not true. (See Example 2.26.)

21 The 8 vertices of a solid wooden cube are numbered, and so are the vertices of a box into which the cube fits exactly. In how many ways can the cube be placed in the box? How many permutations of the vertices are even? Is this set of permutations a group?

22 Describe how the alternating group A_4 is obtained from a group of symmetries of a regular tetrahedron.

23 In Appendix B, Theorem 2.11 is broken down into six theorems: B–2 to B–7. Study these enough to understand the method of proof being used. Then find a book or journal article in your library that gives a different proof of Theorem 2.11(c), and write up the proof.

24 (a) Felix Klein was interested in the relationship between group theory and what other branch of mathematics?

(b) What "concrete" example did Klein find of the symmetric group, S_5? (See Appendix E–18.)

2.6 Abstract Algebras With Two Operations

In this section we define some of the well-known algebras that have two binary operations, and give a few examples of them.

If I is any set of elements, and $P(I)$ is the power set of I, then $(P(I), \cup, \cap)$ is called a *Boolean algebra*, named in honor of George Boole. Boolean algebras have some interesting properties and applications, and so they deserve to be mentioned, even though we shall not be devoting much time to them in this course.

EXAMPLE 2.27

If $I = \{a, b\}$, then $P(I) = \{\emptyset, I, A, B\}$, where $A = \{a\}$, and $B = \{b\}$. The union and intersection tables of the Boolean algebra $P(I)$ are as follows:

$\cup$	$\emptyset$	I	A	B
$\emptyset$	$\emptyset$	I	A	B
I	I	I	I	I
A	A	I	A	I
B	B	I	I	B

$\cap$	$\emptyset$	I	A	B
$\emptyset$	$\emptyset$	$\emptyset$	$\emptyset$	$\emptyset$
I	$\emptyset$	I	A	B
A	$\emptyset$	A	A	$\emptyset$
B	$\emptyset$	B	$\emptyset$	B

□

The two-operation algebras known as *rings* are the algebras with which we are chiefly concerned. By definition, a *ring* $(S, \oplus, \otimes)$ consists of a nonempty set S of elements and two binary operations, called *addition*, $\oplus$, and *multiplication*, $\otimes$, with these properties:

(1) $(S, \oplus)$ is an abelian group.

(2) $\otimes$ is associative.

(3) $a \otimes (b \oplus c) = (a \otimes b) \oplus (a \otimes c)$, and $(b \oplus c) \otimes a = (b \otimes a) \oplus (c \otimes a)$, for all $a, b, c \in S$.

Properties (1), (2), and (3) are called the *ring postulates*. Postulate (3) is a statement of the *left distributive law* and the *right distributive law*, respectively. In

2.6 Abstract Algebras With Two Operations

the definition of a ring, property (3) is significant in that it is the only property that involves *both* operations, and hence ties the two operations together.

If parentheses are omitted in an expression involving both operations, we agree that multiplication is to be performed before addition. The symbol, bc, is an abbreviation of $b \otimes c$. The addition symbol is never omitted. Thus,

$$a \oplus b \otimes c = a \oplus (b \otimes c) = a \oplus bc.$$

The identity of addition is called *zero*, and it will be denoted by z. The symbol, $-a$, denotes the *additive inverse* of a. So

$$a \oplus -a = -a \oplus a = z, \quad \text{for all } a \in S.$$

We obtain various types of rings when we require that additional postulates be satisfied. A *ring with unity* is a ring that has an identity of multiplication. The identity of multiplication is called *unity*, and will be denoted by u. A *commutative ring* is a ring in which $\otimes$ is commutative. Appropriately, a *commutative ring with unity* is a commutative ring that has a unity.

EXAMPLE 2.28

We give examples of three finite rings. In each ring, $S = \{z, a, b, c\}$.

(a) $(S, +, \otimes)$:

+	z	a	b	c
z	z	a	b	c
a	a	z	c	b
b	b	c	z	a
c	c	b	a	z

$\otimes$	z	a	b	c
z	z	z	z	z
a	z	a	b	c
b	z	z	z	z
c	z	a	b	c

There is no unity in $(S, +, \otimes)$, and $\otimes$ is not commutative. We note that a is a left identity of multiplication, but not a right identity, and so is c.

(b) $(S, \oplus, \odot)$:

$\oplus$	z	a	b	c
z	z	a	b	c
a	a	b	c	z
b	b	c	z	a
c	c	z	a	b

$\odot$	z	a	b	c
z	z	z	z	z
a	z	b	z	b
b	z	z	z	z
c	z	b	z	b

This is a commutative ring, since the multiplication table is symmetric with respect to the main diagonal of the table (i.e., the diagonal extending from the upper left-hand corner to the lower right-hand corner). There is no unity.

(c) $(S, \oplus, \cdot)$:

$\oplus$	z	a	b	c
z	z	a	b	c
a	a	b	c	z
b	b	c	z	a
c	c	z	a	b

$\cdot$	z	a	b	c
z	z	z	z	z
a	z	a	b	c
b	z	b	z	b
c	z	c	b	a

This is a commutative ring with unity, a being the unity. □

The symbols 0 and 1 are often used to represent the zero and unity, respectively, in *any* ring. For the present we prefer to use z and u so that the identities will not look like real numbers. Similarly, we frequently represent addition and multiplication by $\oplus$ and $\otimes$ rather than by $+$ and $\cdot$, to make you aware of the fact that the ring operations do not have to be ordinary addition and multiplication of real numbers. In the next example we have two distinct operations of addition, and two distinct operations of multiplication, so we make use of all four symbols: $+, \oplus, \cdot, \otimes$.

EXAMPLE 2.29

Let $M_2(S)$ denote the set of all *two-by-two matrices* over a ring $(S, +, \cdot)$, where elements A and B of $M_2(S)$ are defined as follows:

$$A = \begin{bmatrix} a_{11} & a_{12} \\ a_{21} & a_{22} \end{bmatrix}, \quad B = \begin{bmatrix} b_{11} & b_{12} \\ b_{21} & b_{22} \end{bmatrix}, \text{ with } a_{ij}, b_{ij} \in S.$$

By definition, $A = B$ if and only if $a_{ij} = b_{ij}$, for $i = 1, 2$, and $j = 1, 2$. Addition, $\oplus$, and multiplication, $\otimes$, of matrices are defined as follows:

$$A \oplus B = \begin{bmatrix} a_{11} + b_{11} & a_{12} + b_{12} \\ a_{21} + b_{21} & a_{22} + b_{22} \end{bmatrix},$$

$$A \otimes B = \begin{bmatrix} a_{11} \cdot b_{11} + a_{12} \cdot b_{21} & a_{11} \cdot b_{12} + a_{12} \cdot b_{22} \\ a_{21} \cdot b_{11} + a_{22} \cdot b_{21} & a_{21} \cdot b_{12} + a_{22} \cdot b_{22} \end{bmatrix}.$$

For instance, in $M_2(\mathbf{Z})$ we have:

$$\begin{bmatrix} 2 & 1 \\ -4 & 6 \end{bmatrix} \oplus \begin{bmatrix} 3 & -1 \\ 7 & 3 \end{bmatrix} = \begin{bmatrix} 5 & 0 \\ 3 & 9 \end{bmatrix}, \quad \begin{bmatrix} 2 & 1 \\ -4 & 6 \end{bmatrix} \otimes \begin{bmatrix} 3 & -1 \\ 7 & 3 \end{bmatrix} = \begin{bmatrix} 13 & 1 \\ 30 & 22 \end{bmatrix}.$$

We leave it to you to verify that $(M_2(S), \oplus, \otimes)$ is a ring. The zero matrix is:

$$O = \begin{bmatrix} z & z \\ z & z \end{bmatrix},$$

if z is the zero of S.

If S has a unity u, then $M_2(S)$ has a unity I, where

$$I = \begin{bmatrix} u & z \\ z & u \end{bmatrix}.$$

Hence, the unity of $M_2(\mathbf{Z})$ is:

$$I = \begin{bmatrix} 1 & 0 \\ 0 & 1 \end{bmatrix}.$$

$M_2(S)$ is an example of a ring that is not commutative. If S is infinite, then $M_2(S)$ is an infinite ring, and if S is finite, then $M_2(S)$ is a finite ring.

2.6 Abstract Algebras With Two Operations

The definitions given here can be extended to obtain the ring, $M_n(S)$, of n–by–n matrices. Matrices are studied in detail in a course in *linear algebra*. □

A nonzero element a of a ring S is called a *zero divisor* if there exists a nonzero b in S such that $ab = z$, or a nonzero c in S such that $ca = z$.

EXAMPLE 2.30

Each of the rings of Example 2.28 has zero divisors. In fact, a, b, and c are all zero divisors of ring (a) and of ring (b). However, b is the only zero divisor of ring (c).

The ring of matrices in Example 2.29 also has zero divisors. For instance, in $M_2(\mathbf{Z})$:

$$\begin{bmatrix} 1 & 0 \\ 2 & 0 \end{bmatrix} \otimes \begin{bmatrix} 0 & 0 \\ 3 & 4 \end{bmatrix} = \begin{bmatrix} 0 & 0 \\ 0 & 0 \end{bmatrix},$$

so each matrix on the left side of the equality is a zero divisor. □

A ring is called a *trivial ring* if it contains only one element, and is called a *nontrivial ring* if it contains at least two elements. An *integral domain* is a nontrivial commutative ring with unity that has no zero divisors.

EXAMPLE 2.31

The best known example of an integral domain is $(\mathbf{Z}, +, \cdot)$, the ring of rational integers under the usual operations of addition and multiplication of integers. We shall postulate that $(\mathbf{Z}, +, \cdot)$ is an integral domain. To prove it, we could start with a simple set of postulates concerning the natural numbers, known as *Peano's Postulates*, and eventually verify that the algebra $(\mathbf{Z}, +, \cdot)$ satisfies all the properties of an integral domain. This program is carried out in Appendix C. From your past experience you are probably willing to accept the statement that 0 is the identity of addition, 1 is the identity of multiplication, and other properties, including the fact that $\mathbf{Z}$ has no zero divisors, which can be expressed thus:

$$ab = 0 \Rightarrow a = 0 \text{ or } b = 0, \quad \text{for } a, b \in \mathbf{Z}. \quad \square$$

A *field* is a nontrivial commutative ring with unity with the property that each nonzero element has a multiplicative inverse. The multiplicative inverse of a is denoted by a^{-1}. If $(F, \oplus, \otimes)$ is a field, and F^* is the set of nonzero elements of F, then it is clear that $(F^*, \otimes)$ is an abelian group. We call $(F, \oplus)$ the *additive group*, and $(F^*, \otimes)$ the *multiplicative group* of the field.

EXAMPLE 2.32

At present we postulate that these algebras are fields:

$$(\mathbf{Q}, +, \cdot), \quad (\mathbf{R}, +, \cdot), \quad (\mathbf{C}, +, \cdot),$$

where the operations are the usual addition and multiplication of rational, real, or complex numbers. In previous mathematics courses you have been applying the field properties in working with these numbers, and you may continue to do so in this course. □

EXAMPLE 2.33

$(F, \oplus, \otimes)$ is a finite field, where $F = \{z, u, a\}$, and the operations are defined by the following tables:

$\oplus$	z	u	a		$\otimes$	z	u	a
z	z	u	a		z	z	z	z
u	u	a	z		u	z	u	a
a	a	z	u		a	z	a	u

□

EXAMPLE 2.34

Figure 2–12 is the inclusion diagram of the classes of rings we have discussed. In the diagram we list, after the name of the class, a postulate that distinguishes that class of rings from the class above it. You might well question the fact that we have connected the class of integral domains to the class of fields. If we relied only on the definitions of this section, we could not have done that. However, in the next section we prove that a field has no zero divisors, and hence that every field is an integral domain.

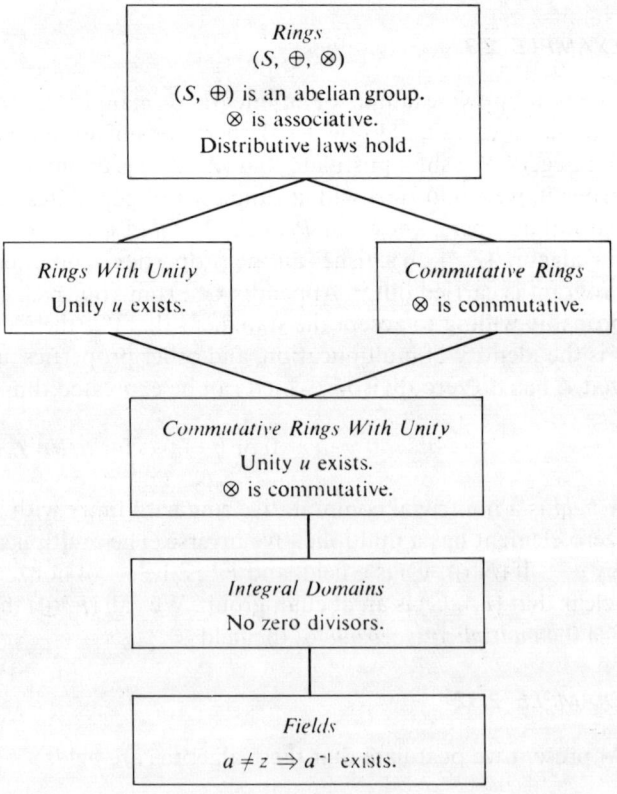

Figure 2–12

2.6 Problems

1 In each part a subset S of $\mathbf{R}$ is given. If the operations on S are the usual addition and multiplication of real numbers, is $(S, +, \cdot)$ a ring? If S is a ring, what kind of ring is it?

(a) $\{5n : n \in \mathbf{Z}\}$.
(b) $\{a + b\sqrt{5} : a, b \in \mathbf{Z}\}$.
(c) $\{a + b\sqrt{5} : a, b \in \mathbf{Q}\}$.
(d) $\{a + b\sqrt{5} + c\sqrt{3} : a, b, c \in \mathbf{Z}\}$.
(e) $\{a + b\sqrt[3]{5} + c\sqrt[3]{25} : a, b, c \in \mathbf{Q}\}$.

2 Let $(S, \oplus)$ be an abelian group with identity z. If $\otimes$ is an operation on S such that $a \otimes b = z$, for all $a, b \in S$, is $(S, \oplus, \otimes)$ a ring?

3 Look at the Boolean algebra of Example 2.27, and decide which ring postulates it satisfies. Is it a ring?

4 These questions refer to the rings of Example 2.28.
(a) Find $-a$, $-b$, and $-c$ in $(S, \oplus)$.
(b) In $(S, \oplus, \otimes)$ compute: $b \otimes (a \oplus b)$, $(a \oplus b) \otimes b$, $(b \otimes a) \oplus (b \otimes b)$, $(a \otimes b) \oplus (b \otimes b)$.
(c) Let $T = \{z, b\}$, and show that $(T, \oplus, \odot)$ is a ring. Is either $(T, +, \otimes)$ or $(T, \oplus, \cdot)$ structurally like $(T, \oplus, \odot)$?

5 Let $A, B \in M_2(\mathbf{Q})$, where

$$A = \begin{bmatrix} 1 & 1 \\ 3 & 5 \end{bmatrix}, \quad B = \begin{bmatrix} 3 & -2 \\ -6 & 4 \end{bmatrix}.$$

Find A^{-1}, and show that B^{-1} does not exist.

6 Let $M_2(S)$ be the set of 2-by-2 matrices over S, where S is a ring with unity. Show that $(M_2(S), \oplus, \otimes)$ is a ring with unity. (See Example 2.29.)

7 Let E be the set of even integers. Is $(M_2(E), \oplus, \otimes)$ a ring? Does $M_2(E)$ have a unity?

8 Let

$$S = \left\{ \begin{bmatrix} a & 0 \\ 0 & b \end{bmatrix} : a, b \in \mathbf{Q} \right\}.$$

Verify that $(S, \oplus, \otimes)$ is a commutative ring with unity. Is S an integral domain?

9 Let $A \in M_2(\mathbf{R})$, where

$$A = \begin{bmatrix} a & b \\ c & d \end{bmatrix}.$$

Prove: A^{-1} exists if and only if $ad - bc \neq 0$.
(Hint: Solve the four simultaneous equations obtained from:

$$\begin{bmatrix} a & b \\ c & d \end{bmatrix} \otimes \begin{bmatrix} x & v \\ y & w \end{bmatrix} = \begin{bmatrix} 1 & 0 \\ 0 & 1 \end{bmatrix}.)$$

If $ad - bc \neq 0$, find A^{-1}.

10 In ring $(M_2(\mathbf{Z}), \oplus, \otimes)$:

(a) Give two matrices, other than unity, that have multiplicative inverses in $M_2(\mathbf{Z})$.

(b) Give the conditions under which a matrix has a multiplicative inverse in $M_2(\mathbf{Z})$.

(c) Find two zero divisors in $M_2(\mathbf{Z})$.

11 For each algebra given, determine which of the field postulates hold.

(a) $(\mathbf{Z}, \oplus, \otimes)$. $a \oplus b = ab, a \otimes b = a + b$.

(b) $(\mathbf{Z}, \oplus, \otimes)$. $a \oplus b = a + b - 1, a \otimes b = a + b - ab$.

(c) $(\mathbf{Q}^*, \oplus, \otimes)$. $a \oplus b = ab, a \otimes b = a + b - ab$.

12 Let $I = \{a\}$, and $P(I) = \{\emptyset, I\}$.

(a) Construct the operation tables of the Boolean algebra $(P(I), \cup, \cap)$.

(b) Show that $(P(I), +, \cap)$ is a field, where the operation, $+$, is the symmetric difference of sets.

13 Prove that $(K, \oplus, \otimes)$ is a field if K is the following subset of $M_2(\mathbf{R})$:

$$K = \left\{ \begin{bmatrix} a & b \\ -b & a \end{bmatrix} : a, b \in \mathbf{R} \right\}.$$

14 Let algebra $(S, \oplus, \otimes)$ have these operation tables:

$\oplus$	z	a	b
z	z	a	b
a	a	b	z
b	b	z	a

$\otimes$	z	a	b
z	z	z	z
a	z	a	b
b	z	a	b

(a) Verify that $\otimes$ is associative. (Hint: Break your proof down into two cases: (1) One of the factors is z. (2) Each factor is either a or b.)

(b) Show that the distributive property does not hold.

15 The addition table and the partially filled-out multiplication table of ring $(S, +, \odot)$ are as follows:

$+$	a	b	c	d
a	a	b	c	d
b	b	a	d	c
c	c	d	a	b
d	d	c	b	a

$\odot$	a	b	c	d
a	a	a	a	a
b	a	—	—	a
c	a	—	c	—
d	a	b	c	—

(a) With the aid of the associative and distributive properties, fill in the five missing elements in the multiplication table.

(b) Which postulates of an integral domain does S fail to possess?

16 Let

$$M_{2,3}(\mathbf{Z}) = \left\{ \begin{bmatrix} a_{11} & a_{12} & a_{13} \\ a_{21} & a_{22} & a_{23} \end{bmatrix} : a_{ij} \in \mathbf{Z} \right\}.$$

Define operations of addition, $\oplus$, and multiplication, $\odot$, on $M_{2,3}(\mathbf{Z})$ so that $(M_{2,3}(\mathbf{Z}), \oplus, \odot)$ is a commutative ring with unity, but is not an integral domain.

17 Let S be the set of all maps (functions) on $\mathbf{R}$, where $(\mathbf{R}, +, \cdot)$ is the real field. If $f, g \in S$, define $f + g$ and $f \cdot g$ as follows:

$$x(f + g) = xf + xg, \qquad x(f \cdot g) = (xf) \cdot (xg), \quad \text{for all } x \in \mathbf{R}.$$

Verify that $(S, +, \cdot)$ is a commutative ring with unity, and that it has zero divisors. (Note: This ring is called a *function ring*. In the left-hand function notation, we would write:

$$(f + g)(x) = f(x) + g(x), \qquad (f \cdot g)(x) = f(x) \cdot g(x).)$$

18 George Boole used abstract algebra for what purpose? (See App. E-13.)

19 Comment on the truth of each of these statements:
(a) $(S, +, \cdot)$ is a ring $\Rightarrow$ $(S, +)$ is an abelian group, $(S, \cdot)$ is a semigroup.
(b) $(S, +)$ is an abelian group, $(S, \cdot)$ is a semigroup $\Rightarrow$ $(S, +, \cdot)$ is a ring.

2.7 Some Basic Properties of Rings

In Problem 1.1–3, you were asked to give reasons for these statements: $3 \cdot 0 = 0$, $(-3)(-7) = 21$, $(\frac{5}{8})(\frac{3}{7}) = \frac{15}{56}$. If you had trouble with your explanations, you need this section! First we establish the fact that, in a ring, zero times any element is always zero.

THEOREM 2.14

If z is the zero of ring $(S, \oplus, \otimes)$, then

$$a \otimes z = z \otimes a = z, \quad \text{for all } a \in S.$$

Proof

Using the definition of zero, and the left distributive law:

$$(a \otimes a) \oplus z = a \otimes a = a \otimes (a \oplus z) = (a \otimes a) \oplus (a \otimes z).$$

By the left cancellation law of addition, which holds in group $(S, \oplus)$, we then have that $z = a \otimes z$. The proof that $z = z \otimes a$ is similar. □

We must emphasize the fact that the symbol, $-a$, is the shorthand for "the additive inverse of a." If you were asked to prove that $s = -t$, for instance, you would want to verify that s is the additive inverse of t, and so you would show that $t \oplus s = z$.

THEOREM 2.15

If a and b are any elements in a ring $(S, \oplus, \otimes)$, then

(a) $a \otimes (-b) = (-a) \otimes b = -(a \otimes b)$, and
(b) $(-a) \otimes (-b) = a \otimes b$.

Proof

(a) We prove that $a \otimes -b$ is the additive inverse of $a \otimes b$, as follows:

$$(a \otimes -b) \oplus (a \otimes b) = a \otimes (-b \oplus b) = a \otimes z = z.$$

So $a \otimes -b$ is *an* inverse of $a \otimes b$. But the inverse is unique, and hence $-(a \otimes b) = a \otimes -b$. Similarly, $-(a \otimes b) = -a \otimes b$.

(b) If g is any element in the group $(S, \oplus)$, then $-(-g) = g$, by Theorem 2.6(a). We apply this remark and part (a) twice, obtaining:

$$(-a) \otimes (-b) = -[a \otimes -b] = -[-(a \otimes b)] = a \otimes b. \quad \square$$

We now define an operation called *subtraction*. By definition, if a and b are any elements in a ring $(S, \oplus, \otimes)$, then

$$a - b = a \oplus (-b).$$

To subtract b from a, one adds to a the additive inverse of b. So the *difference*, $a - b$, always exists in a ring. We read $a - b$ as "a minus b." In the language of maps, subtraction is the map from $S \times S$ to S such that $(a, b) \to a \oplus -b$. Observe that the symbol $-$ is being used in two different ways in the equation: $a - b = a \oplus -b$. The first $-$ denotes the operation of subtraction, while the second $-$ denotes "the additive inverse of" a.

THEOREM 2.16

If a, b, and c are any elements in a ring $(S, \oplus, \otimes)$, then:

(a) $-(a \oplus b) = -a - b$, $\quad -(a - b) = -a \oplus b$.
(b) $a \otimes (b - c) = (a \otimes b) - (a \otimes c)$, $\quad (b - c) \otimes a = (b \otimes a) - (c \otimes a)$.

The proof is left as an exercise.

If a, b, and c are any elements in a ring S, then the *cancellation laws of multiplication* in S are as follows:

Left cancellation law: $a \otimes b = a \otimes c, a \neq z \Rightarrow b = c$.
Right cancellation law: $b \otimes a = c \otimes a, a \neq z \Rightarrow b = c$.

The next theorem should convince you that the cancellation laws are very closely related to the existence or nonexistence of zero divisors.

THEOREM 2.17

Let $(S, \oplus, \otimes)$ be a ring. The cancellation laws of multiplication hold in S if and only if S has no zero divisors.

Proof

(1) Given that S has no zero divisors, suppose that

$$a \otimes b = a \otimes c, \qquad a \neq z.$$

Now, $a \otimes b = a \otimes c \Rightarrow a \otimes b + [-(a \otimes c)] = z$
$\Rightarrow a \otimes b - a \otimes c = z \Rightarrow a \otimes (b - c) = z$.

But S has no zero divisors, and so $b - c = z$. Therefore, $b = c$. The right cancellation law is similarly verified.

(2) Given that the cancellation laws hold, suppose that $a \otimes b = z$. We must show that either $a = z$ or $b = z$. Now $a \otimes z = z$, and so $a \otimes b = a \otimes z$. If $a \neq z$, then $b = z$, by the left cancellation law. $\square$

Because of Theorem 2.17, it would be possible to define an integral domain as a nontrivial commutative ring with unity in which the cancellation laws of multiplication hold.

We define an element x of a ring with unity to be a *unit* if x has a multiplicative inverse in the ring. Of course, unity itself is a unit. In $(\mathbf{Z}, +, \cdot)$ the only units are 1 and -1. In $(\mathbf{Q}, +, \cdot)$ every nonzero number is a unit, since $\mathbf{Q}$ is a field.

THEOREM 2.18

(a) If a is a unit of a ring S, then a is not a zero divisor.

(b) A field has no zero divisors, and so every field is an integral domain.

Proof

(a) Assume that a is a unit of S, and that b is an element of S such that $ab = z$. Then

$$ab = z \Rightarrow a^{-1}(ab) = a^{-1}z \Rightarrow (a^{-1}a)b = z \Rightarrow ub = z \Rightarrow b = z.$$

Similarly, $ca = z \Rightarrow c = z$. Therefore, a is not a zero divisor.

(b) F is a field. $\Rightarrow$ Every nonzero element is a unit. $\Rightarrow$
F has no zero divisors. $\Rightarrow$ F is an integral domain. $\square$

So that you don't waste your time trying to construct a *finite* integral domain that is not a field, we give you the next theorem.

THEOREM 2.19

Every finite integral domain is a field.

Proof

Let D^* be the set of nonzero elements of integral domain D, with

$$D^* = \{a_1, a_2, \ldots, a_m\}.$$

To prove that D is a field, we need show only that every element of D^* has a multiplicative inverse. Let a_k be any element of D^*, and define set S as follows:

$$S = \{a_1 a_k, a_2 a_k, \ldots, a_m a_k\}.$$

Now $z \notin S$, since $a_i a_k \neq z$, for any i. Hence $S \subseteq D^*$. Also, $a_i a_k = a_j a_k \Rightarrow a_i = a_j$, for $a_i, a_j \in D^*$. So $|S| = m$, and $S = D^*$. This means that $u \in S$, and that $a_t a_k = u$, for some t. Therefore a_t is the inverse of a_k. $\square$

If $(F, \oplus, \otimes)$ is a field, we can define a binary operation called *division*. If $a, b \in F$, with $b \neq z$, then *division* of a by b is defined as follows:

$$a \div b = \frac{a}{b} = a \otimes b^{-1} = ab^{-1}.$$

Just as we defined subtraction in terms of addition, we have defined division in terms of multiplication. It is now an easy matter to justify some familiar rules that you have used many times when working with fractions.

THEOREM 2.20

Let $(F, \oplus, \otimes)$ be a field. If $a, c \in F$ and $b, d \in F^*$, where F^* is the set of nonzero elements of F, then:

(a) $\dfrac{ad}{bd} = \dfrac{a}{b},$

(b) $\dfrac{a}{b} = \dfrac{c}{d}$ if and only if $ad = bc$,

(c) $\dfrac{a}{b} \oplus \dfrac{c}{d} = \dfrac{ad \oplus bc}{bd},$

(d) $\dfrac{a}{b} \otimes \dfrac{c}{d} = \dfrac{ac}{bd}.$

Proof

We shall prove part (c), to illustrate the technique, and let you prove the other parts.

$$\frac{ad \oplus bc}{bd} = (ad \oplus bc)(bd)^{-1} = (ad \oplus bc)d^{-1}b^{-1}$$
$$= (ad)(d^{-1}b^{-1}) \oplus (bc)(d^{-1}b^{-1})$$
$$= (ab^{-1})(dd^{-1}) \oplus (cd^{-1})(bb^{-1})$$
$$= ab^{-1}u \oplus cd^{-1}u = ab^{-1} \oplus cd^{-1} = \frac{a}{b} \oplus \frac{c}{d}. \quad \square$$

2.7 Problems

1. In each of the three rings of Example 2.28, compute $a(b - c)$ and $(b - c)c$.

2. Prove: If $(S, \oplus, \otimes)$ is a ring with zero z, then $z \otimes a = z$, for all $a \in S$. (See Theorem 2.14.)

3. Prove Theorem 2.16.

2.7 Some Basic Properties of Rings

4 Given that ring $(S, \oplus, \otimes)$ has no zero divisors, prove that the right cancellation law holds. (See Theorem 2.17.)

5 Let $S = \{(a_1, a_2, a_3): a_i \in \mathbf{Z}\}$. Verify that $(S, \oplus, \otimes)$ is a ring, with the operations as defined below. Is S a commutative ring? Does it have a unity? If it has a unity, find the units.

(a) $(a_1, a_2, a_3) \oplus (b_1, b_2, b_3) = (a_1 + b_1, a_2 + b_2, a_3 + b_3)$,
$(a_1, a_2, a_3) \otimes (b_1, b_2, b_3) = (a_1 b_1, a_2 b_2, a_3 b_3)$.

(b) $(a_1, a_2, a_3) \oplus (b_1, b_2, b_3) = (a_1 + b_1, a_2 + b_2, a_3 + b_3)$,
$(a_1, a_2, a_3) \otimes (b_1, b_2, b_3) = (a_1 b_1, a_2 b_1 + a_3 b_2, a_3 b_3)$.

6 Prove Theorem 2.20 (a, b, d).

7 Let $(F, +, \cdot)$ be the finite field with the following operation tables:

+	z	u	a	b	c
z	z	u	a	b	c
u	u	a	b	c	z
a	a	b	c	z	u
b	b	c	z	u	a
c	c	z	u	a	b

·	z	u	a	b	c
z	z	z	z	z	z
u	z	u	a	b	c
a	z	a	c	u	b
b	z	b	u	c	a
c	z	c	b	a	u

Compute:

(a) $c(-b)$, $\quad -(cb)$.

(b) $-(b - a)$, $\quad -a[u - (c - b)]$.

(c) $a^{-1}(c - b^{-1})$, $\quad c^{-1}b^{-1} + a^{-1}u^{-1}$.

(d) $(ub)/(ac)$, $\quad c/b$, $\quad (u/a) \cdot (c/b)$.

(e) $c/b + b/a$, $\quad (ac + b^2)/ab$, $\quad \dfrac{a/c + u/b}{ac}$, $\quad \dfrac{(a + c)(c + c)}{a + bc}$.

8 Let F be any field. Show that the cancellation laws of multiplication fail to hold in the ring of matrices, $M_2(F)$.

9 Let S be the following subset of the complex field:
$$S = \{a + bi : a, b \in \mathbf{Z}\}.$$

(a) Verify that $(S, +, \cdot)$ is an integral domain, but is not a field.

(b) Find the units of S.

10 Prove: $(\mathbf{Q}(i), +, \cdot)$ is a field, where $\mathbf{Q}(i)$ is the following subset of the complex field: $\mathbf{Q}(i) = \{a + bi : a, b \in \mathbf{Q}\}$.

11 Verify:

(a) If $(S, \oplus, \otimes)$ is a trivial ring, with $S = \{a\}$, then a is both zero and unity.

(b) If S is a nontrivial ring with unity, then zero and unity are distinct elements of S.

12 A *Boolean ring* is defined to be a ring $(S, \oplus, \otimes)$ such that $s \otimes s = s$, for all $s \in S$. Prove: If s and t are any elements in a Boolean ring, then

(a) $-s = s$, and (b) $st = ts$. (Thus, a Boolean ring is always commutative, and each element is its own additive inverse.)

13 Prove: If S is any set, and $P(S)$ is the power set of S, then $(P(S), +, \cap)$ is a Boolean ring. (The operations, $+$ and $\cap$, are symmetric difference and intersection, respectively.)

14 Construct the operation tables of a Boolean ring of order 4. (Hint: See Problem 13.)

15 Let n be a square-free rational integer. (That is, n has no perfect square, other than 1, as a factor.) If

$$\mathbf{Q}(\sqrt{n}) = \{a + b\sqrt{n} : a, b \in \mathbf{Q}\},$$

and if $+$ and $\cdot$ are the usual addition and multiplication of complex, or real, numbers, prove that $(\mathbf{Q}(\sqrt{n}), +, \cdot)$ is a field.

16 If $\mu = a + bi \in \mathbf{C}$, and $\mu^* = a - bi$, define S to be the following subset of $M_2(\mathbf{C})$:

$$S = \left\{ \begin{bmatrix} \mu & \sigma \\ -\sigma^* & \mu^* \end{bmatrix} : \mu, \sigma \in \mathbf{C} \right\}.$$

(a) Prove that $(S, \oplus, \otimes)$ is a ring with unity.
(b) Show that $\otimes$ is not a commutative operation on S.
(c) Verify that each nonzero element of S has a multiplicative inverse.

(Note: A *division ring* is a ring with unity in which each nonzero element has a multiplicative inverse. The division ring of this problem is called a *quaternion ring*, and each matrix in set S is called a *quaternion*. The quaternion ring will be produced in a different form in Problem 6.5–14.)

2.8 Ordered Integral Domains

If a and b are real numbers, we write $a < b$ to express the fact that a is less than b. Can we order the complex numbers and state, for instance, that $2 + i < 5 + 7i$? In this section we explore the matter of inequalities.

An integral domain $(D, \oplus, \otimes)$ is said to be an *ordered integral domain* if the nonzero elements of D are partitioned into sets, D^+ and D^-, such that: (1) D^+ is closed under $\oplus$ and $\otimes$, and (2) $a \in D^+$ if and only if $-a \in D^-$. The elements in D^+ are called the *positive elements* of D, and the elements in D^- are called the *negative elements* of D. Observe that D^- contains the additive inverse of each element of D^+, and that $\{\{z\}, D^+, D^-\}$ is a partition of D.

EXAMPLE 2.35

(a) $(\mathbf{Z}, +, \cdot)$ is an ordered integral domain, with

$$\mathbf{Z}^+ = \{1, 2, 3, \ldots\}, \quad \text{and} \quad \mathbf{Z}^- = \{-1, -2, -3, \ldots\}.$$

Is it possible to choose some other set as the set of positive elements of $\mathbf{Z}$?

If $1 \in \mathbf{Z}^+$, then so are 2, 3, ..., since $\mathbf{Z}^+$ is closed under addition, and we get the usual ordering, as given above. Now assume that $1 \in \mathbf{Z}^-$. Then $-1 \in \mathbf{Z}^+$, and so $(-1)(-1) \in \mathbf{Z}^+$, since $\mathbf{Z}^+$ is closed under multiplication. This means that $1 \in \mathbf{Z}^+$, a contradiction. Therefore, the partition we gave of $\mathbf{Z}^*$ is the only possible one.

(b) The rational field $\mathbf{Q}$ is an ordered integral domain (or, *ordered field*), for if $a/b \in \mathbf{Q}$, with $a, b \in \mathbf{Z}$, define

$$\mathbf{Q}^+ = \{a/b : a, b \in \mathbf{Z}, b \neq 0, ab \in \mathbf{Z}^+\}.$$

(c) The real field $\mathbf{R}$ is also an ordered field, with the set of positive elements $\mathbf{R}^+$ being the set of numbers that are to the right of zero on the usual real number line. □

EXAMPLE 2.36

Some integral domains can be ordered in more than one way. Let $D = \{a + b\sqrt{2} : a, b \in \mathbf{Z}\}$. Then $(D, +, \cdot)$ is an integral domain, if the operations on D are ordinary addition and multiplication of real numbers. If $D^+ = D \cap \mathbf{R}^+$, then the positive numbers of D are simply the positive real numbers that are in D. The conditions for an ordered integral domain are satisfied by the partition $\{\{0\}, D^+, D^-\}$.

Now define

$$D^\oplus = \{a + b\sqrt{2} : (a > 0, a^2 - 2b^2 > 0) \text{ or } (b < 0, a^2 - 2b^2 < 0)\}.$$

It can be shown that $D^\oplus$ is closed under addition and multiplication, and that a nonzero element of D is in $D^\oplus$ if and only if its additive inverse is not. (To show this requires a fair amount of "pencil pushing".) Hence, $D^\oplus$ constitutes a set of positive elements of D, and provides a different ordering than does D^+. For instance, $-\sqrt{2}$ is a negative number relative to D^+, while $-\sqrt{2}$ is a positive number relative to $D^\oplus$, since $b = -1$, and $a^2 - 2b^2 = -2$.

Note that we used the first, or usual, ordering to define the second ordering. Instead of using the sign, $<$, in defining $D^\oplus$, we could have defined $D^\oplus$ as follows:

$$D^\oplus = \{a + b\sqrt{2} : (a \in D^+, a^2 - 2b^2 \in D^+) \text{ or } (b \in D^-, a^2 - 2b^2 \in D^-)\}. \quad □$$

The next theorem cites two basic properties of D^+.

THEOREM 2.21

Let D be an ordered integral domain.

(a) If a is any nonzero element of D, then $a^2 \in D^+$.
(b) If u is the unity of D, then $u \in D^+$.

Proof

(a) Either a or $-a$ is in D^+, and so either a^2 or $(-a)^2$ is in D^+, since D^+ is closed under multiplication. But $(-a)^2 = a^2$, and so $a^2 \in D^+$.

(b) $u \neq z$ (Problem 2.7-11(b)), and so $u^2 \in D^+$. But $u^2 = u$, and hence $u \in D^+$. □

THEOREM 2.22

The complex field $\mathbf{C}$ is not an ordered field.

Proof

Assume that $\mathbf{C}$ is ordered. Then $1 \in \mathbf{C}^+$, and $i^2 \in \mathbf{C}^+$, by Theorem 2.21. But $i^2 \in \mathbf{C}^+ \Rightarrow -1 \in \mathbf{C}^+ \Rightarrow +1 \in \mathbf{C}^-$, a contradiction, since 1 cannot be in both $\mathbf{C}^+$ and $\mathbf{C}^-$. Therefore, $\mathbf{C}$ is not ordered. □

Up to this point we have been using the symbols, $>$, and $<$, based upon past experience. We now are ready to ascertain their exact meaning. Let a and b be any elements in an ordered integral domain D. By definition:

$$a < b \Leftrightarrow b > a \Leftrightarrow b - a \in D^+.$$

We read $a < b$ and $b > a$ as "a is less than b" and "b is greater than a", respectively. The statement, $a < b$, is called an *inequality*. The statement $a \leq b$ means that "either $a = b$ or $a < b$". Similarly, $a \geq b$ means that "either $a = b$ or $a > b$".

The proof of the next theorem will be left as an exercise. The statements are immediate consequences of the definitions, and of Theorem 2.21.

THEOREM 2.23

Let z be the zero of ordered integral domain D, and $a, b \in D$.

(a) $a > z \Leftrightarrow a \in D^+$. $a < z \Leftrightarrow a \in D^-$.
(b) $a > b \Leftrightarrow$ There is a $c \in D^+$ such that $a = b + c$.
(c) $a > z, b > z \Rightarrow a + b > z, ab > z$.
(d) *Trichotomy Law*: Exactly one of the following must hold:

$$a = z, \quad a > z, \quad a < z.$$

(e) $a \neq z \Rightarrow a^2 > z$.
(f) $u > z$, where u is the unity of D.

It is now possible to verify some familiar rules for handling inequalities.

THEOREM 2.24

Let a, b, c, d be in ordered integral domain $(D, +, \cdot)$.

(a) *Transitive property of* $>$: $a > b, b > c \Rightarrow a > c$.
(b) $a > b \Rightarrow a + c > b + c$.

2.8 Ordered Integral Domains

(c) $a > b, c > d \Rightarrow a + c > b + d$.
(d) $a > b, c > z \Rightarrow ac > bc$. $a > b, c < z \Rightarrow ac < bc$.
(e) $a > b \geq z, c > d \geq z \Rightarrow ac > bd$.

Proof

We shall let you prove parts (a), (d) and (e).

(b) $a - b = (a - b) + (c - c) = (a + c) - (b + c)$. So

$$a > b \Rightarrow a - b \in D^+ \Rightarrow (a + c) - (b + c) \in D^+ \Rightarrow a + c > b + c.$$

(c) By (b), $a > b \Rightarrow a + c > b + c$; and $c > d \Rightarrow b + c > b + d$. Therefore, by (a), $a + c > b + d$. $\square$

We can speak of the *absolute value* of a, provided that a is in an ordered integral domain D. The definition of *absolute value of a*, written $|a|$, is:

$$|a| = a \iff a \geq z,$$
$$|a| = -a \iff a < z.$$

Hence, $|a|$ is a positive element whenever a is a nonzero element of D, and is zero if and only if $a = z$.

THEOREM 2.25

If a and b are in an ordered integral domain $(D, +, \cdot)$, then:

(a) $|ab| = |a| \cdot |b|$, and
(b) $|a + b| \leq |a| + |b|$.

Proof

Part (a) will be left as an exercise.

(b) We first note that $x \leq |x|$, and $|x|^2 = x^2$, for all $x \in D$. Also (by Thm. 2.24(e)) $|x| > |y| \Rightarrow |x|^2 > |y|^2$, and so $|x|^2 \leq |y|^2 \Rightarrow |x| \leq |y|$. Hence, $x^2 \leq y^2 \Rightarrow |x| \leq |y|$.

Now $(a + b)^2 = a^2 + ab + ba + b^2 \leq |a^2| + |ab| + |ba| + |b^2|$
$= |a|^2 + |a||b| + |b||a| + |b|^2 = (|a| + |b|)^2$.

Therefore, $|a + b| \leq ||a| + |b|| = |a| + |b|$. $\square$

2.8 Problems

1. Let a, b, c, and d be in an ordered integral domain. Give a counter example to show that if $a < b$ and $c < d$, it does not follow that $a - c < b - d$.

2. Let D be an ordered integral domain. Prove:
   (a) $r \in D^+, s \in D^- \Rightarrow rs \in D^-$.
   (b) $s \in D^-, t \in D^- \Rightarrow st \in D^+$.

3 Prove Theorem 2.23.

4 Prove Theorem 2.24 (a, d).

5 (a) Give a counter example to show that if a, b, c, and d are rational integers such that $a > b$ and $c > d$, then it does not follow that $ac > bd$.
(b) Prove Theorem 2.24(e).

6 Prove Theorem 2.25(a).

7 Prove: If D is any ordered integral domain, and if 0 and 1 are the zero and unity, respectively, in D, then the equation, $x^2 + 1 = 0$, has no solution in D.

8 Verify: If a and b are in an ordered integral domain, and if $a \leq b$ and $b \leq a$, then $a = b$.

9 Show that $<$, $\leq$, $>$, and $\geq$ are four distinct relations on $\mathbf{Z}$.

10 (a) Prove: An ordered integral domain D contains no largest element. (Hint: Assume there is a $b \in D$ such that $x \leq b$, for all $x \in D$. Show that this assumption leads to a contradiction.)

(b) Prove: An ordered field always contains an infinite number of elements. (Hint: Use part (a).)

11 Prove: If $a < b$, where a and b are in the ordered field $\mathbf{Q}$, then there are infinitely many x in $\mathbf{Q}$ such that $a < x < b$.

12 Let $D^\oplus$ be the set defined in Example 2.36. Verify that:
(a) $D^\oplus$ is closed under addition.
(b) $D^\oplus$ is closed under multiplication.
(c) An element x in D satisfies exactly one of the following:

$$x = 0, \qquad x \in D^\oplus, \qquad -x \in D^\oplus.$$

13 Let a and b be in an ordered integral domain D.
(a) Is it true that $a < b \Rightarrow a^2 < b^2$?
(b) Prove: $a < b \Rightarrow a^3 < b^3$.

14 (a) Prove: If $a^3 = b^3$, where a and b are in an ordered integral domain D, then $a = b$.
(b) Cite an example to show that if D is not ordered, then one can have $a^3 = b^3$, with $a \neq b$.
(c) Prove: If D is ordered, $a, b \in D$, and $a^5 = b^5$, then $a = b$.

15 Prove that the rational field can be ordered in one way only.

2.9 Factorization in Integral Domains

One useful partition of an integral domain D is the following:

$$D = O \cup I \cup P \cup C,$$

where

$O = \{z\}$, the set containing just the *zero* of D;

I = the set of *units* of D (that is, the set of elements of D that have multiplicative inverses);

P = the set of *prime elements* of D, where p is called a *prime* if p is neither zero nor a unit, and if every factoring, $p = ab$, implies that either a is a unit or b is a unit;

C = the set of *composite elements* of D, where c is called a *composite* if c is neither zero nor a unit, and if there exists a factoring, $c = ab$, such that neither a nor b is a unit.

EXAMPLE 2.37

If D is the set $\mathbf{Z}$ of rational integers, then the partition just defined enables one to express $\mathbf{Z}$ as a union of four disjoint subsets:

$$\mathbf{Z} = \{0\} \cup \{\pm 1\} \cup \{\pm 2, \pm 3, \pm 5, \ldots\} \cup \{\pm 4, \pm 6, \pm 8, \pm 9, \ldots\}.$$

If D is the rational field $\mathbf{Q}$, then P and C are empty sets, since every nonzero number has an inverse, and hence is a unit. Obviously, there are no primes or composites in *any* field, so the remarks in this section are of little interest if the integral domain in question happens to be a field. ☐

If $c = ab$, we say that *a divides c*, and *b divides c*, and that a and b are *divisors*, or *factors*, of c. In symbols, we write: $a|c$, and $b|c$. If $a|c$, then c is called a *multiple* of a. If $c = av$, where v is a unit, then a is said to be an *associate* of c. If $a|c$, and a is neither a unit nor an associate of c, then a is called a *proper divisor* of c. It is apparent, from the definition, that a prime has no proper divisors. The symbol $\nmid$ is an abbreviation of "does not divide".

EXAMPLE 2.38

The divisors of 12 fall into these categories:

Units: $1, -1$.
Associates: $12, -12$.
Proper divisors: $2, -2, 3, -3, 4, -4, 6, -6$

The divisors of -7 consist of the units, 1 and -1, and the associates, 7 and -7. ☐

The next theorem gives some simple properties of divisibility that can be proved directly from the definition of *divides*.

THEOREM 2.26

If $a, b, c \in D$, where $(D, +, \cdot)$ is an integral domain, then:

(a) $a|b, b|c \Rightarrow a|c$;
(b) $a|b, a|c \Rightarrow a|(bx + cy)$, for all $x, y \in D$;
(c) $a|b, b|a \Rightarrow a$ is an associate of b.

Let's examine some integral domains that are less familiar than $\mathbf{Z}$. If n is a rational integer, with $n \neq 1$, then n is called a *square-free integer* if there is no rational prime p such that $p^2|n$. Thus, -10 is square-free, while -12 is not.

If n is a square-free integer, define $\mathbf{Z}(\sqrt{n}) = \{a + b\sqrt{n} : a, b \in \mathbf{Z}\}$. For instance,

$$\mathbf{Z}(\sqrt{6}) = \{a + b\sqrt{6} : a, b \in \mathbf{Z}\}, \quad \text{and} \quad \mathbf{Z} \subset \mathbf{Z}(\sqrt{6}) \subset \mathbf{R}.$$
$$\mathbf{Z}(\sqrt{-2}) = \{a + b\sqrt{-2} : a, b \in \mathbf{Z}\}, \quad \text{and} \quad \mathbf{Z} \subset \mathbf{Z}(\sqrt{-2}) \subset \mathbf{C}.$$

Addition and multiplication in $\mathbf{Z}(\sqrt{n})$ are the usual operations on complex numbers. That is,

$$(a + b\sqrt{n}) + (c + d\sqrt{n}) = (a + c) + (b + d)\sqrt{n},$$
$$(a + b\sqrt{n}) \cdot (c + d\sqrt{n}) = (ac + bdn) + (ad + bc)\sqrt{n}.$$

THEOREM 2.27

$(\mathbf{Z}(\sqrt{n}), +, \cdot)$ is an integral domain.

Verification of Theorem 2.27 is simple, and is left to the reader.

How do we determine the units and primes in $\mathbf{Z}(\sqrt{n})$? Of great help in this regard is a number known as the *norm* of $a + b\sqrt{n}$. We define the norm in the more general situation in which $\lambda = a + b\sqrt{n}$, with $a, b \in \mathbf{Q}$, and $\lambda^* = a - b\sqrt{n}$. The number λ^* is called the *conjugate* of λ. By definition, the *norm* of λ, written $N(\lambda)$, is the product of λ and λ^*. That is,

$$N(\lambda) = \lambda \cdot \lambda^* = (a + b\sqrt{n})(a - b\sqrt{n}) = a^2 - nb^2.$$

It is evident that if $\lambda \in \mathbf{Z}(\sqrt{n})$; then $N(\lambda) \in \mathbf{Z}$. Also, if $a \in \mathbf{Z}$, then $N(a) = a^2$, and $N(1) = N(-1) = 1$. Observe that if $n < 0$, then $N(\lambda) \geq 0$, for all λ.

THEOREM 2.28

If $\lambda = a + b\sqrt{n}$, and $\eta = c + d\sqrt{n}$, where $n \in \mathbf{Z}$, and $a, b, c, d \in \mathbf{Q}$, then $N(\lambda \cdot \eta) = N(\lambda) \cdot N(\eta)$. Also, $N(\lambda) = 0 \Leftrightarrow \lambda = 0$.

Verification of this is straightforward.
We now give a criterion for finding units in $\mathbf{Z}(\sqrt{n})$.

THEOREM 2.29

λ is a unit of $\mathbf{Z}(\sqrt{n})$ if and only if $N(\lambda)$ is 1 or -1.

Proof

(1) Assume that λ is a unit. The unity is 1, and so there is a number, λ^{-1}, such that $\lambda \cdot \lambda^{-1} = 1$. Hence:

$$1 = N(1) = N(\lambda \cdot \lambda^{-1}) = N(\lambda) \cdot N(\lambda^{-1}).$$

But the only divisors of 1 are 1 and -1. Therefore, $N(\lambda) = \pm 1$.

(2) Assume that $N(\lambda) = \pm 1$. Then one of these two situations must exist: $\lambda \cdot \lambda^* = 1$, or $\lambda \cdot \lambda^* = -1$. Hence, $\lambda^{-1} = \lambda^*$, or $\lambda^{-1} = -\lambda^*$, and so λ is a unit. □

EXAMPLE 2.39

$\mathbf{Z}(\sqrt{-1}) = \mathbf{Z}(i) = \{a + bi : a, b \in \mathbf{Z}\}$, and $N(a + bi) = a^2 + b^2$. Since the norm of each number is nonnegative, a number is a unit iff its norm is 1. So we must find $a, b \in \mathbf{Z}$ such that $a^2 + b^2 = 1$. Clearly $a^2 \leq 1$, $b^2 \leq 1$. If $a = \pm 1$, then b must be 0, and if $b = \pm 1$, then $a = 0$. So $\mathbf{Z}(i)$ has four units: $1, -1, i$, and $-i$. □

EXAMPLE 2.40

$\mathbf{Z}(\sqrt{-6}) = \{a + b\sqrt{-6} : a, b \in \mathbf{Z}\}$, and $N(a + b\sqrt{-6}) = a^2 + 6b^2$. But $a^2 + 6b^2 = 1 \Rightarrow a = \pm 1$ and $b = 0$. Hence this integral domain has but two units: 1 and -1. □

EXAMPLE 2.41

In $\mathbf{Z}(\sqrt{2})$, $N(a + b\sqrt{2}) = a^2 - 2b^2$. Now $N(1 + \sqrt{2}) = -1$, so $1 + \sqrt{2}$ is a unit. We leave it to you to verify that the set $\{(1 + \sqrt{2})^n : n \in \mathbf{Z}^+\}$ is a set of distinct units. Therefore, there are infinitely many units in this integral domain. □

If you were asked to factor 60 into a product of primes in $\mathbf{Z}$, you would probably write $60 = 2 \cdot 2 \cdot 3 \cdot 5$. Of course, $60 = (-5)(2)(-2)(3)$, but this is basically the same factoring, since each factor in the second factoring is an associate of a factor in the first factoring, and the order in which the factors are written is unimportant. In Section 3.4 we shall prove that each rational integer can be factored *uniquely* into a product of primes (apart from order and associates).

If we are to factor a number of $\mathbf{Z}(\sqrt{n})$ into a product of primes, we must be able to determine whether or not a given number is, in fact, a prime.

EXAMPLE 2.42

Is $2 + 3i$ prime or composite in $\mathbf{Z}(i)$? Note that $N(2 + 3i) = 13$. If $2 + 3i = \lambda \cdot \eta$, then

$$N(\lambda) \cdot N(\eta) = N(\lambda\eta) = N(2 + 3i) = 13,$$

so $N(\lambda)$ and $N(\eta)$ are factors of 13. But $13 = (1)(13) = (-1)(-13)$, and hence, either λ or η is a unit. Therefore, $2 + 3i$ is a prime in $\mathbf{Z}(i)$.

It happens that 13 is a prime in $\mathbf{Z}$, but is a composite in $\mathbf{Z}(i)$, since $13 = (2 + 3i)(2 - 3i)$. So 13 factors into a product of two primes in $\mathbf{Z}(i)$. Since we can also write $13 = (3 + 2i)(3 - 2i)$, does this mean that 13 factors into a product of primes in more than one way? No, because $(3 + 2i)(-i) = 2 - 3i$, and $(3 - 2i)(i) = 2 + 3i$. The primes in one factoring are associates of the primes in the other factoring, and so these two factorings are not considered to be essentially different. In Section 4.10 we shall prove that factorization is unique in $\mathbf{Z}(i)$. □

EXAMPLE 2.43

If $\lambda \in \mathbf{Z}(\sqrt{-6})$, then $\lambda = a + b\sqrt{-6}$, and $N(\lambda) = a^2 + 6b^2$. Now 3 and 5 are prime in $\mathbf{Z}(\sqrt{-6})$, since there exist no integers x and y such that either $x^2 + 6y^2 = 3$, or $x^2 + 6y^2 = 5$. Because of these last two equations we conclude that $3 + \sqrt{-6}$ is prime, since $N(3 + \sqrt{-6}) = 15$. For the same reason $3 - \sqrt{-6}$ is also prime.

But
$$15 = 3 \cdot 5 = (3 + \sqrt{-6})(3 - \sqrt{-6}).$$

No two of the primes 3, 5, $3 + \sqrt{-6}$ are associates, because the only units in this integral domain are 1 and -1. Therefore, we have succeeded in factoring 15 into a product of primes in two distinct ways. □

An integral domain is called a *unique factorization domain*, *UFD* for short, if every composite element in it can be factored *uniquely* into a product of primes. In Chapter 3 we examine $(\mathbf{Z}, +, \cdot)$ rather carefully, and determine, among other things, that it is a *UFD*. In Chapter 4 we look at some other integral domains, some of which are *UFD*'s, and some of which are not.

2.9 Problems

1   Verify: A unit in an integral domain has no proper divisors.

2   Prove: If a is any element, and v is a unit of an integral domain D, then $v \mid a$.

3   Prove Theorem 2.26.

4   (a) Prove Theorem 2.27.
   (b) Give an example of an element in $\mathbf{Z}(\sqrt{n})$ that has no multiplicative inverse, thus verifying that $\mathbf{Z}(\sqrt{n})$ is *not* a field.

5   (a) Verify: If m is *not* square-free, then $\mathbf{Z}(\sqrt{m})$ is an integral domain, and is a proper subset of some integral domain $\mathbf{Z}(\sqrt{n})$, where n is square-free.
         (b) Compare the sets, $\mathbf{Z}(\sqrt{12})$ and $\mathbf{Z}(\sqrt{3})$.

6   Define $\mathbf{Q}(\sqrt{n}) = \{a + b\sqrt{n} : a, b \in \mathbf{Q}\}$, with $n \in \mathbf{Z}$. Prove:
         (a) $(\mathbf{Q}(\sqrt{n}), +, \cdot)$ is a field.
         (b) n is square-free, $a + b\sqrt{n} = c + d\sqrt{n} \Rightarrow a = c, b = d$.

7   Prove Theorem 2.28.

8   Find three units of the form, $a + b\sqrt{n}$, with $a, b \in \mathbf{Z}^+$, in each of these integral domains.
         (a) $\mathbf{Z}(\sqrt{3})$ (b) $\mathbf{Z}(\sqrt{6})$ (c) $\mathbf{Z}(\sqrt{7})$

9   In $\mathbf{Z}(i)$, give the associates of:
         (a) $4 - 5i$,   (b) 3.

10   Factor each of these elements of $\mathbf{Z}(i)$ into a product of primes.
         (a) $4 + 3i$ (b) 7   (c) $5i$
         (d) 5   (e) $3 - 5i$ (f) $4 - 5i$

11   (a) In $\mathbf{Z}(\sqrt{-5})$, factor each of these numbers into a product of primes:
         $17 - \sqrt{-5}$, $1 + \sqrt{-5}$,   5,   41,   43.
         (b) Show that $\mathbf{Z}(\sqrt{-5})$ is not a unique factorization domain, by factoring some number into a product of primes in two ways.

12   Prove that there are infinitely many units in $\mathbf{Z}(\sqrt{2})$. (See Example 2.41.)

13   Let a and b be in an integral domain D, and define a relation $\sim$ on D such that: $a \sim b \Leftrightarrow a$ is an associate of b. Prove that $\sim$ is an equivalence relation on D.

14   Let the equivalence relation on integral domain D be that of Problem 13.
         (a) If $D = \mathbf{Z}$, describe the equivalence classes obtained.
         (b) If $D = \mathbf{Z}(\sqrt{-1})$, describe the equivalence classes obtained.

2.10 Morphisms and Isomorphic Algebras

In this chapter we have surveyed some of the basic notions of abstract algebra—notions to which we shall refer in later chapters. Such a survey would not be complete without a discussion of *morphisms*.

     The Greek word, *morphē*, means form, shape, or structure. It also implies a *beautiful* structure, since the Greeks felt that structure was an absolute and inseparable part of beauty. Mathematicians study the structure of algebras by means of *morphisms*, which are maps from one algebra to another—maps which can exist only if the two algebras are structurally similar.

If $\otimes$ is a binary operation on a set S, and $\odot$ is a binary operation on a set T, then, by definition, a *morphism from S to T*, or a *morphic map from S to T*, is a map,

$$\theta: (S, \otimes) \to (T, \odot),$$

such that if
$$s_1 \to t_1,$$
and
$$s_2 \to t_2,$$
then
$$s_1 \otimes s_2 \to t_1 \odot t_2, \quad \text{for all } s_1, s_2 \in S.$$

We say that θ *preserves the operations*. Since $t_1 = s_1\theta$, $t_2 = s_2\theta$, and $t_1 \odot t_2 = (s_1 \otimes s_2)\theta$, we are stating that θ preserves the operations when we say that if s_1 and s_2 are any elements of S, then

$$(s_1 \otimes s_2)\theta = (s_1\theta) \odot (s_2\theta).$$

Morphism is the more recently used abbreviation for *homomorphism*. The Greek word, *homo*, means *similar*, or *resembling*. Thus, S is homomorphic to T (or morphic to T) if and only if the structure of T resembles the structure of S, the resemblance being that defined above.

EXAMPLE 2.44

Let's illustrate what we mean when we say a map "preserves the operations." Let the monoids, $(M_2(\mathbf{Z}), \otimes)$ and $(\mathbf{Z}, \cdot)$, be given. Let θ be a map from $M_2(\mathbf{Z})$ to $\mathbf{Z}$ such that if $A = \begin{bmatrix} a & b \\ c & d \end{bmatrix}$, then $A\theta = ad - bc$. For instance,

$$A = \begin{bmatrix} 2 & 3 \\ 5 & 1 \end{bmatrix} \to -13 = A\theta,$$

$$B = \begin{bmatrix} 6 & 4 \\ 1 & 1 \end{bmatrix} \to 2 = B\theta.$$

Then $A \otimes B = \begin{bmatrix} 15 & 11 \\ 31 & 21 \end{bmatrix} \to 15 \cdot 21 - 11 \cdot 31 = -26 = (A \otimes B)\theta.$

Since $-26 = (-13)(2)$, it is true that $(A \otimes B)\theta = A\theta \cdot B\theta$. Does this always work? We set up the general situation as follows:

$$A = \begin{bmatrix} a & b \\ c & d \end{bmatrix} \to ad - bc = A\theta,$$

$$B = \begin{bmatrix} s & t \\ x & y \end{bmatrix} \to sy - tx = B\theta.$$

2.10 Morphisms and Isomorphic Algebras

Then

$$(A \otimes B)\theta = \begin{bmatrix} as + bx & at + by \\ cs + dx & ct + dy \end{bmatrix}\theta$$
$$= (as + bx)(ct + dy) - (at + by)(cs + dx)$$
$$= asct + bxct + asdy + bxdy - atcs - bycs - atdx - bydx$$
$$= bxct + asdy - bycs - atdx.$$

But

$$(A\theta) \cdot (B\theta) = (ad - bc)(sy - tx) = adsy + bctx - bcsy - adtx.$$

By inspection, we see that $(A \otimes B)\theta = (A\theta) \cdot (B\theta)$.

In words, we have shown that the image of the product of two matrices equals the product of the images of the two matrices. This particular map is called the *determinant function*. It is surjective but not injective. ☐

To say that a two-operation algebra, $(S, \oplus, \otimes)$ is morphic to an algebra $(T, +, \cdot)$ means that there is a map from S to T that preserves *both* operations. That is, there is a map $\theta \colon S \to T$, such that if a and b are any elements in S, then

$$(a \oplus b)\theta = a\theta + b\theta \quad \text{and} \quad (a \otimes b)\theta = a\theta \cdot b\theta.$$

A *ring* morphism is a morphic map from a ring to a ring, and a *group morphism* is a morphic map from a group to a group. Our next example is of a group morphism.

EXAMPLE 2.45

Let $\theta \colon (\mathbf{C}^*, \cdot) \to (\mathbf{R}^+, \cdot)$ be the following map from the nonzero complex numbers to the positive real numbers:

$$\theta \colon \; x = a + bi \to a^2 + b^2 = x\theta,$$
$$y = c + di \to c^2 + d^2 = y\theta.$$

Then $xy = (a + bi)(c + di) = (ac - bd) + (ad + bc)i$, and so $(xy)\theta = (ac - bd)^2 + (ad + bc)^2 = a^2c^2 + b^2d^2 + a^2d^2 + b^2c^2$. But $(x\theta)(y\theta) = (a^2 + b^2)(c^2 + d^2) = a^2c^2 + b^2d^2 + a^2d^2 + b^2c^2$. Therefore, $(x \cdot y)\theta = (x\theta) \cdot (y\theta)$, and so θ is a morphism.

Note that θ is onto, but not 1–1. (Why?) The real number, $a^2 + b^2$, is the *norm* of the complex number, $a + bi$. In showing that θ is a morphism, we have just proved that the norm of the product of two complex numbers equals the product of their norms. Thus, the map θ could have been written as follows, if we define $N(a + bi) = a^2 + b^2$:

$$x \to N(x)$$
$$y \to N(y)$$
$$xy \to N(xy) = N(x) \cdot N(y). \quad \square$$

If the map θ that preserves the (one or more) operations is a bijective map, then the morphism θ is called an *isomorphism*, or an *isomorphic map*. If algebra S is isomorphic to algebra T, we write: $S \cong T$. Keep in mind that to prove that $(S, \oplus, \otimes) \cong (T, +, \cdot)$, we must define $\theta: S \to T$ and verify the following:

(1) $\theta: S \to T$ is a map.
(2) θ is onto.
(3) θ is 1–1.
(4) $(a \oplus b)\theta = (a\theta) + (b\theta)$, for all $a, b \in S$.
(5) $(a \otimes b)\theta = (a\theta) \cdot (b\theta)$, for all $a, b \in S$.

The prefix, *iso*, comes from the Greek *isos*, meaning *equal* or *identical*. Just how identical are two isomorphic groups, for instance? Let's look at some examples.

EXAMPLE 2.46

Given $(G, \cdot)$ and $(H, \circ)$, with $G = \{1, i, -1, -i\}$, G being the set of fourth roots of unity, and $H = \{r_0, r_1, r_2, r_3\}$, where r_j is the permutation obtained by rotating a square about its center through j right angles. (See Examples 2.9(b) and 2.26.) The operation tables of these two groups are as follows:

G:

$\cdot$	1	i	-1	$-i$
1	1	i	-1	$-i$
i	i	-1	$-i$	1
-1	-1	$-i$	1	i
$-i$	$-i$	1	i	-1

H:

$\circ$	r_0	r_1	r_2	r_3
r_0	r_0	r_1	r_2	r_3
r_1	r_1	r_2	r_3	r_0
r_2	r_2	r_3	r_0	r_1
r_3	r_3	r_0	r_1	r_2

Let θ be this map: $1 \to r_0$, $\quad -1 \to r_2$,
$i \to r_1$, $\quad -i \to r_3$.

If we take the group table of G just given and replace each element x of G by its image, $x\theta$, we get precisely the group table of H. This means, of course, that the group operations are preserved under the bijective map θ, and so θ is an isomorphism. Structurally, groups G and H are identical, and we have merely used two different sets of symbols to represent the same group. □

EXAMPLE 2.47

We want to show that the additive group of real numbers, $(\mathbf{R}, +)$, is structurally just like the multiplicative group of positive real numbers, $(\mathbf{R}^+, \cdot)$. There are many different morphic maps that we can set up. We shall give one which, in a calculus course, is called an "exponential function." (See Figure 2–13.) Define θ as follows:

$$\theta: (\mathbf{R}, +) \to (\mathbf{R}^+, \cdot)$$
$$r \to 2^r = r\theta,$$
$$s \to 2^s = s\theta.$$

2.10 Morphisms and Isomorphic Algebras

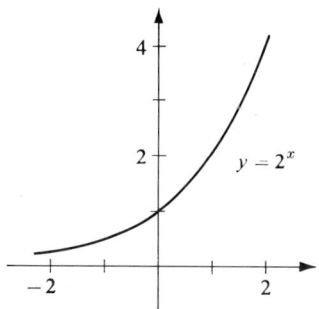

Figure 2–13

Now θ is 1–1, since $r = s$ iff $2^r = 2^s$. Also, θ is onto, since the antecedent r of any positive real number t is $\log_2 t$. That is, $2^r = t \Leftrightarrow r = \log_2 t$. Further,

$$(r + s)\theta = 2^{r+s} = 2^r \cdot 2^s = (r\theta) \cdot (s\theta).$$

So the operations are preserved. Therefore, θ is an isomorphism.

Another way of writing the function θ is $\theta = \{(x, y) : y = 2^x\}$. In a calculus course you would probably write: $\theta(x) = 2^x$.

If b is *any* positive real number, $b \neq 1$, then the map, $\theta : r \to b^r$, is also an isomorphism. This illustrates the fact that frequently there is more than one bijective map that preserves the given operations. Of course, one can also set up bijective maps that do *not* preserve the operations. For instance, ψ is a bijective map, where ψ is defined as follows:

$$\psi : (\mathbf{R}, +) \to (\mathbf{R}^+, \cdot)$$
$$x \to 2^{-x^2}, \text{ if } x \leq 0,$$
$$x \to x + 1, \text{ if } x > 0.$$

However, $r > 0, s > 0 \Rightarrow r\psi = r + 1, s\psi = s + 1, (r + s)\psi = (r + s) + 1$. But $(r\psi)(s\psi) = (r + 1)(s + 1) = rs + r + s + 1 \neq (r + s)\psi$. Therefore, ψ is not an isomorphism. ☐

EXAMPLE 2.48

We now show that the additive group of rational integers $\mathbf{Z}$ is structurally the same as the additive group of even integers, $2\mathbf{Z}$, where $2\mathbf{Z} = \{0, \pm 2, \pm 4, \ldots\}$. Define

$$\theta : (\mathbf{Z}, +) \to (2\mathbf{Z}, +)$$
$$n \to 2n = n\theta,$$
$$k \to 2k = k\theta.$$

So $\qquad n + k \to 2(n + k) = 2n + 2k = n\theta + k\theta.$

Therefore, $(n + k)\theta = n\theta + k\theta$, and the bijective map θ is an isomorphism.

Is the *ring* $(\mathbf{Z}, +, \cdot)$ isomorphic to the *ring* $(2\mathbf{Z}, +, \cdot)$? Under the above map θ we have that $(n \cdot k)\theta = 2(nk)$, while $(n\theta) \cdot (k\theta) = (2n)(2k)$. Therefore,

$(nk)\theta \neq (n\theta)(k\theta)$, and so θ is not a *ring* isomorphism. Would some other map give us an isomorphism of these rings? No, because these two rings are not structurally the same. **Z** is a ring with unity, while ring 2**Z** has no unity. □

In any morphic map θ (bijective or not) from an algebra S to an algebra T, the image set $S\theta$ must possess several important structural features of S itself. These features are given in Theorem 2.30.

THEOREM 2.30

Let $\theta: (S, \otimes) \to (T, \odot)$ be a morphism from algebra S *onto* algebra T.

(a) If $\otimes$ is associative, then so is $\odot$.
(b) If S has identity e with respect to $\otimes$, then T has identity $e\theta$ with respect to $\odot$.
(c) Given that S has identity e, and that $a \in S$; if b is an inverse of a, then $b\theta$ is an inverse of $a\theta$.
(d) If $\otimes$ is commutative, then so is $\odot$.

Proof

Let a, b, and c be any elements of S.

(a) Because the operations are preserved under the map θ, these steps can be taken:

$$[(a\theta) \odot (b\theta)] \odot (c\theta) = (a \otimes b)\theta \odot c\theta = [(a \otimes b) \otimes c]\theta$$
$$= [a \otimes (b \otimes c)]\theta = (a\theta) \odot (b \otimes c)\theta$$
$$= (a\theta) \odot [(b\theta) \odot (c\theta)].$$

(b) Let $a\theta$ be any element in T. Then

$$e\theta \odot a\theta = (e \otimes a)\theta = a\theta, \qquad a\theta \odot e\theta = (a \otimes e)\theta = a\theta.$$

Therefore, $e\theta$ is the identity of T.

(c) Given that b is an inverse of a, then

$$a\theta \odot b\theta = (a \otimes b)\theta = e\theta, \qquad b\theta \odot a\theta = (b \otimes a)\theta = e\theta.$$

Therefore, $b\theta$ is an inverse of $a\theta$.

(d) $b\theta \odot c\theta = (b \otimes c)\theta = (c \otimes b)\theta = c\theta \odot b\theta$.

Therefore, $\odot$ is commutative. □

Theorem 2.30 has given us some guidelines for setting up morphisms. We must map identity elements to identity elements. Also, if two elements correspond, their inverses must correspond. We can use a morphic map to prove that an algebra $(H, \odot)$ is a group. Suppose we know that $(G, \otimes)$ is a group. If we can find a surjective map from G to H that preserves the operations, then we have established the fact that H is also a group. (Why?)

We shall refer to morphisms many times in the chapters which follow.

2.10 Problems

1. Let G and H be the groups of Example 2.46, and let η be the map such that $1\eta = r_0$, $i\eta = r_3$, $(-1)\eta = r_2$, $(-i)\eta = r_1$. Verify that $\eta: G \to H$ is an isomorphism.

2. Assume that $(G, \cdot)$ is a group, and that operation $\otimes$ is defined on set G such that $a \otimes b = b \cdot a$, for all $a, b \in G$.
   (a) Prove: $(G, \cdot)$ is isomorphic to $(G, \otimes)$. (Hint: Let $a \to a'$, where a' is the inverse of a. Show that this is an isomorphic map.)
   (b) By virtue of part (a), can you conclude that $(G, \otimes)$ is a group?

3. Let $P(I)$ be the power set of I, with $I = \{a, b\}$, and let $+$ be the symmetric difference of sets. Show that $(P(I), +)$ is isomorphic to the group of symmetries of a rectangle (that is not a square). (See Example 2.25 and Problem 2.7–14.)

4. (a) Is the Boolean ring of Problem 2.7–14 isomorphic to any of the rings given in Example 2.28?
   (b) Is the group $(P(S), +)$ of Problem 2.7–14 isomorphic to the group $(S, +)$ of Problem 2.6–15?

5. Verify that the map of Example 2.45 is *onto* but not *one-to-one*.

6. Show:
   (a) There is a map of $(\mathbf{Z}(\sqrt{n}), +)$ onto $(\mathbf{Z}, +)$ that is a morphism. (Hint: Prove that $a + b\sqrt{n} \to a$ is a morphism.)
   (b) There is a map of the integral domain $(\mathbf{Z}(\sqrt{n}), +, \cdot)$ onto itself (other than the identity map) that is an isomorphism.
   (Hint: Prove that $a + b\sqrt{n} \to a - b\sqrt{n}$ is an isomorphism.)

7. (a) If G and H are any groups, there exists at least one morphism from G to H. Prove it.
   (b) If S and T are any rings, there exists at least one morphism from S to T. Prove it.

8. Verify that the group of Example 2.25 is not isomorphic to the group of Example 2.46.

9. If $G = \{e, a, a^2, a^3 : a^4 = e\}$, and if $a^i \cdot a^j = a^{i+j}$, construct the Cayley table of $(G, \cdot)$. Find another group that has been discussed that is isomorphic to $(G, \cdot)$.

10. Let $S = \left\{ \begin{bmatrix} r & 0 \\ 0 & r \end{bmatrix} : r \in \mathbf{R} \right\}$, where S is a subset of ring $(M_2(\mathbf{R}), \oplus, \otimes)$. Prove that $(\mathbf{R}, +, \cdot)$ is isomorphic to $(S, \oplus, \otimes)$. (This shows that, structurally, the ring of 2-by-2 matrices with real components contains the real field.)

11. (a) How many structurally different (non-isomorphic) groups are there of order 2, and of order 3?
    (b) Using the set $\{e, a, b, c\}$, where e is the identity, construct group tables

of every possible group of order 4. Show that each group you construct is isomorphic either to the group of Example 2.25 or to the group of Example 2.46, thus proving that there are exactly two non-isomorphic groups of order 4.

12 Construct the operation tables of each ring $(S, +, \cdot)$ of order 2.

13 (a) Prove that $(T, \oplus, \otimes)$ is a subring of $(M_2(\mathbf{Z}), \oplus, \otimes)$ if

$$T = \left\{ \begin{bmatrix} a & 0 \\ b & c \end{bmatrix} : a, b, c \in \mathbf{Z} \right\}.$$

(b) Prove that ring T of part (a) is isomorphic to ring S of Problem 2.7-5(b).

14 (a) Prove that the field $(K, \oplus, \otimes)$ of Problem 2.6-13 is isomorphic to the complex field $(\mathbf{C}, +, \cdot)$. (Hint: Show that the map $\begin{bmatrix} a & b \\ -b & a \end{bmatrix} \to a + bi$ is an isomorphism.)

(b) Show that the complex field is isomorphic to a subring of the quaternion ring of Problem 2.7-16.

15 Let groups $(\mathbf{R}, +)$ and $(G, \otimes)$ be given, with $G = \{(\cos r, \sin r) : r \in \mathbf{R}\}$,

$$(\cos a, \sin a) \otimes (\cos b, \sin b) = (\cos(a + b), \sin(a + b)).$$

Let ψ be the map from $\mathbf{R}$ to G, such that $r\psi = (\cos r, \sin r)$.
(a) Show that ψ is a morphism.
(b) Interpret the sets and the map geometrically (in the Cartesian plane).

16 Prove: If $\theta: (G, \otimes) \to (H, *)$ and $\psi: (H, *) \to (K, \odot)$ are morphisms, then the composite map, $\theta \circ \psi: (G, \otimes) \to (K, \odot)$, is a morphism.

17 Prove: If $G \cong H$ means that "Group G is isomorphic to group H," then $\cong$ is an equivalence relation on a collection of groups.

18 Let $\theta: (S, \otimes) \to (T, \odot)$ be a morphism of S onto T. Use Theorem 2.30 to justify these statements:
(a) If S is a semigroup, then so is T.
(b) If S is a group, then so is T.
(c) If S is an abelian group, then so is T.

19 Let $\theta: (A, \oplus, \otimes) \to (B, +, \cdot)$ be a morphism of A onto B. Use Theorem 2.30 to justify these statements:
(a) If A is a ring, then so is B.
(b) If ring A has a unity, then so does B.
(c) If ring A is commutative, then so is ring B.
(d) If A is an integral domain, and θ is 1–1, then B is an integral domain.
(e) If A is a field and $|B| \geq 2$, then B is a field.

3

The Rational Integers

> *God made the integers, all the rest is the work of man. ... All results of the profoundest mathematical investigation must ultimately be expressible in the simple form of properties of the integers.*—Leopold Kronecker (1823–1891)

3.1 Divisors and the Division Theorem

Certain topics relating to the ordered integral domain $(Z, +, \cdot)$ are not only interesting in themselves, but are useful in studying other algebras. It is for these reasons that we are devoting a chapter to the so-called arithmetic of the rational integers, thus giving you an introduction to the subject known as *theory of numbers*.

Theorem 2.26 and the definitions preceding it will be used in this chapter. A few additional notions concerning divisibility of integers are given in Theorem 3.1.

THEOREM 3.1

Let a, b and c be any elements of $\mathbf{Z}$.
(a) $a \mid b, b \mid a \Rightarrow a = b$ or $a = -b$.
(b) $a \mid b, b \mid a, a, b \in \mathbf{Z}^+ \Rightarrow a = b$.
(c) $b \mid a, a, b \in \mathbf{Z}^+ \Rightarrow 1 \leq b \leq a$.
(d) b is a proper divisor of a, and $a, b \in \mathbf{Z}^+ \Rightarrow 1 < b < a$.

Proof

(a) By Theorem 2.26(c), a is an associate of b. But b and $-b$ are the only associates of b. Hence, $a = b$ or $a = -b$.
(b) By (a), $a = b$ or $a = -b$. But $b \in \mathbf{Z}^+ \Rightarrow -b \in \mathbf{Z}^-$. So $a \neq -b$.
(c) $b \mid a \Rightarrow bc = a$, for some $c \in \mathbf{Z}^+$. If $c = 1$, then $b = a$.
If $c > 1$, then $c = d + 1$, with $d \geq 1$, and $a = bc = b(d + 1) = bd + b$. Since $bd \in \mathbf{Z}^+$, we conclude that $b < a$, by Theorem 2.23(b). Therefore, $1 \leq b \leq a$.
(d) By (c), $1 \leq b \leq a$. If b is a proper divisor of a, then $b \neq 1$ and $b \neq a$. Therefore, $1 < b < a$. □

Let D be an ordered integral domain, and let S be a nonempty subset of D. Set S is said to be *well-ordered* if every nonempty subset T of S contains a smallest element. This means that for each T, where $\emptyset \subset T \subseteq S$, there is an $a \in T$ such that $a \leq t$, for all $t \in T$.

EXAMPLE 3.1

Let $D = \mathbf{Q}$, $S = \mathbf{Q}^+$. We shall verify that the set of positive rationals is *not* well-ordered. Let $T = \{x : x \in \mathbf{Q}^+, x > 5\}$. So T contains all the rational numbers greater than 5. If a is any rational number in T, let $b = (5 + a)/2$. Then b is rational, and $5 < b < a$. No matter which element of T we start with, we can always construct a smaller positive

Figure 3–1

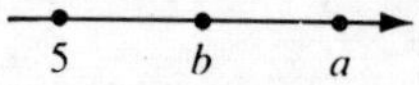

rational number that is in T. So T contains no smallest rational. Therefore, $\mathbf{Q}^+$ is *not* well-ordered. □

Although the set of positive rational numbers is not well-ordered, the set of positive rational integers *is* well-ordered. We do not prove this, but state it as a postulate.

POSTULATE *(Well-Ordering Principle)*

Every nonempty subset of $\mathbf{Z}^+$ contains a smallest positive integer.

We agree that $(\mathbf{Z}, +, \cdot)$ is an ordered integral domain with the property that its set of positive elements is well-ordered. This characterization makes $\mathbf{Z}$ unique among integral domains, for it is not difficult to verify that if $(D, \oplus, \otimes)$ is *any* ordered integral domain with the property that D^+ is well-ordered, then D is isomorphic to $\mathbf{Z}$. (You are asked to prove this in Problem 6.4–14.) We need the Well-Ordering Principle to prove several theorems in this chapter.

You have often divided one positive integer by another, obtaining a quotient and a remainder, with the remainder less than the divisor. We shall prove that this is always possible to do and, further, that the quotient and remainder obtained are unique. But first, we illustrate the theorem.

EXAMPLE 3.2

In Table 3–1, we give various integers a, b, q, and r, with $b \in \mathbf{Z}^+$, that satisfy the equation: $a = bq + r$, where $0 \leq r < b$. We are dividing a by b to obtain the quotient q and remainder r.

Table 3–1 $a = bq + r,$ $0 \leq r < b$

a	b	q	r
17	5	3	2
−17	5	−4	3
0	5	0	0
15	5	3	0
5	17	0	5
−5	17	−1	12

□

THEOREM 3.2 *(Division Theorem)*

If $a \in \mathbf{Z}$, $b \in \mathbf{Z}^+$, then there exist unique rational integers q and r such that
$$a = bq + r, \quad \text{with } 0 \leq r < b.$$

Proof

We prove first (1) that integers q and r exist, and then (2) that they are unique.

(1) Define a set S such that
$$S = \{a - bx : x \in \mathbf{Z}\}.$$
If $0 \in S$, then $a - bx = 0$, for some integer x. So $a = bq + r$, with $q = x$, and $r = 0$.

Now suppose that $0 \notin S$. It is readily verified that S contains some positive integers. (The verification of this is asked for in Problem 3.1–5(b).) Let r be the smallest positive integer in S. By the Well-Ordering Principle, we know that such an r exists. Then, for some integer q, $a - bq = r$, or $a = bq + r$. It remains to show that $r < b$.

Assume that $r \geq b$. Then $r = b + c$, with $0 \leq c < r$. So
$$a = bq + r = bq + (b + c) = b(q + 1) + c.$$

Hence, $c = a - b(q + 1) \in S$. But $c \in S$, $c \geq 0 \Rightarrow c > 0$. This contradicts the choice of r as the *smallest* positive integer in S. Therefore, we must abandon the assumption that $r \geq b$, and so we conclude that
$$a = bq + r, \quad \text{with } 0 < r < b.$$

(2) Suppose there exist integers q, r, q_1, and r_1 which satisfy the conditions:
$$a = bq + r, \quad 0 \leq r < b,$$
$$a = bq_1 + r_1, \quad 0 \leq r_1 < b.$$

There is no loss of generality in choosing the notation so that $q_1 \leq q$. Then $q - q_1 \geq 0$. From the preceding equations:
$$b(q - q_1) = r_1 - r.$$

Now $q - q_1 > 0 \Rightarrow r_1 - r > 0 \Rightarrow b \leq r_1 - r$ (by Theorem 3.1(c)).

However, $r_1 - r \leq r_1 < b$, which would mean that $b < b$, a contradiction. Therefore, $q - q_1 = 0$, and so $q = q_1$. Then $r_1 - r = 0$, and $r = r_1$. □

Mathematical historians tell us that we have a base 10 number system simply because of the biological accident of having 10 fingers (counting thumbs), and that we would probably use base 8 notation if we had but four fingers on each hand. Using a subscript to indicate the base, we know that
$$(2935)_{10} = 2 \cdot (10)^3 + 9 \cdot (10)^2 + 3 \cdot (10) + 5.$$
Similarly, $\quad (3746)_8 = (3 \cdot 8^3 + 7 \cdot 8^2 + 4 \cdot 8 + 6)_{10} = (2022)_{10}.$

So it is a simple matter for a person who is raised in a base-10 environment to convert a number from base 8 notation to base 10 notation, but how does he go from base 10 to base 8? This can be accomplished by applying the Division Theorem repeatedly, as indicated in the following theorem.

THEOREM 3.3

If a is a positive integer in base 10 notation, then
$$a = r_n \cdot b^n + r_{n-1} \cdot b^{n-1} + \cdots + r_2 \cdot b^2 + r_1 \cdot b + r_0$$
$$= (r_n r_{n-1} \cdots r_2 r_1 r_0)_b,$$

where $0 \leq r_i < b$, and the r_i are the remainders obtained by dividing by b as follows:
$$a = q_0 b + r_0,$$
$$q_0 = q_1 b + r_1,$$
$$q_1 = q_2 b + r_2,$$
$$\cdots \cdots$$
$$q_{n-2} = q_{n-1} b + r_{n-1},$$
$$q_{n-1} = 0 \cdot b + r_n;$$

and
$$q_0 > q_1 > \cdots > q_{n-1} > q_n = 0.$$

Proof

By combining the equations given in the theorem, first eliminating q_0 from the first equation (by substituting the value of q_0 given in the second equation), then eliminating q_1, and so forth, one obtains the expression for a in terms of the r_i and powers of b. We leave the details to the reader. □

EXAMPLE 3.3

To change $(29{,}548)_{10}$ to base 8 notation, we can use Theorem 3.3, arranging the repeated division by 8 as follows:

```
                    Remainders
        8 | 29548
          |  3693      4
          |   461      5
          |    57      5
          |     7      1
                0      7
```

Note that we divide each quotient by 8 until we arrive at a quotient of 0. Then
$$(29{,}548)_{10} = (71554)_8. \quad \square$$

3.1 Problems

1. (a) Factor 24 into a product of two or more proper divisors of 24 in all possible ways.
   (b) Classify the divisors of 24 into these three categories: Units, primes, composites.

2. Verify: If p and q are primes in $\mathbf{Z}$ such that $p|q$, then $p = \pm q$.

3. Explain why the rational field $\mathbf{Q}$ contains no primes or composite numbers, even though $\mathbf{Q}$ contains $\mathbf{Z}$.

4. Let $a, b, c, s, t \in \mathbf{Z}$. Prove:
   (a) If $a \neq 0$ and $ab|ac$, then $b|c$.
   (b) If $a|b$ and $s|t$, then $as|bt$.

5. In the proof of Theorem 3.2, the set S was used, with $S = \{a - bx : x \in \mathbf{Z}\}$.
   (a) Given the following values of a and b, find the three smallest positive integers in S and the three largest negative integers in S, and the corresponding values of x.
      (i) $a = 38, b = 7$ (ii) $a = -38, b = 7$
      (iii) $a = 7, b = 38$ (iv) $a = 35, b = 7$
   (b) Prove: if a is any integer and b is any positive integer, then S contains positive integers.

6. The Division Theorem can be stated in a more general form, in which the only stipulation about b is that $b \neq 0$. Then we can say that there exist unique integers q and r such that
   $$a = bq + r, \quad \text{with } 0 \leq r < |b|.$$
   Verify this statement for the case $b < 0$, making use of the results of Theorem 3.2.

7. Prove that the square of any integer is of the form $3k$ or $3k + 1$, but not of the form $3k + 2$.

8. (a) Verify that the relation $|$ of "divides" is a partial order relation on $\mathbf{Z}^+$.
   (b) Is $\mathbf{Z}$ a partially ordered set with respect to "divides"?

9. The following are numbers in base 2 notation. Give these numbers in base 10 notation.
   (a) 10101   (b) 10,000,000,101   (c) 111111

10. The following are numbers in base 10 notation. Change them to base 6 notation, using Theorem 3.3 to do so.
    (a) 298   (b) 3480   (c) 123,456,789

11. (a) If $n = 5$, write the equations of Theorem 3.3, and solve for a in terms of b and the r_i by eliminating the q_i, thus showing that $a = (r_5 r_4 r_3 r_2 r_1 r_0)_b$.
    (b) Generalize part (a), and thus give a proof of Theorem 3.3.

3.2 Greatest Common Divisor and Least Common Multiple

In **Z**, if $d|a$ and $d|b$, then d is called a *common divisor* of a and b. The largest of all the common divisors of a and b is called, appropriately, the *greatest common divisor* (g.c.d.) of a and b. The statement, "g is the greatest common divisor of a and b" is abbreviated as $g = (a, b)$. The context usually makes it clear that one is not talking about the ordered pair (a, b) or the open interval (a, b). It is unfortunate that mathematicians frequently give more than one meaning to the same symbol! Similarly, the greatest of the common divisors of the a_i, for $i = 1, 2, \ldots, n$, is written $g = (a_1, a_2, \ldots, a_n)$.

Note that if $g = (a, b)$, then $g \geq 1$, since 1 is a divisor of every integer. So g is always a positive integer. Since every integer divides zero, $(a, 0) = a$ if $a > 0$, and $(a, 0) = -a$ if $a < 0$. We do not define (a, b) if both a and b are zero. We shall usually be concerned with situations in which neither a nor b is zero. If $(a, b) = 1$, we say that *a is prime to b*, or that a and b are *relatively prime*.

EXAMPLE 3.4

We use the definition to find $(8, 12)$. The set of divisors of 8 is $\{\pm 1, \pm 2, \pm 4, \pm 8\}$, and the set of divisors of 12 is $\{\pm 1, \pm 2, \pm 3, \pm 4, \pm 6, \pm 12\}$. The intersection of these sets is the set, $\{-4, -2, -1, 1, 2, 4\}$, and the largest number in the intersection is 4. Hence, $(8, 12) = 4$.

The set of common divisors of 8 and 15 is $\{-1, 1\}$. Therefore, $(8, 15) = 1$, and so 8 and 15 are said to be relatively prime. □

If $c = ax + by$, where $a, b, x,$ and y are integers, we say that we have expressed c as a *linear combination* of a and b. We now prove that the greatest common divisor of a and b can always be expressed as a linear combination of a and b, and that it is the smallest positive integer that can be so expressed.

THEOREM 3.4

Let $g = (a, b)$, where a and b are integers, not both zero.

(a) If k is the smallest positive integer in the set S, where

$$S = \{ax + by : x, y \in \mathbf{Z}\},$$

then $k = g$. Hence, there exist integers s and t such that

$$g = as + bt.$$

(b) If d is any common divisor of a and b, then $d|g$.

Proof

(a) Set S contains positive integers, since it contains $a^2 + b^2$. (Let $x = a$, $y = b$.) So S contains a smallest positive integer, which we call k. Then for some $x = s$ and $y = t$,

$$k = as + bt.$$

3.2 Greatest Common Divisor and Least Common Multiple

We now proceed to show that $k = g$.

We first verify that $k \mid a$ and $k \mid b$. By the Division Theorem, there exist q and r such that $a = kq + r$, with $0 \le r < k$. Then

$$r = a - kq = a - (as + bt)q = a(1 - sq) + b(-tq) \in S.$$

If $r > 0$, this would contradict the choice of k as the smallest positive integer in S. Hence $r = 0$, and so $a = kq$. Thus, $k \mid a$. The proof that $k \mid b$ is similar. Therefore, k is a common divisor of a and b, and so $k \le g$, by definition of the g.c.d.

Now $g \mid a$ and $g \mid b$, so $g \mid (as + bt)$. Hence $g \mid k$, and so $g \le k$, by Theorem 3.1(c). Therefore, $k = g$, and so $g = as + bt$.

(b) If d is a common divisor of a and b, then $d \mid a$ and $d \mid b$. So $d \mid (as + bt)$. Therefore, $d \mid g$. ☐

We have just proved that the g.c.d. of two integers is divisible by every common divisor of those integers. Frequently the g.c.d. of a and b is *defined* to be the positive common divisor of a and b that is divisible by *every* common divisor of a and b.

EXAMPLE 3.5

Since $(8, 12) = 4$, there exist integers, x and y, such that $8x + 12y = 4$. To find a solution, we apply the cancellation law of multiplication, obtaining: $2x + 3y = 1$. By inspection, we see that one solution is $x = -1, y = 1$. Another is $x = 2, y = -1$. In fact, if $x = s, y = t$ is any solution, then $s + 3n, t - 2n$ is a solution, for each $n \in Z$, since

$$2(s + 3n) + 3(t - 2n) = 2s + 6n + 3t - 6n = 2s + 3t = 1.$$

Therefore, the equation, $8x + 12y = 4$, has infinitely many solutions in integers. ☐

In the above example we have solved a *Diophantine equation*, which is an equation for which solutions in integers are required. Solving various kinds of Diophantine equations has intrigued mathematicians for the past fifteen centuries, and is still an interesting topic in number theory.

EXAMPLE 3.6

Suppose one is asked to find the g.c.d. of 1820 and 231, and to write the g.c.d. as a linear combination of 1820 and 231. We demonstrate a technique for doing this that makes repeated use of the Division Theorem. We divide 1820 by 231, obtaining:

$$1820 = 231(7) + 203.$$

Then we divide 231 by 203, getting:

$$231 = 203(1) + 28.$$

Next $\qquad 203 = 28(7) + 7.$

Finally $\qquad 28 = 7(4) + 0.$

At each step the old divisor became the new dividend, and the old remainder became the new divisor. We continued dividing until we arrived at a zero remainder. The g.c.d. is the last nonzero remainder! (The truth of this statement is not obvious, so we shall soon verify it.) Therefore, $(1820, 231) = 7$.

To complete the exercise, we take the above equations that have nonzero remainders, and solve them for the remainders:

$$7 = 203 - 28(7),$$
$$28 = 231 - 203(1),$$
$$203 = 1820 - 231(7).$$

The object is to eliminate the remainders, 203 and 28, from the first equation. We substitute the right-hand member of the second equation for 28 in the first equation, obtaining:

$$7 = 203 - (231 - 203)7,$$

so $\qquad 7 = 203(8) - 231(7).$

We next eliminate the remainder 203 from this last equation by substitution:

$$7 = [1820 - 231(7)]8 - 231(7),$$

or $\qquad 7 = 1820(8) + 231(-63).$

Therefore, $x = 8, y = -63$, is a solution of the Diophantine equation,

$$1820x + 231y = 7. \quad \square$$

The technique we used in Example 3.6 to find the greatest common divisor of the given integers is known as the *Euclidean Algorithm*. In the language of a mathematician in computer science: "An *algorithm* is a precisely defined sequence of rules telling how to produce specified output information from given input information in a finite number of steps." (D. Knuth, in the *American Scientist*, November–December, 1973.) After we have proved a preliminary theorem (a *lemma*), we shall verify that the algorithm we employed is correct.

THEOREM 3.5

If a, b, c, d are any integers such that $a = bc + d$, then $(a, b) = (b, d)$.

3.2 Greatest Common Divisor and Least Common Multiple

Proof

Let $g = (a, b)$, and $h = (b, d)$. Now

$$g = (a, b) \Rightarrow g \mid a, g \mid b \Rightarrow g \mid (a - bc) \Rightarrow g \mid d.$$

So g is a common divisor of b and d. Hence, $g \leq h$. Also,

$$h = (b, d) \Rightarrow h \mid b, h \mid d \Rightarrow h \mid (bc + d) \Rightarrow h \mid a.$$

So h is a common divisor of a and b, and thus $h \leq g$. Therefore, $g = h$. □

THEOREM 3.6 *(The Euclidean Algorithm)*

Let a and b be integers such that $a > b > 0$.

(a) By repeated applications of the Division Theorem, the following finite set of equations is obtained:

$$\begin{aligned}
a &= bq_1 + r_1, & 0 &< r_1 < b, \\
b &= r_1 q_2 + r_2, & 0 &< r_2 < r_1, \\
r_1 &= r_2 q_3 + r_3, & 0 &< r_3 < r_2, \\
&\;\;\vdots & &\;\;\vdots \\
r_{k-3} &= r_{k-2} q_{k-1} + r_{k-1}, & 0 &< r_{k-1} < r_{k-2}, \\
r_{k-2} &= r_{k-1} q_k + r_k, & 0 &< r_k < r_{k-1}, \\
r_{k-1} &= r_k q_{k+1} + 0. & r_{k+1} &= 0.
\end{aligned}$$

(b) $r_k = (a, b)$.

(c) $r_k = as + bt$, where s and t are found by eliminating, by substitution, the remainders r_i, for $i < k$, from the equation, $r_k = r_{k-2} - r_{k-1} q_k$, the substitutions being:

$$r_i = r_{i-2} - r_{i-1} q_i, \quad \text{for } i = 3, 4, \ldots, k - 1,$$
$$r_2 = b - r_1 q_2, \quad \text{and} \quad r_1 = a - bq_1.$$

Proof

(a) Since each $r_i \geq 0$, and $b > r_1 > r_2 > \cdots$, after a finite number of divisions we must arrive at a zero remainder. Hence, it is always possible to obtain such a set of $k + 1$ equations, with $k \geq 0$.

(b) By Theorem 3.5:

$$(a, b) = (b, r_1) = (r_1, r_2) = (r_2, r_3) = \cdots = (r_{k-1}, r_k).$$

But $r_k \mid r_{k-1}$, and so $(r_{k-1}, r_k) = r_k$. Therefore, $(a, b) = r_k$.

(c) Solving each equation of (a) for the remainder, we get:

$$r_k = r_{k-2} - r_{k-1} q_k, \quad r_{k-1} = r_{k-3} - r_{k-2} q_{k-1}, \ldots.$$

By substitution, we then have that

$$r_k = r_{k-2} - (r_{k-3} - r_{k-2} q_{k-1}) q_k = r_{k-2}(1 + q_{k-1} q_k) + r_{k-3}(-q_k).$$

So we have expressed r_k as a linear combination of r_{k-2} and r_{k-3}. We next eliminate r_{k-2}, and express r_k as a linear combination of r_{k-3} and r_{k-4}. After a finite number of these substitutions, we will have expressed r_k as a linear combination of a and b, and the coefficients of a and b will be s and t, respectively. □

Observe that Theorem 3.6 gives us a *method* for solving the equation, $g = ax + by$, where $g = (a, b)$, while Theorem 3.4 merely informs us of the *existence* of a solution. To the mathematician, however, Theorem 3.4 is the more important of the two theorems because it can be used to obtain other information about divisibility of integers. We illustrate this use in the next two theorems.

THEOREM 3.7

Let a, b, c and p be in **Z**.
(a) If $a|bc$ and $(a, b) = 1$, then $a|c$.
(b) If $p|bc$, where p is a prime, then either $p|b$ or $p|c$.

Proof

(a) Since $(a, b) = 1$, there exist integers s and t such that $as + bt = 1$. Multiplying each side of the equation by c, we get that $a(sc) + (bc)t = c$. Now $a|a$ and $a|bc$, so $a|(asc + bct)$, and $a|c$.

(b) The only positive divisors of p are 1 and $|p|$, and so either $(p, b) = |p|$ or $(p, b) = 1$. If $(p, b) = |p|$, then $p|b$. If $(p, b) = 1$, then $p|c$, by part (a). □

THEOREM 3.8

(a) If $n \in \mathbf{Z}^+$, then $(na, nb) = n \cdot (a, b)$.
(b) If $(a, b) = g$, then $(a/g, b/g) = 1$.

Proof

(a) Let $g = (a, b)$, and $k = (na, nb)$. Since g is a common divisor of a and b, then ng is a common divisor of na and nb. Hence, $ng \le k$. But $g = as + bt$, for some s and t, and so $ng = (na)s + (nb)t$. This means that $k \le ng$, since k is the smallest positive integer that can be expressed as a linear combination of na and nb. Therefore, $k = ng$.

(b) By part (a):

$$g \cdot (a/g, b/g) = (g \cdot (a/g), g \cdot (b/g)) = (a, b) = g.$$

Therefore, $(a/g, b/g) = 1$. □

If $a|m$ and $b|m$, then m is called a *common multiple* of a and b. If h is the smallest positive integer that is a common multiple of a and b, then h is called the *least common multiple* (l.c.m.) of a and b. This is written $h = [a, b]$. We shall prove that *every* common multiple of a and b is divisible by h.

THEOREM 3.9

If $h = [a, b]$, and m is any common multiple of a and b, then $h \mid m$.

Proof

There exist q and r such that $m = hq + r$, with $0 \le r < h$. Now $a \mid m$ and $a \mid h$, so $a \mid (m - hq)$. Thus, $a \mid r$. Similarly, $b \mid r$. So r is a common multiple of a and b. But if $0 < r < h$, we have a contradiction of the definition of l.c.m. Hence $r = 0$, $m = hq$, and so $h \mid m$. □

According to Theorem 3.9, the set of common multiples of a and b is the set $\{0, \pm h, \pm 2h, \pm 3h, \ldots\} = \{nh : n \in \mathbf{Z}\}$, if $h = [a, b]$.

The next theorem enables one to find the l.c.m. of two numbers easily if their g.c.d. is known.

THEOREM 3.10

If $a, b \in \mathbf{Z}^+$, then $[a, b] \cdot (a, b) = ab$.

Proof

Let $h = [a, b]$ and $g = (a, b)$. We must show that $hg = ab$. Now $a \mid a(b/g)$ and $b \mid b(a/g)$, so ab/g is a common multiple of a and b. Hence, $ab/g = nh$, for some $n \in \mathbf{Z}^+$. From this equation we get that

$$a/g = n(h/b), \qquad b/g = n(h/a).$$

So n is a common divisor of a/g and b/g. However, $(a/g, b/g) = 1$, which means that $n = 1$. Therefore, $ab/g = h$, and so $ab = hg$. □

EXAMPLE 3.7

To find $[1950, 1938]$, we apply the Euclidean Algorithm, and find that $(1950, 1938) = 6$. Then

$$[1950, 1938] = \frac{1950 \cdot 1938}{6} = 629{,}850. \quad \square$$

One can speak of the least common multiple of any finite number of integers. If

$$a_i \mid m, \quad \text{for } i = 1, 2, \ldots, n,$$

then m is called a *common multiple* of the n integers: $a_1, a_2, \ldots, a_n$. If h is the smallest positive integer in the set of common multiples of the a_i, then h is called the *least common multiple*, and we write

$$h = [a_1, a_2, \ldots, a_n].$$

In Problem 3.1-8 you were asked to verify that the set of positive integers is a partially ordered set with respect to the relation, |, of divisibility. Because of this, we can construct an inclusion diagram of S, if S is the set of positive divisors of a given positive integer n. It is interesting to note that, in the diagram, we can follow paths upward from any two points, a and b, and the first intersection of these paths is the l.c.m. of a and b. Similarly, the first intersection of paths leading down from a and b is their g.c.d.

EXAMPLE 3.8

The set S of positive divisors of 75 is

$$S = \{1, 3, 5, 15, 25, 75\},$$

and the set T of positive divisors of 30 is

$$T = \{1, 2, 3, 5, 6, 10, 15, 30\}.$$

The inclusion diagrams of S and T, with respect to divisibility, are shown in Figure 3-2.

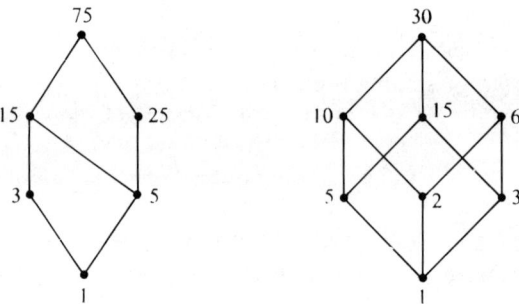

Figure 3-2

3.2 Problems

1. (a) Find two solutions in integers of each equation:

    $$21x + 14y = 7, \quad 8x + 15y = 1.$$

    (b) Can you find solutions in integers of: $6x + 15y = 4$?

2. Complete the proof of Theorem 3.4(a), by verifying that $k \mid b$.

3. Give as much information as you can about (a, b) if there exist s and t such that:
    (a) $as + bt = 6$,   (b) $as + bt = 1$,   (c) $as + bt = -5$.

4. Use the Euclidean Algorithm to find the g.c.d. g of each pair of integers given. Then express g as a linear combination of the integers.
    (a) (867, 810)   (b) (1111, 500)   (c) (397670, 12054)

5   Use Theorem 3.10 to find the l.c.m. of each pair of integers in Problem 4.

6   Give an example to show that if $a \mid bc$, it does not follow that $a \mid b$ or $a \mid c$.

7   (a) $[12, 30, 45] =$ ___   (b) $[10, 21, 35] =$ ___   (c) $[10, 11, 21] =$ ___

8   Construct an inclusion diagram of the set of positive divisors of n, if n is the following integer.
    (a) 5   (b) 6   (c) 8
    (d) 9   (e) 16   (f) 20
    (g) 24

9   If n is an integer such that $2 \leq n \leq 29$, then the inclusion diagram of the set of positive divisors of n is structurally the same as one of the diagrams you made in Problem 8. Partition the set $\{2, 3, \ldots, 29\}$ with respect to the equivalence relation:

    $n \sim m \Leftrightarrow$ The set of positive divisors of n has the same inclusion diagram as the set of positive divisors of m.

    What can you conclude about two integers that are in the same equivalence class?

10 Find all pairs of integers x and y that satisfy both equations simultaneously: $(x, y) = 10$, $[x, y] = 300$.

11 If $n \in \mathbf{Z}^+$, find $(n, n + 1)$ and $[n, n + 1]$, justifying your answers.

12 Prove: $a^2 \mid b^2 \Rightarrow a \mid b$.

13 Prove: 4 does not divide $n^2 + 2$, for any integer n.

14 Verify: $(a, b) = (a, -b)$, and $(a, b) = (a + bn, b)$, for each n.

15 Prove:
    (a) $(a, b) = g, [a, b] = h \Rightarrow g \mid h$.
    (b) $g \mid h \Rightarrow$ There exist a, b such that $(a, b) = g, [a, b] = h$.

16 Prove: If $c > 0$, then $[ac, bc] = c \cdot [a, b]$.

17 Let $ax + by = c$ be a Diophantine equation, with $(a, b) = g$.
    Prove:
    (a) If the equation has a solution, then $g \mid c$.
    (b) If $g \mid c$, then the equation has a solution. (Use Theorem 3.4.)

18 Let x_0, y_0 be a solution of the Diophantine equation, $ax + by = c$, and let $g = (a, b)$. Prove:
    (a) If $s = x_0 + n(b/g), t = y_0 - n(a/g)$, where n is any integer, then s, t is a solution.
    (b) If x_1, y_1 is *any* solution, then

    $$x_1 = x_0 + nb/g, \quad y_1 = y_0 - na/g, \quad \text{for some } n \in \mathbf{Z}.$$

    (Note that Problem 17 has given a necessary and sufficient condition that a linear Diophantine equation in two unknowns have a solution, and Problem 18 has given the complete solution set, once *one* solution is known. The one solution can be found without trial and error by means of the Euclidean Algorithm.)

19 Use the Euclidean Algorithm to obtain a solution of the Diophantine equation, $119x + 70y = 161$. Then give all solutions.

20 (a) What two branches of mathematics was Euclid interested in?

(b) What incident relating to Euclid leads to the remark that "students haven't changed much"? (See Appendix E–1.)

21 If $g = (a_1, a_2, \ldots, a_n)$, state and prove a theorem which is a generalization of Theorem 3.4.

3.3 Mathematical Induction

There are many statements that are true for every natural number n. For instance, one such statement is the following:

$$1^2 + 2^2 + \cdots + n^2 = (n/6)(n + 1)(2n + 1).$$

By substitution, you can verify that this statement is true if $n = 1, n = 2, n = 3, \ldots$, but you certainly cannot employ that technique to prove that the statement is true for *every* n. Fortunately, mathematicians have hit upon a simple principle, called the *Induction Principle*, that enables you to prove a statement like the above in just a few steps. The next example illustrates how the principle operates.

EXAMPLE 3.9

Sometimes you are interested in a statement that is true for all $n \geq c$, where c is some natural number greater than 1. An instance of this is the *Generalized Distributive Property*, which can be stated as follows:

If $a_1, a_2, \ldots, a_n$, and b are in ring $(S, +, \cdot)$, then

$$b(a_1 + a_2 + \cdots + a_n) = ba_1 + ba_2 + \cdots + ba_n, \quad \text{for } n \geq 2.$$

We let P_n represent the preceding statement or proposition. Using summation notation, we have:

$$P_n: \quad b \sum_{i=1}^{n} a_i = \sum_{i=1}^{n} (ba_i).$$

Then P_2 is the statement: $b(a_1 + a_2) = ba_1 + ba_2$.

We know that P_2 is true, by the distributive property of a ring. We don't know if P_3 is true or false, but *if* we take as our hypothesis that P_3 is true, can we *then* conclude that P_4 is true? We show that $P_3 \Rightarrow P_4$ as follows:

$$\begin{aligned}
b(a_1 + a_2 + a_3 + a_4) &= b[(a_1 + a_2 + a_3) + a_4] & \text{[Definition]} \\
&= b(a_1 + a_2 + a_3) + ba_4 & \text{[Distrib. Law]} \\
&= (ba_1 + ba_2 + ba_3) + ba_4 & \text{[By } P_3\text{]} \\
&= ba_1 + ba_2 + ba_3 + ba_4. & \text{[By def.]}
\end{aligned}$$

Let's try to generalize the steps we have just taken. If k is any integer, with $k \geq 2$, we would like to show that *if P_k is true, then P_{k+1} is true.*

$$\begin{aligned}
b(a_1 + a_2 + \cdots + a_k + a_{k+1}) &= b[(a_1 + a_2 + \cdots + a_k) + a_{k+1}] \\
&= b(a_1 + a_2 + \cdots + a_k) + ba_{k+1} \\
&= (ba_1 + ba_2 + \cdots + ba_k) + ba_{k+1} \\
&= ba_1 + ba_2 + \cdots + ba_k + ba_{k+1}.
\end{aligned}$$

We have verified that:

(i) P_2 is true, (ii) $P_k \Rightarrow P_{k+1}$, for all $k \geq 2$.

Is this enough to allow us to conclude that P_n is true for *all* $n \geq 2$? The Induction Principle says it is! □

THEOREM 3.11 *Induction Principle (First Form)*

Let c be a fixed positive integer, and let statement P_n be associated with each positive integer n, for $n \geq c$. *If*:

(i) P_c is true, and

(ii) P_k is true $\Rightarrow P_{k+1}$ is true, for every $k \geq c$,

then P_n is true for every integer $n \geq c$.

Proof

Assume that P_n is false for at least one integer n, where $n > c$. Let

$$F = \{x : P_x \text{ is false}, x > c\}.$$

Now F is a nonempty set of positive integers, and so F contains a smallest integer m. Since $m > c$, we have that $m - 1 \geq c$. But $m - 1 \notin F$, and so P_{m-1} is true. By (ii), if P_{m-1} is true, then P_m is true. Hence, P_m is true. But $m \in F$, which makes P_m false. Because of this contradiction we must abandon the assumption that P_n is false for some $n > c$. Therefore, P_n is true for every integer $n > c$. □

When we prove a theorem by using the Induction Principle, we say that we are proving it by *induction on n*, or by *mathematical induction*.

EXAMPLE 3.10

We illustrate the use of the Induction Principle. Let P_n be the statement:

$$\sum_{i=1}^{n} (a_i + b_i) = \left(\sum_{i=1}^{n} a_i\right) + \left(\sum_{i=1}^{n} b_i\right).$$

P_1 is true, for

$$\sum_{i=1}^{1} (a_i + b_i) = a_1 + b_1 = \left(\sum_{i=1}^{1} a_i\right) + \left(\sum_{i=1}^{1} b_i\right).$$

To prove that $P_k \Rightarrow P_{k+1}$:

$$\begin{aligned}
\sum_{i=1}^{k+1} (a_i + b_i) &= \left[\sum_{i=1}^{k} (a_i + b_i)\right] + (a_{k+1} + b_{k+1}) && \text{[Def. of +]} \\
&= \left(\sum_{i=1}^{k} a_i + \sum_{i=1}^{k} b_i\right) + (a_{k+1} + b_{k+1}) && \text{[}P_k\text{ true, by Hyp.]} \\
&= \left[\left(\sum_{i=1}^{k} a_i\right) + a_{k+1}\right] + \left[\left(\sum_{i=1}^{k} b_i\right) + b_{k+1}\right] && \text{[Assoc., comm. prop. of +]} \\
&= \left(\sum_{i=1}^{k+1} a_i\right) + \left(\sum_{i=1}^{k+1} b_i\right). && \text{[Def. of +]}
\end{aligned}$$

This completes the verification that the two hypotheses of Theorem 3.11 are fulfilled. Therefore, by Theorem 3.11, P_n is true for every $n \geq 1$. □

EXAMPLE 3.11

To show the versatility of use of the Induction Principle, we shall prove the Division Theorem (Theorem 3.2) by induction on a. Let P_a be the statement: If a and b are positive integers, there exist integers q and r such that $a = bq + r$, with $0 \leq r < b$.

P_1: We must find q and r such that $1 = bq + r$, with $0 \leq r < b$.

If $b = 1$, then we let $q = 1, r = 0$. If $b > 1$, then we let $q = 0, r = 1$. Therefore, P_1 is true.

To show that $P_k \Rightarrow P_{k+1}$:
If P_k is true, there exist q_0 and r_0 such that

$$k = bq_0 + r_0, \qquad 0 \leq r_0 < b.$$

Adding 1 to each side of this equation, we get that

$$k + 1 = bq_0 + (r_0 + 1), \qquad 0 \leq r_0 + 1 \leq b.$$

Case 1: $r_0 + 1 < b$.
Let $q = q_0, r = r_0 + 1$. Then $k + 1 = bq + r$, with $0 < r < b$.

Case 2: $r_0 + 1 = b$.
Let $q = q_0 + 1, r = 0$. Then $k + 1 = bq + 0$.
Therefore, in either case, $P_k \Rightarrow P_{k+1}$.

The two conditions of Theorem 3.11 are satisfied, and so P_a is true for every integer $a \geq 1$. Now that we have established the Division Theorem by induction for $a > 0$, it is an easy matter to use this knowledge in order to prove the theorem for $a \leq 0$. (See Problem 3.3–6.) □

A slightly altered version of the Induction Principle is frequently found to be more useful. We shall illustrate its use in proving theorems in the next section.

THEOREM 3.12 *Induction Principle (Second Form)*

Let c be a fixed positive integer, and let statement P_n be associated with each positive integer n, where $n \geq c$. If:
(i) P_c is true, and
(ii) $P_c, P_{c+1}, \ldots, P_{k-1}$ are true $\Rightarrow P_k$ is true, for every $k > c$,

then P_n is true for every $n \geq c$.

Observe that Theorems 3.11 and 3.12 differ only in the wording of part (ii) of the hypothesis. The proof of Theorem 3.12 differs by about one sentence from the proof given for Theorem 3.11, and will be left as an exercise. All problems at the end of this section that require a proof by induction can be done by applying Theorem 3.11.

3.3 Problems

1. Prove by mathematical induction:
   (a) $1^2 + 2^2 + \cdots + n^2 = (n/6)(n + 1)(2n + 1)$, for all $n \in \mathbf{Z}^+$.
   (b) $1^3 + 2^3 + \cdots + n^3 = [n(n + 1)/2]^2$, for all $n \in \mathbf{Z}^+$.

2. Use induction on n to prove that for each positive integer n,
$$\frac{1}{1 \cdot 2} + \frac{1}{2 \cdot 3} + \frac{1}{3 \cdot 4} + \cdots + \frac{1}{n(n+1)} = \frac{n}{n+1}.$$

3. Use induction to prove that if n is any positive integer, then:
   (a) $2|(n^2 + n)$.   (b) $6|(n^3 + 5n)$.

4. Let P_n be the statement: $n^2 + n + 41$ is a prime.
   Verify the statement for $n = 1, 2, 3, \ldots, 10$.
   Is P_n true for all $n \in \mathbf{Z}^+$?

5. Prove: $P_k \Rightarrow P_{k+1}$, where k is any positive integer, and P_n is the statement:
$$1 + 2 + \cdots + n = \frac{(n + 1/2)^2}{2}.$$
   Is P_n true for all $n \in \mathbf{Z}^+$?

6 In Example 3.11 we proved the Division Theorem for $a > 0$ by induction on a. Using the fact that the theorem holds for $a > 0$, prove that it holds for $a < 0$ without employing the Induction Principle. How about $a = 0$?

7 Let p be a prime, and let $a_i \in \mathbf{Z}$. Use induction on n to prove that if $p | a_1 a_2 \cdots a_n$, then $p | a_j$, for some j, with $1 \le j \le n$. (Hint: You may use Theorem 3.7(b) in your proof.)

8 Let $(H, +)$ be a subgroup of $(\mathbf{Z}, +)$.
(a) Prove that if $1 \in H$, then $H = \mathbf{Z}$.
(b) Prove that if $4, 7 \in H$, then $1 \in H$, and so $H = \mathbf{Z}$.
(c) Prove that if $a, b \in H$, where a and b are relatively prime integers, then $H = \mathbf{Z}$.

9 Prove Theorem 3.12.

10 The complex number, $a + bi$, can be represented by the point in the Cartesian plane whose coordinates are (a, b). If $r = \sqrt{a^2 + b^2}$, then

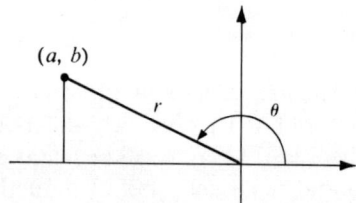

$a + bi = r \cos \theta + (r \sin \theta)i = r(\cos \theta + i \sin \theta)$. Use induction on n to prove *De Moivre's Theorem*:

$$[r(\cos \theta + i \sin \theta)]^n = r^n(\cos n\theta + i \sin n\theta), \quad \text{for all } n \in \mathbf{Z}^+.$$

11 The *Generalized Associative Property* is as follows:
If $*$ is an associative operation on a set S, and if $a_i \in S$, then

$$a_1 * a_2 * \cdots * a_n = (a_1 * a_2 * \cdots * a_r) * (a_{r+1} * \cdots * a_n),$$

for $n \ge 3$, and $1 \le r < n$. Prove this statement by induction on n.

12 Prove: If $(S, +, \cdot)$ is a ring, and $a_i, b_j \in S$, for $i, j \in \mathbf{Z}^+$, then

$$\sum_{j=1}^{m} \left(\sum_{i=1}^{n} a_i b_j \right) = \sum_{i=1}^{n} \left(\sum_{j=1}^{m} a_i b_j \right), \quad \text{for all } m, n \in \mathbf{Z}^+.$$

(Hint: Let m be a given positive integer, and let P_n represent the above equation, which is to be verified by induction on n.)

13 By definition: $k! = k(k - 1)(k - 2) \cdots 3 \cdot 2 \cdot 1$, $0! = 1$, and

$$\binom{k}{r} = \frac{k!}{r!(k-r)!}, \quad \text{for } r = 0, 1, 2, \ldots, k,$$

where $\binom{k}{r}$ represents the number of combinations of k elements, taken r at a time. Without using mathematical induction, prove:

$$\binom{k}{i} + \binom{k}{i-1} = \binom{k+1}{i}.$$

14 The *Binomial Theorem* may be stated as follows:
If a and b are in a commutative ring with unity, then

$$(a+b)^n = \sum_{i=0}^{n} \binom{n}{i} a^{n-i} b^i, \quad \text{for all } n \in \mathbf{Z}^+.$$

(a) Use this equation to write the expansions of $(a+b)^n$, for $n = 1, 2, 3,$ and 4.

(b) Use induction on n to prove the Binomial Theorem.
(Hint: To verify that $P_k \Rightarrow P_{k+1}$: (1) Write the expression for $(a+b)^k$. (2) Multiply each side by $(a+b)$. (3) Use the result of Problem 13 to show that the coefficient of $a^{(k+1)-i}b^i$ is $\binom{k+1}{i}$.)

15 Study Appendix C, and use the Induction Principle to prove these theorems in the appendix: C–5, C–7 through C–11, C–13(b).

16 (a) Besides mathematics and logic, what was another interest of Peano? (See Appendix E–19.)

(b) Compare Peano's Postulates, as given in Appendix C, with their formulation in Appendix E–19. Show that PP–2 in Appendix C combines the notions found in postulates 2, 3, and 4 of Appendix E–19.

17 A board contains 3 vertical rods. On one of these rods is a pile of disks, no two of which have the same diameter. The disks are stacked so that the largest is on the bottom, the smallest is on top, and each disk is smaller than the one directly beneath it. The object is to transfer the entire pile to another rod by moving one disk at a time, and never placing a disk on top of a smaller disk. Prove that a stack of n disks can be transferred in $2^n - 1$ moves.

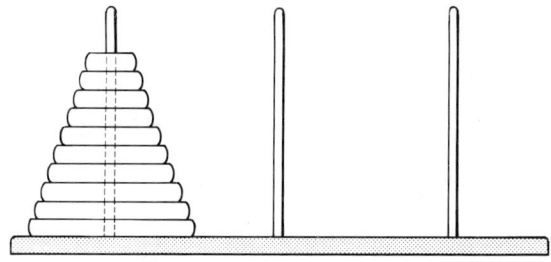

3.4 The Fundamental Theorem of Arithmetic

We shall use the second form of the Induction Principle to prove the main theorems of this section.

THEOREM 3.13

Let $n \geq 2$, with $n \in \mathbf{Z}^+$. Then n can be factored into a product of positive primes.

Proof

Let P_n be the statement: n can be factored into a product of positive primes. We see that P_2 is true, since 2 is a product of one prime. We shall prove that: $P_2, P_3, \ldots, P_{k-1} \Rightarrow P_k$.

Let k be any positive integer greater than 2. If k is a prime, we are done. If k is composite, then there exist integers a and b such that

$$k = ab, \quad \text{where } 2 \leq a < k, \quad 2 \leq b < k.$$

By the induction hypothesis, P_a and P_b are true. So

$$a = p_1 p_2 \cdots p_u, \quad \text{and} \quad b = q_1 q_2 \cdots q_v,$$

where the p_i and q_j are primes. Then

$$k = p_1 p_2 \cdots p_u q_1 q_2 \cdots q_v,$$

and so we have expressed k as a product of primes. Therefore, by Theorem 3.12, every integer $n \geq 2$ can be factored into a product of positive primes. □

It is obvious that we could not use the first form of the Induction Principle to prove Theorem 3.13, since P_{k-1} would not imply P_k. We needed to know that a and b could be factored, and the only information we had about a and b were that they were between 2 and $k - 1$. But since our induction hypothesis stated that P_i is true for $2 \leq i \leq k - 1$, our difficulty was resolved!

Many domains besides the domain of integers have the property that each element can be factored into a product of primes. However, in some of these domains the factorization is not unique. This means that there is an element a in the domain such that

$$a = p_1 p_2 \cdots p_r = q_1 q_2 \cdots q_s,$$

where each p_i and q_j is a prime, but p_1, for instance, does not equal any q_j, nor is it an associate of a q_j. $Z(\sqrt{-6})$ is an example of an integral domain in which factorization is not unique. (See Example 2.43.)

Before we prove that $\mathbf{Z}$ is a *UFD* we state the following lemma, whose proof was asked for in Problem 3.3–7.

THEOREM 3.14

If $p \mid a_1 a_2 \cdots a_n$, where $p, a_i \in \mathbf{Z}$, and p is a prime, then $p \mid a_j$ for some j, with $1 \leq j \leq n$.

THEOREM 3.15 (The Fundamental Theorem of Arithmetic)

Every integer $n > 1$ can be factored *uniquely* into a product of positive primes (apart from the order of the factors).

Proof

By Theorem 3.13, n can be factored into a product of positive primes. Let P_n be the statement: Factorization of n into a product of positive primes is unique. Then P_q is obviously true for each prime q, since q is already a prime, and so cannot equal a product of two or more primes. Hence P_2 is true.

We now prove that $P_2, P_3, \ldots, P_{k-1} \Rightarrow P_k$. That is, we verify that if every positive integer less than k can be factored uniquely, then so can k. Assume that

$$k = p_1 p_2 \cdots p_r = q_1 q_2 \cdots q_s,$$

where the p_i and q_j are positive primes, and $r > 1$. Then $p_1 \mid q_1 q_2 \cdots q_s$, and so $p_1 \mid q_j$ for some j, by Theorem 3.14. For convenience of notation we suppose that $p_1 \mid q_1$. But the only positive divisors of q_1 are 1 and q_1. Hence, $p_1 = q_1$. Then

$$k = p_1(p_2 \cdots p_r) = p_1(q_2 \cdots q_s),$$

and by the cancellation property:

$$m = p_2 \cdots p_r = q_2 \cdots q_s, \quad \text{with } 2 \leq m < k.$$

By the induction hypothesis, factorization of m is unique. So each p_i equals a q_j, and $r = s$. By a re-ordering, if necessary, we have that $p_i = q_i$, for $i = 2, 3, \ldots, r$. So P_k is true, and therefore, by Theorem 3.12, P_n is true for each $n \geq 2$. □

The *standard form* of a positive integer a is defined to be:

$$a = p_1^{n_1} \cdot p_2^{n_2} \cdot \cdots \cdot p_s^{n_s},$$

where the n_i are positive integers, and the p_i are positive primes such that $p_1 < p_2 < \cdots < p_s$. According to the Fundamental Theorem, the standard form of an integer is unique.

EXAMPLE 3.14

If we have the standard forms of two integers, it is then a simple matter to find their g.c.d. and their l.c.m. For instance,

$$2800 = 2^4 \cdot 5^2 \cdot 7^1, \quad 24{,}750 = 2^1 \cdot 3^2 \cdot 5^3 \cdot 11^1.$$

If we write all the prime factors of 24,750 in the factoring of 2800, and all the prime factors of 2800 in the factoring of 24,750, we have:

$$2800 = 2^4 \cdot 3^0 \cdot 5^2 \cdot 7^1 \cdot 11^0, \qquad 24{,}750 = 2^1 \cdot 3^2 \cdot 5^3 \cdot 7^0 \cdot 11^1.$$

The g.c.d. will be the product of these five primes, the exponent of a prime p being the *smaller* of the two exponents in the two factorings. The l.c.m. will be the product of the five primes, the exponent of a prime p being the *larger* of the two exponents in the two factorings. So

$$(2800, 24750) = 2^1 \cdot 3^0 \cdot 5^2 \cdot 7^0 \cdot 11^0 = 50,$$
$$[2800, 24750] = 2^4 \cdot 3^2 \cdot 5^3 \cdot 7^1 \cdot 11^1 = 1{,}386{,}000. \quad \square$$

3.4 Problems

1 Find the g.c.d. and the l.c.m. of each pair of integers by making use of the standard forms of the integers.
(a) 120 and 756 (b) 6600 and 26,000

2 Prove that $\sqrt{7}$ is not a rational number.
(Hint: Assume that $\sqrt{7} = a/b$, where $a, b \in Z$, and $(a, b) = 1$. Use the equation, $7b^2 = a^2$, to arrive at a contradiction.)

3 Let $m > 1$. If $m^2 = p_1^{n_1} \cdot p_2^{n_2} \cdot \cdots \cdot p_s^{n_s}$, where the p_i are distinct positive primes, verify that each n_i is an even integer. (Use the Fundamental Theorem of Arithmetic to prove this.)

4 Given integers a, b, and c such that $ab = c^2$, with $(a, b) = 1$; verify that $a = s^2$ and $b = t^2$, for some integers s and t.

5 (a) Let $p_1, p_2, \ldots, p_n$ be the first n positive primes, with

$$p_1 < p_2 < \cdots < p_n.$$

If $t = 1 + (p_1 \cdot p_2 \cdot \cdots \cdot p_n)$, verify that t is divisible by a prime q, where $q > p_n$.
(b) Use part (a) to prove that there are infinitely many positive primes. (This simple, but elegant, proof that the number of primes is infinite was first given by Euclid.)

6 Let $S = \{2 + (n + 1)!, 3 + (n + 1)!, \ldots, (n + 1) + (n + 1)!\}$.
(a) Verify that S is a set of n consecutive composite integers.
(b) Give the first and last integers in a set of one million consecutive composite integers whose largest integer is not divisible by 2.

7 Let n be expressed in base-10 notation. Prove:
(a) $3 \mid n \Leftrightarrow 3$ divides the sum of the digits of n.
(b) $9 \mid n \Leftrightarrow 9$ divides the sum of the digits of n.

8   Let $(a_1, a_2, \ldots, a_n)$ denote the g.c.d. of the n integers: $a_1, a_2, \ldots, a_n$. Prove: If $n \geq 3$, then

$$(a_1, a_2, \ldots, a_n) = ((a_1, a_2, \ldots, a_{n-1}), a_n).$$

3.5 Congruence Modulo n

Let n be a positive integer. If a and b are integers such that $n \mid (a - b)$, we say that *a is congruent to b modulo n*. Symbolically,

$$a \equiv b \pmod{n}.$$

This is equivalent to saying that there is an integer k such that $a - b = kn$, or that $a = b + kn$.

EXAMPLE 3.15

Let $n = 6$. Then

$$a \equiv b \pmod{6} \Leftrightarrow 6 \mid (a - b) \Leftrightarrow a - b = 6k, \text{ for some } k \in \mathbf{Z}$$
$$\Leftrightarrow a = b + 6k.$$

For instance:

$$4 \equiv 52 \pmod{6}, \quad\quad 52 \equiv 4 \pmod{6}, \quad\quad 17 \equiv 17 \pmod{6},$$
$$-29 \equiv -11 \pmod{6}, \quad -11 \equiv 1 \pmod{6}, \quad -29 \equiv 1 \pmod{6}. \quad \square$$

THEOREM 3.16

Congruence modulo n is an equivalence relation on the set $\mathbf{Z}$.

Proof

(1) $a \equiv a \pmod{n}$ for any $a \in \mathbf{Z}$, since $a - a = 0 \cdot n$.
(2) $a \equiv b \pmod{n} \Rightarrow a - b = kn$, for some $k \in \mathbf{Z} \Rightarrow b - a = (-k)n \Rightarrow b \equiv a \pmod{n}$.
(3) $a \equiv b$, and $b \equiv c \pmod{n} \Rightarrow a - b = sn, b - c = tn$, for some $s, t \in \mathbf{Z} \Rightarrow (a - b) + (b - c) = sn + tn \Rightarrow a - c = (s + t)n \Rightarrow a \equiv c \pmod{n}$. $\square$

In Section 1.4 we proved that an equivalence relation on a set partitions that set into a collection of equivalence classes. Each equivalence class of $\mathbf{Z}$ modulo n is of the form $\bar{a}$, with

$$\bar{a} = \{x : x \equiv a \pmod{n}\}$$
$$= \{x : x = a + kn, \text{ for } k \in \mathbf{Z}\}$$
$$= \{a + kn : k \in \mathbf{Z}\}.$$

So $\bar{a}$ is the set of integers obtained by adding to a each multiple of n. The next theorem indicates how many distinct equivalence classes there are.

THEOREM 3.17

If k is any integer, then k is congruent modulo n to exactly one integer in the set, $\{0, 1, 2, \ldots, n - 1\}$. Hence there are n distinct equivalence classes modulo n: $\bar{0}, \bar{1}, \bar{2}, \ldots, \overline{n-1}$.

Proof

By the Division Theorem there exist *unique* integers, q and r, such that $k = nq + r$, where r is one of the integers: $0, 1, \ldots, n - 1$. But $k = nq + r \Leftrightarrow k \equiv r \pmod{n} \Leftrightarrow k \in \bar{r}$. So each integer lies in one and only one of the n equivalence classes: $\bar{0}, \bar{1}, \bar{2}, \ldots, \overline{n-1}$. □

Recall that if S is any set and $\sim$ is an equivalence relation on S, then the factor set of S with respect to $\sim$ is the set of all equivalence classes of S. The *factor set of* $\mathbf{Z}$ *modulo* n is denoted by either of these symbols: $\mathbf{Z}_n$ or $\dfrac{\mathbf{Z}}{\langle n \rangle}$. So

$$\mathbf{Z} = \bar{0} \cup \bar{1} \cup \bar{2} \cup \cdots \cup \overline{n-1},$$

and
$$\mathbf{Z}_n = \mathbf{Z}/\langle n \rangle = \{\bar{0}, \bar{1}, \bar{2}, \ldots, \overline{n-1}\}.$$

EXAMPLE 3.16

Let $n = 4$. Then $\mathbf{Z}_4 = \{\bar{0}, \bar{1}, \bar{2}, \bar{3}\}$, with

$$\begin{aligned}
\bar{0} &= \{0 + 4k : k \in \mathbf{Z}\} \\
&= \{0, 4, -4, 8, -8, 12, -12, \ldots\} = \bar{4} = \overline{-4} = \bar{8} = \cdots, \\
\bar{1} &= \{1 + 4k : k \in \mathbf{Z}\} \\
&= \{1, 5, -3, 9, -7, 13, -11, \ldots\} = \bar{5} = \overline{-3} = \bar{9} = \cdots, \\
\bar{2} &= \{2 + 4k : k \in \mathbf{Z}\} \\
&= \{2, 6, -2, 10, -6, 14, -10, \ldots\} = \bar{6} = \overline{-2} = \overline{10} = \cdots, \\
\bar{3} &= \{3 + 4k : k \in \mathbf{Z}\} \\
&= \{3, 7, -1, 11, -5, 15, -9, \ldots\} = \bar{7} = \overline{-1} = \overline{11} = \cdots. \quad \square
\end{aligned}$$

We shall have much more to say about the factor set $\mathbf{Z}_n$ in the next section and in later chapters. Before we do that, we must establish a few more facts about congruences.

THEOREM 3.18

If $a \equiv b \pmod{n}$, and $c \equiv d \pmod{n}$, then

$$a + c \equiv b + d \pmod{n}, \quad \text{and} \quad ac \equiv bd \pmod{n}.$$

3.5 Congruence Modulo n

If you understand the definition of congruence modulo n, you should be able to prove Theorem 3.18.

Do we have a cancellation law of multiplication for congruences? Does $ab \equiv ac \pmod{n}$ imply that $b \equiv c \pmod{n}$? A counter example will answer this in the negative. For instance, $6 \cdot 7 \equiv 6 \cdot 5 \pmod{4}$, but $7 \not\equiv 5 \pmod{4}$. The next theorem gives us a kind of modified cancellation law.

THEOREM 3.19

Let $(a, n) = g$. Then $ab \equiv ac \pmod{n}$ iff $b \equiv c \pmod{n/g}$. So if $g = 1$, then $ab \equiv ac \pmod{n}$ iff $b \equiv c \pmod{n}$.

Proof

Given that $ab \equiv ac \pmod{n}$. Then $ab - ac = kn$ for some k, and so $(a/g)(b - c) = k(n/g)$. Hence $(n/g) | (a/g)(b - c)$. But $(a/g, n/g) = 1$, by Thm. 3.8(b), and so $(n/g) | (b - c)$, by Thm. 3.7(a). Therefore $b \equiv c \pmod{n/g}$.

Conversely: $b \equiv c \pmod{n/g} \Rightarrow b = c + (n/g)t$ for some $t \Rightarrow ab = ac + n(a/g)t \Rightarrow ab \equiv ac \pmod{n}$. □

EXAMPLE 3.17

Observe that $6 \cdot 7 \equiv 6 \cdot 3 \pmod{8}$. But $(6, 8) = 2$, so if we are to "cancel" the 6, we must also divide the modulus by 2. Then $7 \equiv 3 \pmod{4}$. We could, however, have divided each side of the given congruence by 3, and left the modulus as 8, since $(3, 8) = 1$. Then we would have: $14 \equiv 6 \pmod{8}$. □

Suppose that n, a, and b are given. If we can find an integer x such that $ax \equiv b \pmod{n}$, then we say that we have found a *solution* of the congruence. This congruence is called a *linear congruence*, just as $ax = b$ is called a linear equation. Similarly, the congruence, $ax^2 + bx + c \equiv 0 \pmod{n}$, is called a *quadratic congruence*, provided that $a \not\equiv 0 \pmod{n}$. The study of congruences and their solutions is an interesting part of a course in number theory. We give here a few examples of solving linear congruences.

EXAMPLE 3.18

We can solve the congruence, $3x \equiv 2 \pmod{10}$, by trial and error, substituting for x the integers: $0, 1, 2, \ldots, 9$. We see that $3(4) \equiv 2 \pmod{10}$, so 4 is a solution. Are there other solutions? We claim that an integer k is a solution iff k is in the equivalence class $\bar{4}$, for:

$$k \text{ is a solution} \Leftrightarrow 3k \equiv 2 \pmod{10}$$
$$\Leftrightarrow 3k \equiv 12 \pmod{10} \Leftrightarrow k \equiv 4 \pmod{10}.$$

So $\bar{4}$ is the solution set, where $\bar{4} = \{4, 14, -6, 24, -16, \ldots\}$. □

EXAMPLE 3.19

To solve $8x - 7 \equiv 5 \pmod{20}$, we add 7 to each side, getting: $8x \equiv 12 \pmod{20}$. We can now divide by 4, obtaining:

$$2x \equiv 3 \pmod 5.$$

By inspection we see that 4 is a solution of the preceding congruence, as is $4 + 5k, k \in \mathbf{Z}$. Since the steps we took in simplifying the given congruence are reversible, the solution set of the given congruence is $\{4 + 5k : k \in \mathbf{Z}\}$. The positive solutions, less than 20, are: 4, 9, 14, 19. Therefore, the solution set is the union of these equivalence classes modulo 20: $\overline{4}, \overline{9}, \overline{14}, \overline{19}$. ☐

EXAMPLE 3.20

Suppose we are asked to solve: $28x \equiv 75 \pmod{102}$. This is equivalent to solving the Diophantine equation, $28x = 75 + 102y$, or $28x - 102y = 75$, or $2(14x - 51y) = 75$. This says, of course, that 2 divides 75, which is false! We conclude that the congruence has no solution. ☐

It is not difficult to prove that the congruence, $ax \equiv b \pmod n$, has solutions if and only if $(a, n) | b$. The general theory of linear congruences is discussed in Problems 3.6-8 and 3.6-9.

EXAMPLE 3.21

Instead of using the Euclidean Algorithm, congruences can be used to simplify the arithmetic when solving a linear Diophantine equation. Suppose we want to find the solution set of

$$587x + 70y = 82.$$

We can proceed as follows:

$$587x + 70y = 82 \Rightarrow 587x = 82 - 70y \Rightarrow 587x \equiv 82 \pmod{70}$$
$$\Rightarrow 27x \equiv 12 \pmod{70} \quad (587 \equiv 27, 82 \equiv 12)$$
$$\Rightarrow 27x = 12 + 70v \Rightarrow 70v = -12 + 27x$$
$$\Rightarrow 70v \equiv -12 \pmod{27} \Rightarrow 16v \equiv -12 \pmod{27}$$
$$\Rightarrow 4v \equiv -3 \pmod{27} \Rightarrow 4v \equiv 24 \pmod{27}$$
$$\Rightarrow v \equiv 6 \pmod{27}.$$

Then $v = 6 \Rightarrow 27x = 12 + 70(6) \Rightarrow x = 16 \Rightarrow 70y = 82 - 587(16) \Rightarrow y = -133$. Therefore, the solution set, according to Problem 3.2-18, is:

$$\{(x, y) : x = 16 + 70n, y = -133 - 587n, \quad \text{for all } n \in \mathbf{Z}\}.$$

3.5 Congruence Modulo n

Note that we converted the given equation to a congruence modulo 70, reduced the integers modulo 70, converted back to an equation, and then to a congruence modulo 27. This process of reducing the size of the integers can be repeated until we arrive at a congruence whose modulus is small enough to allow us to find a solution by trial and error. □

EXAMPLE 3.22

It is desired to find a common solution of these three congruences:

$$x \equiv 2 \pmod 3, \quad x \equiv 1 \pmod 4, \quad x \equiv 3 \pmod 5.$$

First,

$$x \equiv 2 \pmod 3, x \equiv 1 \pmod 4 \Rightarrow x = 2 + 3s, x = 1 + 4t$$
$$\Rightarrow 1 + 4t = 2 + 3s \Rightarrow 4t \equiv 1 \pmod 3$$
$$\Rightarrow t \equiv 1 \pmod 3 \Rightarrow t = 1 + 3k$$
$$\Rightarrow x = 1 + 4(1 + 3k) = 5 + 12k \Rightarrow x \equiv 5 \pmod{12}$$

Therefore, the solution set of the first two congruences is the equivalence class $\bar{5}$ modulo 12.

Next,

$$x \equiv 5 \pmod{12}, x \equiv 3 \pmod 5 \Rightarrow x = 5 + 12h = 3 + 5m$$
$$\Rightarrow 12h \equiv -2 \pmod 5 \Rightarrow 2h \equiv -2 \pmod 5$$
$$\Rightarrow h \equiv 4 \pmod 5 \Rightarrow h = 4 + 5v, \text{ for all } v \in \mathbf{Z}$$
$$\Rightarrow x = 5 + 12(4 + 5v) \Rightarrow x = 53 + 60v$$
$$\Rightarrow x \equiv 53 \pmod{60}.$$

Therefore, the solution set of the three given congruences is $\bar{53}$ modulo 60, and 53 is the smallest positive integer that satisfies all three congruences. □

3.5 Problems

1 List five integers in each equivalence class modulo n if:
  (a) $n = 2$,   (b) $n = 3$,   (c) $n = 5$,   (d) $n = 6$.

2 Complete these statements:
  (a) Integers 29 and 41 are in the same equivalence class modulo n if n is any one of these integers: ___
  (b) $17 \in \bar{24}$ if $n = $ ___.

3 Verify: If $a + b \equiv a + c \pmod n$, then $b \equiv c \pmod n$.

4 Prove Theorem 3.18.

5 Verify that if p is an odd prime, then either $p \equiv 1 \pmod 4$ or $p \equiv -1 \pmod 4$.

6 Prove: If c is a solution of the congruence, $ax \equiv b \pmod{n}$, then any integer in the equivalence class $\bar{c}$ is also a solution.

7 Find integers, if they exist, that satisfy these congruences.
(a) $7x \equiv 21 \pmod 8$
(b) $6x \equiv 10 \pmod{14}$
(c) $5x + 10 \equiv 0 \pmod{15}$
(d) $4x \equiv 5 \pmod 6$
(e) $x^2 + 3x + 2 \equiv 0 \pmod 6$

8 (a) Prove: The congruence, $ax \equiv b \pmod n$, has solutions if and only if $(a, n) | b$. (See Problem 3.2–17.)
(b) Give a congruence modulo 1200 that has no solution, and give one that does have solutions.

9 Prove:
(a) If $(a, n) = 1$, then the solution set of $ax \equiv b \pmod n$ consists of exactly one equivalence class modulo n.
(b) If $(a, n) = g$ and $g | b$, then the solution set of $ax \equiv b \pmod n$ consists of exactly g distinct equivalence classes modulo n.

10 Use the theory of Prob. 8 and 9 to find the solution set of each congruence.
(a) $12x \equiv 45 \pmod{51}$
(b) $32x \equiv 24 \pmod{40}$
(c) $89x \equiv 94 \pmod{97}$

11 Use congruences, as in Example 3.21, to solve each of these Diophantine equations.
(a) $250x - 147y = 150$
(b) $71x + 89y = 1$
(c) $1001x + 30y = 42$

12 Let n be a given positive integer. Prove:
(a) $a \equiv b \pmod n, k \in \mathbf{Z}^+ \Rightarrow a^k \equiv b^k \pmod n$. (Prove by induction on k.)
(b) If $f(t) = c_m t^m + c_{m-1} t^{m-1} + \cdots + c_2 t^2 + c_1 t + c_0$, where $c_i, t \in \mathbf{Z}; m \in \mathbf{Z}^+$, then $a \equiv b \pmod n \Rightarrow f(a) \equiv f(b) \pmod n$.

13 (a) Use the technique of Example 3.22 to find the common solution set of the pair of congruences: $x \equiv 3 \pmod 4$, $x \equiv 2 \pmod 7$.
(b) Use the results of part (a) to find the common solution set of the three congruences:

$$x \equiv 3 \pmod 4, \qquad x \equiv 2 \pmod 7, \qquad x \equiv 8 \pmod 9.$$

14 A band of 13 pirates landed on an island, where they found a chest containing gold coins. When they divided the coins up equally, they found that there were 8 coins left over. The next day two pirates fell overboard and drowned. The coins were all re-distributed, and this time there were 3 remaining. The leader shot three pirates, and the new distribution left a remainder of 5 coins. What is the smallest number of coins that the pirates could have found on the island?

15 Find the common solution set of these congruences:

$$2x \equiv 3 \pmod 5, \qquad 6x \equiv 15 \pmod{21}.$$

16   (a) Show that a common solution does not exist for the pair:

$$x \equiv 1 \pmod{4}, \qquad x \equiv 2 \pmod 6.$$

       (b) Find a common solution of: $x \equiv 1 \pmod 4$, $x \equiv 5 \pmod 6$.

17   Given the congruences: $x \equiv a \pmod n$, $x \equiv b \pmod m$; prove that a common solution exists if and only if $(n, m) | (a - b)$.

18   When Diophantus wanted to solve a second degree equation, how did he express "x^2"? (See Appendix E–2.)

3.6 Factor Ring of Integers Modulo n

In this section we shall define operations, $+$ and $\cdot$, on $\mathbf{Z}_n$ that make $(\mathbf{Z}_n, +, \cdot)$ a commutative ring with unity.

    First, let's examine the possible complications involved in defining an operation on a factor set. Suppose we have a set S and a factor set P, where $P = \{\bar{a} : a \in S\}$. If $*$ is a relation from $P \times P$ to P, then $*$ is a map (and hence a binary operation on P) if and only if $(\bar{a}, \bar{b}) = (\bar{c}, \bar{d})$ implies that $(\bar{a}, \bar{b})* = (\bar{c}, \bar{d})*$, for all $a, b, c, d \in S$. In the usual notation for a binary operation, we say that $*$ is an *operation* on P if: (1) $\bar{a} * \bar{b} \in P$, for all $a, b \in S$; and (2) if $\bar{a} = \bar{c}$ and $\bar{b} = \bar{d}$, then $\bar{a} * \bar{b} = \bar{c} * \bar{d}$. If conditions (1) and (2) are both satisfied, then $*$ is said to be a *well-defined operation* on factor set P. (Actually, there is no such thing as an operation that is *not* well-defined. If $*$ is not well-defined, then it is not an operation!) We now give an example illustrating the fact that one must verify both conditions when one attempts to define an operation on a factor set.

EXAMPLE 3.23

We determine a factor set of $\mathbf{Z}$ as follows: Writing each integer n in the form $n = 10 \cdot k + r$, with $0 \le r \le 9$, we define

$$\bar{n} = \overline{10 \cdot k + r} = \{10k, 10k + 1, 10k + 2, \ldots, 10k + 9\}.$$

Thus, $\overline{30} = \{30, 31, 32, \ldots, 39\} = \overline{31} = \overline{32} = \cdots = \overline{39}$;

$\overline{50} = \{50, 51, 52, \ldots, 59\} = \overline{51} = \overline{52} = \cdots = \overline{59}$.

If $P = \{\overline{10n} : n \in \mathbf{Z}\}$, then P is a factor set of $\mathbf{Z}$. Now define an operation of addition on P as follows: $\bar{n} + \bar{m} = \overline{n + m}$. For instance, $\overline{30} + \overline{50} = \overline{80}$. However, $\overline{30} = \overline{37}$ and $\overline{50} = \overline{56}$, but $\overline{37} + \overline{56} = \overline{93} = \overline{90}$. Thus $\overline{30} + \overline{50} \neq \overline{37} + \overline{56}$, and so we conclude that $+$ is *not* a well-defined operation on P. □

THEOREM 3.20

Let $\bar{a}, \bar{b} \in \mathbf{Z}_n$. If $\bar{a} + \bar{b} = \overline{a + b}$ and $\bar{a} \cdot \bar{b} = \overline{a \cdot b}$, then $+$ and $\cdot$ are well-defined operations on $\mathbf{Z}_n$.

Proof

This result is an immediate consequence of Thm. 3.18 and the definition of an equivalence class. Verification is as follows:

$$\bar{a} = \bar{b}, \bar{c} = \bar{d} \Rightarrow a \equiv b \pmod{n}, c \equiv d \pmod{n}$$
$$\Rightarrow a + c \equiv b + d \pmod{n}, ac \equiv bd \pmod{n}$$
$$\Rightarrow \overline{a + c} = \overline{b + d}, \overline{ac} = \overline{bd}$$
$$\Rightarrow \bar{a} + \bar{c} = \bar{b} + \bar{d}, \bar{a} \cdot \bar{c} = \bar{b} \cdot \bar{d}. \quad \square$$

EXAMPLE 3.24

If we perform addition and multiplication in Z_4, we have, for instance, that $\bar{3} + \bar{2} = \bar{5} = \bar{1}$, and $\bar{3} \cdot \bar{2} = \bar{6} = \bar{2}$. An examination of the equivalence classes given in Example 3.16 reveals that the addition of any integer in set $\bar{3}$ to any integer in set $\bar{2}$ gives an integer that is in set $\bar{1}$. This isn't surprising, in the light of Theorem 3.20. The operation tables for Z_4 are as follows:

+	$\bar{0}$	$\bar{1}$	$\bar{2}$	$\bar{3}$
$\bar{0}$	$\bar{0}$	$\bar{1}$	$\bar{2}$	$\bar{3}$
$\bar{1}$	$\bar{1}$	$\bar{2}$	$\bar{3}$	$\bar{0}$
$\bar{2}$	$\bar{2}$	$\bar{3}$	$\bar{0}$	$\bar{1}$
$\bar{3}$	$\bar{3}$	$\bar{0}$	$\bar{1}$	$\bar{2}$

$\cdot$	$\bar{0}$	$\bar{1}$	$\bar{2}$	$\bar{3}$
$\bar{0}$	$\bar{0}$	$\bar{0}$	$\bar{0}$	$\bar{0}$
$\bar{1}$	$\bar{0}$	$\bar{1}$	$\bar{2}$	$\bar{3}$
$\bar{2}$	$\bar{0}$	$\bar{2}$	$\bar{0}$	$\bar{2}$
$\bar{3}$	$\bar{0}$	$\bar{3}$	$\bar{2}$	$\bar{1}$

Note that $\bar{0}$ is the additive identity, and $\bar{1}$ is the multiplicative identity. Also, $-(\bar{1}) = \bar{3}, -(\bar{2}) = \bar{2}, (\bar{1})^{-1} = \bar{1}, (\bar{3})^{-1} = \bar{3}$, and $\bar{2}$ is a zero divisor. $\square$

THEOREM 3.21

If n is any positive integer, then
(a) $(Z_n, +)$ is an abelian group with identity $\bar{0}$, and
(b) $(Z_n, +, \cdot)$ is a commutative ring with unity $\bar{1}$.

Proof

(a) Operation $+$ is associative, because

$$(\bar{a} + \bar{b}) + \bar{c} = \overline{a+b} + \bar{c} = \overline{(a+b)+c} = \overline{a+(b+c)}$$
$$= \bar{a} + \overline{b+c} = \bar{a} + (\bar{b} + \bar{c}).$$

In the preceding we applied the definition of addition in Z_n four times, and the associative property of addition in Z once.

For all $a \in Z_n$, $\bar{a} + \bar{0} = \overline{a + 0} = \bar{a}$, so $\bar{0}$ is the identity. The additive inverse of $\bar{a}$ is $\overline{n - a}$, since $\bar{a} + \overline{n - a} = \overline{a + (n - a)} = \bar{n} = \bar{0}$. Of course, $-(\bar{a}) = \overline{-a} = \overline{n - a}$. Furthermore, addition is commutative, since $\bar{a} + \bar{b} = \overline{a + b} = \overline{b + a} = \bar{b} + \bar{a}$.

(b) The remaining three properties are left to the reader to establish. $\square$

3.6 Factor Ring of Integers Modulo n

$\mathbf{Z}_n$ is a factor set of the set $\mathbf{Z}$, and since $(\mathbf{Z}_n, +)$ is a group, it is logical to call $(\mathbf{Z}_n, +)$ a *factor group* of the group $(\mathbf{Z}, +)$. Similarly, we shall call $(\mathbf{Z}_n, +, \cdot)$ a *factor ring* of the ring $(\mathbf{Z}, +, \cdot)$.

We now establish a criterion for determining which elements of ring $\mathbf{Z}_n$ are units, and which ones are zero divisors.

THEOREM 3.22

Let $\bar{a}$ be a nonzero element of ring $(\mathbf{Z}_n, +, \cdot)$, and $g = (a, n)$.

(a) If $g > 1$, then $\bar{a}$ is a zero divisor.
(b) If $g = 1$, then $\bar{a}$ is a unit.

Proof

(a) We assume that $1 \leq a < n$. If $(a, n) = g > 1$, then $a = gb, n = gm$, for integers b and m, with $1 < m < n$. Then $\bar{m} \neq \bar{0}$, but $\bar{a} \cdot \bar{m} = \overline{am} = \overline{(gb)m} = \overline{b(gm)} = \overline{bn} = \bar{b} \cdot \bar{n} = \bar{b} \cdot \bar{0} = \bar{0}$. Therefore, $\bar{a}$ is a zero divisor.

(b) If $(a, n) = 1$, there exist integers s and t such that $as + nt = 1$, by Theorem 3.4. Then,

$$\bar{1} = \overline{as + nt} = \overline{as} + \overline{nt} = \bar{a} \cdot \bar{s} + \bar{0} \cdot \bar{t} = \bar{a} \cdot \bar{s},$$

and so $\bar{s}$ is the multiplicative inverse of $\bar{a}$. Hence, $\bar{a}$ is a unit. ☐

The *number* of positive integers that are less than n and prime to n is denoted by $\phi(n)$. Thus, ring $\mathbf{Z}_n$ has $\phi(n)$ units. The map ϕ is called *Euler's phi-function*.

EXAMPLE 3.25

The positive integers that are less than 10 and prime to 10 are: 1, 3, 7, 9, so $\phi(10) = 4$. Then $\bar{1}, \bar{3}, \bar{7}$, and $\bar{9}$ are the units of $\mathbf{Z}_{10}$, while $\bar{2}, \bar{4}, \bar{5}, \bar{6}$, and $\bar{8}$ are the zero divisors of $\mathbf{Z}_{10}$.

Every positive integer less than 13 is prime to 13, so $\phi(13) = 12$. This means that every nonzero element of the ring $\mathbf{Z}_{13}$ has a multiplicative inverse, and so $(\mathbf{Z}_{13}, +, \cdot)$ is actually a field. ☐

EXAMPLE 3.26

How does one compute $\phi(n)$, for large n? If p is a positive prime, then $\phi(p) = p - 1$. If n is composite and small, one can determine $\phi(n)$ simply by inspecting all positive integers less than n.

A theorem from number theory (which we shall not prove) provides us with a simple method for calculating $\phi(n)$. All we have to do is to find the positive prime divisors of n. Suppose they are: $p_1, p_2, \ldots, p_s$. Then, according to the theorem,

$$\phi(n) = n(1 - 1/p_1)(1 - 1/p_2) \cdots (1 - 1/p_s).$$

For instance, the positive primes that divide 3000 are 2, 3, and 5.

Then $\phi(3000) = 3000(1/2)(2/3)(4/5) = 800$.

So the ring $\mathbf{Z}_{3000}$ has 800 units and 2199 zero divisors. □

We now show that ring $\mathbf{Z}_n$ is a field if and only if p is a prime.

THEOREM 3.23

Let n be a positive integer ≥ 2.

(a) If n is composite, then ring $(\mathbf{Z}_n, +, \cdot)$ has zero divisors, and so is not a field.
(b) If n is prime, then $(\mathbf{Z}_n, +, \cdot)$ is a field.

Proof

(a) If n is composite, then $n = ab$, for some a and b such that $1 < a < n$, and $1 < b < n$. Then $\bar{a} \neq \bar{0}, \bar{b} \neq \bar{0}$, but

$$\bar{a} \cdot \bar{b} = \overline{ab} = \bar{n} = \bar{0}.$$

(b) If n is a prime, then $(a, n) = 1$, where $1 \leq a \leq n - 1$. Hence each nonzero element of $\mathbf{Z}_n$ is a unit, by Theorem 3.22, and so $\mathbf{Z}_n$ is a field. □

EXAMPLE 3.27

One can have polynomials whose coefficients are in a ring $(\mathbf{Z}_n, +, \cdot)$. For instance, let $f = \bar{2}x^3 + \bar{3}x + \bar{5}$, where the coefficients are in ring $\mathbf{Z}_6$. A root of f in $\mathbf{Z}_6$ is an element r of $\mathbf{Z}_6$ such that $\bar{2} \cdot r^3 + \bar{3} \cdot r + \bar{5} = \bar{0}$. Thus,

$$\bar{2}(\bar{5})^3 + \bar{3} \cdot \bar{5} + \bar{5} = \bar{2} \cdot \bar{5} + \bar{3} \cdot \bar{5} + \bar{5} = \bar{4} + \bar{3} + \bar{5} = \overline{12} = \bar{0}.$$

It so happens that $\bar{5}$ is the only element of $\mathbf{Z}_6$ that is a root of f.

On the other hand, if $g = x^2 + \bar{5}x$, then g has four roots in $\mathbf{Z}_6$: $\bar{0}, \bar{1}, \bar{3}$, and $\bar{4}$. In the next chapter we look at all sorts of questions relating to polynomials. □

3.6 Problems

1. Construct the addition and multiplication tables of the factor ring $(\mathbf{Z}_n, +, \cdot)$ if:
   (a) $n = 2$,   (b) $n = 3$,   (c) $n = 6$,   (d) $n = 7$.

2. Determine the multiplicative inverse, if it exists, of each element in ring $\mathbf{Z}_n$, if n is:
   (a) 2,   (b) 3,   (c) 6,   (d) 7,
   (e) 8,   (f) 9,   (g) 11.

3   Describe the factor group $(\mathbf{Z}_1, +)$.

4   Prove Theorem 3.21(b).

5   You have probably heard statements like this: "In the 'new math', $2 + 2$ does not always equal 4." In $\mathbf{Z}_3$, we know that $\bar{2} + \bar{2} = \bar{1}$. Does this justify the statement about 'new math'?

6   Tell whether each statement is true or false, and give a reason for your answer.
(a) In ring $\mathbf{Z}_n$, every nonzero element is either a unit or a zero divisor.
(b) If $\mathbf{Z}_n$ is an integral domain, then $\mathbf{Z}_n$ is a field.
(c) In ring $\mathbf{Z}_n$, if $\bar{a} \cdot \bar{b} = \bar{a} \cdot \bar{c}$, where $\bar{a} \neq \bar{0}$, then $\bar{b} = \bar{c}$.

7   Use the formula given in Example 3.26 to find $\phi(n)$, if n is:
(a) 750,   (b) 75,000,000,   (c) 2310,   (d) 243.

8   Let U_n be the set of units of ring $\mathbf{Z}_n$. Verify that $(U_n, \cdot)$ is an abelian group, and that $|U_n| = \phi(n)$.

9   Construct the multiplication table of $(U_{18}, \cdot)$. (See Problem 8.)

10 (a) Give the multiplication tables of these groups: $(U_{10}, \cdot), (U_{12}, \cdot), (U_5, \cdot)$. (See Problem 8.)
(b) In Problem 2.10–11, you showed that there are only two structurally distinct (i.e., non-isomorphic) groups of order 4. Then at least two of the groups of part (a) are isomorphic. Which ones are they?

11 Find all subgroups of $(\mathbf{Z}_n, +)$, and draw the inclusion diagram of subgroups, if:
(a) $n = 8$,   (b) $n = 12$.

12 (a) Explain why groups $(\mathbf{Z}_6, +)$ and $(\mathbf{Z}_{10}, +)$ are not isomorphic.
(b) Show that the inclusion diagrams of subgroups of $\mathbf{Z}_6$ and $\mathbf{Z}_{10}$ look alike.

13 Given the group $(\mathbf{Z}_{24}, +)$, find the smallest subgroup H that contains $\bar{m}$, if $\bar{m}$ is:
(a) $\bar{4}$,   (b) $\overline{12}$,   (c) $\bar{0}$,
(d) $\bar{1}$,   (e) $\bar{5}$,   (f) $\bar{9}$.
(Hint: For instance, if $\bar{8} \in H$, then $\bar{8} + \bar{8} \in H$, and $\bar{8} + \bar{8} + \bar{8} = \bar{0}$. Therefore, $H = \{\bar{0}, \bar{8}, \overline{16}\}$.)

14 Find all roots in $(\mathbf{Z}_n, +, \cdot)$ of the given polynomials.
(a) In $\mathbf{Z}_8: \bar{7}x + \bar{3}$
(b) In $\mathbf{Z}_{14}: \bar{6}x + \bar{4}$
(c) In $\mathbf{Z}_6: \bar{4}x + \bar{1}$
(d) In $\mathbf{Z}_{15}: \bar{5}x + \overline{10}$
(e) In $\mathbf{Z}_6: x^2 + \bar{3}x + \bar{2}$
(f) In $\mathbf{Z}_4: x^3 + x + \bar{1}$
(g) In $\mathbf{Z}_{10}: x^4 - \bar{1}$

15 Given the rings $(\mathbf{Z}, +, \cdot)$ and $(\mathbf{Z}_n, +, \cdot)$; let $\theta: \mathbf{Z} \to \mathbf{Z}_n$ be the map, $a \to \bar{a} = a\theta$. That is, each integer a maps to the equivalence class of which it is a member.

(a) Show what the map θ looks like if $n = 3$. What are the antecedents of $\bar{0}$?

(b) Prove: θ is a morphism of ring $\mathbf{Z}$ onto ring $\mathbf{Z}_n$.

16 Consider group $(\mathbf{Z}_p^*, \cdot)$, where p is an odd prime.

(a) Verify that $\bar{m}$ is not its own inverse, if $1 < m < p - 1$. (Hint: $\bar{m}^2 = \bar{1} \Rightarrow p \mid m^2 - 1 \Rightarrow ?$)

(b) Use part (a) to prove that $\bar{2} \cdot \bar{3} \cdot \cdots \cdot \overline{p - 2} = \bar{1}$.

(c) Use part (b) to prove *Wilson's Theorem*: If p is a prime, then $(p - 1)! \equiv -1 \pmod{p}$.

17 Prove: If $(n - 1)! \equiv -1 \pmod{n}$, then n is a prime.

18 (a) Show that if p is an odd positive integer, then

$$(p - 1)! = \prod_{m=1}^{(p-1)/2} m(p - m) \equiv (-1)^{(p-1)/2} \prod_{m=1}^{(p-1)/2} m^2 \pmod{p}.$$

(b) Let $t = \left(\dfrac{p - 1}{2}\right)!$, where p is an odd positive prime. Use part (a) and Wilson's Theorem (Problem 16(c)) to prove:

$$p \equiv 1 \pmod{4} \Rightarrow t^2 \equiv -1 \pmod{p}.$$
$$p \equiv -1 \pmod{4} \Rightarrow t^2 \equiv 1 \pmod{p}.$$

19 John Wilson (for whom Wilson's Theorem was named) was senior wrangler at Cambridge University. Is that good or bad? (See Appendix E–7.)

4

Polynomials

> The problem of solving polynomial equations led mathematicians to study groups, and is thus one of the historical cornerstones on which contemporary algebra is based. (1973) — Larry Joel Goldstein

4.1 Polynomial Rings

In your secondary school algebra class you learned that a polynomial in x is an expression such as $x^3 - 5x^2 + 6x$. You learned to factor polynomials, and to solve polynomial equations by factoring. Since $x^3 - 5x^2 + 6x = x(x-2)(x-3)$, you would conclude that 0, 2, and 3 are roots of $x^3 - 5x^2 + 6x = 0$.

You are now sophisticated enough, algebraically speaking, to take a second look at polynomials. Your past experience with polynomials will certainly help you to understand the approach we take. For instance, if you are given two polynomials, $f(x)$ and $g(x)$, where

$$f(x) = a_0 + a_1 x + a_2 x^2, \quad \text{and} \quad g(x) = b_0 + b_1 x + b_2 x^2 + b_3 x^3,$$

with $a_i, b_j \in \mathbf{Z}$, you would probably find their sum and product to be as follows:

$$f(x) + g(x) = (a_0 + b_0) + (a_1 + b_1)x + (a_2 + b_2)x^2 + b_3 x^3,$$
$$f(x) \cdot g(x) = a_0 b_0 + (a_0 b_1 + a_1 b_0)x + (a_0 b_2 + a_1 b_1 + a_2 b_0)x^2$$
$$+ (a_0 b_3 + a_1 b_2 + a_2 b_1)x^3 + (a_1 b_3 + a_2 b_2)x^4 + a_2 b_3 x^5.$$

We see that the sum and product depend only on the values of the a_i and b_j, and that the x is acting merely as a placeholder. We could have written our polynomials as sequences of rational integers, and expressed the sum as follows:

$(a_0, a_1, a_2, 0, 0, \ldots) + (b_0, b_1, b_2, b_3, 0, 0, \ldots)$
$$= (a_0 + b_0, a_1 + b_1, a_2 + b_2, b_3, 0, 0, \ldots).$$

This notation enables us to forget about the indeterminate x for the moment, and to concentrate upon the polynomial itself as an element in an abstract algebra.

We begin by defining a polynomial over a ring $(S, +, \cdot)$ to be a sequence,

$$f = (a_0, a_1, a_2, \ldots, a_k, z, z, \ldots),$$

where the a_i are in S, and z is the zero of S. The a_i are called the *coefficients* of f.

Only a finite number of coefficients can be nonzero. If $a_k \neq z$, and $a_m = z$ for all $m > k$, then we say that f is of *degree* k, and write $\deg f = k$. Note that $k \geq 0$. The polynomial, $(z, z, z, \ldots)$ is called the *zero polynomial*, and it has no degree.

If g is another polynomial over S, where

$$g = (b_0, b_1, b_2, \ldots, b_n, z, z, \ldots),$$

then, by definition, $f = g$ if and only if $a_i = b_i$, for $i = 0, 1, 2, \ldots$. The set of all polynomials over ring S will be denoted as $S[x]$. That is, $S[x] = \{(a_1, a_2, a_3, \ldots) : a_i \in S; a_i = z$ for all but a finite number of $i\}$.

Addition, $+$, of polynomials f and g is defined as follows:

$$f + g = (a_0 + b_0, a_1 + b_1, a_2 + b_2, \ldots).$$

Note that we are using $+$ as the symbol for addition in S and also as the symbol for addition in $S[x]$.

Multiplication is a little more complicated. By definition,

$$f \cdot g = (a_0 b_0, a_0 b_1 + a_1 b_0, a_0 b_2 + a_1 b_1 + a_2 b_0, \ldots)$$
$$= (c_0, c_1, c_2, \ldots, c_r, \ldots),$$

with $\qquad c_r = \sum_{i=0}^{r} a_i b_{r-i} = \sum_{i+j=r} a_i b_j, \quad$ for $r = 0, 1, 2, \ldots$.

For instance, $\qquad c_5 = a_0 b_5 + a_1 b_4 + a_2 b_3 + a_3 b_2 + a_4 b_1 + a_5 b_0$.

EXAMPLE 4.1

Suppose we want to compute the product of two polynomials in $\mathbf{Z}[x]$.

$$(2, 3, -1, 0, -4, 0, 0, \ldots) \cdot (7, 5, 6, -2, 3, 9, 0, 0, \ldots) = ?$$

We have diagrammed the procedure for computing c_3. Place your left index finger on the *zero* coefficient of the first polynomial, and your right index finger on the *three* coefficient of the second polynomial. Multiply those two numbers. Then move your left hand to the right, and your right hand to the left, to pick the next two coefficients to multiply. Continue in this fashion until your right hand reaches the zero coefficient of the second polynomial. Thus,

$$c_3 = 2(-2) + 3(6) + (-1)5 + 0(7) = 9.$$

The product polynomial is

$$(14, 31, 20, 9, -34, 9, 0, -1, -12, -36, 0, 0, \ldots). \quad \square$$

THEOREM 4.1

If S is a ring, then $(S[x], +, \cdot)$ is a ring, called a *polynomial ring*.

Proof

From the definitions of addition and multiplication of polynomials, it is obvious that the sum and product are infinite sequences of elements of S, with only a finite number of nonzero coefficients. So $S[x]$ is closed under these operations.

Let $f = (a_0, a_1, a_2, \ldots)$ and $g = (b_0, b_1, b_2, \ldots)$. It is easily verified that addition of polynomials is associative and commutative. The zero is $(z, z, z, \ldots)$, and $-f = (-a_0, -a_1, -a_2, \ldots)$.

We now prove that multiplication is associative. From the definition, fg can be expressed as follows:

$$fg = (d_0, d_1, d_2, \ldots), \quad \text{with } d_m = \sum_{i+j=m} a_i b_j.$$

If $h = (c_0, c_1, c_2, \ldots)$, then $(fg)h = (e_0, e_1, e_2, \ldots)$, with

$$e_s = \sum_{m+k=s} d_m c_k = \sum_{m+k=s} \left(\sum_{i+j=m} a_i b_j \right) c_k = \sum_{i+j+k=s} a_i b_j c_k.$$

Also, $\quad gh = (p_0, p_1, p_2, \ldots), \quad \text{with } p_n = \sum_{j+k=n} b_j c_k.$

Finally, $\quad f(gh) = (q_0, q_1, q_2, \ldots), \quad$ where

$$q_s = \sum_{i+n=s} a_i p_n = \sum_{i+n=s} a_i \left(\sum_{j+k=n} b_j c_k \right) = \sum_{i+j+k=s} a_i b_j c_k.$$

So $e_s = q_s$, for all s. Therefore, $(fg)h = f(gh)$.

The proof of the distributive properties is slightly less messy. We see that

$$f(g + h) = (a_0, a_1, \ldots)(b_0 + c_0, b_1 + c_1, \ldots) = (d_0, d_1, \ldots),$$

with $\quad d_s = \sum_{i+j=s} a_i(b_j + c_j) = \sum_{i+j=s} a_i b_j + \sum_{i+j=s} a_i c_j.$

But $\quad fg = (p_0, p_1, p_2, \ldots), \quad \text{with } p_s = \sum_{i+j=s} a_i b_j,$

and $\quad fh = (q_0, q_1, q_2, \ldots), \quad \text{with } q_s = \sum_{i+j=s} a_i c_j.$

So $d_s = p_s + q_s$, for all s. Therefore, $f(g + h) = fg + fh$. Likewise, $(g + h)f = gf + hf$. □

EXAMPLE 4.2

Let $f, g, h \in \mathbf{Z}_6[x]$, where

$f = (\bar{1}, \bar{4}, \bar{2}, \bar{3}, \bar{0}, \bar{0}, \ldots)$, $g = (\bar{0}, \bar{2}, \bar{2}, \bar{0}, \bar{0}, \ldots)$, and $h = (\bar{3}, \bar{3}, \bar{0}, \bar{0}, \ldots)$.

Then $fg = (\bar{0}, \bar{2}, \bar{4}, \bar{0}, \bar{4}, \bar{0}, \bar{0}, \ldots)$, and $gh = (\bar{0}, \bar{0}, \bar{0}, \ldots) = z$.

Observe that $\deg f = 3$, $\deg g = 2$, and $\deg fg = 4$, so $\deg(fg) < \deg f + \deg g$. Also, g and h are zero divisors, so $\mathbf{Z}_6[x]$ is not an integral domain. □

In Theorem 4.2 we see that certain properties of ring S carry over to ring $S[x]$.

THEOREM 4.2

Let S be a ring.
(a) If S is commutative, then so is $S[x]$.
(b) If S has a unity, then so does $S[x]$.
(c) If S is an integral domain, then so is $S[x]$, and $\deg(fg) = \deg f + \deg g$.
(d) If S is a field, then $S[x]$ is an integral domain whose units are the polynomials of degree 0.

Proof

(a) Let $f = (a_0, a_1, a_2, \ldots)$, $g = (b_0, b_1, b_2, \ldots)$. The kth coefficient of fg is $\sum_{i+j=k} a_i b_j$, and the kth coefficient of gf is $\sum_{i+j=k} b_j a_i$. But if S is commutative, then $a_i b_j = b_j a_i$, for all $a_i, b_j \in S$. Therefore, $fg = gf$, and so $S[x]$ is commutative.

(b) From the definition of multiplication of polynomials, one can verify that if u is the unity of S, then $(u, z, z, \ldots)$ is the unity of $S[x]$.

(c) Let $\deg f = k$, and $\deg g = n$, where $k \geq 0, n \geq 0$. Then $a_k \neq z$, and $b_n \neq z$. If S is an integral domain, then $a_k b_n \neq z$. But $a_k b_n$ is the $(k + n)$ coefficient of fg, and so $\deg fg = k + n = \deg f + \deg g$. Hence $S[x]$ has no zero divisors, and is also a commutative ring with unity, by parts (a) and (b). Therefore, $S[x]$ is an integral domain.

(d) $S[x]$ is an integral domain, by part (c). Let $v = (u, z, z, \ldots)$. Suppose that f has a multiplicative inverse, g. Then $fg = v$. But, by part (c):

$$0 = \deg v = \deg(fg) = \deg f + \deg g.$$

Hence, $\deg f = \deg g = 0$, and any polynomial of degree greater than 0 is not a unit. If $f = (a_0, z, z, \ldots)$, then $f^{-1} = (a_0^{-1}, z, z, \ldots)$. □

If S is an integral domain (or a field), then the integral domain $S[x]$ is called a *polynomial domain*.

Perhaps you feel uncomfortable with the notation we have been using for polynomials, and would like to write a polynomial in its more familiar form. Now that we have convinced you that "x" is of no use whatsoever when the properties of the ring $S[x]$ are being considered, we'll bring x back into the picture. But how? First of all, we verify that the polynomials of zero degree behave just like the elements of ring S.

THEOREM 4.3

If $(S, +, \cdot)$ is a ring, and S' is the following subset of $S[x]$:
$S' = \{(r, z, z, \ldots) : r \in S\}$, then $S \cong S'$.

4.1 Polynomial Rings

Proof

Let $\theta: S \to S'$ be this map:
$$r \to r\theta = (r, z, z, \ldots),$$
$$s \to s\theta = (s, z, z, \ldots).$$

From the definition of equality, $r\theta = s\theta$ iff $r = s$, so θ is 1–1. Each element of S' obviously has an antecedent, so θ is onto. Also,
$$(r + s)\theta = (r + s, z, z, \ldots) = r\theta + s\theta,$$
$$(rs)\theta = (rs, z, z, \ldots) = (r\theta)(s\theta).$$

Therefore, θ is an isomorphism. ☐

For all practical purposes *two* isomorphic rings are but *one* ring, the two rings being distinct only by virtue of the symbols used to represent the elements. Often we agree to use the same symbols to write the elements of both rings. Thus, in the isomorphism of Theorem 4.3, we replace the statement $r\theta = (r, z, z, \ldots)$, by the statement $r = (r, z, z, \ldots)$. That is, the arrow in the map, $r \to r\theta$, is converted to an equals sign, so that $r = r\theta$. You might call this a bit of "notational fudging."

Next, we let x be the name of the polynomial $(z, u, z, z, \ldots)$. With these notational agreements, we can then write a polynomial in the usual way.

THEOREM 4.4

If S is a nontrivial ring with unity, and if the following notation is adopted in the ring $S[x]$:
$$(r, z, z, \ldots) = r, \qquad (z, u, z, z, \ldots) = x,$$
then
$$(a_0, a_1, a_2, \ldots, a_k, z, z, \ldots) = a_0 + a_1 x + a_2 x^2 + \cdots + a_k x^k.$$

Proof

It is readily verified by mathematical induction that
$$x^n = (z, z, \ldots, u, z, z, \ldots),$$
where u is the nth coefficient, and is the only nonzero coefficient. Then
$$a_n x^n = (a_n, z, z, \ldots) \cdot (z, z, \ldots, z, u, z, \ldots)$$
$$= (z, z, \ldots, z, a_n, z, z, \ldots),$$
where a_n is the nth coefficient of this last polynomial. Therefore,
$$(a_0, a_1, a_2, \ldots, a_k, z, z, \ldots)$$
$$= (a_0, z, z, \ldots) + (z, a_1, z, z, \ldots) + \cdots + (z, \ldots, z, a_k, z, \ldots)$$
$$= a_0 + a_1 x + \cdots + a_k x^k. \quad \square$$

Now that we have introduced the indeterminate x, it makes more sense to call the polynomial "$f(x)$", whereas there was no point to using the x when the polynomial was written as a sequence of the coefficients. We can represent a polynomial in any one of these ways:

$$f = (a_0, a_1, a_2, \ldots) = a_0 + a_1 x + a_2 x^2 + \cdots = f(x),$$

where only a finite number of the a_i are not zero. If $a_j = z$, then $a_j x^j = z$. We usually write only those terms for which $a_j \neq z$. In $\mathbf{Z}[x]$, for example, $(3, 0, -4, 0, 0, \ldots) = 3 - 4x^2$, but $(0, 0, 0, \ldots) = 0$.

4.1 Problems

1. In the polynomial ring $\mathbf{Z}[x]$, compute $f + g$ and fg, if $f = (2, 1, 3, 4, -1, 0, 0, \ldots)$, $g = (5, -2, 0, 8, 0, 0, \ldots)$. Then express $f, g, f + g$, and fg in the "usual" notation involving the indeterminate x.

2. Let polynomial domain $\mathbf{Z}_3[x]$ be given.
   (a) Find fg if $f = (\bar{1}, \bar{2}, \bar{0}, \bar{1}, \bar{0}, \bar{2}, \bar{0}, \bar{0}, \ldots)$, and $g = (\bar{2}, \bar{2}, \bar{1}, \bar{1}, \bar{0}, \bar{1}, \bar{2}, \bar{0}, \bar{0}, \ldots)$. What is the degree of fg?
   (b) List the units in $\mathbf{Z}_3[x]$.
   (c) How many polynomials are of degree 2, of degree 3, and of degree n?

3. (a) Give an example to show that if f and g are in $\mathbf{Z}_4[x]$, it is not necessarily true that $\deg fg = \deg f + \deg g$.
   (b) Find a zero divisor of degree ≥ 1 in $\mathbf{Z}_4[x]$.

4. In $\mathbf{Z}_{10}[x]$, find:
   (a) Polynomials f and g such that $\deg fg = \deg f + \deg g$.
   (b) Polynomials f and g such that $\deg fg < \deg f + \deg g$, where f and g are not zero divisors.
   (c) Two zero divisors, f and g.

5. Prove: If S has zero divisors, then so does $S[x]$.

6. Assume that f and g are in ring $M_2(\mathbf{Z})[x]$, where

$$f = \left(\begin{bmatrix} 2 & 1 \\ 0 & 3 \end{bmatrix}, \begin{bmatrix} 0 & 0 \\ 0 & 0 \end{bmatrix}, \begin{bmatrix} 1 & -1 \\ 0 & 0 \end{bmatrix}, O, O, \ldots \right),$$

$$g = \left(\begin{bmatrix} 1 & 0 \\ 0 & 1 \end{bmatrix}, \begin{bmatrix} 4 & 0 \\ 0 & 4 \end{bmatrix}, \begin{bmatrix} 0 & 0 \\ 0 & 0 \end{bmatrix}, \begin{bmatrix} 1 & 3 \\ 1 & 3 \end{bmatrix}, O, O, \ldots \right).$$

   (a) Compute $-f, f + g, fg$, and gf.
   (b) Give the degree of: $f, g, f + g, fg, gf$.
   (c) Give the unity of $M_2(\mathbf{Z})[x]$.

7. Prove: For all $f(x) \in S[x]$ and all $n \in \mathbf{Z}^+$, $f(x) \cdot x^n = x^n \cdot f(x)$, where S is a ring with unity.

8. Prove: If S and T are isomorphic rings, then $S[x]$ and $T[x]$ are isomorphic rings.

9 Prove that $\mathbf{Z}[x]$ is an *ordered* integral domain if one defines the set, $(\mathbf{Z}[x])^+$ of positive elements as follows:
$$a_0 + a_1 x + \cdots + a_k x^k > 0 \Leftrightarrow a_k > 0.$$

10 Given that S is a ring with unity, prove that if $S[x]$ is an integral domain, then S is an integral domain.

11 Another approach to polynomials over a field F is to define a polynomial as a map $f: F \to F$, such that
$$x \to xf = f(x) = \sum_{i=0}^{n} a_i x^i, \quad \text{for all } x \in F.$$
If $F = \mathbf{R}$, verify that $F[x]$ is a subring of the function ring of Problem 2.6–17.

12 If $f = a_0 + a_1 x + a_2 x^2 + a_3 x^3 + \cdots + a_n x^n$, define
$$f^{(1)} = a_1 + 2a_2 x + 3a_3 x^2 + \cdots + na_n x^{n-1},$$
and
$$f^{(m+1)} = [f^{(m)}]^{(1)}, \quad \text{for } m \geq 1.$$

(a) If $\deg f = n$, use induction on n to prove that $\deg f^{(i)} = n - i$, for $i = 1, 2, \ldots, n$, and $f^{(n+1)} = 0$, where 0 is the zero polynomial.

(b) Prove: If $f, g \in \mathbf{R}[x]$, then
$$(f + g)^{(1)} = f^{(1)} + g^{(1)}, \quad \text{and} \quad (fg)^{(1)} = f \cdot g^{(1)} + f^{(1)} \cdot g.$$
(Do *not* use limits, or any formulas from calculus.)

4.2 Division in Polynomial Domains

In Section 4.1 we discovered that if F is a field, then $F[x]$ is an integral domain. Throughout this chapter, F shall always denote a field. An element of $F[x]$ is called a *polynomial over F*. If E is a field such that $F \subseteq E$, then F is called a *subfield* of E, and E is called an *extension field*, or an *extension* of F. If $f(x)$ is a polynomial over F, then $f(x)$ is also a polynomial over E.

As in the case of *any* integral domain, $F[x]$ can be partitioned into these four subsets:

(1) *Zero*, the zero polynomial being denoted by 0.

(2) *Units*, a unit being a polynomial of zero degree. (Thus, a unit is a nonzero element of F.)

(3) *Primes*, or *irreducible polynomials*, an irreducible polynomial $p(x)$ being a polynomial of positive degree such that every factoring, $p(x) = f(x) \cdot g(x)$, implies that either $f(x)$ is a unit or $g(x)$ is a unit, where $f(x), g(x) \in F[x]$.

(4) *Composites, reducible polynomials,* or *factorable polynomials*, a reducible polynomial $h(x)$ being a polynomial of positive degree n such that $h(x) = f(x) \cdot g(x)$, for some polynomials $f(x), g(x) \in F[x]$, where $1 \leq \deg f(x) < n$, $1 \leq \deg g(x) < n$.

If $g(x)$ is a prime in $F[x]$, then $g(x)$ is said to be *irreducible over* F, and if $g(x)$ is a composite element of $F[x]$, then $g(x)$ is said to be *reducible over* F, or *factorable over* F. It can happen that $g(x)$ is irreducible over F, but is reducible over E, where E is an extension field of F. It makes no sense to ask if a given polynomial can be factored unless the field over which the factoring is to be done is specified.

EXAMPLE 4.3

Polynomial $3 - x^2$ is irreducible over $\mathbf{Q}$ but is reducible over $\mathbf{R}$, since $3 - x^2 = (\sqrt{3} - x)(\sqrt{3} + x)$. So $3 - x^2$ is a prime element of $\mathbf{Q}[x]$, but is a composite element of $\mathbf{R}[x]$.

Note that $15 - 5x^2$ is also prime in $\mathbf{Q}[x]$, because every factoring has a unit as a factor. For instance:

$$15 - 5x^2 = 5(3 - x^2) = (-5)(-3 + x^2) = 2(\tfrac{15}{2} - \tfrac{5}{2}x^2). \quad \square$$

In Chapter 3 we proved that $\mathbf{Z}$ is a unique factorization domain. To prove this, we needed first to prove several theorems, beginning with the Division Theorem. (The sequence consisted of these theorems: 3.2, 3.4, 3.7, 3.13, 3.14, 3.15.) We now show that a division theorem exists for polynomials over a field F, and we shall develop a sequence of results from this theorem that will enable us to prove that $F[x]$ also is a unique factorization domain.

EXAMPLE 4.4

In high school you learned a "long division" method for dividing a polynomial $f(x)$ by a polynomial $g(x)$, obtaining a quotient polynomial $q(x)$ and a remainder polynomial $r(x)$. Your work probably took this form:

$$\begin{array}{r} q(x) \\ g(x) \overline{\smash{\big)} f(x)} \\ \cdots\cdots\cdots \\ \hline r(x) \end{array}$$

Recall that you kept on dividing until either $r(x) = 0$, or the degree of $r(x)$ was less than the degree of $g(x)$. You checked your accuracy by showing that $f(x) = g(x) \cdot q(x) + r(x)$.

If one divides $f(x)$ by $g(x)$ in $\mathbf{Z}_7[x]$, for instance, where

$$f(x) = \bar{6}x^4 + \bar{4}x^3 + \bar{2}x^2 + \bar{3}x + \bar{4}, \quad \text{and} \quad g(x) = \bar{2}x^2 + x + \bar{5},$$

one finds that $q(x) = \bar{3}x^2 + \bar{4}x + \bar{2}$, and $r(x) = \bar{2}x + \bar{1}$. We leave it to you to verify this. $\square$

THEOREM 4.5 *(Division Theorem for Polynomials)*

If $f(x), g(x) \in F[x]$, with $g(x) \neq 0$, then there exist unique $q(x), r(x) \in F[x]$ such that

$$f(x) = g(x) \cdot q(x) + r(x), \quad \text{with } r(x) = 0 \text{ or } \deg r(x) < \deg g(x).$$

Proof

Define S to be the following subset of $F[x]$:

$$S = \{f(x) - g(x) \cdot t(x) : t(x) \in F[x]\}.$$

If $0 \in S$, then $0 = f(x) - g(x) \cdot q(x)$, for some $q(x) \in F[x]$, and so

$$f(x) = g(x)q(x) + r(x), \quad \text{where } r(x) = 0.$$

If $0 \notin S$, let $r(x)$ be a polynomial of minimal degree m in S. So $f(x) - g(x)q(x) = r(x)$, for some $q(x) \in F[x]$, and hence

$$f(x) = g(x)q(x) + r(x).$$

We must show that $\deg r(x) < \deg g(x)$. Let

$$g(x) = b_k x^k + \cdots + b_2 x^2 + b_1 x + b_0,$$
$$r(x) = c_m x^m + \cdots + c_2 x^2 + c_1 x + c_0.$$

Assume that $m \geq k$, and define a polynomial $s(x)$ as follows:

$$s(x) = r(x) - [c_m b_k^{-1} x^{m-k}] \cdot g(x).$$

It is readily verified that $\deg s(x) < m$. But

$$\begin{aligned} s(x) &= f(x) - g(x)q(x) - [c_m b_k^{-1} x^{m-k}] \cdot g(x) \\ &= f(x) - g(x) \cdot [q(x) + c_m b_k^{-1} x^{m-k}] \in S, \end{aligned}$$

which is a contradiction of the choice of $r(x)$ as the polynomial of *lowest* degree in S. Therefore, $m < k$.

So far we have established the existence of polynomials $q(x)$ and $r(x)$. We now verify that they are unique. Suppose there exist polynomials, $q_1(x)$, $q_2(x)$, $r_1(x)$, and $r_2(x)$, such that

$$f(x) = g(x)q_1(x) + r_1(x), \quad \text{where } r_1(x) = 0, \ldots \text{ or } \deg r_1(x) < k,$$
$$f(x) = g(x)q_2(x) + r_2(x), \quad \text{where } r_2(x) = 0, \text{ or } \deg r_2(x) < k.$$

Then
$$g(x)[q_1(x) - q_2(x)] = r_2(x) - r_1(x).$$

If $q_1(x) - q_2(x) \neq 0$, then

$$\deg g(x) + \deg [q_1(x) - q_2(x)] = \deg [r_2(x) - r_1(x)].$$

This means that $\deg [r_2(x) - r_1(x)] \geq k$, which is impossible, since $r_1(x)$ and $r_2(x)$ are each of degree less than k. Therefore, $q_1(x) - q_2(x) = 0$, and so $q_1(x) = q_2(x)$. Then $r_2(x) - r_1(x) = g(x) \cdot 0$, and so $r_2(x) = r_1(x)$. $\square$

Suppose you were asked to divide a polynomial $f(x)$ by a first degree polynomial, $x - c$. A quick and easy way to find the quotient and remainder is to use the method known as *synthetic division*. The next example illustrates this method.

EXAMPLE 4.5

If $f(x) = 3x^3 - 5x^2 + 4x - 2$ and $g(x) = x - 2$, where $f(x), g(x) \in \mathbf{Q}[x]$, then there exists a unique polynomial $q(x)$ and a unique rational number r such that

$$3x^3 - 5x^2 + 4x - 2 = (x - 2)q(x) + r.$$

The long division method for finding $q(x)$ and r is as follows:

$$
\begin{array}{r}
3x^2 + x + 6 \\
x - 2 \overline{\smash{\big)} 3x^3 - 5x^2 + 4x - 2} \\
\underline{3x^2 - 6x^2 } \\
x^2 + 4x \\
\underline{x^2 - 2x } \\
6x - 2 \\
\underline{6x - 12} \\
+ 10
\end{array}
$$

Then $q(x) = 3x^2 + x + 6$, and $r = 10$. We note that the x acts merely as a placeholder. The crucial numbers in the process are all found in this abbreviated version:

Line 1: $\quad\quad\quad\quad\quad\quad\quad 3 - 5 + 4 - 2 \;\underline{|\,2}$

Line 2: $\quad\quad\quad\quad\quad\quad\quad\quad\;\; + 6 + 2 + 12$

Line 3: $\quad\quad\quad\quad\quad\quad\quad 3 + 1 + 6 + 10$

We have inserted arrows to indicate the order in which the numbers are obtained. On Line 1 we have written all the coefficients of $f(x)$, starting with the leading coefficient. The divisor, $x - 2$, is replaced by just the 2 placed at the right-hand end of Line 1. The leading coefficient is brought down to Line 3, and the first number, 6, in Line 2 is obtained by multiplying 3 by 2. The two numbers in the second column are added to obtain the 1 in Line 3. In this same way we find, in turn, the numbers in the third, fourth, ..., columns. Finally, Line 3 contains the coefficients of $q(x)$, and the last number in Line 3 is the remainder.

As another example of synthetic division, to divide $2x^4 - x^3 + 5x + 4$ by $x + 3$, we merely perform additions and multiplications, arranging the work as follows:

$$
\begin{array}{r}
2 - 1 + 0 + 5 + 4 \;\underline{|\,-3} \\
\underline{-6 + 21 - 63 + 174} \\
2 - 7 + 21 - 58 + 178
\end{array}
$$

Therefore, the quotient is $2x^3 - 7x^2 + 21x - 58$, and remainder is 178. □

4.2 Division in Polynomial Domains

THEOREM 4.6 *(Synthetic Division)*

Let $f(x) = a_n x^n + a_{n-1} x^{n-1} + \cdots + a_1 x + a_0 \in F[x]$, and $c \in F$. If $f(x)$ is divided by $x - c$, then the quotient is $q(x) = b_{n-1} x^{n-1} + b_{n-2} x^{n-2} + \cdots + b_1 x + b_0$, and the remainder is r,

where
$$b_{n-1} = a_n,$$
$$b_i = a_{i+1} + c \cdot b_{i+1}, \quad \text{if } 0 \leq i \leq n - 2,$$
and
$$r = a_0 + c \cdot b_0.$$

The computation necessary to find the b_i and r can be arranged as follows:

$$
\begin{array}{ccccccc|c}
a_n & a_{n-1} & a_{n-2} \cdots & a_{i+1} \cdots & a_1 & a_0 & & c \\
& cb_{n-1} & cb_{n-2} \cdots & cb_{i+1} \cdots & cb_1 & cb_0 & & \\
\hline
b_{n-1} & b_{n-2} & b_{n-3} \cdots & b_i & \cdots & b_0 & r &
\end{array}
$$

Proof, by induction on n, will be left as an exercise.

4.2 Problems

1. In $\mathbf{Q}[x]$ find a $q(x)$ and an $r(x)$ such that the conditions of the Division Theorem hold if:
   (a) $f(x) = x^4 - 2x^3 + x^2 - \frac{1}{2}$, $\quad g(x) = 2x^3 - x^2 + 5x + 1$.
   (b) $f(x) = 2x + 7$, $\quad g(x) = x^4 + 16x^2 - 5$.
   (c) $f(x) = 0$, $\quad g(x) = 3x^2 + 2$.
   (d) $f(x) = 5$, $\quad g(x) = 3x^2 + 2$.

2. Referring to Example 4.4, perform the long division of $f(x)$ by $g(x)$ in $\mathbf{Z}_7[x]$. Then perform the multiplication and addition necessary to verify that $f(x) = g(x)q(x) + r(x)$.

3. In $\mathbf{Z}_5[x]$, find $q(x)$ and $r(x)$ such that the conditions of the Division Theorem are satisfied, if

   $$f(x) = x^6 + \bar{4}x^5 + \bar{2}x^4 + x^3 + \bar{4}x^2 + x + \bar{3}, \quad g(x) = x^2 + \bar{3}x + \bar{4}.$$

4. Does the Division Theorem hold in the polynomial domain $\mathbf{Z}[x]$? Explain.

5. Let $f(x), g(x) \in F[x]$. Prove:
   (a) If c is a unit of $F[x]$, then $c|f(x)$.
   (b) If $f(x)|g(x)$, and if $h(x)$ is an associate of $f(x)$, then $h(x)|g(x)$.
   (c) If $\deg f(x) = 1$, then $f(x)$ is a prime polynomial.

6. Prove the Division Theorem [4.5] by induction on the degree n of $f(x)$. (Do not prove *uniqueness*, however.)

7. Use synthetic division to find $q(x)$ and r, if $f(x)$ and $g(x)$ are the following polynomials over $\mathbf{Q}$.

(a) $2x^3 - 5x^2 + 6x - 3$, $x - 2$
(b) $3x^4 + 7x^2 + 8x - 2$, $x + 1$
(c) $4x^3 - x^2 - 6x + 7$, $2x - 3$

8   In $\mathbf{Z}_5[x]$, find $q(x)$ and $r(x)$ such that the conditions of the Division Theorem are satisfied, if

$$f(x) = x^6 + \bar{4}x^5 + \bar{2}x^4 + x^3 + \bar{4}x^2 + x + \bar{3}, \qquad g(x) = x - \bar{3}.$$

(Use synthetic division.)

9   Prove Theorem 4.6, using induction on n.

4.3 Unique Factorization of Polynomials Over A Field

In order to continue with our proposed plan of proving that $F[x]$ is a unique factorization domain, we must define a few more terms. If $f(x) = a_k x^k + \cdots + a_1 x + a_0$, with $a_k \neq 0$, then a_k is called the *leading coefficient* of $f(x)$. A polynomial is called a *monic polynomial* if its leading coefficient is the unity 1. Of course, every nonzero polynomial has exactly one associate that is monic, for if $g(x) = a_k^{-1} \cdot f(x)$, where $f(x)$ is the polynomial whose leading coefficient is a_k, then $g(x)$ is a monic polynomial, and is an associate of $f(x)$.

The *greatest common divisor of polynomials $f(x)$ and $g(x)$* in $F[x]$ is a monic polynomial $d(x) \in F[x]$ such that (1) $d(x) \mid f(x)$, $d(x) \mid g(x)$, and (2) if $c(x)$ is any common divisor of $f(x)$ and $g(x)$, where $c(x) \in F[x]$, then $c(x) \mid d(x)$. We write $(f(x), g(x)) = d(x)$.

We leave it as an exercise for you to verify that the g.c.d. is unique. Note that the technique we used for finding the g.c.d. of two rational integers (the Euclidean Algorithm) can be employed to find the g.c.d. of two polynomials, f and g, with this modification: If the last nonzero remainder, $r_k(x)$, has a leading coefficient c, then the g.c.d. of f and g is $d(x)$, where $d(x) = c^{-1} \cdot r_k(x)$, since $d(x)$ must be monic.

The following theorem is the analogue of Theorem 3.4(a).

THEOREM 4.7

If $f(x)$, $g(x)$, and $d(x)$ are polynomials in $F[x]$ such that $d(x) = (f(x), g(x))$, then there exist $s(x), t(x) \in F[x]$ such that

$$d(x) = f(x) \cdot s(x) + g(x) \cdot t(x).$$

Proof

Define $S = \{f(x)h(x) + g(x)k(x) : h(x), k(x) \in F[x]\}$, and let $r(x)$ be a polynomial of lowest degree in S. Then there exist polynomials $s_1(x)$ and $t_1(x)$ such that $r(x) = f(x)s_1(x) + g(x)t_1(x)$. If c is the leading coefficient of $r(x)$, let $d(x) = c^{-1}r(x)$. Then $d(x)$ is monic, and

$$d(x) = f(x)[c^{-1}s_1(x)] + g(x)[c^{-1}t_1(x)] = f(x) \cdot s(x) + g(x) \cdot t(x).$$

4.3 Unique Factorization of Polynomials Over A Field

The proof that the $d(x)$ so obtained is the g.c.d. of f and g will be left as an exercise. It is analogous to the proof of Theorem 3.4(a). □

The next theorem is the analogue of Theorems 3.7(b) and 3.14.

THEOREM 4.8

Assume that $p(x)$ is a prime polynomial of $F[x]$.
(a) If $p(x) \mid f(x)g(x)$, with $f, g \in F[x]$, then $p \mid f$ or $p \mid g$.
(b) If $p(x) \mid f_1(x)f_2(x)\ldots f_n(x)$, with $f_i \in F[x]$, then $p \mid f_i$, for some i.

Proof

(a) If $p(x) \mid f(x)$, we're done. Suppose $p(x) \nmid f(x)$. Then $(p(x), f(x)) = 1$, since the only monic polynomials that divide $p(x)$ are 1 and the monic associate of $p(x)$. By Theorem 4.7, there exist $s(x)$ and $t(x)$ in $F[x]$ such that $p(x)s(x) + f(x)t(x) = 1$. So

$$p(x)[s(x)g(x)] + [f(x)g(x)]t(x) = g(x).$$

From this equation, along with Theorem 2.26(b), we conclude that $p(x) \mid g(x)$.
(b) This generalization of part (a) is proved by induction on n. □

We now have the machinery that is needed to prove the next theorem.

THEOREM 4.9

Let $f(x)$ be a polynomial of degree $n \geq 1$ in $F[x]$.
(a) $f(x)$ can be factored into a product of prime polynomials in $F[x]$.
(b) The factoring of $f(x)$ is unique (except for order and unit factors).

Proof

(a) We prove this by induction on n. If $\deg f(x) = 1$, then $f(x)$ is prime. Now assume that the theorem is true for every polynomial of degree less than n, and let $f(x)$ be of degree n. If $f(x)$ is prime, we are done. If $f(x)$ is reducible, then there exist polynomials $g(x)$ and $h(x)$ in $F[x]$ such that

$$f(x) = g(x)h(x), \quad \text{where } 1 \leq \deg g(x) < n, \quad 1 \leq \deg h(x) < n.$$

By the induction hypothesis, $g(x)$ and $h(x)$ can each be factored into a product of irreducible polynomials. Thus,

$$f(x) = g(x)h(x) = [p_1(x)p_2(x)\ldots p_k(x)][q_1(x)q_2(x)\ldots q_m(x)],$$

where the $p_i(x)$ are the prime factors of $g(x)$, and the $q_j(x)$ are the prime factors of $h(x)$.

(b) We shall prove uniqueness by induction on the degree of $f(x)$. If $\deg f(x) = 1$, then f cannot be expressed as a product of two polynomials of positive degree. Hence the uniqueness is clear.

Assume that every polynomial of degree less than n can be factored uniquely, and let $\deg f(x) = n$. Suppose that

$$f(x) = p_1(x)p_2(x)\ldots p_r(x) = q_1(x)q_2(x)\ldots q_s(x),$$

where each p_i and q_j is a prime polynomial. Then $p_1 \mid q_1 q_2 \ldots q_s$, by definition of divisibility. So $p_1 \mid q_i$, for some i, by Theorem 4.8(b). We choose notation so that $p_1 \mid q_1$. Then $q_1(x) = p_1(x)k(x)$. But $q_1(x)$ is prime, so $k(x)$ must be a unit. Let $k(x) = c_1 \in F$. Then

$$p_1(x)p_2(x)\ldots p_r(x) = p_1(x) \cdot c_1 \cdot q_2(x)\ldots q_s(x).$$

The cancellation law of multiplication holds, since $F[x]$ is an integral domain. Then

$$t(x) = p_2(x)\ldots p_r(x) = c_1 \cdot q_2(x)\ldots q_s(x),$$

with $\deg t(x) < n$. By the induction hypothesis, $t(x)$ can be factored in only one way into a product of prime polynomials. Hence, $r = s$, and $p_i(x) = c_i \cdot q_i(x)$, where $c_i \in F$, for $i = 2, 3, \ldots, r$. Therefore, the factoring of $f(x)$ is unique, except for unit factors. $\square$

EXAMPLE 4.6

If S is a commutative ring with unity, but is not a field, then factorization is not necessarily unique in $S[x]$. For instance, let $f(x) = x^2 + \bar{5}x \in \mathbf{Z}_{12}[x]$. Then

$$f(x) = x(x - \bar{7}) = (x - \bar{3})(x - \bar{4}),$$

so $f(x)$ has been factored into a product of primes in more than one way. $\square$

4.3 Problems

1. Find all associates of $\bar{2}x^2 + \bar{4}x + \bar{3}$ in $\mathbf{Z}_5[x]$.

2. Why does each nonzero polynomial in $F[x]$ have *exactly one* monic polynomial as an associate?

3. Prove: The greatest common divisor of two polynomials is *unique*.

4. Rewrite Theorem 3.6 (the Euclidean Algorithm for rational integers) to obtain the analogue for polynomials over a field F.

5. Apply the Euclidean Algorithm (Problem 4) to the given polynomials $f(x)$ and $g(x)$ to find their g.c.d. $d(x)$, and then find polynomials $s(x)$ and $t(x)$ such that $d(x) = f(x)s(x) + g(x)t(x)$.
   (a) In $\mathbf{Q}[x]$: $f(x) = 2x^5 + 2x^4 + 4x^3 - x^2 - x - 2$,
   $g(x) = 2x^4 + 2x^3 + 5x^2 + x + 2$.
   (b) In $\mathbf{Q}[x]$: $f(x) = x^5 - x^4 + x^3 + x^2 - x - 1$,
   $g(x) = x^2 + x - 2$.

(c) In $\mathbf{Z}_3[x]$: $f(x) = x^4 + \bar{2}x^2 + \bar{2}x + \bar{2}$,
$g(x) = \bar{2}x^3 + \bar{2}x^2 + x + \bar{1}$.

6 Complete the proof of Theorem 4.7.

7 Complete the proof of Theorem 4.8.

8 Given that $f(x) = (\bar{3}x^2 + \bar{1})(\bar{5}x^3 + \bar{2}x - \bar{4})(\bar{4}x - \bar{3}) \in \mathbf{Z}_7[x]$, write $f(x)$ as a product of a unit and three monic polynomials.

9 Prove: If $f(x) \in F[x]$, then

$$f(x) = c \cdot m_1(x) \cdot m_2(x) \cdot \cdots \cdot m_r(x),$$

where c is the leading coefficient of $f(x)$, and the $m_i(x)$ are monic polynomials that are prime over F.

10 In $\mathbf{Z}_{12}[x]$ factor a polynomial (other than the one given in Example 4.6) into a product of primes in two ways.

11 Prove: If n is a composite rational integer, then there exist polynomials in ring $\mathbf{Z}_n[x]$ that do not factor *uniquely* into a product of irreducible polynomials.

12 Find real numbers, a and b, such that the g.c.d. of $x^3 + x^2 - ax + b$ and $x^3 + 2x^2 + bx - a$ is a second-degree polynomial.

13 (This problem develops theory that is the analogue of Sections 3.5 and 3.6.)
Assume that $f(x), g(x), s(x), t(x), m(x) \in F[x]$, with $\deg m(x) = n$.
Define *congruence modulo* $m(x)$ as follows:

$$f(x) \equiv g(x) \pmod{m(x)}$$

if and only if there is a $q(x) \in F[x]$ such that $f(x) = g(x) + m(x) \cdot q(x)$.

Prove:
(a) Congruence modulo $m(x)$ is an equivalence relation on the set $F[x]$.

(b) Each $f(x)$ is congruent to one and only one $r(x) \in F[x]$, where either $r(x) = 0$ or $\deg r(x) < n$.

(c) If $f(x) \equiv g(x) \pmod{m(x)}$, and $s(x) \equiv t(x) \pmod{m(x)}$, then $f(x) + s(x) \equiv g(x) + t(x) \pmod{m(x)}$, and $f(x) \cdot s(x) \equiv g(x) \cdot t(x) \pmod{m(x)}$.

(d) If $f(x)s(x) \equiv f(x)t(x) \pmod{m(x)}$, and if $(f(x), m(x)) = g(x)$, then $s(x) \equiv t(x) \pmod{m(x)/g(x)}$. So if $(f(x), m(x)) = 1$, then $s(x) \equiv t(x) \pmod{m(x)}$.

(e) If

$$\overline{r(x)} = \{r(x) + m(x) \cdot q(x) : q(x) \in F[x]\},$$
$$F[x]/\langle m(x) \rangle = \{\overline{r(x)}\} = \text{factor set of } F[x] \text{ modulo } m(x),$$

and if addition and multiplication are defined as follows:

$$\overline{r(x)} + \overline{s(x)} = \overline{r(x) + s(x)}, \quad \overline{r(x)} \cdot \overline{s(x)} = \overline{r(x) \cdot s(x)};$$

then $+$ and $\cdot$ are well-defined operations on $F[x]/\langle m(x) \rangle$.

(f) $(F[x]/\langle m(x)\rangle, +, \cdot)$ is a commutative ring with unity.

(g) If $m(x)$ is composite over F, then the factor ring $F[x]/\langle m(x)\rangle$ has zero divisors.

(h) If $m(x)$ is prime over F, then $F[x]/\langle m(x)\rangle$ is a field.

4.4 Roots of Polynomials

If $f(x) \in F[x]$, where $f(x) = a_0 + a_1 x + a_2 x^2 + \cdots + a_n x^n$, and if c is an element in some extension field E of F, then, by definition,

$$f(c) = a_0 + a_1 c + a_2 c^2 + \cdots + a_n c^n.$$

Observe that $f(c) \in E$, but $f(c)$ may or may not be in F. Although $f(x)$ is a polynomial of degree n over F, either $f(c)$ is a polynomial of degree 0 over E or $f(c) = 0$, where 0 is the zero element of E. If $f(c) = 0$, then c is called a *root*, or a *zero*, of the polynomial $f(x)$, and we say that $f(x)$ has a root in extension field E. It is possible that $f(x)$ has no root in F, but *does* have one or more roots in some extension field of F.

EXAMPLE 4.7

Let $f(x) = x^4 - 9 \in \mathbf{Q}[x]$. Now f has no roots in $\mathbf{Q}$, but has the two roots, $\sqrt{3}$ and $-\sqrt{3}$, in extension field $\mathbf{R}$. The complex field $\mathbf{C}$ is also an extension of $\mathbf{Q}$, and $f(x)$ has four roots in $\mathbf{C}$: $\sqrt{3}, -\sqrt{3}, \sqrt{3}i$ and $-\sqrt{3}i$. ☐

EXAMPLE 4.8

Suppose that $f(x)$ and $g(x)$ are in $\mathbf{Z}_5[x]$, where

$$f(x) = x^3 + \bar{2}x + \bar{2}, \qquad g(x) = x^2 + x + \bar{1}.$$

By the process of substituting each of the five elements of $\mathbf{Z}_5$ for x, we find that $f(\bar{1}) = \bar{0}$, and that $\bar{1}$ is the only root that f has in $\mathbf{Z}_5$. Furthermore, $g(x)$ has *no* roots in $\mathbf{Z}_5$. But this doesn't mean that $g(x)$ has no roots! In Chapter 6 we devise a scheme for finding an extension field of $\mathbf{Z}_5$ in which $g(x)$ has two roots. We shall also be able to find an extension field of $\mathbf{Z}_5$ in which $f(x)$ has three roots. ☐

If E is an extension of F, and $c \in E$, the map,

$$\theta_c \colon F[x] \to E,$$

such that
$$f(x) \to f(c),$$

is of interest. It turns out that for each c, θ_c is a ring morphism. We first give an example of such a map, and then prove that θ_c is, indeed, a morphism.

EXAMPLE 4.9

Define the map

$$\theta_{2i}: \mathbf{R}[x] \to \mathbf{C}, \quad \text{so that}$$
$$f(x) \to f(2i).$$

This map is not 1–1. For instance:

$$3 + x - x^3 \to 3 + 10i, \quad 28 - 4x + 7x^2 - x^3 \to 0,$$
$$3 + 5x \to 3 + 10i, \quad 0 \to 0,$$
$$1 + \sqrt{3}x^2 \to 1 - 4\sqrt{3}, \quad 4 + x^2 \to 0,$$
$$1 - 4\sqrt{3} \to 1 - 4\sqrt{3}, \quad x \to 2i.$$

Map θ_{2i} is onto, an antecedent of any complex number $a + bi$ being $a + (b/2)x$. Note that each real number maps to itself. □

THEOREM 4.10 (Basic Morphisms of Field Theory)

If F is a subfield of a field E, and if $c \in E$, then the map,

$$\theta_c: F[x] \to E, \quad \text{where } f(x)\theta_c = f(c),$$

is a morphism of $F[x]$ into E.

Proof

Let $f(x)$ and $g(x)$ be any two elements of $F[x]$, where

$$f(x) = \sum_{i=0}^{k} a_i x^i, \quad g(x) = \sum_{i=0}^{n} b_i x^i.$$

Then $f(c) = \sum_{i=0}^{k} a_i c^i \in E$, and so θ_c is a map of $F[x]$ into E.

We must verify that addition and multiplication are preserved. By definition of addition of polynomials,

$$f(x) + g(x) = \sum_{i=0}^{m} (a_i + b_i)x^i,$$

where m is the larger of k and n. Then, by definition of θ_c,

$$[f(x) + g(x)]\theta_c = \sum_{i=0}^{m} (a_i + b_i)c^i.$$

We make use of the distributive property, the associative law of addition, and the commutative law of addition of field E to obtain the following:

$$\sum_{i=0}^{m} (a_i + b_i)c^i = \sum_{i=0}^{m} (a_i c^i + b_i c^i)$$
$$= \sum_{i=0}^{m} a_i c^i + \sum_{i=0}^{m} b_i c^i = f(c) + g(c).$$

Therefore, $[f(x) + g(x)]\theta_c = f(c) + g(c) = f(x)\theta_c + g(x)\theta_c$, and so addition is preserved.

In E we must apply the distributive property, the associative and commutative laws of multiplication, and the associative and commutative laws of addition to obtain the following:

$$f(c)g(c) = (a_0 + a_1 c + \cdots + a_k c^k)(b_0 + b_1 c + \cdots + b_n c^n)$$
$$= a_0 b_0 + (a_0 b_1 + a_1 b_0)c + \cdots + a_k b_n c^{k+n}$$
$$= \sum_{i=0}^{k+n} \left(\sum_{j+t=i} a_j b_t \right) c^i.$$

Hence
$$[f(x) \cdot g(x)]\theta_c = \left[\sum_{i=0}^{k+n} \left(\sum_{j+t=i} a_j b_t \right) x^i \right] \theta_c \quad \text{(Def. of mult. in } F[x]\text{)}$$
$$= \sum_{i=0}^{k+n} \left(\sum_{j+t=i} a_j b_t \right) c^i \quad \text{(Def. of map } \theta_c\text{)}$$
$$= f(c) \cdot g(c) \quad \text{(From above equations)}$$
$$= [f(x)\theta_c] \cdot [g(x)\theta_c]. \quad \text{(Def. of } \theta_c\text{)}$$

So multiplication is preserved and, therefore, θ_c is a ring morphism. □

We shall make good use of Theorem 4.10. For instance, if in $F[x]$ this equation is given: $h(x) = f(x) \cdot g(x) + s(x) \cdot t(x)$, then we are applying Theorem 4.10 when we write: $h(c) = f(c)g(c) + s(c)t(c)$, with c an element of an extension field of F.

We turn now to the manner in which factoring a polynomial over F is related to the roots it may have in F.

THEOREM 4.11 *(Factor Theorem)*

Given that $f(x) \in F[x]$, and $r \in F$, then

$$f(r) = 0 \Leftrightarrow f(x) = (x - r) \cdot q(x),$$

for some $q(x) \in F[x]$.

Proof

By the Division Theorem, there is a $q(x) \in F[x]$ and an $a \in F$ such that

$$f(x) = (x - r) \cdot q(x) + a.$$

Then, by Theorem 4.10,

$$f(r) = (r - r) \cdot q(r) + a = a.$$

So $\quad f(r) = 0 \Leftrightarrow a = 0 \Leftrightarrow f(x) = (x - r) \cdot q(x).$ □

The next theorem is a generalization of the Factor Theorem.

THEOREM 4.12

Let $f(x)$ be a polynomial in $F(x)$ of degree n.

(a) If $f(x)$ has k distinct roots, $r_1, r_2, \ldots, r_k$, in F, then

$$f(x) = (x - r_1)(x - r_2) \ldots (x - r_k) \cdot g(x), \quad \text{with } g(x) \in F[x].$$

(b) $f(x)$ has at most n distinct roots in F.

Proof

(a) The proof will be by induction on k. If $k = 1$, then $f(x) = (x - r_1)g(x)$, by Theorem 4.11. Now assume that the theorem is true for any polynomial that has $k - 1$ distinct roots in F. Let $f(x)$ have the k distinct roots: $r_1, r_2, \ldots, r_k$. Since r_1 is a root of $f(x)$, then $f(x) = (x - r_1)q(x)$, by the Factor Theorem. Now

$$0 = f(r_j) = (r_j - r_1)q(r_j), \quad j = 2, 3, \ldots, k.$$

Since F has no zero divisors, either $r_j - r_1 = 0$ or $q(r_j) = 0$. But $r_j \neq r_1$. So $q(r_j) = 0$, and $q(x)$ has $k - 1$ distinct roots. By the induction hypothesis, there is a $g(x) \in F[x]$ such that

$$q(x) = (x - r_2)(x - r_3) \cdots (x - r_k)g(x).$$

Therefore, $\quad f(x) = (x - r_1)(x - r_2) \cdots (x - r_k)g(x).$

(b) Suppose that deg $g(x) = m$. From the above equation, $n = k + m$, by Theorem 4.2(c). So $k \leq n$. Therefore, $f(x)$ has at most n distinct roots. □

EXAMPLE 4.10

If $f(x)$ is a polynomial of degree n in $S[x]$, where S is not a field, but is a commutative ring with unity, then $f(x)$ can have more than n roots. For instance, let $f(x) = x^2 + 5x \in \mathbf{Z}_{12}[x]$. Then deg $f(x) = 2$, but f has four roots: $\bar{0}, \bar{3}, \bar{4}$, and $\bar{7}$. □

EXAMPLE 4.11

Let $f(x) \in \mathbf{Z}_7[x]$, where

$$f(x) = \bar{3}x^5 - \bar{3}x^4 - x^3 + x^2 + \bar{4}x + \bar{3}.$$

It is readily verified that $f(\bar{1}) = f(\bar{3}) = f(\bar{4}) = \bar{0}$, and $f(c) \neq \bar{0}$ for each other $c \in \mathbf{Z}_7$. Hence $f(x)$ has three roots in $\mathbf{Z}_7$. By Theorem 4.12, $f(x) = (x - \bar{1})(x - \bar{3})(x - \bar{4})g(x)$. To find $g(x)$, one can divide $f(x)$ by the product of the three first-degree polynomials, or one can use synthetic division (three times). It happens that $g(x) = \bar{3}x^2 + \bar{5}$, which has no roots in $\mathbf{Z}_7$, and so is irreducible over $\mathbf{Z}_7$. □

We know that the $g(x)$ of the preceding example is irreducible, by the next theorem.

THEOREM 4.13

Let $f(x)$ be a polynomial over F of degree 2 or 3. Then $f(x)$ is factorable over F if and only if $f(x)$ has a root in F.

Proof

If $f(x)$ has a root r in F, then $f(x) = (x - r)g(x)$, where the degree of $g(x)$ is one or two. Therefore, $f(x)$ is factorable over F.

Conversely, suppose that $f(x)$ is factorable over F. Then $f(x) = g(x)h(x)$, where one of the factors must be of degree one, since $\deg f = 2$ or 3. Let $g(x) = ax + b$, with $a \neq 0$. Then $g(-a^{-1}b) = 0$. Therefore, $-a^{-1}b$ is a root of $f(x)$. □

EXAMPLE 4.12

(a) We want to determine if $x^3 + x + \bar{1}$ is prime over $\mathbf{Z}_5$. By substitution, we see that if $f(x) = x^3 + x + \bar{1}$, then

$$f(\bar{0}) = \bar{1}, \quad f(\bar{1}) = \bar{3}, \quad f(\bar{2}) = \bar{1}, \quad f(\overline{-1}) = \bar{4}, \quad f(\overline{-2}) = \bar{1}.$$

Since $f(x)$ has no root in $\mathbf{Z}_5$, $f(x)$ is prime over $\mathbf{Z}_5$.

(b) We give an example to show that if $\deg f(x) > 3$, then $f(x)$ may be factorable over F and still have no root in F. In $\mathbf{Q}[x]$, for instance, let $f(x) = x^4 - 9$. Then $f(x) = (x^2 - 3)(x^2 + 3)$, so $f(x)$ is factorable over the rational field, but has no rational roots. □

We conclude with a theorem whose simple proof will be left as an exercise.

THEOREM 4.14

Let $g(x)$ be an associate of $f(x)$, where $f(x) \in F[x]$. Then $f(x)$ and $g(x)$ have the same roots in any extension field E of F.

4.4 Problems

1. Find all the roots of $f(x)$ in $\mathbf{Z}_7$, and express $f(x)$ as a product of prime polynomials in $\mathbf{Z}_7[x]$, if
   (a) $f(x) = x^4 + x^3 - x^2 + \bar{3}$.
   (b) $f(x) = \bar{2}x^3 + \bar{3}x^2 + \bar{4}x - \bar{2}$.
   (c) $f(x) = x^7 - x$.
   (d) $f(x) = x^4 + \bar{6}x^3 + \bar{3}x^2 + \bar{6}x + \bar{2}$.

2. (a) Verify that $x^3 - \bar{2}x^2 - x + \bar{1}$ is prime over $\mathbf{Z}_5$.
   (b) For some n find a polynomial $f(x)$ of degree n in $\mathbf{Z}_6[x]$ that has more than n roots, and factor $f(x)$ into a product of prime polynomials in two ways.

3. Apply Theorem 4.13 to find all polynomials of degree 3 in $\mathbf{Z}_2[x]$ that are prime over $\mathbf{Z}_2$.

4 Referring to Example 4.11, use synthetic division to find $g(x)$.

5 Given the morphism, $\theta_{\sqrt{2}}: \mathbf{Q}[x] \to \mathbf{R}$, defined by: $f(x) \to f(\sqrt{2})$; find
(a) a monic polynomial whose image is 0,
(b) three other polynomials whose images are 0.

6 Let c be in an extension field E of field F, and let N be a subset of $F[x]$ that is defined as follows:
$$N = \{f(x): f(c) = 0\}.$$
Prove:
(a) N is a commutative ring.
(b) If $m(x)$ is a polynomial of lowest degree in N, and if $f(x)$ is any polynomial in N, then $m(x) | f(x)$. (Hint: Divide $f(x)$ by $m(x)$, and show that the remainder has to be the zero polynomial.)

7 Let $f(x) \in F[x]$, and let $c \in E$, where E is an extension of F.
Prove: $f(c)$ is the remainder obtained when $f(x)$ is divided by $(x - c)$.

8 Find a polynomial over $\mathbf{R}$ that has no real roots but is factorable over $\mathbf{R}$.

9 Find a complex number, $a + bi$, that is a root of $f(x)$, if $f(x) = x^2 - (3i - 2)x - (5 + i)$, and factor $f(x)$ over $\mathbf{C}$.

10 Prove Theorem 4.14.

11 Let f and g be polynomials over F such that $(f(x), g(x)) = 1$.
Prove:
(a) $f(x)$ and $g(x)$ have no common root.
(b) If f_1 is an associate of f, and g_1 is an associate of g, then $(f_1(x), g_1(x)) = 1$.

12 Let $f^{(i)}(x)$ denote the usual ith derivative of the polynomial $f(x)$ with respect to x, where $f(x) \in \mathbf{R}[x]$. (See Problem 4.1–12.) Define a root r of $f(x)$ to be of *multiplicity* m if $f(x) = (x - r)^m g(x)$, and $g(r) \neq 0$. If $m > 1$, then r is called a *multiple root* of f.
Prove:
(a) $f(x)$ has a root r of multiplicity $m \Leftrightarrow m$ is the smallest positive integer such that $f^{(m)}(r) \neq 0$.
(b) r is a multiple root $\Leftrightarrow (x - r)$ is a common factor of $f(x)$ and $f^{(1)}(x)$.

13 Prove: If $f(x)$ is a polynomial in $F[x]$ of degree 3, and has a nonzero root in F that is twice another root, then all the roots of $f(x)$ are in F.

14 Let $r_1, r_2, \ldots, r_n$ be the roots of $f(x)$, where $r_i \in F$, and
$$f(x) = x^n + a_{n-1}x^{n-1} + \cdots + a_1 x + a_0 \in F[x].$$
Verify that $r_1 + r_2 + \cdots + r_n = -a_{n-1}$, and $r_1 \cdot r_2 \cdot \cdots \cdot r_n = (-1)^n a_0$.

4.5 Polynomials Over The Real and Complex Fields

We have seen that a polynomial over a field F does not necessarily have a root in F. However, if $f(x) \in \mathbf{C}[x]$, it can be proved, by methods beyond the scope of this course (using some advanced topics in analysis), that $f(x)$ has a root in $\mathbf{C}$. We state this important theorem, and ask you to accept it without proof.

THEOREM 4.15 *(Fundamental Theorem of Algebra)*

Every polynomial in $\mathbf{C}[x]$ has a root in $\mathbf{C}$.

The next theorem is a consequence of the Fundamental Theorem.

THEOREM 4.16

(a) Let $f(x) = a_n x^n + a_{n-1} x^{n-1} + \cdots + a_1 x + a_0 \in \mathbf{C}[x]$. Then there exist complex numbers, $c_1, c_2, \ldots, c_n$, not necessarily distinct, such that $f(x) = a_n(x - c_1)(x - c_2) \cdots (x - c_n)$.
(b) The only prime polynomials in $\mathbf{C}[x]$ are polynomials of degree one.

The proof is left as an exercise. It is suggested that you prove part (a) by induction on the degree n of $f(x)$.

We now determine the prime polynomials in $\mathbf{R}[x]$. To do this, we must first say something about complex numbers and their conjugates. By definition, the *conjugate* of the complex number, $a + bi$, is the complex number, $a - bi$. We write: $(a + bi)^* = a - bi$. Note that if r is a real number, then $r^* = r$.

THEOREM 4.17

If s and t are complex numbers, then

$$(s + t)^* = s^* + t^*, \quad \text{and} \quad (s \cdot t)^* = s^* \cdot t^*.$$

Proof

Let $s = a + bi$, and $t = c + di$. The proof is a straightforward application of the definitions of addition and multiplication of complex numbers, and of the conjugate of a complex number. □

Next we show that the complex roots of a polynomial over $\mathbf{R}$ come in complex conjugate pairs.

THEOREM 4.18

If $f(x) \in \mathbf{R}[x]$, $s \in \mathbf{C}$, and $f(s) = 0$, then $f(s^*) = 0$.

4.5 Polynomials Over The Real and Complex Fields

Proof

Let $f(x) = a_n x^n + \cdots + a_1 x + a_0$. If $f(s) = 0$, then $[f(s)]^* = 0^* = 0$. So, by repeated applications of Theorem 4.17:

$$\begin{aligned}
0 &= [a_n s^n + \cdots + a_1 s + a_0]^* \\
&= (a_n s^n)^* + \cdots + (a_1 s)^* + a_0^* \\
&= a_n^* (s^*)^n + \cdots + a_1^* s^* + a_0^* \\
&= a_n (s^*)^n + \cdots + a_1 (s^*) + a_0 = f(s^*). \quad \square
\end{aligned}$$

THEOREM 4.19

Every polynomial in $\mathbf{R}[x]$ that is irreducible over $\mathbf{R}$ is of degree one or two.

Proof

If $c = a + bi \in \mathbf{C}$, with $b \neq 0$, then

$$(x - c)(x - c^*) = x^2 - (c + c^*)x + cc^* = x^2 - 2ax + (a^2 + b^2) \in \mathbf{R}[x].$$

Since its roots are not in $\mathbf{R}$, $(x - c)(x - c^*)$ is a prime polynomial over $\mathbf{R}$ of degree 2.

If $f(x)$ has leading coefficient a_n, and is of degree n, then

$$\begin{aligned}
f(x) \in \mathbf{R}[x] &\Rightarrow f(x) \in \mathbf{C}[x] \\
&\Rightarrow f(x) = a_n(x - c_1)(x - c_2) \cdots (x - c_n),
\end{aligned}$$

where the c_i are complex numbers (by Theorem 4.16). If each c_i is real, then $f(x)$ is a product of prime polynomials of the first degree. If some c_k is not real, then $c_k^* = c_j$, for some $j \neq k$, and $(x - c_k)(x - c_j)$ is a prime polynomial of degree 2 over $\mathbf{R}$. By multiplying each linear factor involving a nonreal number by a linear factor involving the conjugate of that number, we can express $f(x)$ as a product of first and second degree prime polynomials over $\mathbf{R}$. $\quad \square$

THEOREM 4.20

A polynomial in $\mathbf{R}[x]$ of odd degree always has a real root.

Proof

According to Theorem 4.19, if $f(x) \in \mathbf{R}[x]$, then

$$f(x) = p_1(x) \cdot p_2(x) \cdots p_m(x),$$

where each $p_i(x)$ is a prime polynomial over $\mathbf{R}$ of degree 1 or 2. If $\deg p_i(x) = 2$, for each i, then $\deg f(x) = 2m$, a contradiction of the hypothesis. So for some j, $p_j(x) = ax - b$, where $a, b \in \mathbf{R}$, and $a \neq 0$. Then $p_j(b/a) = 0$, which implies that $f(b/a) = 0$. Therefore, b/a is a real root of $f(x)$. $\quad \square$

EXAMPLE 4.13

Suppose we want to find a monic polynomial over F that has $\sqrt[3]{2}$ and $(1 + i)$ as two of its roots. If $F = \mathbf{C}$, then

$$f(x) = (x - \sqrt[3]{2})(x - (1 + i)) = x^2 - (\sqrt[3]{2} + 1 + i)x + \sqrt[3]{2}(1 + i).$$

If $F = \mathbf{R}$, then

$$\begin{aligned} g(x) &= (x - \sqrt[3]{2})(x - (1 + i))(x - (1 - i)) \\ &= (x - \sqrt[3]{2})(x^2 - 2x + 2) \\ &= x^3 - (2 + \sqrt[3]{2})x^2 + (2 + 2\sqrt[3]{2})x - 2\sqrt[3]{2}. \end{aligned}$$

If $F = \mathbf{Q}$, then

$$\begin{aligned} h(x) &= (x^3 - 2)(x^2 - 2x + 2) \\ &= x^5 - 2x^4 + 2x^3 - 2x^2 + 4x - 4. \quad \square \end{aligned}$$

4.5 Problems

1. (a) Find a monic polynomial of lowest possible degree in $\mathbf{R}[x]$ that has $(\sqrt{2} + 3i)$ and $\sqrt{2}$ as two of its roots. Find such a polynomial in $\mathbf{Q}[x]$.
   (b) Find a polynomial in $\mathbf{Q}[x]$ that has $5i$ and $(2 - \sqrt{3})$ as two of its roots.

2. Factor each polynomial into a product of primes: (1) over $\mathbf{Q}$, (2) over $\mathbf{R}$, and (3) over $\mathbf{C}$.
   (a) $x^3 - 3x^2 - 2x + 6$ (b) $x^3 - (5/6)x^2 - (1/2)x + (1/3)$
   (c) $x^4 + 4x^3 + x^2 + 4x$ (d) $x^4 - 3x^2 - 4$

3. Give a monic polynomial of lowest degree in $F[x]$ that has 0, $\sqrt{2}$, and $3i$ as three of its roots if:
   (a) $F = \mathbf{C}$,   (b) $F = \mathbf{R}$,   (c) $F = \mathbf{Q}$.

4. Prove Theorem 4.16.

5. Prove Theorem 4.17.

6. Let $f(x) \in \mathbf{C}[x]$. If $f(a + bi) = 0$, is it true that $f(a - bi) = 0$? Explain.

7. Find the complex roots of $f(x)$ if:
   (a) $f(x) = x^2 - (5 - 12i)$.
   (Hint: Solve two equations in two unknowns to find real numbers, a and b, such that $(a + bi)^2 = 5 - 12i$.)
   (b) $f(x) = x^2 - 2ix + 3$.
   (c) $f(x) = x^2 - 2x + (1 + 2i)$.

8. Show that the map, $\theta: \mathbf{C} \to \mathbf{C}$, such that $(a + bi)\theta = a - bi$, is an isomorphism. (Hint: Use Theorem 4.17.)

9. Verify: If s and t are nonreal complex numbers such that $(x - s)(x - t) \in \mathbf{R}[x]$, then $t = s^*$.

10 (a) Prove: $f(x)$ and $g(x)$ are associates in $\mathbf{C}[x] \Leftrightarrow f(x)$ and $g(x)$ have the same roots in $\mathbf{C}$, each of the same multiplicity. (See Problem 4.4–12 for the definition of *multiplicity*.)

(b) Comment on the statement of part (a) if the complex field $\mathbf{C}$ is replaced by the real field $\mathbf{R}$.

11 Let $f(x) = 3x^5 + x^4 - 2x^3 - 7x^2 - x + 4 \in \mathbf{C}[x]$.

(a) Find a polynomial $g(x)$ whose roots are the additive inverses of the roots of $f(x)$.

(b) Find a polynomial $h(x)$ whose roots are the multiplicative inverses of the roots of $f(x)$.

(Do not attempt to determine the roots of $f(x)$.)

12 (a) Show that -2 cannot be a multiple root of $x^3 + x^2 - ax - 4$, for any real value of a. (See Problem 4.4–12 for the meaning of *multiple root*.)

(b) For what values of r is r a root of

$$2x^4 - 3rx^3 - r^2x^2 + 3r^3x - r^2?$$

What is the multiplicity of r?

(c) Determine a and b so that 1 is a double root of

$$x^4 + ax^3 + (a - b)x^2 + bx + 1.$$

4.6 Polynomials Over The Rational Field

In this section we develop a few results that apply specifically to finding *rational* roots of polynomials, and factoring polynomials in $\mathbf{Q}[x]$.

Suppose that $f(x) \in \mathbf{Q}[x]$, with

$$f(x) = (r_n/s_n)x^n + (r_{n-1}/s_{n-1})x^{n-1} + \cdots + (r_1/s_1)x + (r_0/s_0),$$

and $r_i, s_i \in \mathbf{Z}$. If c is a common multiple of the s_i, and if $g(x) = c \cdot f(x)$, then $g(x) \in \mathbf{Z}[x]$, and $f(x)$ and $g(x)$ have the same roots. Since every polynomial in $\mathbf{Q}[x]$ has an associate in $\mathbf{Z}[x]$, we need only to search for roots of polynomials with integral coefficients when we seek the roots of a polynomial over $\mathbf{Q}$.

The next result is quite useful.

THEOREM 4.21

Let $f(x)$ be a polynomial in $\mathbf{Z}[x]$, with

$$f(x) = a_n x^n + a_{n-1} x^{n-1} + \cdots + a_1 x + a_0.$$

If $f(r/s) = 0$, where r and s are relatively prime integers, then $r \mid a_0$ and $s \mid a_n$.

Proof

$$0 = f(r/s) = a_n(r/s)^n + a_{n-1}(r/s)^{n-1} + \cdots + a_1(r/s) + a_0.$$

If we multiply each side of this equation by s^n, we get:

$$0 = a_n r^n + a_{n-1} r^{n-1} s + \cdots + a_1 r s^{n-1} + a_0 s^n. \qquad [1]$$

Adding $-a_0 s^n$ to each side:

$$-a_0 s^n = r(a_n r^{n-1} + a_{n-1} r^{n-2} s + \cdots + a_1 s^{n-1}).$$

Hence, $r \mid a_0 s^n$. But $(r, s) = 1$, and so $(r, s^n) = 1$. Therefore, $r \mid a_0$.

If we add $-a_n r^n$ to each side of equation [1], we get:

$$-a_n r^n = s(a_{n-1} r^{n-1} + \cdots + a_1 r s^{n-2} + a_0 s^{n-1}).$$

Hence, $s \mid a_n r^n$. But $(r^n, s) = 1$. Therefore, $s \mid a_n$. □

EXAMPLE 4.14

Given that $f(x) = (3/4)x^4 - (1/2)x^3 - (5/4)x^2 - x - 1$, we want to find rational roots of $f(x)$, and to factor $f(x)$ in $\mathbf{Q}[x]$.

First, we find an associate of $f(x)$ with coefficients in $\mathbf{Z}$. If $g(x) = 4 \cdot f(x)$, then

$$g(x) = 3x^4 - 2x^3 - 5x^2 - 4x - 4.$$

According to Theorem 4.21, if r/s is a root, then $r \mid -4$, and $s \mid 3$. So any rational roots must be found among these numbers:

$$1, \quad 2, \quad 4, \quad -1, \quad -2, \quad -4, \quad 1/3, \quad 2/3, \quad 4/3, \quad -1/3, \quad -2/3, \quad -4/3.$$

It is a simple matter to verify that $g(2) = 0$ and $g(-1) = 0$. Then

$$g(x) = (x - 2)(x + 1)(3x^2 + x + 2),$$

and so $\quad f(x) = (\tfrac{1}{4})(x - 2)(x + 1)(3x^2 + x + 2).$

Since $3x^2 + x + 2$ has no rational roots, the factoring of f into the product of a unit and prime *monic* polynomials over $\mathbf{Q}$ is as follows:

$$f(x) = (\tfrac{3}{4})(x - 2)(x + 1)(x^2 + \tfrac{1}{3}x + \tfrac{2}{3}). \quad □$$

The next statement follows immediately from Theorem 4.21.

THEOREM 4.22

Any rational root of a monic polynomial in $\mathbf{Z}[x]$ is a rational integer.

We now show how we can attack the problem of factoring $f(x)$ over $\mathbf{Q}$, where $f(x) \in \mathbf{Q}[x]$, by dealing instead with an associate, $g(x)$, where $g(x) \in \mathbf{Z}[x]$. The next three theorems are concerned with polynomials in $\mathbf{Z}[x]$.

4.6 Polynomials Over The Rational Field

By definition, a *primitive polynomial* is a polynomial in $\mathbf{Z}[x]$ whose coefficients have no common factor other than ± 1. For instance, $6x^2 + 15x + 4$ is primitive, since $(6, 15, 4) = 1$, while $6x^2 + 15x + 12$ is not, since $(6, 15, 12) = 3$.

THEOREM 4.23

The product of two primitive polynomials is a primitive polynomial.

Proof

Let $h(x) = f(x) \cdot g(x)$, where $f(x)$ and $g(x)$ are primitive polynomials, and

$$f(x) = \sum_{i=0}^{m} a_i x^i, \qquad g(x) = \sum_{i=0}^{n} b_i x^i, \qquad h(x) = \sum_{i=0}^{m+n} c_i x^i.$$

Assume that $h(x)$ is not primitive. Then there is a prime p such that $p \mid c_i$, for $i = 0, 1, 2, \ldots, m + n$. Now $c_0 = a_0 b_0$, so $p \mid a_0$ or $p \mid b_0$. $f(x)$ is primitive, so p does not divide every a_i. Let j be the *smallest* subscript such that $p \nmid a_j$. Similarly, let k be the *smallest* subscript such that $p \nmid b_k$. Now

$$c_{j+k} = a_0 b_{j+k} + a_1 b_{j+k-1} + \cdots + a_j b_k + a_{j+1} b_{k-1} + \cdots + a_{j+k} b_0.$$

Since p divides $c_{j+k}, a_0, a_1, \ldots, a_{j-1}, b_{k-1}, \ldots, b_1, b_0$, we must conclude that $p \mid a_j b_k$. This implies that $p \mid a_j$ or $p \mid b_k$, a contradiction. Therefore, $h(x)$ is primitive. □

We now prove that if a polynomial over $\mathbf{Z}$ cannot be factored over $\mathbf{Z}$, then neither can it be factored over $\mathbf{Q}$. Actually, we prove the contrapositive of that statement.

THEOREM 4.24 *(Lemma of Gauss)*

Let $g(x)$ be a polynomial in $\mathbf{Z}[x]$. If $g(x)$ is factorable over $\mathbf{Q}$, then $g(x)$ is factorable over $\mathbf{Z}$.

Proof

Assume that $g(x) = h(x) \cdot k(x)$, with $h(x), k(x) \in \mathbf{Q}[x]$. Writing each coefficient of $h(x)$ as a fraction with the same positive common denominator t,

$$h(x) = (s_0/t) + (s_1/t)x + \cdots + (s_m/t)x^m.$$

If $s = (s_0, s_1, \ldots, s_m)$, then $h(x) = (s/t) \cdot a(x)$, where $a(x)$ is a primitive polynomial. Similarly, there exist positive integers q and r such that $k(x) = (q/r) \cdot b(x)$, where $b(x)$ is primitive. Then

$$g(x) = (s/t) \cdot a(x) \cdot (q/r) \cdot b(x),$$

and so

$$rt \cdot g(x) = sq \cdot [a(x) \cdot b(x)].$$

Then $rt\,|\,sq[a(x)\cdot b(x)]$, which means that $rt\,|\,sq$, since $a(x)\cdot b(x)$ is primitive [Theorem 4.23]. So $sq = rtc$ for some $c \in \mathbf{Z}$. By the cancellation property in domain $\mathbf{Z}[x]$ we have that $g(x) = c\cdot a(x)\cdot b(x)$, with $a(x), b(x) \in \mathbf{Z}[x]$. □

EXAMPLE 4.15

We illustrate Theorem 4.24. Let
$$g(x) = x^5 + 6x^4 + 8x^3 + 22x^2 + 4x - 8.$$
Then $\quad g(x) = [(3/2)x^3 + 6x - 3][(2/3)x^2 + 4x + (8/3)].$

So $g(x)$ is a primitive polynomial that has been factored over $\mathbf{Q}$. Employing the technique used in proving Theorem 4.24, we get:

$$\begin{aligned} g(x) &= [(3/2)x^3 + (12/2)x - (6/2)][(2/3)x^2 + (12/3)x + (8/3)] \\ &= (3/2)(x^3 + 4x - 2)(2/3)(x^2 + 6x + 4) \\ &= (x^3 + 4x - 2)(x^2 + 6x + 4). \end{aligned}$$

Therefore, $g(x)$ can be factored over $\mathbf{Z}$. □

If you were asked to give a polynomial of degree 864 that is prime over $\mathbf{Q}$, for instance, could you do it? The next theorem enables us to accomplish such a feat with a minimum of effort. It also tells us that, given any positive integer n, there exists a polynomial of degree n in $\mathbf{Q}[x]$ that is prime.

THEOREM 4.25   *(Eisenstein's Irreducibility Criterion)*

Let $f(x) = a_0 + a_1 x + \cdots + a_n x^n \in \mathbf{Z}[x]$. If there is a prime $p \in \mathbf{Z}$ such that p divides each a_i except a_n, and $p^2 \nmid a_0$, then $f(x)$ is prime over $\mathbf{Q}$.

Proof

Assume that
$$f(x) = g(x)\cdot h(x) = (b_0 + b_1 x + \cdots + b_k x^k)(c_0 + c_1 x + \cdots + c_m x^m),$$
with $1 \leq k < n$, $1 \leq m < n$. Since $p\,|\,a_0$ and $p^2 \nmid a_0$, then p divides either b_0 or c_0, but not both. Notationally, suppose that $p \nmid b_0$ and $p\,|\,c_0$. Now $a_n = b_k c_m$, so $p \nmid c_m$. Let r be the smallest positive integer such that $p \nmid c_r$. But
$$a_r = b_0 c_r + b_1 c_{r-1} + \cdots + b_t c_{r-t},$$
where t is the smaller of r and k. But p divides $a_r, c_{r-1}, c_{r-2}, \ldots, c_{r-t}$, and so $p\,|\,b_0 c_r$. This implies that either $p\,|\,b_0$ or $p\,|\,c_r$, a contradiction. Therefore, $f(x)$ is prime over $\mathbf{Z}$, and hence is prime over $\mathbf{Q}$, by Theorem 4.24. □

EXAMPLE 4.16

(a) Let $f(x) = 5x^7 + \frac{3}{2}x^4 + 9x + \frac{21}{4}$. Is $f(x)$ prime over $\mathbf{Q}$? We write f as follows:
$$f(x) = (1/4)(20x^7 + 6x^4 + 36x + 21).$$

The prime 3 satisfies Eisenstein's Criterion, and so $4 \cdot f(x)$ is prime over $\mathbf{Q}$. Hence $f(x)$ is prime over $\mathbf{Q}$, We know, therefore, that $f(x)$ has no rational roots.

(b) Earlier we asked you to find a prime polynomial over $\mathbf{Q}$ of degree 864. There are many, of course, one of which is:

$$15x^{864} - 12x^{99} + 14x^{67} + 8x - 50.$$

The prime 2 satisfies the conditions of Theorem 4.25. ☐

4.6 Problems

1. Find all the roots of $f(x)$ in $\mathbf{Q}$, and express $f(x)$ as a product of polynomials in $\mathbf{Q}[x]$ if:
   (a) $f(x) = 3x^5 + 5x^4 - 2x^3 - 3x^2 - 5x + 2$
   (b) $f(x) = x^4 + x^3 - x^2 - 2x - 2$
   (c) $f(x) = x^5 - (3/2)x^4 + x - (3/2)$
   (d) $f(x) = x^4 + x^3 - 3x^2 - 5x - 2$

2. Let n be any positive integer.
   (a) Give a primitive polynomial of degree n, for each n.
   (b) Give a prime polynomial over $\mathbf{Q}$ of degree n, for each n.

3. Determine whether each polynomial is prime over $\mathbf{Q}$, using the Eisenstein Criterion when possible.
   (a) $x^4 - 90x^2 - 2$
   (b) $x^3 + 2x^2 - 6x + 4$
   (c) $6x^8 + 10x^6 - 15x^5 + 25x^3 - 35$
   (d) $x^4 + x^2 - 12$

4. Give an example which illustrates the fact that the Eisenstein criterion for the irreducibility of a polynomial in $\mathbf{Q}[x]$ is sufficient but not necessary.

5. Let $f(x) = ax^2 + bx + c$, with $a, b, c \in \mathbf{Z}$. Verify that $f(x)$ is factorable over $\mathbf{Q}$ if and only if $b^2 - 4ac = m^2$, for some nonnegative integer m.

6. For which primes p and q of $\mathbf{Z}$ is $x^2 + 3x - pq$ prime over $\mathbf{Z}$?

7. Prove that $x^4 - 10x^2 + 1$ is prime over $\mathbf{Q}$. (Hint: Assume that $x^4 - 10x^2 + 1$ is a product of two quadratic polynomials in $\mathbf{Z}[x]$, and show that you get a contradiction.)

8. Let $f(x) = \sum_{i=0}^{n} a_i x^i$, with $a_i \in \mathbf{Q}$, and $g(x) = \sum_{i=0}^{n} a_i(x+1)^i$.
   (a) Verify that $f(x)$ is factorable over $\mathbf{Q}$ if and only if $g(x)$ is factorable over $\mathbf{Q}$.
   (b) Use part (a) and Eisenstein's Criterion to prove that $f(x)$ is prime over $\mathbf{Q}$, where p is a rational prime, and
   $$f(x) = x^{p-1} + x^{p-2} + \cdots + x^2 + x + 1.$$
   (Hint: First note that
   $$f(x) = \frac{x^p - 1}{x - 1}, \quad g(x) = \frac{(x+1)^p - 1}{x} = \cdots.$$

Show that $g(x)$ is a monic polynomial that satisfies the conditions of Theorem 4.25.)

9. Given: $f(x) = \sum_{i=0}^{n} a_i x^i$, $a_i \in \mathbf{Q}$, $a_0 \cdot a_n \neq 0$, $g(x) = x^n \cdot \sum_{i=0}^{n} a_i(1/x)^i$: prove:
   (a) $g(x)$ is a polynomial of degree n.
   (b) $f(x)$ is factorable over $\mathbf{Q}$ $\Leftrightarrow$ $g(x)$ is factorable over $\mathbf{Q}$.

10. Use results of Problem 9 to state another form of Eisenstein's Criterion.

11. Theorems of Gauss and of Eisenstein were given in this section. Did these men know each other? Do you think that Eisenstein could have become as famous as Gauss? Why didn't he? (See Appendices E–8 and E–15.)

4.7 Minimal Polynomials and Algebraic Number Fields

Let s be in some extension field E of a field F. If there is a polynomial $f(x)$ in $F[x]$ such that $f(s) = 0$, then s is said to be *algebraic over field* F. If s is not a root of any polynomial in $F[x]$, then s is said to be *transcendental over* F. A complex number that is algebraic over the rational field is called an *algebraic number*. A complex number that is transcendental over the rational field is called a *transcendental number*.

EXAMPLE 4.17

(a) The number, $\sqrt[3]{5}$, is an algebraic number, since it is a root of a polynomial in $\mathbf{Q}[x]$. One such polynomial is $x^3 - 5$. Another is $2x^4 + 3x^3 - 10x - 15$.

(b) We shall not prove it here, but it can be shown that there is no polynomial over $\mathbf{Q}$ that has π as a root. Hence, π is a transcendental number. But π is algebraic over $\mathbf{R}$, since π is a root of $f(x) \in \mathbf{R}[x]$, where $f(x) = x - \pi$.

(c) If $r = \sqrt{3} + i$, then r is, of course, algebraic over $\mathbf{C}$, since r is a root of $x - (\sqrt{3} + i) \in \mathbf{C}[x]$. Now

$$r = \sqrt{3} + i \Rightarrow r - \sqrt{3} = i \Rightarrow (r - \sqrt{3})^2 = i^2$$
$$\Rightarrow r^2 - 2\sqrt{3}r + 4 = 0.$$

Thus, r is a root of $x^2 - 2\sqrt{3}x + 4 \in \mathbf{R}[x]$, and so r is algebraic over the real field. But we can go farther:

$$r^2 - 2\sqrt{3}r + 4 = 0 \Rightarrow (r^2 + 4)^2 = (2\sqrt{3}r)^2$$
$$\Rightarrow r^4 - 4r^2 + 16 = 0.$$

Hence r is a root of $x^4 - 4x^2 + 16$, and so is algebraic over $\mathbf{Q}$. Therefore, $\sqrt{3} + i$ is an algebraic number. $\square$

If r is a root of *one* polynomial $f(x) \in F[x]$, then r is a root of *infinitely many* polynomials in $F[x]$, for if $f(r) = 0$, then $h(r) = 0$, where $h(x) = f(x) \cdot g(x)$, and $g(x)$

4.7 Minimal Polynomials and Algebraic Number Fields

is any polynomial over infinite field F. However, associated with each r is a *unique* polynomial, which we describe in the next theorem.

THEOREM 4.26

Let r be algebraic over a field F, where r is in an extension field E of F. Then r is a root of a *unique monic* polynomial $m(x) \in F[x]$ that is of *minimal degree* of all polynomials over F that have r as a root. Furthermore, $m(x)$ is *irreducible* over F, and if $f(x)$ is any polynomial in $F[x]$ that has r as a root, then $m(x) \mid f(x)$.

Proof

Since r is algebraic over F, we know that there are polynomials in $F[x]$ that have r as a root. Choose $g(x)$ to be a polynomial of minimal degree n such that $g(r) = 0$. If the leading coefficient of $g(x)$ is b, define $m(x) = b^{-1} \cdot g(x)$. Then $m(x)$ is a monic polynomial whose degree is also n, and $m(r) = 0$. We shall verify that $m(x)$ is prime over F and unique.

Assume that $m(x)$ is not prime. Then $m(x) = s(x)t(x)$, with $1 \leq \deg s(x) < n$, $1 \leq \deg t(x) < n$. But $0 = m(r) = s(r)t(r)$, which implies that either $s(r) = 0$ or $t(r) = 0$, a contradiction of the choice of $m(x)$ with minimal degree n. Therefore, $m(x)$ is prime over F.

Let $m(x) = x^n + a_{n-1}x^{n-1} + \cdots + a_1 \times a_0$. Suppose that $p(x)$ is also a monic polynomial of degree n such that $p(r) = 0$. Let $p(x) = x^n + b_{n-1}x^{n-1} + \cdots + b_1 x + b_0$. Define $h(x) = m(x) - p(x)$. Then

$$h(x) = (a_{n-1} - b_{n-1})x^{n-1} + \cdots + (a_1 - b_1)x + (a_0 - b_0).$$

Either $h(x) = 0$, or $\deg h(x) \leq n - 1$. But $h(r) = m(r) - p(r) = 0$. Therefore, $h(x) = 0$, and so $m(x) = p(x)$. We conclude that the monic polynomial of minimal degree is unique.

Let $f(x)$ be any polynomial over F such that $f(r) = 0$. By the Division Theorem, there exist polynomials $w(x)$ and $v(x)$ such that

$$f(x) = m(x)w(x) + v(x), \quad \text{with } v(x) = 0, \text{ or } \deg v(x) < n.$$

But $\quad v(r) = f(r) - m(r)w(r) = 0 - 0 \cdot w(r) = 0.$

We conclude that $v(x) = 0$. Therefore, $m(x) \mid f(x)$. □

The polynomial $m(x)$ of Theorem 4.26 is called the *minimal polynomial* of r over F. The *degree of an element that is algebraic over a field* F is the degree of its minimal polynomial over F.

EXAMPLE 4.18

The minimal polynomial of $\sqrt[3]{5}/2$ over $\mathbf{Q}$ is $x^3 - 5/8$, so $\sqrt[3]{5}/2$ is an algebraic number of degree 3. The minimal polynomial of $\sqrt{3} + i$ over $\mathbf{Q}$ is $x^4 - 4x^2 + 16$, so $\sqrt{3} + i$ is an algebraic number of degree 4. Note that

$\sqrt{3} + i$ is a root of $x^2 - 2\sqrt{3}x + 4$, which means that $\sqrt{3} + i$ is algebraic and of degree 2 over the real field. Each rational number p/q is an algebraic number of degree one, since it is a root of $x - p/q \in \mathbf{Q}[x]$. □

If r is algebraic and of degree n over F, then its minimal polynomial takes the form:

$$m(x) = x^n + b_{n-1}x^{n-1} + \cdots + b_1 x + b_0 \in F[x].$$

Since $m(r) = 0$, one has that

$$r^n = -b_0 - b_1 r - \cdots - b_{n-1} r^{n-1}.$$

So if $f(x)$ is *any* polynomial in $F[x]$, then $f(r)$ can be written as a sum involving powers of r that are less than n. We illustrate this.

EXAMPLE 4.19

Let $m(x) = x^4 + 3x^2 - 9x - 6$. By the Eisenstein Criterion, $m(x)$ is prime over $\mathbf{Q}$. If r is a root of $m(x)$, then

$$r^4 = 6 + 9r - 3r^2.$$

If $f(x)$ is any polynomial over $\mathbf{Q}$, then

$$f(r) = a_0 + a_1 r + a_2 r^2 + a_3 r^3,$$

for some $a_i \in \mathbf{Q}$. For instance, if $f(x) = 2x^5 + 3x^4 - x^3 + 7$, then

$$\begin{aligned} f(r) &= 2r^5 + 3r^4 - r^3 + 7 \\ &= 2r(6 + 9r - 3r^2) + 3(6 + 9r - 3r^2) - r^3 + 7 \\ &= 25 + 39r + 9r^2 - 7r^3. \quad \square \end{aligned}$$

THEOREM 4.27

If r is algebraic and of degree n over F, and if $f(x)$ is any polynomial in $F[x]$, then $f(r)$ is uniquely representable in the form:

$$f(r) = a_0 + a_1 r + \cdots + a_{n-1} r^{n-1}, \qquad a_i \in F.$$

Proof

Let $m(x)$ be the minimal polynomial of r. Then $\deg m(x) = n$. By the Division Theorem for polynomials, there exist unique $q(x)$ and $s(x)$ in $F[x]$ such that

$$f(x) = m(x) \cdot q(x) + s(x), \quad \text{with } s(x) = 0, \quad \text{or} \quad \deg s(x) < n.$$

So there are unique a_i in F such that

$$s(x) = a_0 + a_1 x + \cdots + a_{n-1} x^{n-1}.$$

4.7 Minimal Polynomials and Algebraic Number Fields

Then
$$f(r) = m(r) \cdot q(r) + s(r)$$
$$= 0 \cdot q(r) + s(r)$$
$$= s(r) = a_0 + a_1 r + \cdots + a_{n-1} r^{n-1}.$$

Proof of the uniqueness of the representation is asked for in Problem 4.7–12. □

If r and t have the same minimal polynomial over F, then t is called a *conjugate* of r, and r and t are said to be *conjugates over F*. We now make use of a number algebraic over F to obtain an extension field of F.

THEOREM 4.28

(a) Let r be algebraic and of degree n over a field F, where r is in an extension field E of F. If

$$F(r) = \{a_0 + a_1 r + a_2 r^2 + \cdots + a_{n-1} r^{n-1} : a_i \in F\},$$

then $(F(r), +, \cdot)$ is a field, the operations being those in E.

(b) If t is a conjugate of r, then fields $F(r)$ and $F(t)$ are isomorphic.

Proof

(a) Since $a_i \in F \subseteq E$, and $r \in E$, set $F(r)$ is a subset of E. So to show that $F(r)$ is a field, we need prove only that $F(r)$ is closed under the operations, and that the additive and multiplicative inverses of elements of $F(r)$ are in $F(r)$.

Let $a(r)$ and $b(r)$ be any elements in $F(r)$, where $a(r) = \sum_{i=0}^{n-1} a_i r^i$, $b(r) = \sum_{i=0}^{n-1} b_i r^i$. Define $a(x) = \sum_{i=0}^{n-1} a_i x^i$ and $b(x) = \sum_{i=0}^{n-1} b_i x^i$. If $f(x) = a(x) + b(x)$, and $g(x) = a(x) \cdot b(x)$, then $f(x)$ and $g(x)$ are in $F[x]$, and so $f(r), g(r) \in F(r)$, by Theorem 4.27. Therefore, $F(r)$ is closed under addition and multiplication, since $f(r) = a(r) + b(r)$, and $g(r) = a(r) \cdot b(r)$.

The additive inverse of $a(r)$ is $(-a_0) + (-a_1)r + \cdots + (-a_{n-1})r^{n-1}$, which is in $F(r)$. If $a(r) \neq 0$, we must show that $[a(r)]^{-1} \in F(r)$. Let $m(x)$ be the minimal polynomial of r. The only monic polynomials that divide $m(x)$ are 1 and $m(x)$, so $(m(x), a(x)) = 1$. (For if $m(x)$ divides $a(x)$, then $a(r) = 0$.) By Theorem 4.7, there are polynomials $v(x)$ and $w(x)$ over F such that

$$1 = m(x) \cdot v(x) + a(x) \cdot w(x).$$

So
$$1 = m(r) \cdot v(r) + a(r) \cdot w(r) = a(r) \cdot w(r),$$

which means that $w(r) = [a(r)]^{-1}$. Therefore, $F(r)$ is a field.

(b) If r and t are conjugates, then the map,

$$\theta : F(r) \rightarrow F(t),$$

such that $a_0 + a_1 r + \cdots + a_{n-1} r^{n-1} \rightarrow a_0 + a_1 t + \cdots + a_{n-1} t^{n-1}$,

is obviously an isomorphism. □

It can be shown that if $F(r) = F(s)$, then $\deg r = \deg s$ over F. (See Problem 4.7–26.) So if $\deg r = n$, it makes sense to call $F(r)$ an *algebraic extension field of F of degree n*. If $\deg(r) = 2$, then $F(r)$ is called a *quadratic extension* of F, and if $\deg(r) = 3$, then $F(r)$ is called a *cubic extension* of F. When F is the rational field, then $\mathbf{Q}(r)$ is called an *algebraic number field*. Also $\mathbf{Q}(r)$ is called a *quadratic field* if $\deg r = 2$, and a *cubic field* if $\deg r = 3$. Following are some examples of these fields.

EXAMPLE 4.20

If $r = 2 + \sqrt{-3}$, then $\mathbf{Q}(r)$ is a quadratic field, for
$r - 2 = \sqrt{-3} \Rightarrow (r - 2)^2 = (\sqrt{-3})^2 \Rightarrow r^2 - 4r + 7 = 0 \Rightarrow$ The minimal polynomial of r is $x^2 - 4x + 7$. We express this quadratic field as follows:

$$\mathbf{Q}(r) = \{a + br : a, b \in \mathbf{Q}, \text{ and } r^2 = 4r - 7\}.$$

We can derive a formula for the product of any two numbers of $\mathbf{Q}(r)$:

$$(a + br)(c + dr) = ac + (ad + bc)r + bdr^2$$

$$= ac + (ad + bc)r + bd(4r - 7)$$

$$= (ac - 7bd) + (ad + bc + 4bd)r.$$

A formula for $(a + br)^{-1}$ can be obtained as follows:

$$(a + br)(c + dr) = 1 \Rightarrow ac - 7bd = 1, \quad bc + (a + 4b)d = 0.$$

The solution of this last pair of equations is:

$$c = \frac{a + 4b}{a^2 + 4ab + 7b^2}, \quad d = \frac{-b}{a^2 + 4ab + 7b^2}.$$

Then $c + dr = (a + br)^{-1}$. For instance,

$$(3 - 2r)^{-1} = (-5/13) + (2/13)r. \quad \square$$

EXAMPLE 4.21

We now examine a cubic field. Let $m(x) = x^3 - 3x + 1$. We apply Theorems 4.21 and 4.13 to verify that $m(x)$ is prime over $\mathbf{Q}$. The graph of the equation, $y = x^3 - 3x + 1$, shows that $m(x)$ has three real, irrational roots: two positive roots and one negative root (see Figure 4–1). Let r be any one of those roots. Then $\mathbf{Q}(r) = \{a_0 + a_1 r + a_2 r^2 : r^3 = 3r - 1\}$, where the $a_i \in \mathbf{Q}$.

We shall illustrate two different methods for finding the multiplicative inverse of $2 - r^2$.

Method 1: We seek a, b, and c such that

$$(2 - r^2)(a + br + cr^2) = 1.$$

4.7 Minimal Polynomials and Algebraic Number Fields

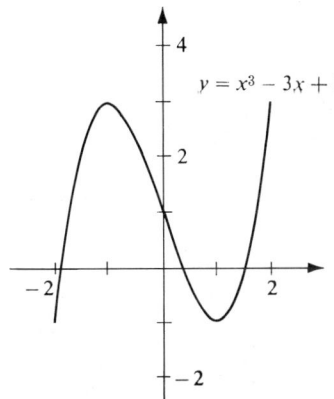

Figure 4–1

Multiplying the left members, we get:

$$2a + 2br + (2c - a)r^2 - br^3 - cr^4 = 1.$$

So $\quad 2a + 2br + (2c - a)r^2 - b(3r - 1) - cr(3r - 1) = 1.$

Then $\quad (2a + b) + (-b - c)r + (-a + 5c)r^2 = 1 + (0)r + (0)r^2.$

From this we obtain the three equations:

$$2a + b = 1, \quad -b - c = 0, \quad -a + 5c = 0,$$

whose common solution is: $a = 1, b = -1, c = -1$. Therefore, $(2 - r^2)^{-1} = 1 - r - r^2$.

Method 2: This is the method employed in the proof of Theorem 4.28. Let $f(x) = 2 - x^2$. Then $f(r) = 2 - r^2$. The only monic polynomials dividing $m(x)$ are 1 and $m(x)$. Hence, $(m(x), f(x)) = 1$. We use the Euclidean Algorithm to find $g(x)$ and $h(x)$ such that

$$m(x)g(x) + f(x)h(x) = 1.$$

Then $\quad m(r)g(r) + f(r)h(r) = 1.$

But $m(r) = 0$, and so $h(r)$ is the multiplicative inverse of $f(r)$. Putting this program into operation, we get:

$$x^3 - 3x + 1 = (-x^2 + 2)(-x) + (-x + 1),$$
$$-x^2 + 2 = (-x + 1)(x + 1) + 1.$$

So $\quad 1 = (-x^2 + 2) - (-x + 1)(x + 1)$
$\quad\quad = (-x^2 + 2) - [x^3 - 3x + 1 - (-x^2 + 2)(-x)](x + 1).$

Finally,

$$1 = (2 - x^2)(1 - x - x^2) + (x^3 - 3x + 1)(-1).$$

Therefore,

$$1 = (2 - r^2)(1 - r - r^2). \quad \square$$

EXAMPLE 4.22

We illustrate the fact that if r and t are conjugate algebraic numbers, it may happen that $\mathbf{Q}(r) \neq \mathbf{Q}(t)$, even though it is always true that $\mathbf{Q}(r) \cong \mathbf{Q}(t)$. Let $m(x) = x^3 - 2$, which is prime over $\mathbf{Q}$. If w is any root of $x^3 - 2$, then

$$\mathbf{Q}(w) = \{a_0 + a_1 w + a_2 w^2 : w^3 = 2\}.$$

Now $x^3 - 2 = (x - r)(x - s)(x - t)$,

where

$$r = \sqrt[3]{2}, \quad s = (\sqrt[3]{2}/2)(-1 + \sqrt{-3}), \quad t = (\sqrt[3]{2}/2)(-1 - \sqrt{-3}).$$

According to Theorem 4.28, fields $\mathbf{Q}(r)$, $\mathbf{Q}(s)$, and $\mathbf{Q}(t)$ are isomorphic. However, $\mathbf{Q}(r)$ contains only real numbers, while $\mathbf{Q}(s)$ contains some nonreal complex numbers. Therefore, $\mathbf{Q}(r) \neq \mathbf{Q}(s)$. It can also be shown that $\mathbf{Q}(s) \neq \mathbf{Q}(t)$. Field $\mathbf{Q}(r)$ is called a *real cubic field*, while $\mathbf{Q}(s)$ and $\mathbf{Q}(t)$ are called *imaginary cubic fields*. □

By the Eisenstein Criterion, if n is any positive integer, and p is any rational prime, then $x^n + p$ is prime over $\mathbf{Q}$. Hence we know that, for any positive integer n, there exists an algebraic number field $\mathbf{Q}(r)$ of degree n.

We have been calling $\mathbf{Z}$ the set of *rational* integers. You might well ask the question: Is there a set of *irrational* integers? We now proceed to describe such a set. An *algebraic integer* is, by definition, an algebraic number whose minimal polynomial is in $\mathbf{Z}[x]$. Every rational integer n is an algebraic integer, since the minimal polynomial of n is $x - n$, which is in $\mathbf{Z}[x]$. An algebraic integer that is *not* a rational integer will be called an *irrational integer*. If you fail to find an irrational integer defined in some other algebra book, don't be concerned. Remember that a word can be defined in any way that one chooses. As Humpty Dumpty said to Alice, in Lewis Carroll's *Through The Looking Glass*: "When I use a word, it means just what I choose it to mean—neither more nor less."

EXAMPLE 4.23

The minimal polynomials of $\sqrt[3]{2/5}$, $\frac{3}{4} + i$, and $\frac{9}{7}$ are, respectively: $x^3 - \frac{2}{5}, x^2 - \frac{3}{2}x + \frac{25}{16}, x - \frac{9}{7}$. So these three complex numbers are algebraic numbers, but are not algebraic integers. On the other hand, $\sqrt[3]{2}i, \frac{7}{2} + \frac{3}{2}\sqrt{5}$, and $\sqrt{-5} + \sqrt{3}$ are algebraic integers, since their minimal polynomials are, respectively:

$$x^6 + 4, \quad x^2 - 7x + 1, \quad x^4 + 4x^2 + 64.$$

The second of these irrational integers is real, while the other two are imaginary. □

EXAMPLE 4.24

The diagram in Figure 4–2 categorizes the complex numbers with respect to their minimal polynomials $m(x)$ over $\mathbf{Q}$.

4.7 Minimal Polynomials and Algebraic Number Fields

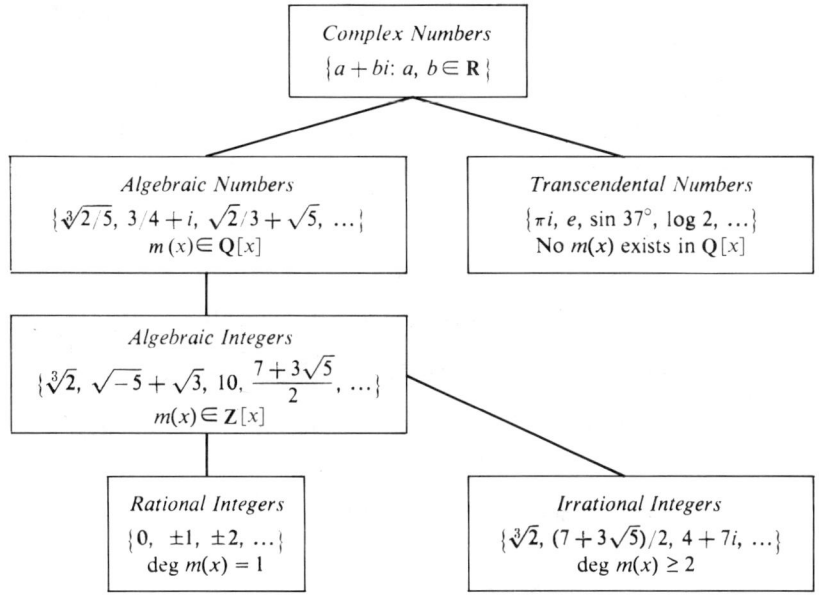

Figure 4-2

4.7 Problems

1 Find the minimal polynomials over **Q** of these numbers, and indicate which numbers are algebraic integers.

(a) $\frac{2}{3}$
(b) $2 + \sqrt{3}i$
(c) $\sqrt[3]{7}/2$
(d) $\sqrt{2} - \sqrt{3}$
(e) $\sqrt[3]{2} + \sqrt{5}$
(f) $\sqrt[3]{6} - 1/\sqrt{2}$
(g) $\dfrac{5 - \sqrt{-3}}{2}$

2 Give the possible degrees of a complex number r that is algebraic:

(a) over **Q** (b) over **R** (c) over **C**

3 List four elements of $F(r)$, not all of which are in F. Give the minimal polynomial of r over F.

(a) $\mathbf{Q}(i)$
(b) $\mathbf{R}(i)$
(c) $\mathbf{Q}(\sqrt[3]{7})$
(d) $\mathbf{R}(\sqrt{-2})$
(e) $F(\sqrt{5})$, if $F = \mathbf{Q}\sqrt{6})$
(f) $\mathbf{Q}(\sqrt{3} + 2\sqrt{2})$

4 Let $f(x) = x^4 - 3$.

(a) Factor $f(x)$ over $\mathbf{Q}(\sqrt{3})$, over **R**, and then over **C**.
(b) Show that $f(x)$ has two roots, r and s, such that $\mathbf{Q}(r) = \mathbf{Q}(s)$, and that there are two roots, s and t, such that $\mathbf{Q}(s) \neq \mathbf{Q}(t)$.

5 A typical element in $\mathbf{Q}(\sqrt{6})$ is $a + b\sqrt{6}$, with $a, b \in \mathbf{Q}$. Also, $(a + b\sqrt{6})^{-1} = c + d\sqrt{6}$, with $c = a/(a^2 - 6b^2)$, $d = (-b)/(a^2 - 6b^2)$.

Give a typical element in these fields, and find the multiplicative inverse of that element.

(a) $\mathbf{Q}(\sqrt{10})$ (b) $\mathbf{Q}(\sqrt[3]{4})$ (c) $\mathbf{Q}\sqrt{-2})$

6   Prove: If $r = \sqrt{6}$, and $s = 2/3 + \sqrt{6}/2$, then $\mathbf{Q}(r) = \mathbf{Q}(s)$.

7   Show that $\mathbf{Q}(\sqrt{3})$ is not isomorphic to $\mathbf{Q}(\sqrt{5})$.

8   (a) If $s = \sqrt{5} - \sqrt{2}$, prove that

$$\mathbf{Q}(s) = \{a_0 + a_1 s + a_2 s^2 + a_3 s^3 : a_i \in \mathbf{Q}, s^4 = 14s^2 - 9\}$$
$$= \{b_0 + b_1\sqrt{2} + b_2\sqrt{5} + b_3\sqrt{10} : b_i \in \mathbf{Q}\}.$$

(b) If $F = \mathbf{Q}(\sqrt{2})$, show that $F(\sqrt{5}) = \mathbf{Q}(\sqrt{5} - \sqrt{2})$.

9   If the order of field F is k, and if r is algebraic over F of degree n, what is the order of field $F(r)$?

10   If $m(x) \in \mathbf{Q}[x]$, $\deg m(x) = 1$, and $m(r) = 0$, what can you say about field $\mathbf{Q}(r)$?

11   Give an example of a field F, a field E, and an element r such that:

(a) $F = F(r) \subset E$.   (b) $F \subset F(r) = E$.
(c) $F = F(r) = E$.   (d) $F \subset F(r) \subset E$.

12   Prove: If r is of degree n over F, and if

$$a_0 + a_1 r + \cdots + a_{n-1} r^{n-1} = b_0 + b_1 r + \cdots + b_{n-1} r^{n-1},$$

with $a_i, b_j \in F$, then $a_i = b_i$, for $i = 0, 1, 2, \ldots, n-1$.

13   In Theorem 4.28(b), why is θ "obviously" an isomorphism?

14   Referring to the cubic field of Example 4.21:

(a) Develop the formula for the product of any two numbers of $\mathbf{Q}(r)$. That is, $(a + br + cr^2) \cdot (d + er + fr^2) = ?$

(b) Find $(1 - 3r + 2r^2)^{-1}$ by two distinct methods.

15   If r is a root of $x^2 + x + 1 \in \mathbf{Q}[x]$, derive a formula for:

(a) The product of any two elements of $\mathbf{Q}(r)$.
(b) The multiplicative inverse of any element of $\mathbf{Q}(r)$.

16   If $f(x) \in \mathbf{Q}[x]$, with $f(x) = (x^3 + 5x^2 - 2)(x^2 - 4x + 1)$, give two extension fields of $\mathbf{Q}$ in which $f(x)$ has a root.

17   Each of the given polynomials $m(x)$ is prime over $\mathbf{Q}$. If r is a root of $m(x)$, factor $m(x)$ over $\mathbf{Q}(r)$.

(a) $m(x) = x^3 - 2$ (b) $m(x) = x^3 - 3x + 1$
(c) $m(x) = x^2 + bx + c$ (d) $m(x) = x^4 + 3x^3 - 9x + 6$

(Hint: Divide $m(x)$ by $x - r$ to find the other factor.)

18   In Section 2.9 we defined the *conjugate* of $a + b\sqrt{n}$ to be $a - b\sqrt{n}$, where $a, b \in \mathbf{Q}$, and n is a square-free rational integer. Show that this definition is compatible with the definition of conjugate given in Section 4.7.

4.7 Minimal Polynomials and Algebraic Number Fields

19 If $\deg r = 2$ over $\mathbf{Q}$, prove that $\mathbf{Q}(r) = \mathbf{Q}(r^*)$, where r^* is the conjugate of r.

20 If s is a root of $x^2 + 1$, where $x^2 + 1 \in \mathbf{R}[x]$, show that $\mathbf{R}(s)$ is isomorphic to the complex field $\mathbf{C}$.

21 Referring to Example 4.22, verify that $\mathbf{Q}(s) \neq \mathbf{Q}(t)$.

22 Let r be an algebraic number of degree n. The *norm* of r is defined to be the product of the n conjugates of r. Prove: The norm of r is the constant term of the minimal polynomial of r, if n is even.

23 Let s be in an extension field E of F, and let $\theta_s : F[x] \to E$, be the morphism of Theorem 4.10. Prove that θ_s is 1–1 iff s is transcendental over F.

24 With the assistance of a reference book, if necessary, write up proofs of the following:
(a) If λ and η are algebraic numbers, then so are $\lambda + \eta$ and $\lambda \cdot \eta$.
(b) If λ and η are algebraic integers, then so are $\lambda + \eta$ and $\lambda \cdot \eta$.
(c) The set of all algebraic numbers is a field. The set of all algebraic integers is an integral domain.

Let $m(x)$ be the minimal polynomial of r over a field F. Prove: $F(r)$ is isomorphic to $F[x]/\langle m(x) \rangle$, where $F(r)$ is the field of Theorem 4.28, and $F[x]/\langle m(x) \rangle$ is the field of Problem 4.3–13(h). (Hint: Show that the map:

$$a_0 + a_1 r + a_2 r^2 + \cdots + a_{n-1} r^{n-1} \to \overline{a_0 + a_1 x + a_2 x^2 + \cdots + a_{n-1} x^{n-1}},$$

is an isomorphism.)

26 An abelian group $(V, \oplus)$ is called a *vector space* over a field $(F, +, \cdot)$ iff there exists a map $* : F \times V \to V$, called *scalar multiplication*, such that for all $a, b \in F$ and $\mu, \sigma \in V$:

(1) $a * (\mu + \sigma) = (a * \mu) \oplus (b * \mu)$,
(2) $(a + b) * \mu = (a * \mu) \oplus (b * \mu)$,
(3) $(a \cdot b) * \mu = a * (b * \mu)$,
(4) $1 * \mu = \mu$, where 1 is the unity of F.

The elements of V are called *vectors*, and the elements of F are called *scalars*. If there is a set $\{\mu_1, \mu_2, \ldots, \mu_n\}$ of vectors such that each vector σ can be represented uniquely as follows:

$$\sigma = a_1 \mu_1 \oplus a_2 \mu_2 \oplus \cdots \oplus a_n \mu_n, \qquad a_i \in F;$$

then $\{\mu_1, \mu_2, \ldots, \mu_n\}$ is called a *basis* for V. (Note: By $a\mu$ we mean $a * \mu$.)

(a) If r is an element of degree n over a field F, show that $F(r)$ is a vector space over F, and that $F(r)$ has a basis consisting of n vectors.
(b) In a linear algebra course one proves that if $\{\mu_1, \mu_2, \ldots, \mu_n\}$ is a basis for a vector space V, and $\{\sigma_1, \sigma_2, \ldots, \sigma_m\}$ is also a basis for V, then $n = m$. Use this fact to prove that if $F(r) = F(s)$, then $\deg r = \deg s$.

4.8 Quadratic Fields and Quadratic Domains

The arithmetic of the algebraic integers in quadratic fields is not only interesting in itself, but it provides examples that shed light on other aspects of algebra. For these reasons, we shall devote the remainder of this chapter to the topic of quadratic numbers, extending some ideas that were touched upon briefly in Section 2.9. First we must decide on what distinguishes one quadratic field from another.

EXAMPLE 4.25

Let $p(x) = 3x^2 + 6x + 7$. Then $p(x)$ is prime over $\mathbf{Q}$, and so if r is either root of $p(x)$, then $\mathbf{Q}(r)$ is a quadratic field. One of the roots is:
$r = -1 + (2/3)\sqrt{-3}$. Then

$$\mathbf{Q}(r) = \{x + yr : x, y \in \mathbf{Q}\} = \{(x - y) + (2y/3)\sqrt{-3}\}.$$

But $\quad \mathbf{Q}(\sqrt{-3}) = \{h + k\sqrt{-3} : h, k \in \mathbf{Q}\}$.

If we let $x - y = h$, and $2y/3 = k$, we can readily show that $\mathbf{Q}(r) = \mathbf{Q}(\sqrt{-3})$. It would appear that $\sqrt{-3}$ is the vital part of r in determining the quadratic field. Note that the minimal polynomial of r is $x^2 + 2x + \frac{7}{3}$, while the minimal polynomial of $\sqrt{-3}$ is $x^2 + 3$. Thus, distinct monic polynomials can give rise to the same quadratic field. □

THEOREM 4.29

(a) If $\mathbf{Q}(r)$ is any quadratic field, then there is a square-free integer n such that $\mathbf{Q}(r) = \mathbf{Q}(\sqrt{n})$.
(b) If k and n are distinct square-free integers, then fields $\mathbf{Q}(\sqrt{k})$ and $\mathbf{Q}(\sqrt{n})$ are not isomorphic.

Proof

(a) Let $m(x) = x^2 + b_1 x + b_0$ be the minimal polynomial of r. If $b_1 = b/a$, and $b_0 = c/a$, with $a, b, c \in \mathbf{Z}$, then r is a root of $p(x)$, where

$$p(x) = a \cdot m(x) = ax^2 + bx + c.$$

So r is one of the two numbers: $(-b \pm \sqrt{b^2 - 4ac})/2a$, where $b^2 - 4ac$ is not a perfect square, but may contain some factors that are perfect squares. Let $b^2 - 4ac = d^2 n$, where $d \geq 1$, and n is square-free. Hence,

$$r = (-b \pm \sqrt{d^2 n}/2a = (-b/2a) \pm (d/2a)\sqrt{n} = s + t\sqrt{n},$$

with $t \neq 0$, and $s = -b/2a$, $t = \pm d/2a$. Then,

$$\mathbf{Q}(r) = \{x + yr : x, y \in \mathbf{Q}\} = \{(x + ys) + yt\sqrt{n} : x, y \in \mathbf{Q}\},$$
$$\mathbf{Q}(\sqrt{n}) = \{u + v\sqrt{n} : u, v \in \mathbf{Q}\}.$$

Now $\quad x + ys = u, yt = v \Leftrightarrow x = u - (s/t)v, y = (1/t)v$.

Therefore, $\mathbf{Q}(r) = \mathbf{Q}(\sqrt{n})$.

(b) Assume there is an isomorphism, $\theta: \mathbf{Q}(\sqrt{k}) \to \mathbf{Q}(\sqrt{n})$. Then $\sqrt{k}\theta = x + y\sqrt{n}$ for some $x, y \in \mathbf{Q}$. We leave it to you (Problem 4.8–2) to prove that this last equation leads to a contradiction. Hence there is no possible image of $\sqrt{k}$, and so an isomorphism θ does not exist. (Hint: You must show that $a \in \mathbf{Q} \Rightarrow a\theta = a$, and that there are no rational x and y such that $(x + y\sqrt{n})^2 = k$.) □

If n is a *positive* square-free integer, then $\mathbf{Q}(\sqrt{n})$ is called a *real quadratic field*, and if n is a *negative* square-free integer, then $\mathbf{Q}(\sqrt{n})$ is called an *imaginary quadratic field*. So the distinct real quadratic fields are given by $\mathbf{Q}(\sqrt{n})$, where

$$n = 2, 3, 5, 6, 7, 10, 11, 13, 14, 15, 17, \ldots,$$

and the distinct imaginary quadratic fields are given by $\mathbf{Q}(\sqrt{n})$, where

$$n = -1, -2, -3, -5, -6, -7, -10, -11, -13, \ldots.$$

If $r \in \mathbf{Q}(\sqrt{n})$, $r \notin \mathbf{Q}$, and if r^* is the conjugate of r, then $r = a + b\sqrt{n}$, and $r^* = a - b\sqrt{n}$, for some $a, b \in \mathbf{Q}$, and the minimal polynomial of both r and r^* is $m(x)$, where

$$\begin{aligned} m(x) &= (x - r)(x - r^*) = x^2 - (r + r^*)x + r \cdot r^* \\ &= x^2 - 2ax + (a^2 - nb^2) \\ &= x^2 - 2ax + N(r), \end{aligned}$$

since the norm of r is $r \cdot r^*$. We know that every quadratic field contains irrational integers as well as rational integers, since if $a, b \in \mathbf{Z}$, then $a + b\sqrt{n}$ is an algebraic integer, its minimal polynomial being $x^2 - 2ax + (a^2 - nb^2)$. The symbol, $\mathbf{Q}[\sqrt{n}]$, will be used to denote the set of algebraic integers that are in the field of $\mathbf{Q}(\sqrt{n})$. Then

$$\mathbf{Z} \subset \mathbf{Q}[\sqrt{n}] \subset \mathbf{Q}(\sqrt{n}) \subset \mathbf{C}.$$

THEOREM 4.30

If r is an algebraic integer, then $N(r)$ is a rational integer; but not conversely.

Proof

If $r \in \mathbf{Q}[\sqrt{n}]$, then its minimal polynomial $m(x)$ is in $\mathbf{Z}[x]$. But $m(x) = x^2 - (r + r^*)x + N(r)$. Therefore, $N(r) \in \mathbf{Z}$.

To show that the converse is not true, we note that if $r = \frac{6}{5} + \frac{8}{5}\sqrt{-1}$, then $N(r) = 4$. However, r is not an algebraic integer, because it has $x^2 - \frac{12}{5}x + 4$ as its minimal polynomial. □

In our next proof we show that if $n \equiv 1 \pmod 4$, then $\mathbf{Q}(\sqrt{n})$ contains not only algebraic integers of the form $a + b\sqrt{n}$ with $a, b \in \mathbf{Z}$, but also algebraic integers of the form $(a + b\sqrt{n})/2$ with a and b both odd rational integers. For instance,

$$\mathbf{Q}[\sqrt{5}] = \{2 - 3\sqrt{5}, -7, 2\sqrt{5}, (1 + 3\sqrt{5})/2, (7 - 19\sqrt{5})/2, \ldots\}.$$

First note that if n is any square-free integer, then n must be congruent modulo 4 to either 1, 2, or 3.

THEOREM 4.31

Let n be a square-free integer.

$n \equiv 2$ or $3 \pmod{4} \Rightarrow \mathbf{Q}[\sqrt{n}] = \{a + b\sqrt{n} : a, b \in \mathbf{Z}\}$.

$n \equiv 1 \pmod{4} \Rightarrow \mathbf{Q}[\sqrt{n}] = \left\{\dfrac{a + b\sqrt{n}}{2} : a, b \in \mathbf{Z}, a \text{ and } b \text{ have same parity}\right\}$.

Proof

We can write the elements of the quadratic field $\mathbf{Q}(\sqrt{n})$ as follows:

$$\mathbf{Q}(\sqrt{n}) = \{s + t\sqrt{n} : s, t \in \mathbf{Q}\}$$
$$= \left\{\dfrac{a + b\sqrt{n}}{c} : a, b, c \in \mathbf{Z}, c \geq 1, (a, b, c) = 1\right\}.$$

If $r = (a + b\sqrt{n})/c$, then the minimal polynomial of r is $m(x)$, where

$$m(x) = x^2 - (r + r^*)x + rr^* = x^2 - \dfrac{2a}{c}x + \dfrac{a^2 - nb^2}{c^2}.$$

We must find all a, b, and c such that:

$$c \mid 2a, \quad c^2 \mid (a^2 - nb^2), \quad (a, b, c) = 1.$$

First, $b = 0 \Rightarrow c^2 \mid a^2 \Rightarrow c \mid a$, and $(a, b, c) = c$. So $c = 1$, and algebraic integer r is simply the rational integer a. In the remainder of the proof, we assume that $b \neq 0$.

We now prove that $(a, c) = 1$. Assume there is a prime p such that $p \mid a$ and $p \mid c$. Then $p^2 \mid c^2$, and $p^2 \mid a^2$. Hence, $p^2 \mid [(a^2 - nb^2) - a^2]$, or $p^2 \mid (-nb^2)$. But n is square-free, which means that $p \mid b$. Hence, p is a common divisor of a, b, and c, which is a contradiction. Therefore, $(a, c) = 1$. But $(a, c) = 1$, and $c \mid 2a \Rightarrow c \mid 2 \Rightarrow c = 1$ or $c = 2$.

If $c = 1$, then $r = a + b\sqrt{n}$, which is in $\mathbf{Q}[\sqrt{n}]$, for all n.

We now let $c = 2$, and determine the values that a and b must have, if $(a + b\sqrt{n})/2$ is to be an algebraic integer. We know that a must be an odd integer. $[(a, 2) = 1.]$ Assume that b is even. If $a = 2k + 1$, and $b = 2h$, then

$$N(r) = \dfrac{a^2 - nb^2}{4} = \dfrac{(2k + 1)^2 - n(2h)^2}{4} = (k^2 + k - nh^2) + 1/4,$$

which is not a rational integer. So if b is even, then r is not an algebraic integer.

4.8 Quadratic Fields and Quadratic Domains

We have arrived at the one remaining possibility. Let $b = 2h + 1$. Then

$$\frac{a^2 - nb^2}{4} = \frac{(2k+1)^2 - n(2h+1)^2}{4} = (k^2 + k - nh^2) + \frac{1-n}{4},$$

which is a rational integer if and only if $n \equiv 1 \pmod{4}$.

Given that $r = \dfrac{a + b\sqrt{n}}{2}$, results can be summarized as follows:

(1) a and b are both even $\Rightarrow r \in \mathbf{Q}[\sqrt{n}]$, for all n.
(2) a and b are both odd $\Rightarrow r \in \mathbf{Q}[\sqrt{n}]$ iff $n \equiv 1 \pmod{4}$.
(3) a and b are of opposite parity $\Rightarrow r \notin \mathbf{Q}[\sqrt{n}]$, for any n. $\square$

In Section 2.9 we discussed the integral domain $\mathbf{Z}(\sqrt{n})$, where $\mathbf{Z}(\sqrt{n}) = \{a + b\sqrt{n} : a, b \in \mathbf{Z}\}$. We now see that if $n \equiv 2$ or $3 \pmod{4}$, then $\mathbf{Z}(\sqrt{n}) = \mathbf{Q}[\sqrt{n}]$, but if $n \equiv 1 \pmod{4}$, then $\mathbf{Z}(\sqrt{n}) \subset \mathbf{Q}[\sqrt{n}]$. So by Theorem 2.27 we know that if $n \equiv 2$ or $3 \pmod{4}$, then $\mathbf{Q}[\sqrt{n}]$ is an integral domain.

THEOREM 4.32

If n is any square-free integer, then $\mathbf{Q}[\sqrt{n}]$ is an integral domain.

Proof

We leave it to the reader to prove the theorem for the case of $n \equiv 1 \pmod{4}$. Since $\mathbf{Q}[\sqrt{n}]$ is a subset of field $\mathbf{Q}(\sqrt{n})$, one needs to verify only that $\mathbf{Q}[\sqrt{n}]$ is closed under addition and multiplication, and that if $r \in \mathbf{Q}[\sqrt{n}]$, then $-r \in \mathbf{Q}[\sqrt{n}]$. It is suggested that one let $r = (a + b\sqrt{n})/2$, and $s = (c + d\sqrt{n})/2$, where a and b have the same parity, and c and d have the same parity. It is a simple matter to show that $r + s$, rs, $-r \in \mathbf{Q}[\sqrt{n}]$. $\square$

4.8 Problems

1. Give the norm and the minimal polynomial of each of these numbers of $\mathbf{Q}(\sqrt{5})$.

   (a) $(\tfrac{2}{3})\sqrt{5}$ (b) $\tfrac{9}{7}$ (c) $\dfrac{7 - 3\sqrt{5}}{2}$
   (d) $4 - \sqrt{5}$ (e) $(1 + \sqrt{5})/3$

2. Complete the proof of Theorem 4.29(b).

3. Let n be a given square-free integer. Verify that there are infinitely many polynomials, $x^2 + bx + c$, with $b, c \in \mathbf{Q}$, such that if r is a root of $x^2 + bx + c$, then $\mathbf{Q}(r) = \mathbf{Q}(\sqrt{n})$.

4. Find a number r in $\mathbf{Q}(\sqrt{-2})$ such that $N(r)$ is a rational integer, but r is not an algebraic integer.

5 Prove Theorem 4.32.

6 Prove by mathematical induction: If $r \in \mathbf{Q}(\sqrt{n})$, then $N(r^n) = [N(r)]^n$.

7 Let $f(x) = x^2 - (37 + 12\sqrt{7})$.
  (a) Give an integral domain in which $f(x)$ lies.
  (b) Is $f(x)$ prime or composite over $\mathbf{Q}(\sqrt{7})$?
  (c) If r is a root of $f(x)$, give the minimum polynomial of r over $\mathbf{Q}$.

8 (a) Prove: If $r \mid s$, then $N(r) \mid N(s)$, where $r, s \in \mathbf{Q}[\sqrt{n}]$.
  (b) Is the converse of part (a) true?

9 It is desired to find all isomorphisms θ of quadratic field $\mathbf{Q}(\sqrt{n})$ onto $\mathbf{Q}(\sqrt{n})$. Prove: $a\theta = a$, for all $a \in \mathbf{Q}$, and $\sqrt{n}\theta = \pm\sqrt{n}$. Hence there are but two isomorphisms:

$$a + b\sqrt{n} \to a + b\sqrt{n}, \quad \text{and} \quad a + b\sqrt{n} \to a - b\sqrt{n}.$$

10 Prove:
  (a) If $r = (a + b\sqrt{5})/2$ and $s = c + d\sqrt{3}$, where $a, b, c, d \in \mathbf{Z}$, and a and b have the same parity, then $r + s$ and rs are algebraic integers.
  (b) If an element of $\mathbf{Q}[\sqrt{5}]$ is added to or multiplied by an element of $\mathbf{Q}[\sqrt{3}]$, the resulting element is an algebraic integer of degree 1, 2, or 4.

11 Let $F_1 = \mathbf{Q}(\sqrt{n_1})$, and $F_2 = \mathbf{Q}(\sqrt{n_2})$, where n_1 and n_2 are distinct square-free integers. Verify that:
  (a) $F_1(\sqrt{n_2}) = F_2(\sqrt{n_1})$, so that this field can be denoted as $\mathbf{Q}(\sqrt{n_1}, \sqrt{n_2})$.
  (b) $\mathbf{Q}(\sqrt{n_1}, \sqrt{n_2})$ is of degree 2 over F_i, and of degree 4 over $\mathbf{Q}$.

4.9 Units and Primes in Quadratic Domains

For the record, we restate Theorem 2.9 in its slightly broader form. The proof of Theorem 4.33 is exactly the same as the proof of Theorem 2.9.

THEOREM 4.33

Let $r \in \mathbf{Q}[\sqrt{n}]$. Then r is a unit $\Leftrightarrow N(r) = \pm 1$.

EXAMPLE 4.26

Suppose we want to find all units v in $\mathbf{Q}[\sqrt{-7}]$. Note that $-7 \equiv 1$ (mod 4), and so let $v = (a + b\sqrt{-7})/2$, where a and b have the same parity. Now $N(v) = (a^2 + 7b^2)/4$. Hence, v is a unit iff $N(v) = 1$. We seek integers a and b such that $a^2 + 7b^2 = 4$. Thus, $b = 0$, $a = \pm 2$. Therefore, $\mathbf{Q}[\sqrt{-7}]$ has but two units: 1 and -1. □

We shall now determine all the units in each imaginary quadratic domain.

THEOREM 4.34

$\mathbf{Q}[\sqrt{-1}]$ has four units: $\pm 1, \pm i$.
$\mathbf{Q}[\sqrt{-3}]$ has six units: $\pm 1, (\pm 1 \pm \sqrt{-3})/2$.
$n < -1, n \neq -3, n$ is square-free $\Rightarrow \mathbf{Q}[\sqrt{n}]$ has two units: 1 and -1.

Proof

First, we find all units of the form, $v = a + b\sqrt{n}$, with $a, b \in \mathbf{Z}$. Let $n = -k$. Then

$$N(v) = a^2 - nb^2 = a^2 + kb^2 = 1, \quad \text{with } k > 0.$$

If $k = 1$, we must solve the Diophantine equation, $a^2 + b^2 = 1$. If $a = \pm 1$, then $b = 0$, and if $b = \pm 1$, then $a = 0$. Hence, the units of $\mathbf{Q}[\sqrt{-1}]$ are: $\pm 1, \pm i$.

Now $$k \geq 2 \Rightarrow b = 0 \Rightarrow a = 1 \text{ or } -1.$$

So 1 and -1 are units in every imaginary quadratic domain, $\mathbf{Q}[\sqrt{n}]$, and are the only units if $n \equiv 2$ or $3 \pmod 4$, $n \neq -1$.

Second, if $n \equiv 1 \pmod 4$, we must consider the possibility of having units of the form, $v = (a + b\sqrt{n})/2$, with a and b both odd. Now

$$N(v) = 1 \Rightarrow a^2 + kb^2 = 4, \quad \text{with } k = 3, 7, 11, \ldots.$$

If $k = 3$, the equation becomes: $a^2 + 3b^2 = 4$, whose solutions in odd rational integers are: $a = \pm 1, b = \pm 1$. Thus, $\mathbf{Q}[\sqrt{-3}]$ has the additional four units: $(\pm 1 \pm \sqrt{-3})/2$. It is evident that the equation, $a^2 + kb^2 = 4$, has no solution in odd integers if $k \geq 7$. □

Although all but two of the *imaginary* quadratic domains have only the same two units that $\mathbf{Z}$ has, the situation is quite different for the *real* quadratic domains. We illustrate this next.

EXAMPLE 4.27

In $\mathbf{Q}[\sqrt{2}]$, if $v = 1 + \sqrt{2}$, then $N(v) = -1$, and so v is a unit. Then, if k is a positive integer,

$$N(v^k) = [N(v)]^k = \begin{cases} -1, & \text{if } k \text{ is odd,} \\ +1, & \text{if } k \text{ is even.} \end{cases}$$

Hence, every positive integral power of v is also a unit. Some of these units are as follows:

$$v^2 = 3 + 2\sqrt{2}, \quad v^3 = 7 + 5\sqrt{2},$$
$$v^4 = 17 + 12\sqrt{2}, \quad v^5 = 41 + 29\sqrt{2}.$$

It is not difficult to verify that these powers of v are all distinct. Of course v^{-k} is also a unit, for $k \in \mathbf{Z}^+$. So $\mathbf{Q}[\sqrt{2}]$ has infinitely many units. □

It can be proved, using number theory topics that we have not discussed, that if $n > 1$, and n is square-free, then the Diophantine equation, $x^2 - ny^2 = 1$, always has a solution, (x_1, y_1), with $x_1, y_1 \in \mathbf{Z}^+$. But $N(x_1 + y_1\sqrt{n}) = x_1^2 - ny_1^2 = 1$, so $x_1 + y_1\sqrt{n}$ is a unit in $\mathbf{Q}[\sqrt{n}]$. If we accept this fact, we can then prove that a real quadratic domain has infinitely many units.

THEOREM 4.35

Let $n \geq 2$ be a square-free integer. If v is a unit of $\mathbf{Q}[\sqrt{n}]$, where $v = a + b\sqrt{n}$, and $a, b \in \mathbf{Z}^+$, then $\{v, v^2, v^3, \ldots\}$ is a set of distinct units of $\mathbf{Q}[\sqrt{n}]$. Hence, every real quadratic domain contains infinitely many units.

You are asked to prove this theorem as an exercise.

If we are to factor a number into a product of primes, we must be able to recognize a prime when we see one. The next theorem is of some help.

THEOREM 4.36

(a) If $r \in \mathbf{Q}[\sqrt{n}]$, and $N(r) = p$, where p is a rational prime, then r is a quadratic prime.

(b) Every rational prime p is either prime in $\mathbf{Q}[n]$ or is a product of two irrational primes of $\mathbf{Q}[\sqrt{n}]$.

Proof

(a) Given that $N(r) = p$, suppose that $r = st$. Then $N(s) \cdot N(t) = p$. So either $N(s)$ or $N(t)$ must be 1 or -1, which means that either s is a unit or t is a unit. Therefore, r is prime.

(b) Let p be any rational prime. Then $N(p) = p^2$. If p is not prime in $\mathbf{Q}[\sqrt{n}]$, then $p = rs$, for some r and s in $\mathbf{Q}[\sqrt{n}]$ such that $N(r) = N(s) = \pm p$. Therefore, r and s are quadratic primes, by part (a). □

We show, by an example, that although the condition of Theorem 4.36(a) is sufficient for r to be a prime, it is not necessary.

EXAMPLE 4.28

Let $r = 4 + \sqrt{-5}$. Is r prime or composite in $\mathbf{Q}[\sqrt{-5}]$? Now $N(a + b\sqrt{-5}) = a^2 + 5b^2$, so $N(r) = 21$. If r is not prime, then there exist algebraic integers s and t in $\mathbf{Q}[\sqrt{-5}]$ such that $r = st$, with $N(s) = 3$, and $N(t) = 7$. But the Diophantine equation, $x^2 + 5y^2 = 3$, has no solution. Therefore, r is a prime, even though its norm is a composite rational integer. □

Every composite quadratic integer can be factored into a product of primes in at least one way, according to the next theorem.

THEOREM 4.37

Each element in $\mathbf{Q}[\sqrt{n}]$ that is not zero or a unit can be factored into a product of primes.

Proof

We prove this by induction on the absolute value of the norm. If r is neither zero nor a unit, then $|N(r)| = m \geq 2$. Let P_m be the statement:

$|N(r)| = m \Rightarrow r = \pi_1 \pi_2 \cdots \pi_s$, where the π_i are primes in $\mathbf{Q}[\sqrt{n}]$.

The proof is the analogue of the proof of Theorem 3.13. We leave it to you to carry it out. □

4.9 Problems

1. In $\mathbf{Q}[\sqrt{-3}]$, factor each of the following into a product of primes.
   (a) $3 - 4\sqrt{-3}$
   (b) 7
   (c) $1 + \sqrt{-3}$
   (d) $15 + 8\sqrt{-3}$
   (e) $(9 - \sqrt{-3})/2$
   (f) $(1 + \sqrt{-3})/2$

2. In $\mathbf{Q}[\sqrt{-5}]$, factor each of these integers into a product of primes.
   (a) $17 - \sqrt{-5}$
   (b) $1 + \sqrt{-5}$
   (c) 5
   (d) 41
   (e) 43
   (f) $25 + 35\sqrt{-5}$

3. Assume that $r = st$, with $r, s, t \in \mathbf{Q}[\sqrt{n}]$.
   (a) If $N(r) = 6$, what can you conclude about $N(s)$ if $n > 0$, and if $n < 0$?
   (b) If $N(r) = 7$, classify r, s, and t in terms of whether they are primes, composites, or units.

4. Give the associates of $4 + 5\sqrt{n}$ in domain $\mathbf{Q}[\sqrt{n}]$, if:
   (a) $n = -1$,
   (b) $n = -2$,
   (c) $n = -3$.

5. Prove: If $r, v \in \mathbf{Q}[\sqrt{n}]$, and v is a unit, then $v \mid r$.

6. Verify that if $a + b\sqrt{n}$ is a unit of $\mathbf{Q}[\sqrt{n}]$, where $a, b \in \mathbf{Z}^+$, then the four numbers, $\pm a \pm b\sqrt{n}$, are units, and one of them is the multiplicative inverse of $a + b\sqrt{n}$.

7. Find the pair (x_1, y_1) of smallest positive rational integers that give a solution of the Diophantine equation, $x^2 - ny^2 = 1$, if:
   (a) $n = 2$,
   (b) $n = 3$,
   (c) $n = 5$,
   (d) $n = 6$,
   (e) $n = 7$.

8. Use the results of Problem 7 to find three units of the form $a + b\sqrt{n}$, with $a, b \in \mathbf{Z}^+$, in the integral domain $\mathbf{Q}[\sqrt{n}]$, for each n given in Problem 7.

9. Prove Theorem 4.35(a). (Hint: Let $v^k = a_k + b_k\sqrt{n}$, for $k \in \mathbf{Z}^+$. Verify that $v^{k+1} > v^k$, by showing that $a_{k+1} > a_k$, and $b_{k+1} > b_k$.)

10. Prove Theorem 4.37.

4.10 Quadratic Unique Factorization Domains

In Example 2.41 we showed that $\mathbf{Q}[\sqrt{-6}]$ is not a UFD. The next theorem gives a necessary and sufficient condition that a quadratic domain $\mathbf{Q}[\sqrt{n}]$ be a unique factorization domain.

THEOREM 4.38

$\mathbf{Q}[\sqrt{n}]$ is a unique factorization domain if and only if the following condition holds: For all $s_i \in \mathbf{Q}[\sqrt{n}]$ and every prime π in $\mathbf{Q}[\sqrt{n}]$, if $\pi \mid s_1 s_2 \cdots s_k$, then $\pi \mid s_j$, for some j.

Proof

(1) Assume that $\pi \mid s_1 s_2 \cdots s_k \Rightarrow \pi \mid s_j$, for some j. Let r be in $\mathbf{Q}[\sqrt{n}]$, where r is not zero or a unit. The proof that factorization of r is unique is analogous to the proof of Theorem 3.15, and can be carried out by mathematical induction on m, where $m \geq 2$, and P_m is the statement: If $|N(r)| = m$, then the factorization of r into a product of primes in $\mathbf{Q}[\sqrt{n}]$ is unique. We leave the proof to you.

(2) Assume that $\mathbf{Q}[\sqrt{n}]$ is a UFD. Let π be any prime in $\mathbf{Q}[\sqrt{n}]$ such that $\pi \mid s_1 s_2 \cdots s_h$. Then

$$\pi t = s_1 s_2 \cdots s_h, \quad \text{for some } t \in \mathbf{Q}[\sqrt{n}].$$

But each s_i factors into a product of primes, by Theorem 4.37. Hence,

$$\pi t = \pi_1 \pi_2 \cdots \pi_w,$$

the π_j are primes, and each $\pi_j \mid s_i$, for some i. Since factorization is unique, π is an associate of one of the π_k, so $\pi_k = v\pi$, where v is a unit. But $\pi \mid \pi_k$, and $\pi_k \mid s_j$ for some $j \Rightarrow \pi \mid s_j$. □

To compare similarities and differences, let's review the steps we took when we proved that $\mathbf{Z}$ is a UFD. In $\mathbf{Z}$ we first verified the Division Theorem [3.2]. We used the Division Theorem to prove that if $g = (a, b)$, then there exist rational integers, s and t, such that $g = as + bt$ [3.4]. This enabled us to prove that if a rational prime divides a product of rational integers, it must divide one of the factors [3.14]. This result was then used to prove the uniqueness of factorization in $\mathbf{Z}$ [3.15]. Thus, the Division Theorem is the key theorem that leads to unique factorization of a rational integer into a product of primes.

We now introduce a form of division for quadratic domains. A quadratic domain $\mathbf{Q}[\sqrt{n}]$ is called a *Euclidean quadratic domain*, if, for each $s, t \in \mathbf{Q}[\sqrt{n}]$, with $t \neq 0$, there exist $q, r \in \mathbf{Q}[\sqrt{n}]$ such that

$$s = tq + r, \quad \text{where } 0 \leq |N(r)| < |N(t)|.$$

EXAMPLE 4.29

Let $s, t \in \mathbf{Q}[\sqrt{3}]$, where

$$s = 2 + 3\sqrt{3}, \quad t = 1 - 2\sqrt{3}.$$

Now $N(t) = -11$. We want to find a q and an r such that

$$2 + 3\sqrt{3} = (1 - 2\sqrt{3}) \cdot q + r, \quad \text{with } |N(r)| < 11.$$

The easiest way to find q is as follows:

$$\frac{s}{t} = \frac{2 + 3\sqrt{3}}{1 - 2\sqrt{3}} = \frac{(2 + 3\sqrt{3})(1 + 2\sqrt{3})}{-11} = \left(-\frac{21}{11}\right) + \left(-\frac{7}{11}\right)\sqrt{3}.$$

Let $q = x + y\sqrt{3}$, where x is the rational integer closest to $-\frac{21}{11}$, and y is the rational integer closest to $-\frac{7}{11}$. Then $q = -2 - \sqrt{3}$, and

$$r = s - tq = (2 + 3\sqrt{3}) - (1 - 2\sqrt{3})(-2 - \sqrt{3}) = -2.$$

The q and r so chosen satisfy the desired conditions, since $|N(r)| = 4$.

However, the quotient and remainder are not unique, for if we let $q_1 = -1 - \sqrt{3}$, we get that $r_1 = s - tq_1 = -3 + 2\sqrt{3}$, and $|N(r_1)| = |-3| = 3$. $\square$

We now illustrate how the technique employed in the preceding example can be used to prove that $Q[\sqrt{-2}]$, $Q[\sqrt{-1}]$, $Q[\sqrt{2}]$, and $Q[\sqrt{3}]$ are Euclidean domains.

THEOREM 4.39

$Q[\sqrt{n}]$ is a Euclidean domain if $n = -2, -1, 2, 3$.

Proof

Given the quadratic domain $Q[\sqrt{n}]$, where n is one of these integers: $-2, -1, 2, 3$. Let s and t, $t \neq 0$, be any two elements in $Q[\sqrt{n}]$. We must find q and r such that $s = tq + r$, with $|N(r)| < |N(t)|$.

Now $$\frac{s}{t} = \frac{st^*}{tt^*} = h + k\sqrt{n}, \quad h, k \in Q.$$

Choose $x, y \in Z$ so that $|x - h| \leq \frac{1}{2}$, and $|y - k| \leq \frac{1}{2}$. Let $q = x + y\sqrt{n}$ and $r = s - tq$. Then

$$r = t(h + k\sqrt{n}) - t(x + y\sqrt{n}) = t[(h - x) + (k - y)\sqrt{n}],$$

which means that

$$|N(r)| = |N(t)| \cdot |(x - h)^2 - n(y - k)^2|.$$
Hence $$|N(r)| < |N(t)| \Leftrightarrow |(x - h)^2 - n(y - k)^2| < 1.$$

We now consider our four values of n:

$$n = -2 \Rightarrow |(x - h)^2 + 2(y - k)^2| < \tfrac{1}{4} + 2(\tfrac{1}{4}) = \tfrac{3}{4} < 1.$$
$$n = -1 \Rightarrow |(x - h)^2 + (y - k)^2| < \tfrac{1}{4} + \tfrac{1}{4} = \tfrac{1}{2} < 1.$$

$$n = 2 \Rightarrow |(x - h)^2 - 2(y - k)^2| \leq (x - h)^2 + 2(y - k)^2 < 1.$$
$$n = 3 \Rightarrow |(x - h)^2 - 3(y - k)^2| \leq (x - h)^2 + 3(y - k)^2 < 1.$$

Therefore $\mathbf{Q}[\sqrt{n}]$ is Euclidean for $n = -2, -1, 2, 3$. □

EXAMPLE 4.30

We show how to go about verifying that $\mathbf{Q}[\sqrt{-6}]$ is *not* Euclidean. Let $s = -3$, and $t = -1 + \sqrt{-6}$. So $N(t) = 7$. If $\mathbf{Q}[\sqrt{-6}]$ is Euclidean, then we can find a quadratic integer $q = x + y\sqrt{-6}$ and an $r = s - tq$, with $N(r) < 7$. But

$$r = -3 - (-1 + \sqrt{-6})(x + y\sqrt{-6})$$
$$= (-3 + x + 6y) + (-x + y)\sqrt{-6}.$$

This means that we must solve the Diophantine inequality:

$$(x + 6y - 3)^2 + 6(x - y)^2 < 7.$$

Then $(x - y)^2$ must be either 0 or 1. Now

$$(x - y)^2 = 0 \Rightarrow x = y \Rightarrow (7y - 3)^2 < 7 \Rightarrow 3 - \sqrt{7} < 7y < 3 + \sqrt{7},$$

which has no solution. Hence $(x - y)^2 \neq 0$. Also,

$$(x - y)^2 = 1 \Rightarrow x + 6y - 3 = 0 \Rightarrow x = 3 - 6y \Rightarrow (7y - 3)^2 = 1,$$

which has no solution. Therefore, no q and r exist to satisfy the equation, $s = tq + r$, with $N(r) < 7$, and so $\mathbf{Q}[\sqrt{-6}]$ is not a Euclidean domain. □

The next two theorems give some basic properties of Euclidean domains.

THEOREM 4.40

If $\mathbf{Q}[\sqrt{n}]$ is a Euclidean domain, and $s, t, \pi, r, r_i \in \mathbf{Q}[\sqrt{n}]$, then:
(a) $(s, t) = 1 \Rightarrow sh + tk = 1$, for some $h, k \in \mathbf{Q}[\sqrt{n}]$;
(b) $\pi | rs$, where π is a prime $\Rightarrow \pi | r$, or $\pi | s$;
(c) $\pi | r_1 r_2 \cdots r_k$, where π is a prime $\Rightarrow \pi | r_j$, for some j.

Proof

(a) Define set S as follows, given that $(s, t) = 1$:

$$S = \{sx + ty : x, y \in \mathbf{Q}[\sqrt{n}]\}.$$

Let m be the smallest positive rational integer such that there is a $v \in S$ for which $|N(v)| = m$. Then there exist $x_0, y_0 \in \mathbf{Q}[\sqrt{n}]$ such that

$$v = sx_0 + ty_0, \quad \text{with } |N(v)| = m.$$

4.10 Quadratic Unique Factorization Domains

We shall prove that v is a unit by showing that $v\,|\,s$ and $v\,|\,t$.

Since $\mathbf{Q}[\sqrt{n}]$ is a Euclidean domain, there exist q and r in $\mathbf{Q}[\sqrt{n}]$ such that $s = vq + r$, with $0 \leq |N(r)| < |N(v)|$. But

$$r = s - vq = s - (sx_0 + ty_0)q = s(1 - x_0) + t(-y_0 q) \in S.$$

This means that $|N(r)| = 0$, because of the minimality of $|N(v)|$. So $r = 0$, and $v\,|\,s$. Similarly, it can be shown that $v\,|\,t$. But the only common divisors of s and t are the units, and hence v is a unit. Then

$$1 = vv^{-1} = s(x_0 v^{-1}) + t(y_0 v^{-1}) = sh + tk, \quad \text{with } h = x_0 v^{-1}, \quad k = y_0 v^{-1}.$$

We leave it to you to use part (a) to prove (b), and then to use (b) and induction on k to prove (c). □

The next theorem is an immediate consequence of Theorems 4.40(c) and 4.38.

THEOREM 4.41

If $\mathbf{Q}[\sqrt{n}]$ is a Euclidean domain, then $\mathbf{Q}[\sqrt{n}]$ is a unique factorization domain.

The converse of Theorem 4.41 is not true, as there exist unique factorization domains that are not Euclidean.

EXAMPLE 4.31

The situation with regard to unique factorization domains, at the time of writing, is summarized in Table 4–1.

Table 4–1 Quadratic Unique Factorization Domains

Type of UFD	Values of Square-free n, if $\mathbf{Q}[\sqrt{n}]$ is a UFD	
	$\mathbf{Q}[\sqrt{n}]$ is Euclidean	$\mathbf{Q}[\sqrt{n}]$ is not Euclidean
Imaginary $[n < 0]$	$-1, -2, -3, -7, -11$	$-19, -43, -67, -163$ (In 1966, H. M. Stark proved there are only these four.)
Real, with $0 < n < 100$	2, 3, 5, 6, 7, 11, 13, 17, 19, 21, 29, 33, 37, 41, 57, 73.	14, 22, 23, 31, 38, 43, 46, 47, 53, 59, 61, 62, 67, 69, 71, 77, 83, 86, 89, 93, 94, 97.
Real, with $n > 100$	None. (H. Davenport published this conclusion in 1950.)	101, 103, 107, 109, 113, ... (Research not completed. Are there a finite or an infinite number of n?)

Note that there are only 21 Euclidean domains: 5 imaginary and 16 real. Also, there are a mere *nine imaginary* unique factorization domains.

Mathematicians are still working on the question of how many *real* unique factorization domains there are. Several number theorists conjecture that there are infinitely many. ☐

The rational primes play an important role in helping us to determine *all* the algebraic primes in a UFD, $\mathbf{Q}[\sqrt{n}]$. Their role is made clear in the following result.

THEOREM 4.42

If π is a prime in a unique factorization domain $\mathbf{Q}[\sqrt{n}]$, then there is exactly one positive rational prime p such that $\pi \,|\, p$.

Proof

We know that π divides *some* rational integer, since $\pi \cdot \pi^*$ is $N(\pi)$, which is in $\mathbf{Z}$. Let p be the smallest positive rational integer such that $\pi\,|\,p$. If $p = 1$, then $\pi t = 1$, for some t, and π would be a unit. Thus, $p > 1$. If p is composite, then $p = ab$, where $a, b \in \mathbf{Z}^+$, and $1 < a < p$, $1 < b < p$. But

$$\pi\,|\,ab \Rightarrow \pi\,|\,a, \quad \text{or} \quad \pi\,|\,b,$$

a contradiction of the minimality of p. Hence, p is a rational prime.

Now assume that $\pi\,|\,q$, where q is also a positive rational prime, and $p \neq q$. Since $(p, q) = 1$, there exist rational integers, x and y, such that $px + qy = 1$. But

$$\pi\,|\,p \quad \text{and} \quad \pi\,|\,q \Rightarrow \pi\,|\,px + qy \Rightarrow \pi\,|\,1 \Rightarrow \pi \text{ is a unit},$$

which is not so. Therefore, π divides one and only one positive rational prime, p. ☐

EXAMPLE 4.32

In order to find all primes in $\mathbf{Q}[\sqrt{-2}]$ with norms less than 20, one simply factors each *rational* prime p, with $2 \leq p \leq 19$. It turns out that 5, 7, and 13 are all prime in $\mathbf{Q}[\sqrt{-2}]$, and that the other rational primes factor as follows:

$$2 = (\sqrt{-2})(-\sqrt{-2}), \qquad 17 = (3 + 2\sqrt{-2})(3 - 2\sqrt{-2}),$$
$$3 = (1 + \sqrt{-2})(1 - \sqrt{-2}), \qquad 19 = (1 + 3\sqrt{-2})(1 - 3\sqrt{-2}).$$
$$11 = (3 + \sqrt{-2})(3 - \sqrt{-2}),$$

These factors, together with their associates, give us all primes with norms less than 20. Since the only units in this domain are 1 and -1, there are 18 such irrational primes:

$$\pm\sqrt{-2}, \quad \pm 1 \pm \sqrt{-2}, \quad \pm 3 \pm \sqrt{-2}, \quad \pm 3 \pm 2\sqrt{-2},$$
$$\pm 1 \pm 3\sqrt{-2}. \quad \square$$

4.10 Quadratic Unique Factorization Domains

Given any UFD $\mathbf{Q}[\sqrt{n}]$, there is a test (which will not be given here) that one can apply to determine whether or not a rational prime p is prime or composite in $\mathbf{Q}[\sqrt{n}]$. For the case in which $n = -1$, the test reduces to these simple conditions: If p is a positive rational prime of the form $4k - 1$, then p is a prime in $\mathbf{Q}[i]$, and if p is a positive rational prime of the form $4k + 1$, then p is composite in $\mathbf{Q}[i]$; 2 is composite in $\mathbf{Q}[i]$.

Incidentally, mathematicians of over a century ago used quadratic domains to help them solve certain Diophantine equations. Some men obtained incorrect results, because they erroneously assumed that factorization was unique in every quadratic domain. It was Gauss who looked at these matters more carefully, and he proved that $\mathbf{Q}[i]$ is a UFD. In his honor, the elements of $\mathbf{Q}[i]$ are called *Gaussian integers*.

EXAMPLE 4.33

Let us examine two rational primes, 19 and 29. Now $19 = 4(5) - 1$, and so 19 is a Gaussian prime. But $29 = 4(7) + 1$, and so 29 is factorable in $\mathbf{Q}[i]$. By trial and error one finds that $29 = (5 + 2i)(5 - 2i)$.

Since $\mathbf{Q}[i]$ has four units, each nonzero Gaussian integer has four associates. Therefore, there are eight Gaussian primes whose norm is 29:

$$\pm 5 \pm 2i, \quad \pm 2 \pm 5i.$$

Each of these eight Gaussian primes is, of course, a divisor of 29. □

4.10 Problems

1. Show that $\mathbf{Q}[\sqrt{-5}]$ is not a UFD, by factoring some integer into a product of primes in two ways.

2. Let $s = 5 + 7i$, and $t = 2 - i$, with $s, t \in \mathbf{Q}[i]$. Find two distinct sets of q and r in $\mathbf{Q}[i]$ such that $s = qt + r$, with $|N(r)| < |N(t)|$, thus showing that the quotient and remainder are not unique.

3. (a) Factor each rational prime p into a product of Gaussian primes, if $1 < p < 40$.
   (b) How many Gaussian primes are there with norm less than 40?

4. (a) Factor these rational primes in $\mathbf{Q}[\sqrt{6}]$: 3, 5, 19, 23.
   (b) In $\mathbf{Q}[\sqrt{6}]$, note that
   $$2 = (2 + \sqrt{6})(-2 + \sqrt{6}) = (22 + 9\sqrt{6})(-22 + 9\sqrt{6}).$$
   But you were told that $\mathbf{Q}[\sqrt{6}]$ is a UFD. Is there a contradiction here? Explain.

5. Show that $1 + \sqrt{-7}$ and 2 are prime in $\mathbf{Z}(\sqrt{-7})$, but are composite in $\mathbf{Q}[\sqrt{-7}]$.

6 Verify that $\mathbf{Q}[\sqrt{-10}]$ is not a Euclidean domain, by employing the technique of Example 4.30 on suitable s and t. (Hint: Try $s = 3 - \sqrt{-10}$, $t = 1 + \sqrt{-10}$.)

7 In $\mathbf{Q}[\sqrt{-10}]$, find a quadratic prime π which divides no rational prime. Use this result and Theorem 4.42 to verify that $\mathbf{Q}[\sqrt{-10}]$ is not a UFD.

8 (a) Show that each number in $\mathbf{Q}[\sqrt{-3}]$ has an associate in $\mathbf{Z}(\sqrt{-3})$.
(b) It was noted in Example 4.31 that $\mathbf{Q}[\sqrt{-3}]$ is a UFD. Show that $\mathbf{Z}(\sqrt{-3})$ is *not* a UFD.

9 (a) For n in the interval, $2 \le n < 50$, what percent of the quadratic domains are unique factorization domains?
(b) What percent of $\mathbf{Q}[\sqrt{n}]$ are UFD, if $-50 < n \le -1$?

10 Give an example to show that in a $\mathbf{Q}[\sqrt{n}]$ that is not a UFD, one can have a prime π such that $\pi \mid rs$, and yet π divides neither r nor s.

11 Prove Theorem 4.38, part (1).

12 Prove Theorem 4.40 (b, c).

13 (a) Verify: If π is prime in $\mathbf{Q}[\sqrt{n}]$, then so is π^*.
(b) Show that if π is a prime in a UFD $\mathbf{Q}[\sqrt{n}]$, then $|N(\pi)| = p$ or p^2, for some positive rational prime p.

14 Prove that $\mathbf{Q}[\sqrt{5}]$ is a Euclidean domain. (Hint: Use the technique used in proving Theorem 4.39, making use of the fact that $\mathbf{Q}[\sqrt{5}] = \{(a + b\sqrt{5})/2 : a \text{ and } b \text{ have same parity}\}$. Let $s/t = (h + k\sqrt{5})/2$, and $q = (x + y\sqrt{5})/2$, where rational integers x and y are chosen to have the same parity and to satisfy the inequalities: $|x - h| \le 1$, $|y - k| \le \frac{1}{2}$.)

15 (a) If $n \equiv 2$ or $3 \pmod 4$, and $n \le -5$, show that $\mathbf{Q}[\sqrt{n}]$ is not Euclidean. (Hint: Let $s = 1 + \sqrt{n}$, $t = 2$, and show that no q and r exist such that $s = tq + r$, with $N(r) < 4$.)
(b) If $n \equiv 1 \pmod 4$, and $n \le -15$, show that $\mathbf{Q}[\sqrt{n}]$ is not Euclidean. (Hint: Let $s = (1 + \sqrt{n})/2$, $t = 2$.)
(c) Assuming that $\mathbf{Q}[\sqrt{n}]$ is Euclidean if $n = -1, -2, -3, -7, -11$, have you now proved that these are the only imaginary Euclidean domains?

16 The year 1966 was a very good year for the mathematician, Harold Stark. Why? (See Appendix E–22.)

5
Groups

> *In a little more than a century, it (group theory) has effected a remarkable unification of mathematics, revealing connections between parts of algebra and geometry that were long considered distinct and unrelated. (1956)*—James R. Newman

5.1 Integral Powers of Group Elements

In previous chapters you have learned a little about groups, and have worked with many different examples of groups. In the present chapter you will be introduced to some interesting and useful ideas concerning the basic structure of a group.

In studying group structures, we find it helpful to use exponential notation such as a^4, a^{-7}, a^0. If a is an element in a group $(G, \otimes)$, then a^{-1} shall denote the *inverse* of a. In Section 1.7 we defined:

$$a^1 = a, \qquad a^{k+1} = a^k \otimes a \quad \text{if } k \in \mathbf{Z}^+.$$

We need these additional definitions;

$$a^0 = e, \qquad a^{-k} = (a^{-1})^k,$$

where e is the identity of G. The integer n in the expression, a^n, is called the *exponent* of a, and a^n is said to be an *integral power* of a. It is evident that if $a \in G$, then $a^n \in G$, for all $n \in \mathbf{Z}$. (Why?)

The following three theorems give the *laws of exponents*. To prove them rigorously we make use of mathematical induction. The proof of Theorem 5.2 looks particularly messy, simply because there are so many cases to consider. We apologize for beginning a new chapter with such uninspiring theorems, but they are mathematical "housekeeping chores" of which we wish to dispose as quickly as possible.

THEOREM 5.1

If a is any element in group $(G, \otimes)$, and if $n \in \mathbf{Z}^+$, then
(a) $a^n = a \otimes a \otimes \cdots \otimes a$, where a appears as a factor n times.
(b) $a^{-n} = (a^{-1})^n = (a^n)^{-1}$.

Proof

(a) This is readily verified by induction on n, using the definitions: $a^1 = a$, and $a^{k+1} = a^k \otimes a$.

(b) By definition, $a^{-n} = (a^{-1})^n$. We must show that $(a^{-1})^n$ is the inverse of a^n. Now

$$a^n(a^{-1})^n = (a \otimes a \otimes \cdots \otimes a)(a^{-1} \otimes a^{-1} \otimes \cdots \otimes a^{-1}),$$

where there are n factors inside each set of parentheses, by part (a). By the generalized associative property (Problem 3.3–11), this product is e. Therefore, $(a^n)^{-1} = (a^{-1})^n$. □

THEOREM 5.2

Let a be any element in group $(G, \otimes)$. Then
(a) $a^m \otimes a^n = a^{m+n}$, where $m, n \in \mathbf{Z}$.
(b) $a^{m_1} \otimes a^{m_2} \otimes \cdots \otimes a^{m_n} = a^{m_1 + m_2 + \cdots + m_n}$, where $m_i \in \mathbf{Z}$, and $n \in \mathbf{Z}^+$.

Proof

(a) We break the proof down into cases, based on whether n is positive, negative, or zero, with subcases involving the size of m when $n = 1$.

Case 1: $n = 0$. Then

$$a^m a^n = a^m a^0 = a^m e = a^m = a^{m+0} = a^{m+n}.$$

Case 2: $n \geq 1$. We prove this by induction on n. Let m be any integer, and define P_n to be the statement: $a^m a^n = a^{m+n}$. We first prove that P_1 is true. Now

$$m = 0 \Rightarrow a^m a^1 = a^0 a^1 = ea^1 = a^1 = a^{0+1} = a^{m+1}.$$
$$m > 0 \Rightarrow a^m a^1 = a^m a = a^{m+1}, \quad \text{by definition.}$$
$$m < 0 \Rightarrow m = -s, \quad s > 0 \Rightarrow a^m a^1 = a^{-s} a = (a^{-1})^s a$$
$$= (a^{-1})^{(s-1)+1} a = (a^{-1})^{s-1}(a^{-1})a = (a^{-1})^{s-1} e$$
$$= a^{-(s-1)} = a^{-s+1} = a^{m+1}.$$

Therefore, P_1 is true.

We prove that $P_k \Rightarrow P_{k+1}$, as follows:

$$a^m a^{k+1} = a^m(a^k a) = (a^m a^k)a = a^{m+k} a = a^{(m+k)+1} = a^{m+(k+1)}.$$

Case 3: $n \leq -1$. Let $n = -t$. Then $t \geq 1$, and Case 2 can be applied as follows:

$$a^m a^n = a^m a^{-t} = a^{-(-m)} a^{-t} = (a^{-1})^{-m}(a^{-1})^t$$
$$= (a^{-1})^{-m+t} = (a^{-1})^{-m+(-n)} = a^{-[-(m+n)]} = a^{m+n}.$$

(b) The proof is immediate from (a), by induction on n. □

THEOREM 5.3

Let a and b be any elements in group $(G, \otimes)$, and let $m, n \in \mathbf{Z}$. Then
(a) $(a^m)^n = a^{mn}$, and
(b) $(a \otimes b)^n = a^n \otimes b^n$, if G is abelian.

Proof

(a) We examine the three cases: $n = 0, n > 0, n < 0$.

$n = 0 \Rightarrow (a^m)^n = (a^m)^0 = e = a^0 = a^{m \cdot 0} = a^{mn}$.

$n \geq 1 \Rightarrow (a^m)^n = a^m \otimes a^m \otimes \cdots \otimes a^m$, for n factors (Theorem 5.11(a))
$= a^{m+m+\cdots+m} = a^{mn}$. (Theorem 5.2(b))

$n \leq -1 \Rightarrow n = -t, t \geq 1 \Rightarrow (a^m)^n = (a^m)^{-t} = [(a^m)^t]^{-1} = (a^{mt})^{-1} = a^{-mt}$
$= a^{mn}$.

(b) The proof will be left as an exercise. □

Remember that the properties of exponents that we have verified in this section can be applied to *any* group, no matter what the group operation may be. The abstract notation used looks like the notation you have always used when the operation is multiplication of, say, real numbers. If the operation is $+$, we translate as follows, for instance:

$$a^2 = a \otimes a = a + a.$$

In a field $(F, +, \cdot)$, we need to distinguish between the additive group $(F, +)$ and the multiplicative group $(F^*, \cdot)$ in applying Theorems 5.1–5.3. So we must use two kinds of notation. Thus, if $\otimes$ is multiplication, $\cdot$, we denote $a \cdot a$ as a^2, and if $\otimes$ is addition, $+$, we denote $a + a$ as $2a$ (moving the exponent down to the same level as a and to the left of a). However, in dealing with abstract groups in this chapter, we let a^2 represent $a \otimes a$, regardless of the operation. Thus, in $(\mathbf{Q}, +)$, $3^2 = 3 + 3 = 6$, and in $(\mathbf{Q}^*, \cdot)$, $3^2 = 3 \cdot 3 = 9$.

Let a be an element in a group $(G, \otimes)$ with identity e. If m is the smallest positive integer such that $a^m = e$, then a is said to be an *element of order m*. We write $|a| = m$, which is our shorthand for "The order of a is m." If there is no positive integer m such that $a^m = e$, then a is said to be of *infinite order*, and we write: $|a| = \infty$. It is apparent that if $|a| = m$, then $a^{m-1} \otimes a = e$, and so $a^{-1} = a^{m-1}$.

EXAMPLE 5.1

The distinct powers of elements of $(\mathbf{Z}_7^*, \cdot)$ are given:

$\bar{1}^1 = \bar{1}$	$\bar{2}^1 = \bar{2}$	$\bar{3}^1 = \bar{3}$	$\bar{4}^1 = \bar{4}$	$\bar{5}^1 = \bar{5}$	$\bar{6}^1 = \bar{6}$
	$\bar{2}^2 = \bar{4}$	$\bar{3}^2 = \bar{2}$	$\bar{4}^2 = \bar{2}$	$\bar{5}^2 = \bar{4}$	$\bar{6}^2 = \bar{1}$
	$\bar{2}^3 = \bar{1}$	$\bar{3}^3 = \bar{6}$	$\bar{4}^3 = \bar{1}$	$\bar{5}^3 = \bar{6}$	
		$\bar{3}^4 = \bar{4}$		$\bar{5}^4 = \bar{2}$	
		$\bar{3}^5 = \bar{5}$		$\bar{5}^5 = \bar{3}$	
		$\bar{3}^6 = \bar{1}$		$\bar{5}^6 = \bar{1}$	

Therefore, $|\bar{1}| = 1$, $|\bar{2}| = 3$, $|\bar{3}| = 6$,
$|\bar{4}| = 3$, $|\bar{5}| = 6$, $|\bar{6}| = 2$. ☐

EXAMPLE 5.2

It is easily verified that the orders of the elements of $(\mathbf{Z}_6, +)$ are as follows:
$|\bar{0}| = 1$, $|\bar{1}| = 6$, $|\bar{2}| = 3$, $|\bar{3}| = 2$, $|\bar{4}| = 3$, $|\bar{5}| = 6$. ☐

EXAMPLE 5.3

In group $(\mathbf{Z}, +)$, if a is any nonzero integer, and $n \in \mathbf{Z}^+$, then $a^n = a + a + \cdots + a \neq 0$. So $|a| = \infty$. Also, $|0| = 1$. ☐

THEOREM 5.4

Assume that a is an element of a group G with identity e, and that $|a| = m$. Then:

(a) $a^k = e \Leftrightarrow m \mid k$.

(b) $a^s = a^t \Leftrightarrow s \equiv t \pmod{m}$.

(c) If $H = \{a, a^2, a^3, \ldots, a^{m-1}, a^m\}$, then H is a set of m distinct elements of G. If s is any integer, then $a^s = a^i$, for some $a^i \in H$.

Proof

(a) We are given that $a^m = e$. Then
$m \mid k \Rightarrow k = mv$, for some $v \in \mathbf{Z} \Rightarrow a^k = a^{mv} = (a^m)^v = e^v = e$.

Conversely, suppose that $a^k = e$. By the Division Theorem there exist q and r such that $k = mq + r$, with $0 \leq r < m$. Then

$$a^r = a^{k-mq} = a^k \cdot a^{m(-q)} = e \cdot (a^m)^{-q} = e^{-q} = e.$$

But if $r > 0$, we have a contradiction, since $|a| = m > r$. So $r = 0$. Therefore, $m \mid k$.

(b) $a^s = a^t \Leftrightarrow a^{s-t} = e \Leftrightarrow m \mid (s - t) \Leftrightarrow s \equiv t \pmod{m}$.

(c) This follows from part (b), and the fact that every integer in $\mathbf{Z}$ is congruent modulo m to exactly one integer in the set $\{0, 1, \ldots, m - 1\}$, by Theorem 3.17. ☐

5.1 Problems

1. Find the order of each element in the group:
   (a) $(\mathbf{Z}_4, +)$. (See Example 3.24.)
   (b) Klein Four-group, K_4. (See Example 2.25.)
   (c) Symmetric group, S_3. (See Problem 2.3–4.)
   (d) Dihedral group, D_4. (See Example 2.26.)

5.1 Integral Powers of Group Elements

2. Find the order of each element of $(\mathbf{Z}_{24}, +)$.

3. Find the order of each of these elements in the symmetric group, S_6.
   (a) (134)   (b) (134)(256)   (c) (25)
   (d) (134)(25)   (e) (13)(24)(56)   (f) (2465)
   (g) (24651)   (h) (123456)

4. If $a = \bar{3}$, find a^0, a^1, a^2, a^{-1}, and a^{-2} in (a) $(\mathbf{Z}_{11}^*, \cdot)$, and in (b) $(\mathbf{Z}_{11}, +)$.

5. Rewrite the statements of Theorems 5.1, 5.2, and 5.3, using the additive notation.

6. Let group $(\mathbf{Q}^*, \cdot)$ be given.
   (a) Find: $3^0, 3^1, 3^2, 3^{-1}, 3^{-2}$.
   (b) If $H = \{3^n : n \in \mathbf{Z}\}$, prove that $(H, \cdot)$ is a subgroup of $(\mathbf{Q}^*, \cdot)$.

7. Given group $(\mathbf{Z}, +)$, identify the subset $\{3^n : n \in \mathbf{Z}\}$.

8. Prove Theorem 5.1(a).

9. Prove by induction on n: $(a^{-1})^n = (a^n)^{-1}$, for all $n \in \mathbf{Z}^+$. (This is Theorem 5.1(b).)

10. Prove Theorem 5.2(b).

11. Prove Theorem 5.3(b).

12. (a) Give the orders of these elements of $(\mathbf{Z}_{12}, +)$: $\bar{2}, \bar{3}, \bar{4}, \bar{5}, \bar{6}$.
    (b) Identify each of these five subsets of $(\mathbf{Z}_{12}, +)$:
    $$\{a^n : n \in \mathbf{Z}\}, \quad \text{for } a = \bar{2}, \bar{3}, \bar{4}, \bar{5}, \bar{6}.$$

13. Prove: If a is in group $(G, \otimes)$, then $a^n \in G$, for each $n \in \mathbf{Z}$.

14. Prove: If a is a group element of order k, then a^{-1} is of order k.

15. Prove: If $\theta = (a_1, a_2, \ldots, a_k)$, where θ is a k-cycle of the symmetric group S_n, then $|\theta| = k$.

16. Prove: If θ is an isomorphism of group G onto group H, and $a \in G$, then $|a| = |a\theta|$.

17. Given that a and b are in a group $(G, \otimes)$, and that $b \otimes a = a \otimes b^3$, prove that $b^n \otimes a = a \otimes b^{3n}$, for all $n \in \mathbf{Z}^+$.

18. Prove: If a, b, c are in group G, and if $c = aba^{-1}$, then $c^n = ab^n a^{-1}$, for all $n \in \mathbf{Z}$.

19. Let a and b be elements in an abelian group G. If $|a| = r$ and $|b| = s$, where $(r, s) = 1$, prove that $|ab| = rs$.

5.2 Cyclic Groups

Suppose a subset S of $(\mathbf{Q}^+, \cdot)$ is defined as follows:

$$S = \{2^n : n \in \mathbf{Z}\} = \{1, 2, \tfrac{1}{2}, 4, \tfrac{1}{4}, \ldots\}.$$

It is readily verified that S is a group. The next theorem generalizes this idea.

THEOREM 5.5

If a is any element of a group $(G, \otimes)$, and if $H = \{a^n : n \in \mathbf{Z}\}$, then H is a subgroup of G.

Proof

(1) $a^0 = e$, so $e \in H$.
(2) $m, k \in \mathbf{Z}, a \in H \Rightarrow a^{m+k} \in H$. Hence, H is closed, since $a^m \cdot a^k = a^{m+k}$.
(3) $(a^n)^{-1} = a^{-n}$, and $a^{-n} \in H$. Therefore, by Theorem 2.8, $H \leq G$. □

Subgroup H of Theorem 5.5 is called the *cyclic subgroup of G generated by a*, and will be denoted by $\langle a \rangle$. That is,

$$\langle a \rangle = \{a^n : n \in \mathbf{Z}\}.$$

Thus, each element of a group G generates a cyclic subgroup of G.

To say that a *group G is cyclic* means that there is an element a in G such that $G = \langle a \rangle$. Element a is called a *generator* of G.

EXAMPLE 5.4

Consider the group $(G, \cdot)$, with $G = \{1, -1, i, -i\}$. Now, $i^1 = i, i^2 = -1$, $i^3 = -i, i^4 = 1$. So $|i| = 4$, and each element in G is a power of i. Each element can also be written as a power of $-i$. So $G = \langle i \rangle = \langle -i \rangle$. Note that -1 generates a proper cyclic subgroup, since $\langle -1 \rangle = \{1, -1\}$. Of course, 1 generates the identity subgroup, since $\langle 1 \rangle = \{1\}$. □

EXAMPLE 5.5

Given the multiplicative group $(\mathbf{Q}^+, \cdot)$. If $r \in \mathbf{Q}^+, r \neq 1$, then $\langle r \rangle$ is an infinite cyclic subgroup, where

$$\langle r \rangle = \{r^n : n \in \mathbf{Z}\} = \{r^0, r^1, r^{-1}, r^2, r^{-2}, r^3, \ldots\}.$$

Thus, $\langle 2 \rangle = \{1, 2, \tfrac{1}{2}, 4, \tfrac{1}{4}, 8, \tfrac{1}{8}, \ldots\}$,
$\langle \tfrac{2}{3} \rangle = \{1, \tfrac{2}{3}, \tfrac{3}{2}, \tfrac{4}{9}, \tfrac{9}{4}, \tfrac{8}{27}, \tfrac{27}{8}, \ldots\}$.

It is easily verified that $\mathbf{Q}^+$ itself is not a cyclic group. You are asked to do that in Problem 5.2–15. □

5.2 Cyclic Groups

EXAMPLE 5.6

The order of each element of $(\mathbf{Z}_6, +)$, and the cyclic group generated by each element are as follows:

$$|\bar{0}| = 1, \qquad \langle \bar{0} \rangle = \{\bar{0}\},$$
$$|\bar{3}| = 2, \qquad \langle \bar{3} \rangle = \{\bar{0}, \bar{3}\},$$
$$|\bar{2}| = |\bar{4}| = 3, \qquad \langle \bar{2} \rangle = \langle \bar{4} \rangle = \{\bar{0}, \bar{2}, \bar{4}\},$$
$$|\bar{1}| = |\bar{5}| = 6, \qquad \langle \bar{1} \rangle = \langle \bar{5} \rangle = \mathbf{Z}_6.$$

Hence $(\mathbf{Z}_6, +)$ is a cyclic group, and it has two proper cyclic subgroups. Observe that the order of an element is the same as the order of the subgroup which that element generates. □

THEOREM 5.6

Let a be any element of a group.
(a) If $|a| = m$, then $\langle a \rangle = \{e, a, a^2, \ldots, a^{m-1}\}$.
(b) If $|a| = \infty$, then $\langle a \rangle = \{a^n : n \in \mathbf{Z}\}$, where the a^n are all distinct.

Proof

(a) This is immediate, from Theorem 5.4(c) and the definition of $\langle a \rangle$.

(b) Given that $|a| = \infty$, assume that $a^s = a^t$, for some $s > t$. Then $a^{s-t} = e$, with $s - t > 0$, a contradiction of the hypothesis that a is of infinite order. Therefore, $\langle a \rangle = \{a^n : n \in \mathbf{Z}\}$, where the a^n are all distinct. □

Cyclic groups are easy algebras to deal with. For one thing, a cyclic group is always abelian. Also, the subgroups are not difficult to discover, since a subgroup of a cyclic group is itself cyclic. We now prove these assertions.

THEOREM 5.7

Let $(G, \otimes)$ be a cyclic group. Then:
(a) G is abelian.
(b) If H is a subgroup of G, then H is cyclic.

Proof

(a) By hypothesis, $G = \langle a \rangle$, for some $a \in G$. If a^m and a^n are any two elements of G, then

$$a^m \otimes a^n = a^{m+n} = a^{n+m} = a^n \otimes a^m.$$

Therefore, G is abelian.

(b) If $H = \{e\}$, then H is certainly cyclic. Now suppose that H is a *proper* subgroup of G, and let m be the smallest positive integer such that $a^m \in H$. We shall prove that $H = \langle a^m \rangle$.

If a^s is any element in H, then there exist integers q and r such that $s = mq + r$, with $0 \leq r < m$. Then $a^r = a^{s-mq} = a^s \otimes (a^m)^{-q} \in H$. If $r > 0$, the minimality of m is contradicted. So $r = 0$, and $s = mq$. Then $a^s = a^{mq} = (a^m)^q$, and so $a^s \in \langle a^m \rangle$. Therefore, $H = \langle a^m \rangle$, which means that H is a cyclic subgroup of G. □

We shall now record the fact that $(\mathbf{Z}, +)$ and $(\mathbf{Z}_n, +)$ are cyclic groups. These are rather significant groups, for in a later section we show that *every* infinite cyclic group is isomorphic to $(\mathbf{Z}, +)$, and *every* cyclic group of order n is isomorphic to $(\mathbf{Z}_n, +)$.

THEOREM 5.8

(a) $(\mathbf{Z}, +) = \langle 1 \rangle$. The distinct subgroups of $\mathbf{Z}$ are

$$\langle k \rangle = k\mathbf{Z} = \{kn : n \in \mathbf{Z}\}, \quad \text{with } k = 0, 1, 2, \ldots.$$

(b) $(\mathbf{Z}_n, +) = \langle \bar{1} \rangle$.

Proof

(a) If n is any positive integer, then, in abstract notation:

$$n = 1 + 1 + \cdots + 1 = 1^n,$$
$$-n = (-1) + (-1) + \cdots + (-1) = (-1)^n = (1^{-1})^n = 1^{-n},$$
$$0 = 1^0.$$

Hence, $(\mathbf{Z}, +) = \langle 1 \rangle$.

Each subgroup of $\mathbf{Z}$ is of the form $\langle k \rangle$, and $\langle k \rangle = \langle -k \rangle$, with

$$\langle k \rangle = k\mathbf{Z} = \{0, \pm k, \pm 2k, \pm 3k, \ldots\}.$$

Also, if k and m are two distinct positive integers, then $\langle k \rangle \neq \langle m \rangle$. (Why?) So the distinct subgroups of $\mathbf{Z}$ are: $0\mathbf{Z}, \mathbf{Z}, 2\mathbf{Z}, 3\mathbf{Z}, \ldots$.

(b) We leave this as an exercise. □

Given a cyclic group $\langle a \rangle$ of order n, we might ask these questions: Does any element other than a generate the group? How many subgroups are there? What are the orders of the subgroups? By which elements are these subgroups generated? The next theorem provides some answers.

THEOREM 5.9

Let $G = \langle a \rangle$, where $|G| = n$.
(a) If $(n, k) = d$, then $|a^k| = |\langle a^k \rangle| = n/d$.
(b) If $(n, k) = 1$, then $\langle a^k \rangle = G$.

5.2 Cyclic Groups

Proof

(a) Let $H = \langle a^k \rangle = \{(a^k)^m : m \in \mathbf{Z}\}$. We must show that if $(n, k) = d$, then $|H| = n/d$. There exist integers s and t such that $n = sd$, and $k = td$. Now

$$(a^k)^s = a^{ks} = a^{tds} = a^{tn} = (a^n)^t = e.$$

So $|a^k|$ divides s, and thus $|a^k| \leq s$. Hence,

$$H = \{a^k, a^{2k}, \ldots, a^{sk}\}.$$

Assume that $a^{ik} = a^{jk}$, for some i and j such that $1 \leq j < i \leq s$. Then $ik \equiv jk \pmod{n}$, by Theorem 5.4(b), and $i \equiv j \pmod{n/d}$, which means that $i \equiv j \pmod{s}$, which is impossible, since $0 < i - j < s$. Therefore, $|a^k| = s = n/d = |H|$.

(b) By part (a), if $(n, k) = 1$, then $\langle a^k \rangle$ is a subgroup of G of order n, and so $\langle a^k \rangle = G$. □

EXAMPLE 5.7

Let $a \in G$, and $|a| = 10$. Then

$$\langle a \rangle = \{e, a, a^2, \ldots, a^9\}.$$

The integers prime to 10 are: 1, 3, 7, and 9. By Theorem 5.9,

$$\langle a \rangle = \langle a^3 \rangle = \langle a^7 \rangle = \langle a^9 \rangle.$$

If $k = 2, 4, 6$, or 8, then $(k, 10) = 2$, and so

$$|a^2| = |a^4| = |a^6| = |a^8| = \tfrac{10}{2} = 5,$$

and $\quad \langle a^2 \rangle = \langle a^4 \rangle = \langle a^6 \rangle = \langle a^8 \rangle = \{e, a^2, a^4, a^6, a^8\}.$

If $k = 5$, then $(k, 10) = 5$, $\quad |a^5| = \tfrac{10}{5} = 2,\quad$ and $\quad \langle a^5 \rangle = \{e, a^5\}$.
If $k = 0$, then $(k, 10) = 10$, $\quad |a^0| = \tfrac{10}{10} = 1,\quad$ and $\quad \langle a^0 \rangle = \{e\}$. □

The next two theorems are corollaries to Theorem 5.9.

THEOREM 5.10

Let $G = \langle a \rangle$ be a cyclic group of order n; then G has a subgroup H of order m iff $m \mid n$.

Proof

(1) $m \mid n \Rightarrow n = mk \Rightarrow (n, k) = k \Rightarrow |\langle a^k \rangle| = n/k = m$. Therefore, $\langle a^k \rangle$ is the subgroup H of order m.

(2) Conversely, let $H = \langle a^k \rangle$ be a subgroup of G, where $|H| = m$, and let $(n, k) = d$. Then $|\langle a^k \rangle| = n/d$, so $m = n/d$. Thus, $m \mid n$. □

THEOREM 5.11

If $\bar{k}$ is an element of $(\mathbf{Z}_n, +)$, and if $(n, k) = d$, then $|\bar{k}| = |\langle \bar{k} \rangle| = n/d$.

Proof

$\bar{k} = \bar{1} + \bar{1} + \cdots + \bar{1} = \bar{1}^k$. So the result is immediate, by Theorem 5.9. □

EXAMPLE 5.8

Suppose we want to find the order of each element of $(\mathbf{Z}_{12}, +)$, and the subgroup that each element generates. The results are summarized in Table 5–1.

Table 5–1 Subgroups of $(\mathbf{Z}_{12}, +)$

$(12, k)$	$\|\langle \bar{k} \rangle\|$	$\bar{k}$	$\langle \bar{k} \rangle$
1	12	$\bar{1}, \bar{5}, \bar{7}, \bar{11}$	$\mathbf{Z}_{12}$
2	6	$\bar{2}, \bar{10}$	$\{\bar{0}, \bar{2}, \bar{4}, \bar{6}, \bar{8}, \bar{10}\}$
3	4	$\bar{3}, \bar{9}$	$\{\bar{0}, \bar{3}, \bar{6}, \bar{9}\}$
4	3	$\bar{4}, \bar{8}$	$\{\bar{0}, \bar{4}, \bar{8}\}$
6	2	$\bar{6}$	$\{\bar{0}, \bar{6}\}$
12	1	$\bar{0}$	$\{\bar{0}\}$

□

We have seen that a cyclic group can be generated by just one element. The question arises as to how a noncyclic group is generated. Let X be a subset of a group G, where $X = \{a_1, a_2, \ldots\}$. If each element of G is a product of a finite number of integral powers of the a_i, where powers of an element of X may occur more than once in the product, then we say that *G is generated by X*, or X is *a set of generators* of G. If X is a finite set, then we say that G is *finitely generated*. Of course, a finite group is always finitely generated. If group G is generated by X, where $X = \{a_1, a_2, \ldots, a_s\}$, then we write: $G = \langle a_1, a_2, \ldots, a_s \rangle$. If $(G, \otimes)$ is not abelian it may be necessary, for instance, to write an element of G in the form: $a_1^2 \otimes a_3 \otimes a_1 \otimes a_3^{-1} \otimes a_1^{-4}$. If G is cyclic, than there is a generator set X consisting of just one element. If G is not cyclic, then every set of generators of G must contain at least two elements.

EXAMPLE 5.9

Let $G = (\mathbf{Z}_{10}, +)$. If $X = \{\bar{7}\}$, then X is a set of generators of G. But $\{\bar{7}, \bar{4}, \bar{5}\}$ also generates G, and even the whole set $\mathbf{Z}_{10}$ itself generates G. In fact, a set generating $\mathbf{Z}_{10}$ can contain anywhere from one to ten elements. □

EXAMPLE 5.10

The dihedral group D_n is a noncyclic group of order $2n$. (See Example 2.26 and the discussion preceding the example.) It can be defined abstractly as

5.2 Cyclic Groups

a set of formal symbols as follows:

$$D_n = \langle a, b \rangle = \{a^i b^j : i = 0, 1, \ldots, n-1, j = 0, 1\},$$

where $a^i b^j = a^u b^v$ if and only if $i = u, j = v$; $a^n = b^2 = e$; $n \geq 2$; and $ba = a^{n-1}b$.

If $n \geq 3$, then D_n represents the group of symmetries of a regular n-gon. A rotation through the angle $2\pi/n$ about the center of the polygon can be represented by a, and one of the reflections about a line of symmetry can be represented by b. If $n = 4$, we have:

$$D_4 = \{e, a, a^2, a^3, b, ab, a^2b, a^3b\},$$

with $|a| = 4$, $|b| = 2$, and $ba = a^3 b$. The equation, $ba = a^3 b$, is used to find the product of two elements of D_4. By induction it can be shown that $ba^n = a^{3n}b$. So

$$ba^2 = a^6 b = a^2 b, \qquad ba^3 = a^9 b = ab.$$

In the following, we find the order of $a^3 b$:

$$(a^3 b)^2 = (a^3 b)(a^3 b) = a^3 (ba^3) b = a^3 (ab) b = a^4 b^2 = ee = e.$$

Hence, $|a^3 b| = 2$. D_4 has five elements of order 2, each of which generates a cyclic group of order 2. There are subgroups of order 4, since

$$\langle a \rangle = \langle a^3 \rangle = \{e, a, a^2, a^3\},$$
$$\langle a^2, b \rangle = \langle a^2, a^2 b \rangle = \{e, a^2, b, a^2 b\}, \quad \text{and}$$
$$\langle a^2, ab \rangle = \langle a^2, a^3 b \rangle = \{e, a^2, ab, a^3 b\}.$$

Note that $\langle a \rangle \cong \mathbf{Z}_4$, while $\langle a^2, b \rangle \cong \langle a^2, ab \rangle \cong K_4$. □

The group tables of D_n, for $n = 3, 4, 5,$ and 6 are given in Appendix D. These are tables 6.2, 8.4, 10.2, and 12.3, respectively. We shall make use of the tables of Appendix D to illustrate the theory that is developed in this chapter.

EXAMPLE 5.11

If $n = 2$, then the group D_n that was defined in Example 5.10 becomes:

$$D_2 = \langle a, b \rangle = \{e, a, b, ab\}, \quad \text{with} \quad |a| = |b| = 2, \quad \text{and} \quad ba = ab.$$

It is readily seen that this is the Klein group K_4 that was discussed in Example 2.25. Group K_4 has three proper cyclic subgroups: $\{e, a\}$, $\{e, b\}$, and $\{e, ab\}$. □

5.2 Problems

1. List a few elements in each of the following cyclic subgroups of $(\mathbf{R}^*, \cdot)$.
   (a) $\langle \sqrt{2} \rangle$ (b) $\langle -1 \rangle$ (c) $\langle 4/5 \rangle$ (d) $\langle \pi \rangle$

2   List each proper subgroup of $(\mathbf{Z}_n, +)$, and indicate which elements generate the subgroup if:

(a) $n = 15$,   (b) $n = 16$,   (c) $n = 17$,   (d) $n = 24$.

3   Construct the inclusion diagram of subgroups of $(\mathbf{Z}_n, +)$, for each group of Problem 2.

4   Let G be a cyclic group of order 300, where $G = \{a^i : |a| = 300\}$.

(a) How many different elements, including a, can generate G?

(b) Give a generator of each proper subgroup of G.

5   Prove Theorem 5.8(b).

6   Show that if U_{18} is the set of units of ring $(\mathbf{Z}_{18}, +, \cdot)$, then $(U_{18}, \cdot)$ is isomorphic to $(\mathbf{Z}_6, +)$.

7   List all proper cyclic subgroups of each of these groups.

(a) D_3 (See Appendix D–6.2.)

(b) D_4 (See Appendix D–8.4.)

(c) D_5 (See Appendix D–10.2.)

(d) D_6 (See Appendix D–12.3.)

8   If it exists, find a proper noncyclic subgroup of:

(a) D_3,   (b) D_4,   (c) D_5,   (d) D_6.

(See Appendix D.)

9   If every proper subgroup of G is cyclic, does this insure that G is cyclic?

10   Verify that each non-identity element of group $(M_2(\mathbf{Z}_2), \oplus)$ is of order 2. How many cyclic subgroups does this group of matrices have?

11   The *dicyclic group*, C_{4m}, is defined as follows:

$$C_{4m} = \{a^i b^j : |a| = 2m, |b| = 4, b^2 = a^m, ba = a^{2m-1}b\}.$$

(a) Show that C_4 is isomorphic to $\mathbf{Z}_4$.

(b) Show that C_8 is isomorphic to the quaternion group, Q_8. (See Problem 2.4–4.)

(c) List the elements in C_{12}, and find the cyclic subgroup generated by each element.

(d) Show that C_{12} has no noncyclic proper subgroups. (Hint: Refer to Appendix D–12.4, and use the orders of the elements in your argument.)

12   Verify: If k and m are two positive integers, and $k \neq m$, then $k\mathbf{Z} \neq m\mathbf{Z}$. (See Theorem 5.8.)

13   Let $k\mathbf{Z}$ and $m\mathbf{Z}$ be subgroups of $(\mathbf{Z}, +)$, where k and m are distinct positive integers. Prove:

(a) $k\mathbf{Z} < m\mathbf{Z}$ if and only if $m \mid k$.

(b) Let $k\mathbf{Z} < m\mathbf{Z}$. There is no subgroup $t\mathbf{Z}$ such that $k\mathbf{Z} < t\mathbf{Z} < m\mathbf{Z}$ if and only if $k/m = p$, where p is a prime.

14 (a) Given group G; verify that if K is a subgroup of G, and if $a \in K$, then $\langle a \rangle$ is a subgroup of K.

(b) The statement of part (a) says, in effect, that $\langle a \rangle$ is the "smallest" subgroup containing a. In what sense is it the "smallest"?

15 (a) Let a/b be any positive rational number in lowest terms. That is, $(a, b) = 1$. Suppose c is an integer such that $c > 1$, $(a, c) = 1$, and $(b, c) = 1$. Prove that c is not in the cyclic subgroup, $\langle a/b \rangle$, of $(\mathbf{Q}^+, \cdot)$.

(b) Is $(\mathbf{Q}^+, \cdot)$ a cyclic group? Explain.

16 If w is a complex number, and $w^m = 1$, then w is called an *m*th *root of unity*. Let $w = \cos(2\pi/m) + i \sin(2\pi/m)$, where m is a positive integer. Prove:

(a) $\langle w \rangle$ is a cyclic subgroup of order m of the multiplicative group of nonzero complex numbers. (Hint: Use De Moivre's Theorem, given in Problem 3.3–10.)

(b) Each element of group $(\langle w \rangle, \cdot)$ is an *m*th root of unity.

17 Write the m *m*th roots of unity, and plot the m points in the plane that correspond to these roots if:

(a) $m = 2$,   (b) $m = 3$,   (c) $m = 4$,
(d) $m = 6$,   (e) $m = 8$.

(See Problem 16.)

5.3 Cosets and Lagrange's Theorem

In the preceding section we proved that if a cyclic group G is of finite order n, then the order of any subgroup of G must be a divisor of n. Furthermore, if k is a divisor of n, then cyclic group G has a subgroup of order k. For instance, a cyclic group of order 30 has proper subgroups of orders 2, 3, 5, 6, 10, and 15, while a cyclic group of order 31 has no proper subgroups.

We propose to examine the situation with respect to finite groups in general. In this section we shall answer these two questions concerning a group G of order n:

(1) If H is a subgroup of order k, must k divide n?

(2) If k divides n, must G have a subgroup of order k?

We shall now set up the machinery necessary to answer question (1).

Let H be a subgroup of a group $(G, \otimes)$, where G is finite or infinite. A *left coset of* H is a set $g \otimes H$, abbreviated as gH, where $g \in G$, defined by

$$gH = \{g \otimes h : h \in H\}.$$

Thus we get a coset of H if we take an element g of G and multiply it by each element of H.

Similarly, a *right coset of H* is Hg, where

$$H \otimes g = Hg = \{h \otimes g : h \in H\}.$$

EXAMPLE 5.12

We shall find cosets of a subgroup of A_4. (In Problem 2.5–14 you were asked to construct the group table of A_4.) To do this we refer to Figure 5–1.

Figure 5–1 *Alternating group,* A_4

Element	Order
$e = (1)$	1
$a = (12)(34)$	2
$b = (13)(24)$	2
$c = (14)(23)$	2
$g = (123)$	3
$h = (134)$	3
$i = (243)$	3
$j = (142)$	3
$r = (132)$	3
$s = (234)$	3
$t = (124)$	3
$u = (143)$	3

	e a b c	g h i j	r s t u
e	e a b c	g h i j	r s t u
a	a e c b	h g j i	s r u t
b	b c e a	i j g h	t u r s
c	c b a e	j i h g	u t s r
g	g i j h	r t u s	e b c a
h	h j i g	s u t r	a c b e
i	i g h j	t r s u	b e a c
j	j h g i	u s r t	c a e b
r	r u s t	e c a b	g j h i
s	s t r u	a b e c	h i g j
t	t s u r	b a c e	i h j g
u	u r t s	c e b a	j g i h

Let $H = \langle t \rangle = \{e, t, j\}$. The group table can be used to verify that the following are left cosets of H:

$$aH = \{ae, at, aj\} = \{a, u, i\}, \quad cH = \{ce, ct, cj\} = \{c, s, g\},$$
$$bH = \{be, bt, bj\} = \{b, r, h\}, \quad eH = \{ee, et, ej\} = \{e, t, j\}.$$

Note that $A_4 = H \cup aH \cup bH \cup cH$, and that $\{H, aH, bH, cH\}$ is a factor set of A_4. The following are right cosets of H:

$$Ha = \{ea, ta, ja\} = \{a, s, h\}, \quad Hc = \{ec, tc, jc\} = \{c, r, i\},$$
$$Hb = \{eb, tb, jb\} = \{b, u, g\}, \quad He = \{ee, te, je\} = \{e, t, j\}.$$

Then $A_4 = H \cup Ha \cup Hb \cup Hc$, and $\{H, Ha, Hb, Hc\}$ is also a factor set of A_4, but is different than the factor set of left cosets. □

EXAMPLE 5.13

Given the group $(\mathbf{Z}_{12}, +)$ and subgroup $H = \{\bar{0}, \bar{4}, \bar{8}\}$; cosets of H are the following:

$$\bar{1}H = \bar{1} + H = \{\bar{1}, \bar{5}, \bar{9}\} = H + \bar{1} = H\bar{1},$$
$$\bar{2}H = \bar{2} + H = \{\bar{2}, \bar{6}, \overline{10}\} = H + \bar{2} = H\bar{2},$$

$$\overline{3}H = \overline{3} + H = \{\overline{3},\overline{7},\overline{11}\} = H + \overline{3} = H\overline{3},$$
$$\overline{0}H = \overline{0} + H = \{\overline{0},\overline{4},\overline{8}\} = H + \overline{0} = H\overline{0} = H.$$

Because $\mathbf{Z}_{12}$ is abelian, $aH = Ha$, for all $a \in \mathbf{Z}_{12}$. Also, $\{H, \overline{1}H, \overline{2}H, \overline{3}H\}$ is a factor set of $\mathbf{Z}_{12}$. ☐

The results we are after can be obtained by using left cosets only. However, statements involving left cosets are also true if "left" is replaced by "right."

Let H be a subgroup of a group $(G, \otimes)$, where G may be either finite or infinite. If a and b are any elements in G, we say that *a is left congruent to b modulo H* if and only if there is an element h in H such that $a = b \otimes h$. In symbols:

$$a \equiv b \pmod{H} \Leftrightarrow a = b \otimes h, \quad \text{for some } h \in H.$$

THEOREM 5.12

(a) If H is a subgroup of a group G, then left congruence modulo H is an equivalence relation on G.

(b) If $\bar{a}$ is an equivalence class with respect to left congruence modulo H, then $\bar{a} = aH$, and so the set of distinct left cosets of H is a partition of G.

Proof

(a) We prove the symmetric property as follows:

$$a \equiv b \pmod{H} \Rightarrow a = b \otimes h \Rightarrow a \otimes h^{-1} = (b \otimes h) \otimes h^{-1}$$
$$\Rightarrow a \otimes h^{-1} = b \Rightarrow b = a \otimes h^{-1} \Rightarrow b \equiv a \pmod{H}.$$

We leave it to you to verify that left congruence is reflexive and transitive.

(b) If $\bar{a}$ is one of the equivalence classes, then

$$\bar{a} = \{x : x \equiv a \pmod{H}\} = \{x : x = a \otimes h, h \in H\} = \{ah : h \in H\} = aH.$$

So the equivalence class $\bar{a}$ is precisely the left coset aH. Therefore, by Theorem 1.4, the set of distinct left cosets of H is a partition, or factor set, of G. ☐

In Chapter 3 we discussed congruence modulo a *rational integer*, and now we're talking about congruence modulo a *subgroup* of a group. The next example shows that there is no ambiguity or contradiction in this.

EXAMPLE 5.14

Let $(G, \otimes) = (\mathbf{Z}, +)$, and let

$$H = n\mathbf{Z} = \{0, \pm n, \pm 2n, \ldots\} = \langle n \rangle.$$

Then
$$a \equiv b \pmod{H} \Leftrightarrow a = b \otimes h, \quad \text{for some } h \in H$$
$$\Leftrightarrow a = b + kn, \quad \text{for some } k \in \mathbf{Z}$$
$$\Leftrightarrow a \equiv b \pmod{n}.$$

Hence, congruence modulo an integer n is actually congruence modulo the additive subgroup of $\mathbf{Z}$ consisting of all multiples of n. Therefore, the congruence of Section 3.5 is merely a special case of the congruence of the present section. If a is any rational integer, then the equivalence class $\bar{a}$ is the following:

$$\bar{a} = a + H = a + \{kn : k \in \mathbf{Z}\} = \{a + kn : k \in \mathbf{Z}\}.$$

The factor set $\mathbf{Z}_n$ is the collection of distinct left cosets of $n\mathbf{Z}$. ☐

In Example 5.12, we saw that each coset of H contains the same number of elements. We want to prove that this is always the case, even if H and G are infinite sets.

To say that *two sets, A and B, contain the same number of elements* means that there is a bijective map from A to B. That is, sets A and B are the same size if and only if a one-to-one correspondence can be established between the elements of the two sets.

THEOREM 5.13

If H is a subgroup of a group G, then any two left cosets of H contain the same number of elements.

Proof

Let aH and bH be any two left cosets of H. Define map $\mu: aH \to bH$ so that, for each $h \in H$, $(ah)\mu = bh$. Obviously μ is onto. Let $h, k \in H$. Then

$$(ah)\mu = (ak)\mu \Leftrightarrow bh = bk \Leftrightarrow h = k \Leftrightarrow ah = ak.$$

Therefore, μ is 1–1, and so is a bijective map. ☐

We have now arrived at the main theorem of this section, the proof of which is very simple.

THEOREM 5.14 *(Lagrange)*

If G is a group of order n, and H is a subgroup of order k, then k divides n.

Proof

We verified that the left cosets of H constitute a partition of G. There can be only a finite number of distinct cosets, since G is finite. Let the distinct cosets be denoted as $g_1 H, g_2 H, \ldots, g_m H$, where $g_i \in G$. Then

$$G = g_1 H \cup g_2 H \cup \cdots \cup g_m H.$$

Each element of G is in exactly one of these cosets. Now $eH = H$, so subgroup H is one of the cosets. But H contains k elements. Hence, each coset contains k elements. Then G is a union of m pairwise disjoint sets,

5.3 Cosets and Lagrange's Theorem

each of which contains k elements. This means that $n = km$. Therefore, k divides n. □

The next theorem is a corollary to Lagrange's Theorem.

THEOREM 5.15

(a) If a is any element in a finite group G, then the order of a divides the order of G.

(b) If the order of group G is a prime p, then G is a cyclic group.

Proof

(a) Let $|a| = k$, $|G| = n$. Then $|\langle a \rangle| = k$, by Thm. 5.6, and so $k \mid n$, by Thm. 5.14.

(b) $a \in G, a \neq e \Rightarrow |a| = p \Rightarrow \langle a \rangle = G \Rightarrow G$ is cyclic. □

The number of distinct cosets of a subgroup H is called the *index* of H in G, and is denoted by $[G:H]$. In the notation of Theorem 5.14:

$$[G:H] = |G|/|H| = n/k = m.$$

We are now ready to discuss question (2): If $|G| = n$, and $k \mid n$, is there a subgroup of G of order k? The answer is in the affirmative if G is an abelian group, and we go into this later on. Right now we show, by a counter example, that a nonabelian group of order n does not have to have a subgroup of order k, where $k \mid n$.

THEOREM 5.16

A_4, the alternating group of order 12, has no subgroup of order 6. Hence, a nonabelian group of order n does not necessarily have a subgroup of order k, even if k divides n.

Proof

We refer to the group table of A_4, given in Example 5.12. Each element of A_4 is of order 1, 2, or 3, so A_4 is not cyclic. It is a simple matter to verify that $A_4 = \langle a, g \rangle$, where $a = (12)(34)$, and $g = (123)$. The elements can be expressed as products of a and g as follows:

$$a = a, \quad g^2 = r, \quad ga = i, \quad ag^2a = t,$$
$$a^2 = e, \quad ag = h, \quad g^2a = u, \quad gag^2 = b,$$
$$g = g, \quad ag^2 = s, \quad aga = j, \quad g^2ag = c.$$

In A_4, each element of order 2 is of the form $(mn)(pq)$, where m, n, p, and q are the integers 1, 2, 3, 4 in some order, and each element of order 3 is of the form (mnp). Any two elements of order 2 generate the Klein four-group, and so two or more elements of order 2 generate a subgroup whose order is divisible by 4. It will be left as an exercise to verify that $\langle (mnp), (mn)(pq) \rangle = A_4$, and $\langle (mnp), (mnq) \rangle = A_4$. Since this exhausts the possibilities, it is impossible to generate a subgroup of order 6. □

EXAMPLE 5.15

A word is in order about the generators of A_4. In the proof of Theorem 5.16 we saw that a set of generators of A_4 can consist of either two elements of order 3, or one element of order 3 and one element of order 2. But although it takes only *two* elements to generate A_4, it is not unusual to define A_4 in terms of *three* elements. This is done in Appendix D–12.5, where $A_4 = \langle a, b, c \rangle$. Using the notation of that appendix, it can be shown that the following map is an isomorphism:

$$
\begin{array}{lll}
e \to (1) & c \to (123) & c^2 \to (132) \\
a \to (12)(34) & ac \to (134) & ac^2 \to (234) \\
b \to (13)(24) & bc \to (243) & bc^2 \to (124) \\
ab \to (14)(23) & abc \to (142) & abc^2 \to (143)
\end{array}
$$
□

If G is a finite nonabelian group, can one guarantee the existence of *any* proper subgroups? Yes, for in Section 5.9 we shall prove that if $|G| = n$ and $p^k \mid n$, where p is a positive prime and $k \in \mathbf{Z}^+$, then G has a subgroup of order p^k.

EXAMPLE 5.16

If $|G| = 200 = 2^3 \cdot 5^2$, then, according to the above statement, we are assured that G has proper subgroups of orders 2, 4, 8, 5, and 25. □

5.3 Problems

1. Using the notation of Example 5.12, find the left cosets and the right cosets of each of these subgroups of A_4:

   (b) $M = \{e, b\}$

   (c) $P = \{e, i, s\}$ (d) $N = \{e, a, b, c\}$

2. List all subgroups of A_4, and draw the inclusion diagram of subgroups.

3. Given group $(\mathbf{Z}, +)$ and subgroup $\langle 5 \rangle$; find the cosets of $\langle 5 \rangle$, and the factor set of $\mathbf{Z}$ with respect to congruence modulo $\langle 5 \rangle$.

4. Complete the proof of Theorem 5.12(a).

5. Given that H is a subgroup of a group G, and $gH = \{g, s, t, v\}$; explain why $gH = sH = tH = vH$.

6   Find the cosets of the subgroup $\langle \overline{4} \rangle$ of group $(\mathbf{Z}_{24}, +)$, and give the order of the factor set.

7   An operation, $+$, on $\mathbf{R} \times \mathbf{R}$ is defined as follows:

$$(a, b) + (c, d) = (a + c, b + d).$$

(a) Prove: (1) $(\mathbf{R} \times \mathbf{R}, +)$ is a group, and (2) if $H = \{(x, 0): x \in \mathbf{R}\}$, then $H < \mathbf{R} \times \mathbf{R}$.

(b) Describe geometrically the cosets of H.

8   Show that $\{(123), (124)\}$ is a set of generators of A_4 by writing each element of A_4 as a product of (123) and (124).

9   Complete the proof of Theorem 5.16 by showing that

$$\langle (mnp), (mn)(pq) \rangle = A_4, \quad \langle (mnp), (mnq) \rangle = A_4,$$

where $m, n, p,$ and q are the integers 1, 2, 3, 4, in any given order.

10   Find a set of generators for each of these groups, choosing the smallest possible set.
(a) S_3 (b) D_4 (c) $(M_2(\mathbf{Z}_2), \oplus)$ (d) $(\mathbf{Q}^+, \cdot)$

11   Prove: If G is any group of order n, and if $a \in G$, then $a^n = e$.

12   Prove:

(a) (*Euler's Theorem*) If n is a positive integer, and a is an integer such that $(a, n) = 1$, then $a^{\phi(n)} \equiv 1 \pmod{n}$. (Hint: See Problems 3.6–8 and 5.3–11.)

(b) (*Fermat's Theorem*) If p is a positive prime, $a \in \mathbf{Z}$, then $a^p \equiv a \pmod{p}$. (Show that this is virtually a special case of part (a).)

13   Let H and K be subgroups of group G with identity e. Prove:
(a) If $|H| = 6$ and $|K| = 35$, then $H \cap K = \{e\}$.
(b) If $|H| = m, |K| = n$, with $(m, n) = 1$, then $H \cap K = \{e\}$.

14   Prove: Every infinite group G has proper subgroups.

15   (a) If G is a group of order 60, discuss the existence of proper subgroups, and the orders of those subgroups.
(b) Answer part (a), given that G is abelian.

16   Prove: If $|G| = 2p$, where p is a prime, then every proper subgroup of G is cyclic.

17   Given that G is a finite group, and that $|G| > 1$; show that if G has no proper subgroups, then $|G| = p$, where p is a prime.

18   (a) If G is a group of order p^2, where p is a prime, prove that G has at least one subgroup of order p.
(b) Prove: If $|G| = p^n$, where p is prime and $n \in \mathbf{Z}^+$, then G contains a subgroup of order p.

19 Let H be any subgroup of a group G, $L(H)$ be the collection of left cosets of H, and let $*$ be defined as follows:

$$(xH) * (yH) = (xy)H, \quad \text{for all } x, y \in G.$$

Using the notation of Example 5.12, if $G = A_4$ and $H = \{e, t, j\}$, show that $*$ is not a *well-defined* operation on $L(H)$.

20 (a) Why has Fermat been called the "prince of amateurs"?

(b) What is Fermat's Last Theorem, and where was it found? (See Appendix E–4.)

21 (a) Although Swiss, Euler spent most of his life in either Russia or Prussia. What was he doing in those countries?

(b) List some of the mathematical symbols that Euler invented. (See Appendix E–5.)

22 (a) Did Napoleon like Lagrange?

(b) What well-known notation was invented by Lagrange? (See Appendix E–6.)

5.4 Normal Subgroups and Factor Groups

In this section we shall convert the collection of cosets of a special kind of subgroup into an algebra by defining a binary operation on the cosets.

If A and B are subsets of an algebra $(S, \otimes)$, we define *set multiplication* as follows: $A \otimes B = AB = \{a \otimes b : a \in A, b \in B\}$.

EXAMPLE 5.17

An examination of the group table of Example 5.12 reveals that N is a subgroup of A_4, where $N = \{e, a, b, c\}$. The left cosets of N are:

$$N = \{e, a, b, c\},$$
$$gN = \{g, i, j, h\},$$
$$rN = \{r, u, s, t\}.$$

Then $\{N, gN, rN\}$ is a factor set. Applying the definition of set multiplication to these cosets, we get, for instance:

$$(gN)(rN) = \{gr, gu, gs, gt, ir, \ldots, ht\}$$
$$= \{e, a, b, c, b, c, e, a, c, b, a, e, a, e, c, b\}$$
$$= \{e, a, b, c\} = N.$$

So the product of these two cosets of N is a coset of N. By multiplying cosets, we arrive at the following multiplication table:

5.4 Normal Subgroups and Factor Groups

	N	gN	rN
N	N	gN	rN
gN	gN	rN	N
rN	rN	N	gN

Note that $(gN)^2 = rN$, and $(gN)^3 = N$. Therefore, the factor set $\{N, gN, rN\}$ is the cyclic group of order 3 under the operation of coset multiplication. The group table of Example 5.12 is partitioned into nine blocks that show this coset multiplication. In a sense, the above table is a condensed version of the table of Example 5.12.

In Example 5.12 we obtained the factor set $\{H, aH, bH, cH\}$, with $H = \{e, t, j\}$. Is this set closed under coset multiplication? We multiply aH by bH, and get: $(aH)(bH) = \{c, s, g, t, j, e, h, b, r\}$, which is not a coset at all! So this factor set is certainly not a group under coset multiplication.

This tells us that the structures of subgroups H and N must differ in some significant way. If we compute the right cosets of N we find that $gN = Ng$, and $rN = Nr$. We have found a structural difference:

$$\{N, gN, rN\} = \{N, Ng, Nr\}, \quad \text{but} \quad \{H, aH, bH, cH\} \neq \{H, Ha, Hb, Hc\}. \quad \square$$

The statement that a left coset equals its right coset, $gH = Hg$, can be made in several equivalent ways. We list these ways in the next theorem.

THEOREM 5.17

Let g be any element in a group G, and let H be a subgroup of G. Then these statements are equivalent:

(1) $gH = Hg$.
(2) If $h \in H$, then $gh = h_1 g$, and $hg = gh_2$, for some $h_1, h_2 \in H$.
(3) $gHg^{-1} = H$.
(4) $ghg^{-1} \in H$, for all $h \in H$.

Proof

We can show that all four statements are equivalent by verifying the following:

$$(1) \Rightarrow (2), \quad (2) \Rightarrow (3), \quad (3) \Rightarrow (4), \quad (4) \Rightarrow (1).$$

We shall prove the first implication, and leave the others as an exercise. Given (1): $gH = Hg$. Now

$$h \in H \Rightarrow gh \in gH \Rightarrow gh \in Hg \Rightarrow gh = h_1 g, \text{ for some } h_1 \in H, \text{ and}$$
$$h \in H \Rightarrow hg \in Hg \Rightarrow hg \in gH \Rightarrow hg = gh_2, \text{ for some } h_2 \in H.$$

Therefore, $(1) \Rightarrow (2)$. $\square$

A subgroup H of a group G is called a *normal subgroup* of G if and only if, for every $g \in G$, $gH = Hg$. If H is a normal subgroup of G we write $H \triangleleft G$. The factor set of G modulo H is denoted by G/H. If $|G| = n$ and $|H| = k$, it is obvious that $|G/H| = n/k$. We now prove the important fact that if $H \triangleleft G$, then G/H is a group.

THEOREM 5.18

Let H be a normal subgroup of $(G, \otimes)$.
(a) Under coset multiplication, $(aH)(bH) = (ab)H$, for all $a, b \in G$.
(b) $(G/H, \otimes)$ is a group.

Proof
(a) Let $x \in (aH)(bH)$. Then, with obvious notation,

$$x = (ah_1)(bh_2) = a(h_1 b)h_2 = a(bh_3)h_2 = (ab)(h_3 h_2) \in (ab)H.$$

So $(aH)(bH) \subseteq abH$. Now let $y \in abH$. Then $y = abh = (ae)(bh) \in (aH)(bH)$. So $abH \subseteq (aH)(bH)$. Therefore, $(aH)(bH) = abH$.

(b) The associative property may be verified as follows:

$$[(aH)(bH)]cH = [(ab)H]cH = [(ab)c]H = [a(bc)]H = aH[(bc)H]$$
$$= aH[(bH)(cH)].$$

Now H is the identity of G/H, since $H(gH) = (eH)(gH) = (eg)H = gH$, for all $g \in G$. As for inverses, $(g^{-1}H)(gH) = g^{-1}gH = eH = H$, and so $(gH)^{-1} = g^{-1}H$. Therefore, G/H is a group. $\square$

Group G/H is called the *factor group* of G modulo H. Note that the order of G/H is the index of H. That is, $|G/H| = [G:H]$.

THEOREM 5.19

If group G is abelian, then each subgroup H is normal, and factor group G/H is abelian.

The proof is left as an exercise.

EXAMPLE 5.18

Let $G = \mathbf{R} \times \mathbf{R}$, and define addition, $+$, on G as follows:
$(a, b) + (c, d) = (a + c, b + d)$. If $H = \{(x, 0): x \in \mathbf{R}\}$, then $H \triangleleft G$. (See Problem 5.3-7.)

For each $(a, b) \in G$, $(a, b) + H$ is a coset of H, with

$$(a, b) + H = \{(x + a, b): x \in \mathbf{R}\} = \{(r, b): r \in \mathbf{R}\} = (0, b) + H.$$

Hence, $\qquad G/H = \{(0, y) + H: y \in \mathbf{R}\},$

5.4 Normal Subgroups and Factor Groups

with the coset operation being:

$$[(0, y_1) + H] + [(0, y_2) + H] = (0, y_1 + y_2) + H.$$

The geometric interpretation is interesting (see Figure 5–2). G is the set of points in the Cartesian plane. The origin is the identity of G, and a point

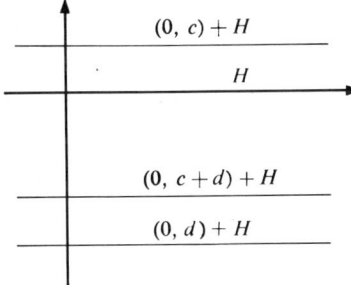

Figure 5–2

and its inverse are symmetric about the origin. Subgroup H is the x–axis. Each horizontal line is a coset of H, and, conversely, each coset is a horizontal line. So the plane is partitioned into a collection, G/H, of horizontal lines. The "sum" of two horizontal lines is the horizontal line whose y–intercept is the sum of the y–intercepts of the two given lines. In group G/H the x–axis is the identity, and the inverse of a line above the x–axis is a line below the x–axis, and the same undirected distance from the x–axis. ☐

EXAMPLE 5.19

The group $(\mathbf{Z}_n, +)$ of integers modulo n is actually the factor group of $(\mathbf{Z}, +)$ modulo the subgroup $n\mathbf{Z}$. For if a is any integer, then

$$\bar{a} = a + n\mathbf{Z}, \quad \text{and} \quad \mathbf{Z}_n = \mathbf{Z}/n\mathbf{Z} = \{\bar{0}, \bar{1}, \bar{2}, \ldots, \overline{n-1}\}. \quad \square$$

EXAMPLE 5.20

The group table of a group G of order 9 is given in Appendix D–9.2. Observe that each of the 9 blocks contains 9 symbols, but only 3 distinct elements. It is apparent from the table that if $H = \{e, a, a^2\}$, then $H \triangleleft G$. The cosets of H can be written as follows:

$$\bar{e} = eH = \{e, a, a^2\},$$
$$\bar{b} = bH = \{b, ab, a^2b\},$$
$$\overline{b^2} = b^2H = \{b^2, ab^2, a^2b^2\}.$$

The following is the group table of G/H.

	$\bar{e}$	$\bar{b}$	$\overline{b^2}$
$\bar{e}$	$\bar{e}$	$\bar{b}$	$\overline{b^2}$
$\bar{b}$	$\bar{b}$	$\overline{b^2}$	$\bar{e}$
$\overline{b^2}$	$\overline{b^2}$	$\bar{e}$	$\bar{b}$

It turns out that G/H is isomorphic to $(\mathbf{Z}_3, +)$. ☐

Let M be a normal subgroup of G. Then M is called a *maximal normal subgroup* of G iff there is no subgroup N, $N \ne M$, $N \ne G$, such that $M \triangleleft N \triangleleft G$. If the G_i are subgroups of G such that

$$E = G_k \triangleleft G_{k-1} \triangleleft \cdots \triangleleft G_2 \triangleleft G_1 \triangleleft G_0 = G,$$

where E is the identity subgroup, and G_i is a maximal normal subgroup of G_{i-1}, for $i = 1, 2, \ldots, k$, then the collection, $\{G_1, G_2, \ldots, G_k\}$ is called a *composition series of G of length k*. The k factor groups,

$$G/G_1, G_1/G_2, \ldots, G_{k-1}/G_k,$$

are called the *composition factors* of G.

EXAMPLE 5.21

Let's find a composition series of S_4. Certainly, $A_4 \triangleleft S_4$, since $|A_4| = \frac{1}{2}|S_4|$. (You are asked to verify this in Problem 5.4–6.) Using the notation of Examples 5.12 and 5.17, we have that $N \triangleleft A_4$, where $N = \{e, a, b, c\}$. If $H = \{e, a\}$, and $E = \{e\}$, we then have this composition series:

$$E \triangleleft H \triangleleft N \triangleleft A_4 \triangleleft S_4.$$

The composition factors are these factor groups:

$$S_4/A_4, A_4/N, N/H, H/E,$$

whose orders are, respectively: 2, 3, 2, 2. □

Although a group G can have more than one composition series, it can be proved that, in terms of group structures, any two composition series of G produce the same set of composition factors. That is, factor groups obtained from one composition series can be put into a 1–1 correspondence with the factor groups obtained from another composition series so that the corresponding factor groups are isomorphic. We have seen in Example 5.21 that the composition factors of S_4 consist of three cyclic groups of order 2 and one cyclic group of order 3.

A finite group is said to be a *solvable group* if its composition factors have prime orders. Thus, S_4 is a solvable group.

A *simple group* is one that has no proper normal subgroups. Obviously, a group of prime order is always simple. The fact that the alternating group A_5 is simple has some important consequences, so we take time to prove this fact.

THEOREM 5.20

A_5 is a simple group, and hence A_5 is not solvable. The symmetric group S_5 is not solvable.

Proof

We must prove that A_5 has no proper normal subgroups. Let N be a normal subgroup of A_5, with $|N| \ge 2$. We now show that $N = A_5$.

5.4 Normal Subgroups and Factor Groups

Written as a product of disjoint cycles, each non-identity element of A_5 must be in one of these forms:

$$(abc), \quad (abcde), \quad (ab)(cd).$$

First we verify that N must contain at least one 3–cycle. Observe that if $h \in N$ and $g \in A_5$, then $ghg^{-1} \in N$, and so $h(ghg^{-1}) \in N$. Now if $h = (abcde) \in N$, we choose $g = (ab)(cd)$, and conclude that $hghg^{-1} = (bed)$. Also, if $h = (ab)(cd) \in N$, we choose $g = (abe)$, and find that $hghg^{-1} = (abe)$. Therefore, N contains a 3–cycle; call it (abc).

Second, we show that if (rst) is *any* 3–cycle, then $(rst) \in N$. Note that if

$$g_1 = \begin{pmatrix} r & s & t & u & v \\ a & b & c & d & e \end{pmatrix}, \quad g_2 = \begin{pmatrix} r & s & t & u & v \\ a & b & c & e & d \end{pmatrix}, \quad \text{and} \quad x = (de),$$

then $g_1 x = g_2$, and so g_1 and g_2 are of opposite parity. Choose the g_i that is an even permutation, and call it g. Then $g \in A_5$. But if $k = (abc)$, then $gkg^{-1} \in N$. However, $gkg^{-1} = (rst)$. Therefore, N contains *every* 3–cycle.

Since $\quad (abcde) = (abc)(ade), \quad$ and $\quad (ab)(cd) = (abc)(adc),$

we see that every even permutation can be expressed as a product of 3–cycles. We conclude that every even permutation is in N, and so $N = A_5$. Therefore, A_5 contains no proper normal subgroup, and hence A_5 is a simple group.

If E is the identity subgroup, the only composition series of A_5 is:

$$E \triangleleft A_5.$$

The composition factor, A_5/E, is of order 60, and so A_5 is not solvable.

Since $|S_5| = 120$, and $|A_5| = 60$, A_5 is a maximal normal subgroup of S_5. A composition series of S_5 is:

$$E \triangleleft A_5 \triangleleft S_5,$$

and the composition factors are S_5/A_5 and A_5/E, which are isomorphic, respectively, to $\mathbf{Z}_2$ and A_5, which are of orders 2 and 60. Therefore, S_5 is not solvable. □

A remarkable result concerning the solution by radicals of polynomials of degree 5 makes use of the fact that the symmetric group S_5 is not solvable. We discuss this in Section 6.8. Incidentally, it has been proved that the alternating group, A_5, is the group of lowest order that is not solvable. Hence, G is solvable if $|G| < 60$. As recently as 1962 it was first proved that every group of *odd* order is solvable.

5.4 Problems

1 Referring to group A_4, Example 5.12:
(a) If $K = \{e, i, s\}$, is K a normal subgroup of A_4?
(b) Is $\{e, b\}$ a normal subgroup of A_4?

2 The dihedral group D_6 can be defined as follows:
$$D_6 = \langle a, b \rangle = \{a^i b^j : |a| = 6, |b| = 2, ba = a^5 b\}.$$
(See Appendix D–12.3.)
(a) If $H = \{e, a^2, a^4\}$, verify that $H \triangleleft D_6$.
(b) Construct the group table of D_6/H, and identify the group. Is D_6/H a simple group?

3 Rewrite the group table of D_6, Appendix D–12.3, listing the elements in the heading in an order that will exhibit blocks of elements that are all in the same coset of H, where $H = \{e, a^2, a^4\}$. (Let the first three elements be those of H, the next three those of Ha, then those of Hb, and finally, those of Hab.)

4 (a) Show that $H \triangleleft C_{12}$, if $H = \{e, a^3\}$. (See Appendix D–12.4.)
(b) Find the elements of C_{12}/H, and the order of each element. Is C_{12}/H simple?

5 Rewrite the group table of C_{12} to show the block multiplication of cosets of $\{e, a^3\}$. (See Problem 4.)

6 Prove: If $H < G$, where $|G| = n$ and $|H| = n/2$, then $H \triangleleft G$.

7 Explain these statements:
(a) The group table of Example 5.17 is a condensation of the group table of Example 5.12.
(b) If G is the group of Appendix D–9.2, and $H = \{e, a, a^2\}$, it is apparent from the form of the group table that H is a normal subgroup of G.

8 (a) Verify: If C is the center of a group G, then C is a normal subgroup of G. (See Problem 2.4–6.)
(b) If C is the center of G, is G/C abelian?

9 Find the center C of D_6, and show that D_6/C is isomorphic to S_3.

10 (a) Show that each subgroup of the quaternion group, $\mathbf{Q}_8$, is normal. (See Problem 2.4–4.)
(b) Verify that $\{1, -1\}$ is the center C of $\mathbf{Q}_8$, and give the group table of $\mathbf{Q}_8/C$.

11 Let $(\mathbf{R} \times \mathbf{R}, +)$ be the group of Example 5.18.
(a) Verify that $K \triangleleft \mathbf{R} \times \mathbf{R}$, if $K = \{(r, r) : r \in \mathbf{R}\}$.
(b) Interpret geometrically the factor group, $(\mathbf{R} \times \mathbf{R})/K$.

12 Let group $(\mathbf{Q}, +)$ and subgroup $(\mathbf{Z}, +)$ be given.
   (a) Describe a typical coset of $\mathbf{Z}$ in $\mathbf{Q}$.
   (b) Verify that each element of $\mathbf{Q}/\mathbf{Z}$ is of finite order.
   (c) What is the order of $\mathbf{Q}/\mathbf{Z}$?

13 Assume that H is a normal subgroup of G. Prove:
   (a) If $a \in G$, then $(aH)^n = a^n H$, for all $n \in \mathbf{Z}$.
   (b) If G is a cyclic group, then G/H is a cyclic group.

14 Verify: If $H < K < G$, and if $H \triangleleft G$, then $H \triangleleft K$.

15 If $H \triangleleft K$ and $K \triangleleft G$, is it true that $H \triangleleft G$?

16 Prove Theorem 5.17.

17 Prove Theorem 5.19.

18 Which finite abelian groups are simple?

19 Prove: $\{e, a\}$ is a normal subgroup of G if and only if a is in the center of G.

20 Verify the permutation multiplications used in the proof of Theorem 5.20.

21 (a) Find all maximal normal subgroups of $(\mathbf{Z}_{30}, +)$.
   (b) List each subgroup H of $\mathbf{Z}_{30}$ for which $\mathbf{Z}_{30}/H$ is a simple group.
   (c) Give three different composition series of $\mathbf{Z}_{30}$, and identify the composition factors.

22 Give two different composition series of $(\mathbf{Z}_{24}, +)$, and identify the composition factors.

23 Determine whether S_4/D_4 is a group. (See Example 2.26.)

24 Prove: If $H \triangleleft G$, $H' \leq G/H$, and $S = \cup \, g_i H$, the union being over all $g_i H \in H'$, then $S \leq G$.

5.5 Group Morphisms

In group theory, many useful results are obtained by means of morphisms. As a corollary to Theorem 2.30, we state the following:

THEOREM 5.21

Let $\theta: (G, \otimes) \to (H, \odot)$ be a morphism. If G is a group, then so is $G\theta$. If G is an abelian group, then so is $G\theta$.

It is a simple matter, by induction on n, to verify the next theorem.

THEOREM 5.22

If $\theta: (S, \otimes) \to (T, \odot)$ is a morphism, and $s_i \in S$, then
$$(s_1 \otimes s_2 \otimes \cdots \otimes s_n)\theta = s_1\theta \odot s_2\theta \odot \cdots \odot s_n\theta,$$
and
$$(s_1^n)\theta = (s_1\theta)^n, \quad \text{for all } n \in \mathbf{Z}^+.$$

EXAMPLE 5.22

The following are morphisms from $(\mathbf{Z}_8, +)$ to $(\mathbf{Z}_{12}, +)$:

$\lambda:$	$\eta:$	$\theta:$	$\psi:$
$\bar{0} \to \bar{0}$	$\bar{0} \to \bar{0}$	$\bar{0} \to \bar{0}$	$\bar{0} \to \bar{0}$
$\bar{1} \to \bar{0}$	$\bar{1} \to \bar{6}$	$\bar{1} \to \bar{3}$	$\bar{1} \to \bar{9}$
$\bar{2} \to \bar{0}$	$\bar{2} \to \bar{0}$	$\bar{2} \to \bar{6}$	$\bar{2} \to \bar{6}$
$\bar{3} \to \bar{0}$	$\bar{3} \to \bar{6}$	$\bar{3} \to \bar{9}$	$\bar{3} \to \bar{3}$
$\bar{4} \to \bar{0}$	$\bar{4} \to \bar{0}$	$\bar{4} \to \bar{0}$	$\bar{4} \to \bar{0}$
$\bar{5} \to \bar{0}$	$\bar{5} \to \bar{6}$	$\bar{5} \to \bar{3}$	$\bar{5} \to \bar{9}$
$\bar{6} \to \bar{0}$	$\bar{6} \to \bar{0}$	$\bar{6} \to \bar{6}$	$\bar{6} \to \bar{6}$
$\bar{7} \to \bar{0}$	$\bar{7} \to \bar{6}$	$\bar{7} \to \bar{9}$	$\bar{7} \to \bar{3}$

In each map the image set is a proper subgroup of $\mathbf{Z}_{12}$, with the exception of λ, since $\{\bar{0}\}$ is not a *proper* subgroup. Because $\mathbf{Z}_8 = \langle \bar{1} \rangle$, the image of $\bar{1}$ determines all the other images.

We verify that θ is a morphism as follows:
$$\bar{1}\theta = \bar{3} \Rightarrow \bar{k}\theta = \bar{3} + \bar{3} + \cdots + \bar{3} = \overline{3k}.$$
So $\quad (\bar{k} + \bar{m})\theta = \overline{3(k+m)} = \overline{3k + 3m} = \overline{3k} + \overline{3m} = \bar{k}\theta + \bar{m}\theta.$

Therefore, θ is a morphism.

Maps $\lambda, \eta, \theta,$ and ψ are the only morphisms from $\mathbf{Z}_8$ to $\mathbf{Z}_{12}$. Note that in each map, the order of the image of $\bar{1}$ divides the order of $\bar{1}$. That is, in $\mathbf{Z}_8$, $|\bar{1}| = 8$, while in $\mathbf{Z}_{12}$, $|\bar{0}| = 1$, $|\bar{6}| = 2$, $|\bar{3}| = |\bar{9}| = 4$. ◻

THEOREM 5.23

Let $\theta: G \to H$ be a morphism from group G to group H, and let $g \in G$, with $|g| = k$ and $|g\theta| = m$.

(a) $m \mid k$.
(b) If θ is an isomorphism, then $m = k$.

Proof

(a) $|g| = k \Rightarrow (g\theta)^k = (g^k)\theta = e\theta$, where e is the identity of G. But $|g\theta| = m$, and so $m \mid k$, by Theorem 4.5.

(b) $|g\theta| = m \Rightarrow e\theta = (g\theta)^m = (g^m)\theta$. But θ is 1–1, and so $e = g^m$. Hence, $k \mid m$, by Theorem 4.5. Therefore, $m = k$. ◻

The fact that subgroups correspond under a morphism is dealt with in the next theorem.

5.5 Group Morphisms

THEOREM 5.24

Let $\theta: G \to G'$ be a morphism from group $(G, \otimes)$ to group $(G', \odot)$. Then:

(a) $H \leq G \Rightarrow H\theta \leq G'$.
(b) $H \triangleleft G \Rightarrow H\theta \triangleleft G\theta$.
(c) $L \leq G' \Rightarrow L\theta^{-1} \leq G$.
(d) $L \triangleleft G\theta \Rightarrow L\theta^{-1} \triangleleft G$.

Proof

(a) If $H \leq G$, then H is a group. Then $H\theta$ is a group, by Theorem 5.21. But $H\theta$ is also a subset of G', and so $H\theta \leq G'$.

(b) $\quad H \triangleleft G \Rightarrow g \otimes h \otimes g^{-1} \in H, \quad$ for all $h \in H, \quad g \in G$

$\Rightarrow (g \otimes h \otimes g^{-1})\theta \in H\theta$

$\Rightarrow g\theta \odot h\theta \odot g^{-1}\theta \in H\theta$

$\Rightarrow g\theta \odot h\theta \odot (g\theta)^{-1} \in H\theta, \quad$ for all $h\theta \in H\theta, \quad g\theta \in G\theta$

$\Rightarrow H\theta \triangleleft G\theta$.

(c) If $L \leq G'$, then, by definition, $L\theta^{-1} = \{x : x \in G, x\theta \in L\}$. We need to verify that $L\theta^{-1}$ satisfies the three conditions of Theorem 2.8.

(1) $e\theta \in L$, since L is a group, and so $e \in L\theta^{-1}$.
(2) Closure:

$a, b \in L\theta^{-1} \Rightarrow a\theta, b\theta \in L \Rightarrow a\theta \odot b\theta \in L \Rightarrow (a \otimes b)\theta \in L \Rightarrow a \otimes b \in L\theta^{-1}$.

(3) Inverses:

$a \in L\theta^{-1} \Rightarrow a\theta \in L \Rightarrow (a\theta)^{-1} \in L \Rightarrow a^{-1}\theta \in L \Rightarrow a^{-1} \in L\theta^{-1}$.

Therefore, if L is a subgroup of G', then $L\theta^{-1}$ is a subgroup of G.

(d) $\quad L \triangleleft G\theta \Rightarrow g\theta \odot x\theta \odot (g\theta)^{-1} \in L, \quad$ for all $g \in G, \quad x \in L\theta^{-1}$

$\Rightarrow (g \otimes x \otimes g^{-1})\theta \in L \Rightarrow g \otimes x \otimes g^{-1} \in L\theta^{-1}$

$\Rightarrow L\theta^{-1} \triangleleft G. \quad \square$

According to Theorem 5.24, not only do subgroups correspond under a morphism, but *normal* subgroups correspond to *normal* subgroups.

If the morphism is from any algebra $(S, \otimes)$ to *itself*, the morphism is called an *endomorphism of* S, and if the endomorphism is bijective, it is called an *automorphism of* S. Thus, an automorphism of S is a permutation of S that preserves the binary operations.

EXAMPLE 5.23

The inclusion diagram for the four types of morphisms is shown in Figure 5–3.

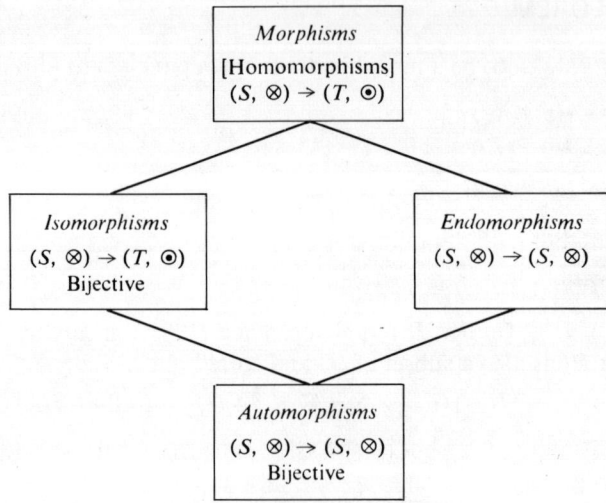

Figure 5–3

Equivalently, an automorphism can be defined as an isomorphism from algebra S to algebra S.

EXAMPLE 5.24

There are four endomorphisms of $(\mathbf{Z}_4, +)$:

$$
\begin{array}{llll}
\theta: \begin{array}{l}\bar{0} \to \bar{0}\\ \bar{1} \to \bar{0}\\ \bar{2} \to \bar{0}\\ \bar{3} \to \bar{0}\end{array} &
\psi: \begin{array}{l}\bar{0} \to \bar{0}\\ \bar{1} \to \bar{2}\\ \bar{2} \to \bar{0}\\ \bar{3} \to \bar{2}\end{array} &
\lambda: \begin{array}{l}\bar{0} \to \bar{0}\\ \bar{1} \to \bar{1}\\ \bar{2} \to \bar{2}\\ \bar{3} \to \bar{3}\end{array} &
\eta: \begin{array}{l}\bar{0} \to \bar{0}\\ \bar{1} \to \bar{3}\\ \bar{2} \to \bar{2}\\ \bar{3} \to \bar{1}\end{array}
\end{array}
$$

In each morphism the image is, of course, a subgroup of $\mathbf{Z}_4$. Also, λ and η are automorphisms, λ being the identity automorphism. Under the operation of composition of maps, $\circ$, $\{\lambda, \eta\}$ is a group. □

THEOREM 5.25

If A_G is the set of all automorphisms of a group $(G, \otimes)$, and if $\circ$ is the operation of composition of maps, then $(A_G, \circ)$ is a group.

Proof

We have already proved that the set S_G of all permutations of a set G is a group under composition of maps (Theorem 2.9). Since A_G is a subset of S_G, we need only to show that:

(1) ε is in A_G,
(2) θ, ψ in $A_G \Rightarrow \theta \circ \psi$ is in A_G,
(3) θ in $A_G \Rightarrow \theta^{-1}$ in A_G.

(1) $\varepsilon: g \to g\varepsilon = g$ is obviously an automorphism, and so ε is in A_G.

(2) Suppose that $\theta, \psi \in A_G$. Let a and b be any elements of G. Then

$$(ab)(\theta \circ \psi) = \lfloor (ab)\theta \rfloor \psi = [(a\theta)(b\theta)]\psi = [(a\theta)\psi][(b\theta)\psi] = [a(\theta \circ \psi)][b(\theta \circ \psi)],$$

and so $\theta \circ \psi$ is in A_G.

(3) The inverse of a permutation is a permutation. Suppose that $\theta: G \to G$ is any automorphism. So

$$\theta: a \to a' = a\theta,$$
$$b \to b' = b\theta.$$

Now $\quad a'\theta^{-1} = a, b'\theta^{-1} = b \Rightarrow ab = (a'\theta^{-1})(b'\theta^{-1}),$

and $\quad (ab)\theta = (a\theta)(b\theta) = a'b' \Rightarrow ab = (a'b')\theta^{-1}.$

Therefore, $(a'\theta^{-1})(b'\theta^{-1}) = (a'b')\theta^{-1}$, which means that permutation θ^{-1} preserves the operation, and so θ^{-1} is in A_G.

In conclusion, $A_G < S_G$, which makes A_G a group. □

EXAMPLE 5.25

Let $(G, \otimes) = (\mathbf{Z}_{10}, +)$. Since $G = \langle \bar{1} \rangle$, each automorphism is completely determined by specifying the image of $\bar{1}$. The positive integers that are less than 10 and prime to 10 are: 1, 3, 7, and 9. Hence $\bar{1}, \bar{3}, \bar{7},$ and $\bar{9}$ are each of order 10, and can be the only images of $\bar{1}$ under an automorphism. If we define $\bar{1}\varepsilon = \bar{1}, \bar{1}\theta_1 = \bar{3}, \bar{1}\theta_2 = \bar{7}, \bar{1}\theta_3 = \bar{9}$, we can easily verify that each of these choices for an image of $\bar{1}$ leads to an automorphism, and so $A_G = \{\varepsilon, \theta_1, \theta_2, \theta_3\}$.

To identify A_G, we begin by finding the order of θ_1, as follows:

$$\bar{1}\theta_1^2 = (\bar{1}\theta_1)\theta_1 = \bar{3}\theta_1 = \bar{9}, \quad \text{and so } \theta_1^2 = \theta_3.$$

Therefore, θ_1 must be of order 4, and so A_G is the cyclic group of order 4. □

EXAMPLE 5.26

Let us find the automorphism group of D_4. (See Appendix D–8.4.) Now

$$D_4 = \langle a, b \rangle = \{a^i b^j : a^4 = b^2 = e, ba = a^3 b\}.$$

By Theorem 5.23, elements that correspond must have the same order. Also an automorphism is completely determined when the images of the generators are specified. It can be shown that the following maps of the generators yield automorphisms:

$$\begin{aligned}
\varepsilon: &\ a \to a & \theta_1: &\ a \to a & \theta_2: &\ a \to a & \theta_3: &\ a \to a \\
&\ b \to b & &\ b \to ab & &\ b \to a^2 b & &\ b \to a^3 b \\
\theta_4: &\ a \to a^3 & \theta_5: &\ a \to a^3 & \theta_6: &\ a \to a^3 & \theta_7: &\ a \to a^3 \\
&\ b \to b & &\ b \to ab & &\ b \to a^2 b & &\ b \to a^3 b
\end{aligned}$$

The group table of A_{D_4} is shown in Figure 5–4.

	ε	θ_1	θ_2	θ_3	θ_4	θ_5	θ_6	θ_7
ε	ε	θ_1	θ_2	θ_3	θ_4	θ_5	θ_6	θ_7
θ_1	θ_1	θ_2	θ_3	ε	θ_7	θ_4	θ_5	θ_6
θ_2	θ_2	θ_3	ε	θ_1	θ_6	θ_7	θ_4	θ_5
θ_3	θ_3	ε	θ_1	θ_2	θ_5	θ_6	θ_7	θ_4
θ_4	θ_4	θ_5	θ_6	θ_7	ε	θ_1	θ_2	θ_3
θ_5	θ_5	θ_7	θ_4	θ_6	θ_3	ε	θ_1	θ_2
θ_6	θ_6	θ_4	θ_7	θ_5	θ_2	θ_3	ε	θ_1
θ_7	θ_7	θ_6	θ_5	θ_4	θ_1	θ_2	θ_3	ε

Figure 5–4

Now, $A_{D_4} = \langle \theta_1, \theta_4 \rangle = \{\theta_1^i \theta_4^j : \theta_1^4 = \theta_4^2 = \varepsilon, \theta_4 \theta_1 = \theta_1^3 \theta_4\}$.
Therefore, A_{D_4} is isomorphic to D_4. It is not usual for the automorphism group of a group G to be structurally the same as G, or even to be the same size as G, but it so happened in this example. □

There is a certain subgroup of A_G that is of interest. We examine this next.

THEOREM 5.26

(a) If g is a fixed element of a group $(G, \otimes)$, then the map $i_g : G \to G$, such that for all $x \in G$, $xi_g = gxg^{-1}$, is an automorphism of G, called an *inner automorphism* of G.

(b) The set I_G of inner automorphisms of G is a normal subgroup of the group A_G, of all automorphisms of G.

Proof

(a) It is readily verified that i_g is a bijective map, and that $(xy)i_g = (xi_g)(yi_g)$, for all $x, y \in G$.

(b) One first proves that $i_a \circ i_b = i_{ba}$, as follows:
$x(i_a \circ i_b) = (xi_a)i_b = (axa^{-1})i_b = b(axa^{-1})b^{-1} = (ba)x(ba)^{-1} = xi_{ba}$.
So I_G is closed under composition of maps, i_e is the identity map, where e is the identity of G, and $i_a^{-1} = i_{a^{-1}}$. Therefore, $I_G \leq A_G$.

To prove that I_G is a *normal* subgroup of A_G, let ψ be any automorphism of G, and let i_a be any inner automorphism. We must show that $\psi i_a \psi^{-1}$ is an inner automorphism. For each $x \in G$:

$$x(\psi i_a \psi^{-1}) = [(x\psi)i_a]\psi^{-1} = [a(x\psi)a^{-1}]\psi^{-1}$$
$$= [a\psi^{-1}][(x\psi)\psi^{-1}][a^{-1}\psi^{-1}] \qquad \text{(since } \psi^{-1} \text{ is a morphism)}$$
$$= [a\psi^{-1}]x[a\psi^{-1}]^{-1} = bxb^{-1} = xi_b,$$

if $b = a\psi^{-1}$. So $\psi i_a \psi^{-1}$ is the inner automorphism i_b, and thus $I_G \triangleleft A_G$. □

5.5 Group Morphisms

Note that if G is abelian, then the only inner automorphism is the identity map.

EXAMPLE 5.27

We find the inner automorphism group of D_4. (See Example 5.26.) For each $g \in D_4$ the inner automorphism is completely determined by the images of the set $\{a, b\}$ of generators of D_4. So we need only to compute gag^{-1} and gbg^{-1}, for each $g \in D_4$.

Letting g range over all 8 elements, we obtain:

i_e: $a \to a,$
       $b \to b.$

i_a: $a \to a,$
       $b \to aba^{-1} = aba^3 = a^2b.$

i_{a^2}: $a \to a,$
           $b \to a^2ba^2 = b.$

i_{a^3}: $a \to a,$
           $b \to a^3ba = a^2b.$

i_b: $a \to bab = a^3,$
       $b \to bbb = b.$

i_{ab}: $a \to (ab)a(ab) = a^3,$
          $b \to (ab)b(ab) = a^2b.$

i_{a^2b}: $a \to (a^2b)a(a^2b) = a^3,$
            $b \to (a^2b)b(a^2b) = b.$

i_{a^3b}: $a \to (a^3b)a(a^3b) = a^3,$
            $b \to (a^3b)b(a^3b) = a^2b.$

So $i_e = i_{a^2} = \varepsilon,$ $i_a = i_{a^3} = \theta_2,$ $i_b = i_{a^2b} = \theta_4,$ $i_{ab} = i_{a^3b} = \theta_6,$

using the notation of Example 5.26, and $I_G = \{\varepsilon, \theta_2, \theta_4, \theta_6\}$. Since each of these θ_i is of order 2, I_G is isomorphic to the Klein four-group. □

5.5 Problems

1 Determine all images of $\bar{1}$ under each morphism, $(\mathbf{Z}_n, +) \to (\mathbf{Z}_m, +)$.
   (a) $\mathbf{Z}_6 \to \mathbf{Z}_{12}$
   (b) $\mathbf{Z}_8 \to \mathbf{Z}_4$
   (c) $\mathbf{Z}_8 \to \mathbf{Z}_5$
   (d) $\mathbf{Z}_{12} \to \mathbf{Z}_{18}$
   (e) $\mathbf{Z}_{12} \to \mathbf{Z}_{13}$

2 Determine the range of each endomorphism of $(\mathbf{Z}_6, +)$. How many of these endomorphisms are automorphisms?

3 For each algebra, determine whether the given map is an endomorphism.
   (a) $(\mathbf{Z}, +),$ $n \to 2n$
   (b) $(\mathbf{Z}, +),$ $n \to n + 1$
   (c) $(\mathbf{R}^+, \cdot),$ $r \to r$
   (d) $(\mathbf{Q}, +),$ $r \to |r|$
   (e) $(\mathbf{Q}, +),$ $r \to -r$
   (f) $(\mathbf{Z}, \cdot),$ $k \to r$, where $k \equiv r \pmod{10}$, and $0 \le r \le 9$.

4 Prove Theorem 5.22.

5 Verify: $\theta: \bar{1} \to \bar{a}$, is an automorphism of $(\mathbf{Z}_n, +)$ if and only if $(a, n) = 1$.

6 Prove: For each integer n there is an endomorphism θ of $(\mathbf{Z}, +)$ such that $1\theta = n$.

7   Let $(G, \otimes)$ be any group, and let a be a fixed element of G. If $\theta: (\mathbf{Z}, +) \to (G, \otimes)$ is the map such that $n\theta = a^n$, prove that θ is a morphism.

8   Find all morphisms of $(\mathbf{Z}, +)$ into $(\mathbf{Z}_8, +)$. Which ones are surjective?

9   Prove: If $\theta: (G, \otimes) \to (H, *)$ and $\psi: (H, *) \to (K, \odot)$ are morphisms, then the composite map, $\theta \circ \psi: (G, \otimes) \to (K, \odot)$ is a morphism.

10 Give the image of $\bar{1}$ under each automorphism of $(\mathbf{Z}_n, +)$ if:
    (a) $n = 12$,   (b) $n = 13$,   (c) $n = 14$.
    Also, identify each automorphism group.

11 Show that the automorphism group of K_4 is isomorphic to S_3.

12 (a) Determine the number of automorphisms of the quaternion group, $\mathbf{Q}_8$, and the order of each automorphism in the automorphism group.
    (b) Is $A_{\mathbf{Q}_8}$ isomorphic to S_4?

13 Verify that if $G = S_3$, then $A_G = I_G \cong S_3$.

14 Complete the proof of Theorem 5.26.

15 Prove: If $N \triangleleft G$, and i_g is an inner automorphism of G, then $Ni_g = N$.

16 Let H be a subgroup of G, and define K to be a *conjugate* of H if and only if $K = gHg^{-1}$, for some $g \in G$. Prove:
    (a) K is a subgroup of G. (Hint: See Theorems 5.24 and 5.26.)
    (b) The relation, "is a conjugate of", is an equivalence relation on the collection of subgroups of a group G, and thus partitions the subgroups into equivalence classes.
    (c) Subgroup H is a normal subgroup of G if and only if the only conjugate of H is H itself.

17 Determine which subgroups of D_4 are conjugate subgroups. (See Problem 16.)

18 (a) Show that the inner automorphism group of Q_8 is isomorphic to K_4.
    (b) Determine which subgroups of Q_8 are conjugate subgroups. (See Problem 16.)

19 Find the inner automorphism group of A_4, and partition the subgroups of A_4 into sets of conjugate subgroups. (See Problem 16.)

20 Prove: If G is a nonabelian group, then its inner automorphism group contains at least two elements.

5.6 Direct Products

A group $(H, \odot)$ is said to be *embedded* in group $(G, \otimes)$ if G contains a subgroup H' that is isomorphic to H. G is said to be an *extension* of H. (See Figure 5–5.) In symbols:

$$(H, \odot) \cong (H', \otimes) \leq (G, \otimes).$$

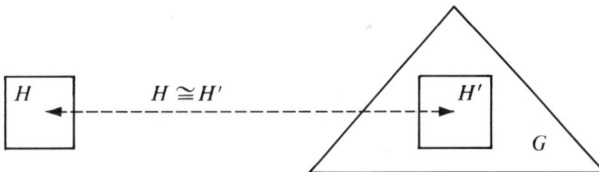

Figure 5–5

If $H' < G$, then G is a "larger" group that contains a subgroup just like H. In this section we examine one kind of embedding.

By definition, the *Cartesian product* of sets $G_1, G_2, \ldots, G_m$ is:

$$G_1 \times G_2 \times \cdots \times G_m = \{(g_1, g_2, \ldots, g_m) : g_i \in G_i, i = 1, 2, \ldots, m\}.$$

This is a generalization of the definition of Section 1.3.

THEOREM 5.27

Let $(G_i, \otimes_i)$ be a group, for $i = 1, 2, \ldots, m$. If $G = G_1 \times G_2 \times \cdots \times G_m$, and if operation $\otimes$ on G is defined by:

$$(g_1, \ldots, g_m) \otimes (h_1, \ldots, h_m) = (g_1 \otimes_1 h_1, \ldots, g_m \otimes_m h_m),$$

then $(G, \otimes)$ is a group. If each G_i is abelian, then G is abelian.

Proof

We'll let you fill in the details. Note that if the identity of G_i is e_i, then the identity of G is: $e = (e_1, \ldots, e_m)$. Also, $(g_1, g_2, \ldots, g_m)^{-1} = (g_1^{-1}, g_2^{-1}, \ldots, g_m^{-1})$. □

Group G of Theorem 5.27 is called the *direct product* of groups $G_1, G_2, \ldots, G_m$. The next result informs us that we have succeeded in finding a G in which each G_i is embedded.

THEOREM 5.28

If $(G_j, \otimes_j)$ is a group with identity e_j, for $j = 1, 2, \ldots, m$, $(G, \otimes)$ is the direct product of groups $G_1, G_2, \ldots, G_m$, and

$$G'_i = \{(e_1, e_2, \ldots, e_{i-1}, a_i, e_{i+1}, \ldots, e_m) : a_i \in G_i\},$$

then

(a) $(G_i, \otimes_i) \cong (G_i', \otimes)$, and
(b) $(G_i', \otimes) \triangleleft (G, \otimes)$.

Proof

(a) It is a simple matter to verify that the map, $\theta: G_i \to G_i'$, is an isomorphism, and hence that $(G_i', \otimes)$ is a group, where θ is defined as follows:

$$a_i \to (e_1, e_2, \ldots, e_{i-1}, a_i, e_{i+1}, \ldots, e_m) = a_i\theta, \quad \text{for all } a_i \in G_i.$$

(b) If g is any element of G, where $g = (g_1, \ldots, g_i, \ldots, g_m)$, and h is any element of G_i', where $h = (e_1, \ldots, e_{i-1}, a_i, \ldots, e_m)$, then $g \otimes h \otimes g^{-1} = (e_1, \ldots, e_{i-1}, g_i a_i g_i^{-1}, e_{i+1}, \ldots, e_m) \in G_i'$, and so G_i' is a normal subgroup of G. □

EXAMPLE 5.28

Consider these groups:

$$K_4 = \{e, a, b, c\}, \quad Z_2 = \{\bar{0}, \bar{1}\}.$$

The elements of $K_4 \times Z_2$, and the order of each element are:

Element:	$(e, \bar{0})$	$(a, \bar{0})$	$(b, \bar{0})$	$(c, \bar{0})$
Order:	1	2	2	2
Element:	$(e, \bar{1})$	$(a, \bar{1})$	$(b, \bar{1})$	$(c, \bar{1})$
Order:	2	2	2	2

So $K_4 \times Z_2$ is an abelian group of order 8. Also,

$$K_4 \cong \{(e, \bar{0}), (a, \bar{0}), (b, \bar{0}), (c, \bar{0})\}, \quad Z_2 \cong \{(e, \bar{0}), (e, \bar{1})\}. \quad \square$$

EXAMPLE 5.29

Another abelian group of order 8 is $Z_4 \times Z_2$. Its elements are as follows:

Element:	$(\bar{0}, \bar{0})$	$(\bar{1}, \bar{0})$	$(\bar{2}, \bar{0})$	$(\bar{3}, \bar{0})$
Order:	1	4	2	4
Element:	$(\bar{0}, \bar{1})$	$(\bar{1}, \bar{1})$	$(\bar{2}, \bar{1})$	$(\bar{3}, \bar{1})$
Order:	2	4	2	4

By examining the orders of the elements, we can see that $K_4 \times Z_2$ is structurally different than $Z_4 \times Z_2$. Observe that $Z_4 \times Z_2$ contains more than one subgroup isomorphic to Z_4, and a subgroup isomorphic to K_4. Can you find these subgroups? □

EXAMPLE 5.30

We ask this question: Are Z_{12} and $Z_4 \times Z_3$ structurally different groups? Consider an element (a, b) of $Z_4 \times Z_3$ such that $|a| = 4$ and $|b| = 3$. If

$(a, b)^k = (\bar{0}, \bar{0})$, then $a^k = \bar{0}$ and $b^k = \bar{0}$. Hence, $4 \mid k$ and $3 \mid k$, by Theorem 5.4. So k must be a multiple of both 4 and 3. But $[4, 3] = 12$, and so (a, b) is of order 12. This means that $\mathbf{Z}_4 \times \mathbf{Z}_3$ is the cyclic group of order 12. Therefore, $\mathbf{Z}_4 \times \mathbf{Z}_3 \cong \mathbf{Z}_{12}$. □

THEOREM 5.29

If m and n are relatively prime integers, then $\mathbf{Z}_m \times \mathbf{Z}_n \cong \mathbf{Z}_{mn}$.

The proof parallels the special case that was discussed in Example 5.30.

THEOREM 5.30

If G, H, and K are any groups, then
(a) $G \times H \cong H \times G$, and
(b) $G \times H \times K \cong (G \times H) \times K$.

The proof is left as an exercise. Note that the elements of $G \times H \times K$ are ordered triplets, (g, h, k), while the elements of $(G \times H) \times K$ are ordered pairs, $((g, h), k)$, with the first component also being an ordered pair.

EXAMPLE 5.31

Suppose we want to find all possible nonisomorphic groups of order 40 that are direct products of cyclic groups. We do this by factoring 40 in various ways. Since $40 = 2 \cdot 2 \cdot 2 \cdot 5$, then $\mathbf{Z}_2 \times \mathbf{Z}_2 \times \mathbf{Z}_2 \times \mathbf{Z}_5$ is a group of order 40. However, $\mathbf{Z}_2 \times \mathbf{Z}_5 \cong \mathbf{Z}_{10}$, so it is clear that $\mathbf{Z}_2 \times \mathbf{Z}_2 \times \mathbf{Z}_2 \times \mathbf{Z}_5 \cong \mathbf{Z}_2 \times \mathbf{Z}_2 \times \mathbf{Z}_{10}$. To avoid listing two groups that are structurally the same, we agree to replace $\mathbf{Z}_m \times \mathbf{Z}_n$ by $\mathbf{Z}_{mn}$ whenever $(m, n) = 1$. So the nonisomorphic groups are:

$$\mathbf{Z}_2 \times \mathbf{Z}_2 \times \mathbf{Z}_{10}, \quad \mathbf{Z}_2 \times \mathbf{Z}_{20}, \quad \mathbf{Z}_{40}.$$

Any other factoring of 40 will yield a direct product that is isomorphic to one of these three groups. We know that no two of these three groups are isomorphic, since the second group contains an element of order 20, and there is no element of order higher than 10 in the first group, while the third group has an element of order 40, and there is no element of order higher than 20 in the second group. The simplest set of generators of each group is as follows:

$$\mathbf{Z}_2 \times \mathbf{Z}_2 \times \mathbf{Z}_{10} = \langle (\bar{1}, \bar{0}, \bar{0}), (\bar{0}, \bar{1}, \bar{0}), (\bar{0}, \bar{0}, \bar{1}) \rangle,$$
$$\mathbf{Z}_2 \times \mathbf{Z}_{20} = \langle (\bar{1}, \bar{0}), (\bar{0}, \bar{1}) \rangle,$$
$$\mathbf{Z}_{40} = \langle \bar{1} \rangle. \quad □$$

5.6 Problems

1. Let $G = D_3 \times \mathbf{Z}_2$.
   (a) List the elements of G, and give the order of each element.
   (b) Find all subgroups of G, and indicate which one is isomorphic to D_3.

2. (a) List all subgroups of $K_4 \times \mathbf{Z}_2$. Verify that each subgroup of order 4 is isomorphic to K_4. (See Example 5.28.)
   (b) List all subgroups of $\mathbf{Z}_4 \times \mathbf{Z}_2$. Show that there are two nonisomorphic subgroups of order 4. (See Example 5.29.)

3. Prove Theorem 5.27.

4. Complete the proof of Theorem 5.28.

5. (a) List the elements in $\mathbf{Z}_2 \times \mathbf{Z}_6$, and find the order of each element.
   (b) List the elements in $\mathbf{Z}_3 \times \mathbf{Z}_2 \times \mathbf{Z}_2$, and find the order of each element.
   (c) Are groups $\mathbf{Z}_2 \times \mathbf{Z}_6$ and $\mathbf{Z}_3 \times \mathbf{Z}_2 \times \mathbf{Z}_2$ isomorphic? Justify your answer.

6. Let G and H be these cyclic groups: $G = \{e_1, a, a^2\}$, $H = \{e_2, b\}$. List the elements of $G \times H$, and find the order of each element.

7. Prove Theorem 5.29.

8. Prove Theorem 5.30.

9. Comment on the truth of the statement: A group is an extension of each of its proper subgroups.

10. Give an element of greatest order in each of the following groups.
    (a) $\mathbf{Z}_4 \times \mathbf{Z}_6$ (b) $\mathbf{Z}_3 \times \mathbf{Z}_8$ (c) $\mathbf{Z}_2 \times \mathbf{Z}_{12}$ (d) $\mathbf{Z}_2 \times \mathbf{Z}_2 \times \mathbf{Z}_6$

11. List each group G that is a direct product of cyclic groups $(\mathbf{Z}_n, +)$, if the order of G is:
    (a) 8,   (b) 9,   (c) 10,   (d) 12,   (e) 16.

12. Referring to Appendix D, set up the isomorphic maps that show the following:
    (a) $\mathbf{Z}_4 \times \mathbf{Z}_2 \cong \langle a, b \rangle = \{a^i b^j : |a| = 4, |b| = 2, ba = ab\}$.
    (b) $\mathbf{Z}_2 \times \mathbf{Z}_2 \times \mathbf{Z}_2 \cong \langle a, b, c \rangle = \{a^i b^j c^k : |a| = |b| = |c| = 2, ab = ba, ac = ca, bc = cb\}$.

13. Given that $G = G_1 \times G_2 \times \cdots \times G_n$, where $G = \langle a_i \rangle$, prove:
    $$G = \langle (a_1, e_2, \ldots, e_n), (e_1, a_2, \ldots, e_n), \ldots, (e_1, e_2, \ldots, a_n) \rangle.$$

14. Set up a map, $\theta: K_4 \times \mathbf{Z}_2 \to \mathbf{Z}_2 \times \mathbf{Z}_2 \times \mathbf{Z}_2$, that is an isomorphism. (Write the elements of $K_4 \times \mathbf{Z}_2$ as in Example 5.29.)

15. (a) If $G = G_1 \times G_2 \times \cdots \times G_m$, show that the map,
    $$\pi_i: (g_1, g_2, \ldots, g_m) \to g_i,$$
    is a morphism of G onto G_i, for each i.

(b) The map of part (a) is called the *projection* of G onto G_i. If **R** is the additive group of real numbers, and $G = \mathbf{R} \times \mathbf{R}$, interpret the two projections geometrically.

16   Prove: If $G = H \times K$, there is an $H' \leq G$ so that $G/H' \cong K$, and $H \cong H'$.

17   (a) Prove: If H and K are normal subgroups of group $(G, \odot)$, $H \cap K = \{e\}$, and $G = H \odot K$, then $G \cong H \times K$. (Note: To say that $G = H \odot K$ means that if g is any element of G, then $g = h \odot k$, where $h \in H$, and $k \in K$.)

(b) Show that the Klein four-group satisfies the conditions of (a).

(c) Give an example which shows that the statement of part (a) is false if either of the subgroups is not normal.

5.7 Group Representations

A few decades ago algebraists made a big point of distinguishing between *abstract* groups and *special* groups. They would say that a group is an *abstract group* if the elements are just symbols, with no particular interpretation attached to those symbols. (All abstract groups of order ≤ 12 are listed in Appendix D.) If the group refers to integers, to real numbers, to matrices, to rotations or reflections in the plane, etc., then the group was called a *special group*. If an abstract group G is isomorphic to a special group G', then G' is called a *representation* of G. For instance, if

$$G = \{a^i b^j : |a| = 3, |b| = 2, ba = a^2 b\},$$

then G is an abstract group, while the group of rigid motions of an equilateral triangle is a representation of G. We don't insist on using the words, *abstract* and *special*, but merely define G' to be a *representation* of G if G is isomorphic to G'. In this section we look at some convenient and interesting ways of representing groups.

If you are asked to set up an isomorphic map from a group G to a group G', you have certain guidelines to follow. These three have already been established:

(1) The identity of G must map to the identity of G'.

(2) If a maps to b, then a and b must have the same order.

(3) If a maps to b, then a^{-1} must map to b^{-1}.

The next theorem should come as no surprise. We'll let you prove it.

THEOREM 5.31

Isomorphism, $\cong$, is an equivalence relation on a given set of groups.

By Theorem 5.31, $\cong$ partitions the collection of groups into equivalence classes. In studying the structure of groups, we need to look at only one representative of each equivalence class. The next theorem informs us that, structurally, $(\mathbf{Z}, +)$ and $(\mathbf{Z}_n, +)$ are the only *cyclic* groups.

THEOREM 5.32

Let $(G, \otimes)$ be a cyclic group.
(a) If G is infinite, then $(G, \otimes) \cong (\mathbf{Z}, +)$.
(b) If $|G| = n$, then $(G, \otimes) \cong (\mathbf{Z}_n, +)$.

Proof

(a) There is an element $a \in G$ such that $G = \langle a \rangle$. Define $\theta: \mathbf{Z} \to G$ to be the map such that $k\theta = a^k$, for all $k \in \mathbf{Z}$. According to Theorem 5.6(b), $a^r = a^s$ only if $r = s$, so θ is injective. Also, θ is clearly surjective, and

$$(k + t)\theta = a^{k+t} = a^k \otimes a^t = (k\theta) \otimes (t\theta).$$

Therefore, θ is an isomorphism.

(b) Group $G = \langle a \rangle$ for some $a \in G$, where a is of order n. Define $\psi: \mathbf{Z}_n \to G$ to be the map such that $\bar{k}\psi = a^k$, for all $k \in \mathbf{Z}$. By Theorem 5.4, $a^r = a^s \Rightarrow r \equiv s \pmod{n} \Rightarrow \bar{r} = \bar{s}$, and so ψ is injective. Also, ψ is clearly surjective, and

$$(\bar{k} + \bar{t})\psi = \overline{(k + t)}\psi = a^{k+t} = a^k \otimes a^t = (\bar{k}\psi) \otimes (\bar{t}\psi).$$

Therefore, ψ is an isomorphism. □

According to the proof of the preceding theorem, if G is a cyclic group of order n, then a morphism $\psi: \mathbf{Z}_n \to G$ is an isomorphism whenever the image of $\bar{1}$ is a generator g of G.

EXAMPLE 5.32

If $G = \langle b \rangle$, and $|b| = 10$, then G can be generated by any one of these elements: b, b^3, b^7, b^9. So $\theta_i: \mathbf{Z}_n \to G$ is an isomorphism, for $i = 1, 2, 3, 4$, if $\bar{1}\theta_1 = b, \bar{1}\theta_2 = b^3, \bar{1}\theta_3 = b^7$, and $\bar{1}\theta_4 = b^9$.

Similarly, if G is an infinite cyclic group, then $(\mathbf{Z}, +)$ is isomorphic to G, and the image of 1 can be any generator of G. But 1 and -1 are the only generators of $\mathbf{Z}$, so G can have but two different generators. Therefore, only two distinct isomorphic maps are possible. □

Of great interest to mathematicians is the fact that the cyclic groups are the "building blocks" with which *all* abelian groups are constructed, since it can be proved that *every* abelian group has a representation as a direct product of cyclic groups. We state, without proof, the powerful theorem that enables us to find all finite abelian groups. We do not have time to develop the machinery to prove the theorem in this course.

THEOREM 5.33

If G is an abelian group of order n, then there is a unique set of integers, $\{n_1, n_2, \ldots, n_s\}$, such that

$$G \cong \mathbf{Z}_{n_1} \times \mathbf{Z}_{n_2} \times \cdots \times \mathbf{Z}_{n_s},$$

where $n = n_1 \cdot n_2 \cdot \cdots \cdot n_s$, and $n_i | n_{i+1}$, for $i = 1, 2, \ldots, s - 1$.

EXAMPLE 5.33

The three ways in which 40 can be factored so that the factors satisfy the conditions of Theorem 5.33 are:

$$(2)(2)(10), \quad (2)(20), \quad (40).$$

Hence, every abelian group of order 40 is isomorphic to one and only one of the three distinct groups:

$$\mathbf{Z}_2 \times \mathbf{Z}_2 \times \mathbf{Z}_{10}, \quad \mathbf{Z}_2 \times \mathbf{Z}_{20}, \quad \mathbf{Z}_{40}.$$

Group $\mathbf{Z}_{10} \times \mathbf{Z}_4$, for instance, is isomorphic to one of these groups. We use Theorems 5.29 and 5.30 to write:

$$\mathbf{Z}_{10} \times \mathbf{Z}_4 \cong (\mathbf{Z}_2 \times \mathbf{Z}_5) \times \mathbf{Z}_4 \cong \mathbf{Z}_2 \times (\mathbf{Z}_5 \times \mathbf{Z}_4) \cong \mathbf{Z}_2 \times \mathbf{Z}_{20}. \quad \square$$

The fact that every abelian group is isomorphic to a direct product of cyclic groups enables us to prove easily a kind of converse, for abelian groups, to the Lagrange Theorem (Theorem 5.14).

THEOREM 5.34

If G is an abelian group of order n, and if $k | n$, then G has a subgroup of order k.

Proof

By Theorem 5.33, $G \cong \mathbf{Z}_{n_1} \times \mathbf{Z}_{n_2} \times \cdots \times \mathbf{Z}_{n_s}$, where $n = n_1 n_2 \cdots n_s$. If $k | n$, then k can be factored as follows:

$$k = k_1 k_2 \cdots k_s, \quad \text{with } k_i | n_i, \quad \text{for } i = 1, 2, \ldots, s.$$

By Theorem 5.10, $\mathbf{Z}_{n_i}$ has a subgroup $\langle \bar{a}_i \rangle$ or order k_i, $i = 1, 2, \ldots, s$. (If $k_i = 1$, then $\bar{a}_i = \bar{0}$.) Let

$$H' = \langle \bar{a}_1 \rangle \times \langle \bar{a}_2 \rangle \times \cdots \times \langle \bar{a}_s \rangle.$$

So H' is a subgroup of order k of the direct product. Under the isomorphism, G has a subgroup H that is isomorphic to H'. Therefore, H is of order k. $\quad \square$

In Section 5.9 we construct another proof of Theorem 5.34, using only theorems that we have proved.

Theorem 5.33 gives us a scheme for determining all *abelian* groups of a given order, but tells us nothing about the non-abelian groups. It is not an easy task to determine all non-isomorphic groups of a given order n, even for a relatively small n.

EXAMPLE 5.34

In Table 5–2 we list the number of distinct groups of order n, for small values of n. From the work we have done already, you will be able to identify some of these groups.

Table 5–2

Number of Groups of Order n, for $1 \leq n \leq 30$														
Order of Group	1	2	3	4	5	6	7	8	9	10	11	12	13	14	15
No. of Groups	1	1	1	2	1	2	1	5	2	2	1	5	1	2	1
Order of Group	16	17	18	19	20	21	22	23	24	25	26	27	28	29	30
No. of Groups	14	1	5	1	5	1	2	1	15	2	2	5	4	1	4

Obviously, the two groups of order 4 are $\mathbf{Z}_4$ and $\mathbf{Z}_2 \times \mathbf{Z}_2$ (which is isomorphic to K_4), and the two groups of order 6 are $\mathbf{Z}_6$ and S_3. □

We have seen that we can construct every possible finite abelian group by taking direct products of the groups $(\mathbf{Z}_m, +)$, for $m \in \mathbf{Z}^+$. We shall now look at a remarkable theorem, known as Cayley's Theorem, which tells us that every group of order n, be it abelian or non-abelian, can be represented by a subgroup of the symmetric group, S_n. Furthermore, every infinite group can also be represented by a group of permutations. Structurally speaking, this means that the only groups in existence are groups of permutations! Before we state and prove the theorem, we give an example of what we are talking about.

EXAMPLE 5.35

It is desired to find a permutation group to which $(G, \otimes)$ is isomorphic, if $G = \{e, a, b, r, s, t\}$, where G has the following group table.

$\otimes$	e	a	b	r	s	t
e	e	a	b	r	s	t
a	a	b	e	t	r	s
b	b	e	a	s	t	r
r	r	s	t	e	a	b
s	s	t	r	b	e	a
t	t	r	s	a	b	e

We know that S_6 consists of the 720 permutations of the set G. (The choice of the six symbols to use is immaterial in determining the elements of S_6.) If g is an element of G, we define the map,

$$\psi_g: \quad G \to G,$$

as follows:
$$\psi_g = \begin{pmatrix} e & a & b & r & s & t \\ eg & ag & bg & rg & sg & tg \end{pmatrix}.$$

Note that the second line of the expression for ψ_g is obtained by multiplying each element of G by g. This second line would then correspond to the column in the group table that is headed by g. For example,

$$\psi_a = \begin{pmatrix} e & a & b & r & s & t \\ a & b & e & s & t & r \end{pmatrix}, \quad \psi_r = \begin{pmatrix} e & a & b & r & s & t \\ r & t & s & e & b & a \end{pmatrix},$$

$$\psi_t = \begin{pmatrix} e & a & b & r & s & t \\ t & s & r & b & a & e \end{pmatrix}.$$

If $H = \{\psi_e, \psi_a, \psi_b, \psi_r, \psi_s, \psi_t\}$, then H is a set of six permutations, and it turns out that G is isomorphic to H. The isomorphic map $\theta: G \to H$, is a rather logical one:

$$\theta: \quad g \to \psi_g, \quad \text{for all } g \in G.$$

If you multiply the permutations, ψ_a and ψ_r, you will find that $\psi_a \circ \psi_r = \psi_t$. But $ar = t$, and so $\psi_a \circ \psi_r = \psi_{ar}$. Thus,

$$(a\theta)(r\theta) = \psi_a \circ \psi_r = \psi_{ar} = (ar)\theta.$$

Of course, to prove that the bijective map θ is an isomorphism, one would need to show that $(x\theta)(y\theta) = (xy)\theta$, for all $x, y \in G$. $\square$

THEOREM 5.35 *(Cayley)*

If $(G, \otimes)$ is any group, then group G is isomorphic to a group of permutations of the set G.

Proof

If $a \in G$, define a map, ψ_a, on G as follows:

$$\psi_a: \quad x \to x\psi_a = xa, \quad \text{for all } x \in G.$$

Let $H = \{\psi_a : a \in G\}$. We subdivide our proof into three parts. We shall prove that:

(1) ψ_a is a permutation, for every $a \in G$.
(2) $\psi_a \circ \psi_b = \psi_{ab}$, for all $a, b \in G$.
(3) The map, $\theta: G \to H$, is an isomorphism, if $a\theta = \psi_a$.

(1) If y is any element in G, then it is the image of ya^{-1} under ψ_a, since $(ya^{-1})\psi_a = (ya^{-1})a = y$.

So ψ_a is onto. Also, $x\psi_a = v\psi_a \Rightarrow xa = va \Rightarrow x = v$, so ψ_a is 1–1. Therefore, ψ_a is a permutation of G.

(2) If x is any element in G, then

$$x(\psi_a\psi_b) = (x\psi_a)\psi_b = (xa)\psi_b = x(ab) = x\psi_{ab}.$$

Therefore, $\psi_a\psi_b = \psi_{ab}$.

(3) Define map θ from G to H to be the following:

$$\theta: \quad a \to a\theta = \psi_a, \quad \text{for all } a \in G.$$

This is obviously a surjective map. Now suppose that $\psi_a = \psi_b$. Then $xa = xb$, for all $x \in G$, and so $a = b$. Hence, θ is bijective. Further,

$$(ab)\theta = \psi_{ab} = \psi_a\psi_b = (a\theta)(b\theta).$$

We conclude that θ is an isomorphism. By Theorem 5.21, H is a group. Therefore, G is isomorphic to a permutation group. □

Cayley's Theorem is also called the *Representation Theorem for Groups*, and group $H = \{\psi_a : a \in G\}$ is called the *right regular representation* of G. The *left regular representation* is discussed in Problem 5.7–21.

We have shown, by Cayley's Theorem, that every group of order n is embedded in the symmetric group S_n. This fact, although interesting, does not simplify the difficult task of determining all groups of order n.

5.7 Problems

1. Let cyclic group $G = \{e, a, a^2, \ldots, a^{11}\} = \langle a \rangle$, with $|a| = 12$.
   (a) List all possible isomorphic maps of G onto $\mathbf{Z}_{12}$.
   (b) Explain this statement: Each of the isomorphisms of part (a) corresponds to an automorphism of $\mathbf{Z}_{12}$.

2. Prove Theorem 5.31.

3. List, as direct products of the form of Theorem 5.33, all abelian groups of these orders: 8, 16, 18, 24, 25, 72, 93.

4. It has been calculated that there are 267 groups of order 64, and 238 groups of order 160. How many of these groups are abelian?

5. For each abelian group of order 40, give the generators of a subgroup of each of these orders: 2, 4, 5, 8, 10, 20.

6. Using the data in Example 5.34, make a table in which you break down the number of groups of order n into the number of abelian and the number of nonabelian groups, for $1 \le n \le 30$.

7. Give the five distinct groups of order 8, and list the orders of the elements in each group.

8 Construct the inclusion diagram of subgroups of each group of order 8. (See Problem 7.)

9 (a) Give the elements in each of the two groups of order 9, and list the order of each element.

(b) Construct the inclusion diagrams of the subgroups of each group of order 9.

10 You have become familiar with these five groups of order 12: Z_{12}, A_4, D_6, $Z_2 \times Z_6$, C_{12}. (See Appendix D.)

(a) For each group, list the 12 numbers that are the orders of its elements.

(b) Use the results of part (a) as verification that no two of these groups are isomorphic.

11 Which group of Problem 10 is a representation of $D_3 \times Z_2$?

12 How many distinct equivalence classes (with respect to isomorphism of groups) contain groups of orders less than or equal to 20?

13 List the elements in the right regular representation of each of the following groups, writing each permutation as a product of disjoint cycles.

(a) Z_4 (b) K_4 (c) S_3 (d) D_4

14 Using the isomorphism of Cayley's Theorem, determine the permutation to which 3 corresponds if:

(a) $3 \in (Z, +)$.   (b) $3 \in (Q^*, \cdot)$.

15 Given groups $(Z, +)$ and $(M_2(Z), +)$, show that $M_2(Z)$ is isomorphic to $Z \times Z \times Z \times Z$.

16 (a) Prove: If p and q are distinct positive primes, then there is exactly one abelian group of order pq.

(b) Look at the table of Example 5.34, and make a conjecture as to the number of nonabelian groups of order pq if: (1) p and q are distinct odd primes, (2) $p = 2$, and q is an odd prime.

17 Given that there are s abelian groups of order m and t abelian groups of order n, where $(m, n) = 1$; verify that there are st abelian groups of order mn. (Hint: Consider direct product groups.)

18 Find the number of abelian groups of order n, if n is 200, 400, 450, 600, 1800, 2325. (Hint: Factor each number, and apply Prob. 17.)

19 Prove:

(a) S_n is embedded in S_{n+1}, for all $n \in Z^+$.

(b) If $|G| = n$, then group G is embedded in S_k, for all $k \geq n$.

20 Prove:

(a) If G is abelian and of order 6, then G is the cyclic group.

(b) There is only one nonabelian group of order 6.

21 Let G be a group. If $a \in G$, define a map, λ_a, on G as follows:

$$\lambda_a: \quad x \to ax = x\lambda_a, \quad \text{for all } x \in G.$$

(a) Prove that λ_a is a permutation of G, for each $a \in G$.

(b) Prove that, under composition of maps, $\lambda_a \lambda_b = \lambda_{ba}$.

(c) If $L = \{\lambda_a : a \in G\}$, let $\theta: G \to L$ be the map such that $a\theta = \lambda_{a^{-1}}$. Prove that θ is an isomorphism. (Note: This is the analogue of Theorem 5.32. Group L is called the *left regular representation* of G.)

22 Comment on this statement: If G is abelian, then the left regular representation of G equals the right regular representation of G. (See Problem 21.)

23 List the elements in the left regular representation L of D_4, writing each permutation as a product of disjoint cycles. Set up the isomorphic map, $D_4 \to L$, indicating the image of each element of D_4. (See Problem 21.)

24 (a) What was Cayley's contribution to the Women's Liberation Movement of his day?

(b) Cayley is credited with inventing some mathematics that is known to most undergraduate mathematics majors today. Name one of his inventions. (See Appendix E-14.)

5.8 The Kernel and the Fundamental Morphism Theorem

Let $\theta: G \to G'$ be a morphism of group G into group G'. If

$$K = \{x : x \in G, x\theta = e'\},$$

where e' is the identity of G', then K is called the *kernel of the morphism* θ. That is, the kernel is the set of elements of G that map to the identity of G'.

EXAMPLE 5.36

If $\theta: (\mathbf{Z}_8, +) \to (\mathbf{Z}_{12}, +)$ is the morphism such that $\bar{1}\theta = \bar{6}$, then

$$\begin{array}{llll} \bar{0} \to \bar{0}, & \bar{1} \to \bar{6}, & \bar{2} \to \bar{0}, & \bar{3} \to \bar{6}, \\ \bar{4} \to \bar{0}, & \bar{5} \to \bar{6}, & \bar{6} \to \bar{0}, & \bar{7} \to \bar{6}, \end{array}$$

and so the kernel $K = \{\bar{0}, \bar{2}, \bar{4}, \bar{6}\}$. □

THEOREM 5.36

Let $\theta: G \to G'$ be a morphism from group G to group G', with kernel K. Then:

(a) K is a normal subgroup of G, and

(b) θ is 1-1 iff $K = \{e\}$, where e is the identity of G.

5.8 The Kernel and the Fundamental Morphism Theorem

Proof

(a) Let e' be the identity of G'. Then $\{e'\}$ is a normal subgroup of $G\theta$, since $xe'x^{-1} = e'$, for all $x \in G\theta$. But $\{e'\}\theta^{-1} = K$. By Theorem 5.24(d), K is a normal subgroup of G.

(b) Given that $K = \{e\}$; then

$$x\theta = y\theta \Rightarrow (x\theta)(y\theta)^{-1} = e' \Rightarrow (xy^{-1})\theta = e' \Rightarrow xy^{-1} = e \Rightarrow x = y.$$

Therefore, θ is 1–1.

Conversely, suppose that θ is 1–1. We know that $e\theta = e'$. So

$$a \in K \Rightarrow a\theta = e' \Rightarrow a\theta = e\theta \Rightarrow a = e \Rightarrow K = \{e\}. \quad \square$$

We have learned that if N is a normal subgroup of G, then G/N is a group. In a sense, G/N is a "smaller version" of G. We define the *natural map* of G onto G/N to be the map such that $g \to gN$, for all $g \in G$. That is, each element of G maps to the coset in which it lies.

EXAMPLE 5.37

Let $\lambda: (\mathbf{Z}, +) \to (\mathbf{Z}_5, +)$ be the map such that $m\lambda = \bar{m}$. Kernel $K = \{0, \pm 5, \pm 10, \ldots\} = 5\mathbf{Z}$. Now $5\mathbf{Z} \triangleleft \mathbf{Z}$, and $\mathbf{Z}/5\mathbf{Z} = \mathbf{Z}_5$. So λ is the natural map of each integer to the coset of $5\mathbf{Z}$ in which that integer lies (or of each integer to the equivalence class modulo n in which the integer lies). $\quad \square$

THEOREM 5.37

If N is a normal subgroup of G, then the natural map,

$$\lambda: \quad G \to G/N,$$

is a morphism with kernel N.

Proof

For $g, h \in G$, we have the natural map:

$$\lambda: \quad g \to gN,$$
$$h \to hN,$$
$$gh \to (gh)N.$$

But $(g\lambda)(h\lambda) = (gN)(hN) = (gh)N = (gh)\lambda$. So map λ is a morphism. Now N is the identity of the factor group G/N, and $g\lambda = N$ iff $g \in N$. Therefore, N is the kernel of λ. $\quad \square$

Note the two roles that N plays in Theorem 5.37. N is an *element* of group G/N, but N is a *subset* of group G. N is the identity element of one group, but it is a normal subgroup of the other, and $N\lambda = \{N\}$.

EXAMPLE 5.38

N is a normal subgroup of $(\mathbf{Z}_{12}, +)$, where $N = \{\bar{0}, \bar{4}, \bar{8}\}$. So

$$\mathbf{Z}_{12}/N = \{\{\bar{0}, \bar{4}, \bar{8}\}, \{\bar{1}, \bar{5}, \bar{9}\}, \{\bar{2}, \bar{6}, \overline{10}\}, \{\bar{3}, \bar{7}, \overline{11}\}\}$$
$$= \{N, \bar{1} + N, \bar{2} + N, \bar{3} + N\},$$

and the natural map of $\mathbf{Z}_{12}$ onto $\mathbf{Z}_{12}/N$ is as follows:

$\bar{0} \to \{\bar{0}, \bar{4}, \bar{8}\} = N,$ $\quad\bar{6} \to \{\bar{2}, \bar{6}, \overline{10}\} = \bar{2} + N,$
$\bar{1} \to \{\bar{1}, \bar{5}, \bar{9}\} = \bar{1} + N,$ $\quad\bar{7} \to \{\bar{3}, \bar{7}, \overline{11}\} = \bar{3} + N,$
$\bar{2} \to \{\bar{2}, \bar{6}, \overline{10}\} = \bar{2} + N,$ $\quad\bar{8} \to \{\bar{0}, \bar{4}, \bar{8}\} = N,$
$\bar{3} \to \{\bar{3}, \bar{7}, \overline{11}\} = \bar{3} + N,$ $\quad\bar{9} \to \{\bar{1}, \bar{5}, \bar{9}\} = \bar{1} + N,$
$\bar{4} \to \{\bar{0}, \bar{4}, \bar{8}\} = N,$ $\quad\overline{10} \to \{\bar{2}, \bar{6}, \overline{10}\} = \bar{2} + N,$
$\bar{5} \to \{\bar{1}, \bar{5}, \bar{9}\} = \bar{1} + N,$ $\quad\overline{11} \to \{\bar{3}, \bar{7}, \overline{11}\} = \bar{3} + N.$ □

EXAMPLE 5.39

In Appendix D–9.2, the group table of $\mathbf{Z}_3 \times \mathbf{Z}_3 = G$ is given. If $H = \{e, a, a^2\}$, then it is apparent from the group table that H is a normal subgroup of G. (Why?) Also, $Hb = \{b, ab, a^2b\}$, $Hb^2 = \{b^2, ab^2, a^2b^2\}$, and $G/H = \{H, Hb, Hb^2\}$. The group table of G is subdivided into nine 3-by-3 blocks, and each block contains only the elements in precisely one of the cosets of H. It is as follows:

	H	Hb	Hb^2
H	H	Hb	Hb^2
Hb	Hb	Hb^2	H
Hb^2	Hb^2	H	Hb

The group table of G/H, given in this example, is a condensation, or "mini-version", of the group table of G. The natural map $\lambda: G \to G/H$ is a three–to–one map. □

We use the natural map to prove the next theorem, whose usefulness will become apparent in Chapter 6.

THEOREM 5.38

Let N be a normal subgroup of G. Then N is maximal in G if and only if G/N is a simple group.

Proof

We shall prove the equivalent statement:

N is not maximal in $G \Leftrightarrow G/N$ is not simple.

(1) Assume that N is not maximal in G. Then there is a subgroup M such

that $N \triangleleft M \triangleleft G$, where $M \neq N$, and $M \neq G$. Consider the natural morphism,

$$\lambda: \quad G \to G/N,$$
$$M \to M\lambda,$$
$$N \to N\lambda = \{N\}.$$

We know that $M\lambda \triangleleft G/N$, by Theorem 5.24(b). We must show that $M\lambda$ is a *proper* subgroup of G/N. There is an $h \in M$ such that $h \notin N$. Then $hN \neq N$, and $hN \in M\lambda$. Hence, $\{N\} < M\lambda$. There is a $g \in G$ such that $g \notin M$. Now if $gN \in M\lambda$, then $g \in M$. So $gN \notin M\lambda$, and $M\lambda < G/N$. Therefore, $M\lambda$ is a proper normal subgroup of G/N, and thus G/N is not simple.

(2) Assume that G/N is not simple. Then G/N has a proper normal subgroup H', and $H'\lambda^{-1}$ is a normal subgroup of G. Consider the map,

$$\lambda: \quad G \to G/N,$$
$$H'\lambda^{-1} \to H',$$
$$N \to \{N\}.$$

We now show that $H'\lambda^{-1}$ is a *proper* subgroup of G. There is a coset $hN \in H'$ such that $hN \neq N$. Hence $h \notin N$, but $h \in H'\lambda^{-1}$. So $N < H'\lambda^{-1}$. There is a coset $gN \in G/N$ such that $gN \notin H'$. Then $g \notin H'\lambda^{-1}$, and so $H'\lambda^{-1} < G$. Therefore, N is not maximal in G. □

EXAMPLE 5.40

Continuing with Example 5.38, we see that N is not a *maximal* normal subgroup of $\mathbf{Z}_{12}$, since $N < M < \mathbf{Z}_{12}$, where $M = \{\overline{0}, \overline{2}, \overline{4}, \overline{6}, \overline{8}, \overline{10}\}$. Now $\{N, \overline{2} + N\}$ is a proper normal subgroup of $\mathbf{Z}_{12}/N$, where $\mathbf{Z}_{12}/N = \{N, \overline{1} + N, \overline{2} + N, \overline{3} + N\}$. Therefore, $\mathbf{Z}_{12}/N$ is not simple. □

Suppose that $\theta: G \to H$ is a morphism of a group G onto a group H. If K is the kernel of θ, then the natural map, $\lambda: G \to G/K$, is a morphism. The interesting fact is that, under these two morphisms, the image groups, H and G/K, turn out to be isomorphic. This relationship is depicted in Figure 5–6.

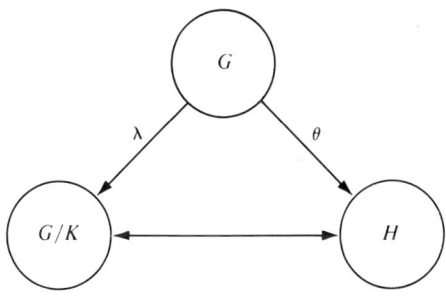

Figure 5–6

THEOREM 5.39 *(Fundamental Theorem on Group Morphisms)*

If $\theta: G \to H$ is a morphism of group G onto group H with kernel K, then $G/K \cong H$.

Proof

We define ψ as follows:

$$\psi: G/K \to H$$

where
$$gK \to (gK)\psi = g\theta.$$

By hypothesis, θ is a morphism. We must verify that ψ is an isomorphism. Let $a, b \in G$, and let e' be the identity of H. Then:

$$aK = bK \Leftrightarrow a \in bK \Leftrightarrow a = bk, \quad \text{for some } k \in K$$
$$\Leftrightarrow b^{-1}a = k \Leftrightarrow (b^{-1}a)\theta = e'$$
$$\Leftrightarrow (b\theta)^{-1}(a\theta) = e' \Leftrightarrow a\theta = b\theta.$$

So ψ is a map, and is 1–1. Map ψ is obviously surjective, since each element $g\theta$ of H has coset gK as its antecedent. To show that the operations are preserved:

$$[(aK)(bK)]\psi = [(ab)K]\psi = (ab)\theta = (a\theta)(b\theta) = [(aK)\psi][(bK)\psi].$$

Therefore, ψ is an isomorphism. □

EXAMPLE 5.41

We illustrate Theorem 5.39 by letting $(G, +) = (\mathbf{R} \times \mathbf{R}, +)$, $(H, +) = (\mathbf{R}, +)$, and defining θ as follows:

$$\theta: \quad G \to H,$$
$$(x, y) \to x.$$

Then kernel $K = \{(0, y): y \in \mathbf{R}\}$, and $G/K = \{(a, b) + K: a, b \in \mathbf{R}\}$. We see that

$$(a, b) + K = \{(a, b + y): y \in \mathbf{R}\} = \{(a, r): r \in \mathbf{R}\} = (a, 0) + K.$$

Geometrically (Figure 5–7): G is the plane, H is the x–axis, K is the y–axis, coset $(a, 0) + K$ is the vertical line through $(a, 0)$, G/K is the set of all vertical lines, λ maps each point P in the plane to the vertical line through P, and ψ maps each vertical line to its point of intersection with the x–axis.

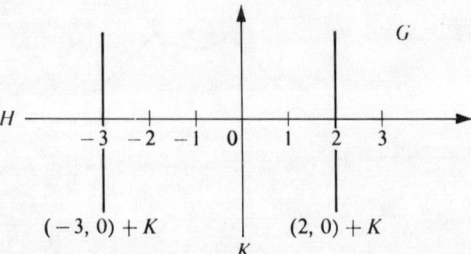

Figure 5–7

□

5.8 Problems

1 (a) Find a morphism, $\theta: 4\mathbf{Z} \to \mathbf{Z}_5$, whose kernel is $20\mathbf{Z}$.
   (b) Find a morphism, $\theta: \mathbf{Z}_{12} \to \mathbf{Z}_4$, whose kernel is $\langle \bar{4} \rangle$.

2 For each morphism of Problem 5.5–1, determine: (1) the kernel K, (2) the factor group, $\mathbf{Z}_n/K$, and (3) whether $\mathbf{Z}_n/K$ is simple.

3 Let $(G, \cdot) = (\{1, -1\}, \cdot)$, where the operation is multiplication of real numbers. Define $\theta: S_n \to G$ to be the map such that $a\theta = 1$ if a is an even permutation, and $a\theta = -1$ if a is an odd permutation.
   (a) Verify that θ is a morphism.
   (b) Describe the kernel K of θ, and the elements of G/K.

4 (a) Show that $\theta: (\mathbf{R} \times \mathbf{R}, +) \to (\mathbf{R}, +)$ is a morphism if $(a, b)\theta = b$.
   (b) Use the map θ of part (a) to illustrate Theorem 5.39. Describe the various maps and groups as was done in Example 5.41.

5 Verify that $\theta: (\mathbf{R} \times \mathbf{R}, +) \to (\mathbf{R}, +)$ is a morphism, if $(a, b)\theta = 2a - b$. Describe, both algebraically and geometrically, the kernel K of θ, a coset $(a, b) + K$, the factor group, $(\mathbf{R} \times \mathbf{R})/K$, and the isomorphism, $\psi: (\mathbf{R} \times \mathbf{R})/K \to \mathbf{R}$.

6 Let $G = \mathbf{Z}_2 \times \mathbf{Z}_4$, and let H and L be these subgroups:

$$H = \langle (\bar{0}, \bar{2}) \rangle, \qquad L = \langle (\bar{1}, \bar{0}) \rangle.$$

   (a) Give the elements in G/H and G/L, and identify the groups.
   (b) Give the natural morphism of G onto G/H, and of G onto G/L. Find any other morphisms that exist from G onto G/H, and from G onto G/L.
   (c) Suppose that M and N are normal subgroups of a group G, and that $M \cong N$. Does it follow that $G/M \cong G/N$?

7 Suppose you have the chain of subgroups:

$$E < K_4 < L < G,$$

with $E = \{e\}$, $G/L \cong \mathbf{Z}_7$, and $L/K_4 \cong \mathbf{Z}_6$.
   (a) Give the orders of L and G.
   (b) Can any more normal subgroups be inserted in the chain? If so, insert them, and give their orders.

8 Given: $H \triangleleft G, K \triangleleft G, H \cap K = \{e\}, HK = \{hk : h \in H, k \in K\}$. Prove:
   (a) $hk = kh$ if $h \in H, k \in K$, and
   (b) $HK \triangleleft G$.

9 Assume that $G = HK, H \leq G, K \triangleleft G, H \cap K = \{e\}$. Make use of Theorem 5.39 to prove that $G/K \cong H$.

10 Given: Group $(\mathbf{Z}, +)$ and subgroup $(n\mathbf{Z}, +)$.
Prove: $\mathbf{Z}/n\mathbf{Z}$ is a simple group $\Leftrightarrow n$ is a prime.

5.9 A Sylow Theorem

In Section 5.3 we raised the question of the existence of proper subgroups of a given finite group, and we stated this result: If G is a group of order n, and p is a prime such that $p^k \mid n$, where $k \geq 1$, then G contains a subgroup of order p^k. We now take time to prove this. But first we must develop some preliminary theorems that will enable us to arrive, more or less painlessly, at the main result.

THEOREM 5.40 (Cauchy's Theorem for Abelian Groups)

If G is an abelian group of order mp, where p is a prime, then G contains an element of order p.

Proof

We prove this by induction on m. The theorem is obviously true for $m = 1$. Suppose now that the theorem holds for every group of order less than mp. Let $|G| = mp$, $m > 1$, and let H be a proper subgroup of G. (See Problem 5.3–17.) If $|H| = n$, then $1 < n < mp$. If $p \mid n$, then H contains an element of order p (by the induction hypothesis), and hence, so does G.

If $p \nmid n$, then $|G/H| = mp/n = kp$, with $1 \leq k < m$. Since G/H is an abelian group of order less than mp, G/H contains an element aH of order p, for some $a \in G$. That is, $H = (aH)^p = a^p H$. Thus $a^p \in H$, and so $e = (a^p)^n = (a^n)^p$, which means that $|a^n| = 1$ or p. However,

$$|a^n| = 1 \Rightarrow a^n = e \Rightarrow (aH)^n = a^n H = H \Rightarrow p \mid n,$$

since $|aH| = p$. But $p \nmid n$. Therefore, $|a^n| = p$, and so G contains an element of order p. □

We need a few new definitions. Let a and b be elements of a group G. To say that *a is conjugate to b* means that $gag^{-1} = b$, for some $g \in G$. If G is abelian, then $gag^{-1} = a$, for all $g \in G$, and so each element is conjugate to itself only. If $N_a = \{x : x \in G, xax^{-1} = a\}$, the set N_a is called the *normalizer of a*. Obviously, G is the normalizer of the identity e. In any case, N_a is a subgroup of G.

THEOREM 5.41

If a is an element of a group G, then N_a is a subgroup of G.

Theorem 5.41 was proved in Example 2.14. The proof of the next theorem will be left an an exercise.

THEOREM 5.42

The relation of conjugacy of elements of a group G is an equivalence relation on G.

The equivalence classes of G that are determined by conjugacy are called the *conjugacy classes* of G. We shall let C_a represent the conjugacy class in which element a lies. It is apparent that if a is in the center of G, then $C_a = \{a\}$.

5.9 A Sylow Theorem

EXAMPLE 5.42

We obtain the conjugacy classes of A_4 by using the group table of Example 5.12. First, $C_e = \{e\}$. Next, we obtain the conjugates of a as follows:

$$xax^{-1} = \begin{cases} a & \text{if } x = e, a, b, c, \\ b & \text{if } x = g, h, i, j, \\ c & \text{if } x = r, s, t, u. \end{cases}$$

So $C_a = \{a, b, c\}$. Note also that $N_a = \{e, a, b, c\}$, a subgroup of A_4. In similar fashion we find that $C_g = \{g, h, i, j\}$, $C_r = \{r, s, t, u\}$. The conjugacy classes partition A_4 as follows:

$$A_4 = \{e\} \cup \{a, b, c\} \cup \{g, h, i, j\} \cup \{r, s, t, u\}. \quad \square$$

If $C_{g_1}, C_{g_2}, \ldots, C_{g_k}$ are the distinct conjugacy classes of a group G of order n, then the equation, $n = |C_{g_1}| + |C_{g_2}| + \cdots + |C_{g_k}|$, is called the *class equation* of G. From Example 5.42 we see that the class equation of A_4 is: $12 = 1 + 3 + 4 + 4$. It is no mere coincidence that the order of each conjugacy class divides 12.

THEOREM 5.43

Let G be a group of order n. If C_a is a conjugacy class of order m, then $m \mid n$.

Proof

The theorem is obviously true if $m = 1$. Suppose $m > 1$, and let a and b be two distinct elements of C_a. Then

Define
$$N_a = \{x : xax^{-1} = a\} = \{x_1, x_2, \ldots, x_r\}.$$
$$B = \{y : yay^{-1} = b\} = \{y_1, y_2, \ldots, y_s\}.$$

We now show that $r = s$. Define sets T and V as follows:

$$T = y_1^{-1}B = \{y_1^{-1}y_1, y_1^{-1}y_2, \ldots, y_1^{-1}y_s\},$$
$$V = y_1 N_a = \{y_1 x_1, y_1 x_2, \ldots, y_1 x_r\}.$$

Since $(y_1^{-1}y_i)a(y_1^{-1}y_i)^{-1} = y_1^{-1}(y_i a y_i^{-1})y_1 = a$ for $i = 1, 2, \ldots, s$, then $T \subseteq N_a$, and thus $s \leq r$. Furthermore, $(y_1 x_j)a(y_1 x_j)^{-1} = y_1(x_j a x_j^{-1})y_1^{-1} = y_1 a y_1^{-1} = b$, for $j = 1, 2, \ldots, r$, so $V \subseteq B$, and $r \leq s$. Thus, $r = s$. Now if $C_a = \{a, b_1, b_2, \ldots, b_{m-1}\}$, and $B_i = \{x : xax^{-1} = b_i\}$, for $i = 1, 2, \ldots, m - 1$, then $r = |N_a| = |B_i|$. But $G = N_a \cup B_1 \cup B_2 \cup \cdots \cup B_{m-1}$, and it is clear that the subsets in this union are pairwise disjoint. Therefore, $n = rm$, and so $m \mid n$. $\square$

In the proof of Theorem 5.43 we saw that $n = rm$, with $|G| = n$, $|N_a| = r$, and $|C_a| = m$. Hence, the following statement is immediate.

THEOREM 5.44

If a is any element of a finite group G, then

$$|C_a| = [G:N_a].$$

We are ready to return to the mainstream of the development.

THEOREM 5.45

Let G be a group of order p^m, where p is a prime, and $m \geq 1$. Then the center C of G contains an element of order p.

Proof

If e is the identity of G, then $C_e = \{e\}$. So the class equation of G can be written:

$$p^m = 1 + |C_{g_1}| + |C_{g_2}| + \cdots + |C_{g_k}|.$$

Now $|C_{g_i}|$ divides p^m, by Theorem 5.43. If p divides $|C_{g_i}|$, for $i = 1, 2, \ldots, k$, then $p \mid 1$, a contradiction. Hence, $|C_{g_j}| = 1$, for at least one $g_j \neq e$, which means that $g_j \in C$. Now $C \leq G$, so $|C| = p^r$, where $1 \leq r \leq m$. By Theorem 5.40, C contains an element of order p. □

THEOREM 5.46

If G is a group of order p^m, where p is a prime, $m \geq 1$, then G contains subgroups of order p^r, for $r = 1, 2, \ldots, m$.

Proof

Proof is by induction on m. The theorem is obviously true if $m = 1$. Let $|G| = p^m$, $m > 1$, and assume the theorem is true for every group of order p^{m-1}.

Let a be an element of order p that is in the center of G (possible, by Theorem 5.45), and let $H = \langle a \rangle$. Then $|H| = p$, and H is a normal subgroup of G. So G/H is a factor group of order p^{m-1}. By the induction hypothesis, G/H contains a subgroup K of order p^s, for $s = 1, 2, \ldots, m - 1$. Then

$$K = \{H, g_2H, g_3H, \ldots, g_{p^s}H\},$$

for some set $\{g_i\}$ of elements of G. If

$$S = H \cup g_2H \cup g_3H \cup \cdots \cup g_{p^s}H,$$

then $|S| = p^{s+1}$, and S is a subgroup of G, by Problem 5.4-24. Therefore, S is a subgroup of order p^r of G, where $r = 2, 3, \ldots, m$. The fact that G contains a subgroup of order p was established in Theorem 5.45. □

We are now in a position to prove the main result, which was first established by the Norwegian algebraist Ludwig Sylow [1832–1918].

THEOREM 5.47 (Sylow)

If G is a group of order kp^m, where p is a prime and $(k, p) = 1$, then G contains a subgroup of order p^m.

Proof

The theorem is clearly true if $k = 1$ or $m = 0$. Assume there is at least one group for which the theorem is false. Let n be the smallest positive integer such that there is a group G with $|G| = n = kp^m$, $(k, p) = 1$, and G has no subgroup of order p^m. We know that $k > 1$ and $m \geq 1$. If H is any proper subgroup of G, then $|H| = h$, where $1 < h < kp^m$, and $kp^m = ht = |H| \cdot [G:H]$. If $p^m | h$, then H has a subgroup of order p^m (since $h < n$), and, therefore, so does G. We are forced to conclude that $p^m \nmid h$, which means that $p | t$. Thus, the index of every proper subgroup of G is divisible by p.

Next, we turn our attention to the class equation of G. If $|C| = w$, where C is the center of G, then $kp^m = w + \Sigma |C_x|$, the summation being over all the distinct conjugacy classes of order greater than 1. But $|C_x|$ is the index of the proper subgroup N_x of G (by Theorem 5.44), and so p divides each $|C_x|$. We conclude that $p | w$. By Theorem 5.40, C contains an element v of order p. If $N = \langle v \rangle$, then N is a normal subgroup of G, and $|G/N| = kp^{m-1} < n$. Hence G/N has a subgroup K of order p^{m-1}. If

and
$$K = \{N, a_2N, a_3N, \ldots, a_{p^{m-1}}N\},$$
$$S = N \cup a_2N \cup a_3N \cup \cdots \cup a_{p^{m-1}}N,$$

then $|S| = p^m$, and S is a subgroup of G (by Problem 5.4-24). Again, this contradicts our choice of G. We conclude that the theorem is true for every finite group. □

If $|G| = kp^m$, with $(k, p) = 1$, then each subgroup of order p^m is called a *Sylow p-subgroup of G*.

As a corollary to Theorems 5.46 and 5.47, we have the following result.

THEOREM 5.48

If G is a group, and $|G| = p_1^{e_1} p_2^{e_2} \cdots p_m^{e_m}$, where the p_i are distinct primes and $e_i \geq 1$, then G contains subgroups of order $p_i^{n_i}$, for $i = 1, 2, \ldots, m$, and $n_i = 1, 2, \ldots, e_i$.

Proof

By Theorem 5.47, G contains a subgroup H_i of order $p_i^{e_i}$, for $i = 1, 2, \ldots, m$, and by Theorem 5.46, H_i contains a subgroup of order $p_i^{n_i}$, for $n_i = 1, 2, \ldots, e_i$. □

We can use Theorem 5.48 to say more about G if G is abelian.

THEOREM 5.49

If G is an abelian group of order n, and $d \mid n$, then G has a subgroup of order d.

Proof

Order n can be written: $n = p_1^{e_1} p_2^{e_2} \cdots p_m^{e_m}$, where the p_i are distinct primes. Since $d \mid n$, then

$$d = p_1^{n_1} p_2^{n_2} \cdots p_m^{n_m}, \quad \text{with } 0 \le n_i \le e_i, \quad \text{for } i = 1, 2, \ldots, m.$$

By Theorem 5.48, G has a subgroup H_i of order $p_i^{n_i}$, for $i = 1, 2, \ldots, m$.

Define $\quad H = H_1 H_2 \cdots H_m = \{h_1 h_2 \cdots h_m : h_i \in H_i\}.$

It is readily verified that H is a subgroup of G. We show that $|H| = d$ by proving the following by induction on m:

$$h_1 h_2 \cdots h_m = k_1 k_2 \cdots k_m \Rightarrow h_i = k_i, \quad \text{for } i = 1, 2, \ldots, m,$$

where $h_i, k_i \in H_i$. This statement is certainly true if $m = 1$. Suppose it holds for $m - 1$. Now

$$h_1 h_2 \cdots h_m = k_1 k_2 \cdots k_m \Rightarrow h_m k_m^{-1} = (k_1 h_1^{-1}) \cdots (k_{m-1} h_{m-1}^{-1})$$
$$\Rightarrow h_m k_m^{-1} \in H_m \cap (H_1 H_2 \cdots H_{m-1})$$
$$\Rightarrow h_m k_m^{-1} = e \quad \text{(why?)}$$
$$\Rightarrow h_m = k_m, h_1 h_2 \cdots h_{m-1} = k_1 k_2 \cdots k_{m-1}$$
$$\Rightarrow h_i = k_i, \quad \text{for } i = 1, 2, \ldots, m.$$

Therefore, $|H| = d$. □

5.9 Problems

1. (a) Find the normalizer of each element of D_6. (See Appendix D–12.3.)
   (b) Partition D_6 into conjugacy classes.
   (c) Give the class equation of D_6.

2. Find all Sylow p-subgroups of each group of order 12. (Use the notation given in Appendix D.)

3. Two subgroups, H and K, are said to be *conjugate subgroups* of G if $K = gHg^{-1}$, for some $g \in G$. (See Problem 5.5–16.) A second Sylow theorem states: For a given prime p, all Sylow p-subgroups of G are conjugate subgroups. Verify this for the Sylow subgroups of D_6, C_{12}, and A_4. (Use results from Problem 2.)

4. A third Sylow theorem states that if m is the number of Sylow p-subgroups in G, then $m \equiv 1 \pmod{p}$. Verify this for the Sylow subgroups of D_6, C_{12}, and A_4. (See Problem 2.)

5.9 A Sylow Theorem

5 Prove Theorem 5.42.

6 Prove: An element x of a group G is in a conjugacy class of order one if and only if x is in the center of G.

7 Prove: If G is an abelian group, and

$$H = H_1 H_2 \cdots H_m = \{h_1 h_2 \cdots h_m : h_i \in H_i\},$$

where each H_i is a subgroup of G, then H is a subgroup of G.

8 Prove: If G is an abelian group of order mn, with $(m, n) = 1$, then there exist subgroups H and K of G such that $G = HK \cong H \times K$.

9 Prove: If $|G| = p^2$, where p is a prime, then group G is abelian.

10 Prove: If $|G| = p^m$, where p is a prime and $m \geq 1$, then G contains a normal subgroup of order p.

11 (a) Prove: If G is a finite abelian group, then G is isomorphic to a direct product of Sylow subgroups of G.

(b) Show, by an example, that the statement of part (a) is false if the word "abelian" is omitted.

6
Rings and Field Theory

Algebra is generous; she often gives more than is asked of her.—Jean Le Rond D'Alembert (1717–1783)

6.1 Subrings and Ideals

Chapter 2 introduced some basic ideas about the structure of rings, and Chapter 3 examined, in some detail, the ring of rational integers and the ring of integers modulo n. Chapter 4 discussed polynomial rings, and the roots of polynomials over various fields. In this chapter we shall develop a few more significant properties of rings, and shall look again at the problem of solving polynomials over a field. We shall frequently parallel, or make use of, the group structure that we developed in Chapter 5.

Let $(S, +, \cdot)$ be a ring. If T is a nonempty subset of S such that $(T, +, \cdot)$ is a ring, then T is called a *subring* of S, and we write $T \leq S$. If $\{z\} \subset T \subset S$, then subring T is called a *proper subring* of S, and we write $T < S$. Appropriately, if S and T are integral domains, then subring T is called a *subdomain* of S, and if S and T are fields, then subring T is called a *subfield* of S.

To determine whether or not a subset of a ring is a subring, it is not necessary to check every ring postulate, since the commutative, associative, and distributive properties of a set must always hold in *any* subset. The next theorem gives the conditions that must be checked.

THEOREM 6.1

Given that $(S, +, \cdot)$ is a ring, T is a nonempty subset of S, T is closed under $+$ and $\cdot$, and $x \in T$ implies that $-x \in T$; then $(T, +, \cdot)$ is a subring of $(S, +, \cdot)$.

We leave the verification of this to you.

It is interesting to note that if S is a ring with unity, then subring T does not necessarily have a unity. Also, if S has unity u, subring T may have unity v, where u and v are not the same element.

EXAMPLE 6.1

The even integers, $2\mathbf{Z}$, form a subring of the ring $\mathbf{Z}$, but $2\mathbf{Z}$ has no unity, while $\mathbf{Z}$ has 1 as unity. □

EXAMPLE 6.2

Let $S = \mathbf{Z}_6$, and $T = \{\bar{0}, \bar{2}, \bar{4}\}$. The addition and multiplication tables of T are as follows:

+	$\bar{0}$	$\bar{2}$	$\bar{4}$
$\bar{0}$	$\bar{0}$	$\bar{2}$	$\bar{4}$
$\bar{2}$	$\bar{2}$	$\bar{4}$	$\bar{0}$
$\bar{4}$	$\bar{4}$	$\bar{0}$	$\bar{2}$

·	$\bar{0}$	$\bar{2}$	$\bar{4}$
$\bar{0}$	$\bar{0}$	$\bar{0}$	$\bar{0}$
$\bar{2}$	$\bar{0}$	$\bar{4}$	$\bar{2}$
$\bar{4}$	$\bar{0}$	$\bar{2}$	$\bar{4}$

Observe that $T < \mathbf{Z}_6$. However, $\bar{1}$ is the unity of $\mathbf{Z}_6$, while $\bar{4}$ is the unity of T. □

Of interest to us is a special kind of subring, known as an *ideal*. A subring N of a ring S is called an *ideal* of S if

$$sN \subset N, \quad \text{and} \quad Ns \subset N, \quad \text{for all } s \in S.$$

This means, of course, that the product of any element n of N and any element s of S is an element of N. This situation is depicted in Figure 6-1.

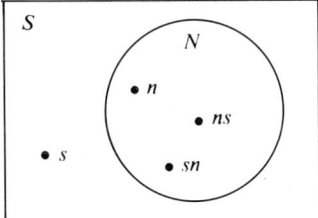

Figure 6-1

EXAMPLE 6.3

Let $(S, +, \cdot)$ be a ring whose addition and multiplication tables are the following:

+	z	a	b	c
z	z	a	b	c
a	a	z	c	b
b	b	c	z	a
c	c	b	a	z

·	z	a	b	c
z	z	z	z	z
a	z	a	b	c
b	z	z	z	z
c	z	a	b	c

By examining these tables we see that $\{z, a\}$ and $\{z, c\}$ are subrings of S, but are not ideals, while $\{z, b\}$ is an ideal of S. In any ring S, the zero subring, $\{z\}$, and S itself are ideals of S. □

EXAMPLE 6.4

We know that $(5\mathbf{Z}, +)$ is a subgroup of $(\mathbf{Z}, +)$. Not only is $5\mathbf{Z}$ closed under multiplication, but the product of any multiple of 5 and *any* integer is again a multiple of 5. Hence $(5\mathbf{Z}, +, \cdot)$ is an ideal of ring $(\mathbf{Z}, +, \cdot)$. Similarly, $n\mathbf{Z}$ is an ideal of ring $\mathbf{Z}$, for any integer n. ☐

THEOREM 6.2

Let A and B be ideals of a ring $(S, +, \cdot)$. If

$$A + B = \{a + b : a \in A, \text{ and } b \in B\},$$

then $A + B$ is an ideal of S. Also, $A \subseteq A + B$, and $B \subseteq A + B$.

The proof of this is left as an exercise. Keep in mind that to prove that a subset N of a ring S is an ideal, you must prove the following: (1) N is not empty, (2) $x \in N \Rightarrow -x \in N$, (3) $x, y \in N \Rightarrow x + y \in N$, and (4) $x \in N, s \in S \Rightarrow xs, sx \in N$.

THEOREM 6.3

Let $(S, +, \cdot)$ be a commutative ring with unity, and let $a \in S$. If $N = \{as : s \in S\}$, then N is an ideal of S.

Proof

(1) $a = au \in N$, so N is not empty.
(2) $-(as) = a(-s) \in N$.
(3) $as, at \in N \Rightarrow as + at = a(s + t) \in N$, since $s + t \in S$.
(4) $as \in N, t \in S \Rightarrow (as)t = a(st) \in N$, since $st \in S$. ☐

The ideal N of Theorem 6.3 is called a *principal ideal*, and is denoted by $\langle a \rangle$. We say that a *generates* the ideal. Note that each ideal that contains element a also contains ideal $\langle a \rangle$.

The following theorem makes use of Theorems 6.2 and 6.3, and is verified by induction on n.

THEOREM 6.4

Let $\{a_1, a_2, \ldots, a_n\} \subseteq S$, where $(S, +, \cdot)$ is a commutative ring with unity. If

$$N = \langle a_1, a_2, \ldots, a_n \rangle = \{a_1 s_1 + a_2 s_2 + \cdots + a_n s_n : s_i \in S\},$$

then N is an ideal of S.

We say that the ideal of Theorem 6.4 is *finitely generated*, and that $\{a_1, a_2, \ldots, a_n\}$ is a *set of generators* of N. We shall be concerned only with finitely generated ideals.

A commutative ring with unity in which every ideal is principal is called a *principal ideal ring* (PIR), and an integral domain in which every ideal is principal is called a *principal ideal domain* (PID).

THEOREM 6.5

The ring $(\mathbf{Z}, +, \cdot)$ is a principal ideal domain.

Proof

By Theorem 5.8, the only additive subgroups of $(\mathbf{Z}, +)$ are $(k\mathbf{Z}, +)$, for $k = 0, 1, 2, \ldots$, with $k\mathbf{Z} = \{kn : n \in \mathbf{Z}\}$. If $(N, +, \cdot)$ is an ideal of $\mathbf{Z}$, then $(N, +)$ is a subgroup of $(\mathbf{Z}, +)$. Therefore, for some k, $N = k\mathbf{Z} = \langle k \rangle$. □

Observe that *every* subring of $\mathbf{Z}$ is a principal ideal.

THEOREM 6.6

Let $F[x]$ be the set of polynomials over a field F. Then the integral domain, $(F[x], +, \cdot)$ is a principal ideal domain.

Proof

Let N be any nonzero ideal in $F[x]$, and let $g(x)$ be a polynomial of lowest degree in N. We shall prove that $N = \langle g(x) \rangle$.

Let $f(x)$ be any polynomial in N. By the Division Theorem, there are polynomials $q(x)$ and $r(x)$ in $F[x]$ such that

$$f(x) = g(x)q(x) + r(x), \quad \text{with } r(x) = 0, \quad \text{or} \quad \deg r(x) < \deg g(x).$$

So $r(x) = f(x) - g(x)q(x)$. But $g(x)q(x) \in N$, since N is an ideal, and hence $r(x) \in N$. Because of the minimality of the degree of $g(x)$, we conclude that $r(x) = 0$. Thus, $f(x) = g(x)q(x)$, and so $f(x)$ is in $\langle g(x) \rangle$. Therefore, $N = \langle g(x) \rangle$, and N is a principal ideal. □

An ideal N of a ring S is called a *proper ideal* if $\{z\} \subset N \subset S$. Ideals $\{z\}$ and S are called *trivial ideals* of S. The next theorem tells us that the subject of ideals is rather dull if the ring happens to be a field.

THEOREM 6.7

(a) Let S be a ring with unity. If N is an ideal containing a unit, then $N = S$.
(b) A field has no proper ideals.

Proof

(a) Let N contain the unit a. Then a^{-1} exists in S, and so $a^{-1}a \in N$, since N is an ideal. Hence $u \in N$. Therefore,

$$x \in S \Rightarrow xu \in N \Rightarrow x \in N \Rightarrow N = S.$$

(b) Let N be an ideal of a field F. If $N \neq \{z\}$, then N contains a unit, and so $N = F$, by part (a). □

EXAMPLE 6.5

We now show that $Z(\sqrt{-5})$ is *not* a principal ideal domain. (According to Theorem 2.27, $Z(\sqrt{n})$ is an integral domain.) Let

$$I = \langle 3, 1 + \sqrt{-5} \rangle = \{3s_1 + (1 + \sqrt{-5})s_2 : s_i \in Z(\sqrt{-5})\}.$$

Assume that I is a principal ideal. Then $I = \langle a + b\sqrt{-5}\rangle$, for some fixed $a, b \in Z$, and $a + b\sqrt{-5}$ must divide both 3 and $1 + \sqrt{-5}$. Now $N(3) = 9$, and $N(1 + \sqrt{-5}) = 6$, so $N(a + b\sqrt{-5})$ must divide both 6 and 9. This means that the norm of $a + b\sqrt{-5}$ is either 1 or 3. But if $N(a + b\sqrt{-5}) = 1$, then $a + b\sqrt{-5}$ is 1 or -1. Now

$$1 \in I \Rightarrow 1 = 3(s + t\sqrt{-5}) + (1 + \sqrt{-5})(x + y\sqrt{-5})$$
$$\Rightarrow 1 = (3s + x - 5y) + (3t + x + y)\sqrt{-5}$$
$$\Rightarrow 1 = 3s + x - 5y, \quad 0 = 3t + x + y$$
$$\Rightarrow 1 = 3(s - t) - 6y \Rightarrow 3\,|\,1, \quad \text{a contradiction.}$$

Hence $1 \notin I$, and so $-1 \notin I$. Thus I contains no element with norm of 1. Further, if $N(a + b\sqrt{-5}) = 3$, then $a^2 + 5b^2 = 3$. But this last equation has no solution in rational integers. Therefore, I is not principal. □

In Chapter 4 we talked about factorization in quadratic domains. You recall that factorization is not unique in $Q[\sqrt{-5}]$. It can be proved that $Q[\sqrt{n}]$ is a unique factorization domain if and only if $Q[\sqrt{n}]$ is a principal ideal domain. Relationships such as this hint at the pattern and beauty of the mathematics upon which we're touching in this all-too-brief survey. The invention of ideals by the mathematician Dedekind came about as a result of his observation that factorization is not unique in some algebraic domains. He developed an algebra of ideals in which factorization *is* unique!

6.1 Problems

1. Find all proper ideals of $(Z_{12}, +, \cdot)$, and give a generator of each ideal. Determine the unity (if it exists) of each ideal.

2. Prove Theorem 6.1.

3. Let T and N be these subsets of ring $M_2(Z)$:

$$T = \left\{\begin{bmatrix} a & 0 \\ b & c \end{bmatrix}\right\}, \quad N = \left\{\begin{bmatrix} 2a & 2b \\ 2c & 2d \end{bmatrix}\right\}, \quad \text{where } a, b, c, d \in Z.$$

   (a) Show that T is a subring of $M_2(Z)$, but is not an ideal.
   (b) Show that N is an ideal of $M_2(Z)$. Is N a principal ideal?

4. If $T = \left\{\begin{bmatrix} a & b \\ -b & a \end{bmatrix} : a, b \in R\right\}$, verify that T is a subring of $M_2(R)$, but is not an ideal of $M_2(R)$.

6.1 Subrings and Ideals

5. Suppose that $(S_i, +, \cdot)$ is a ring, for $i = 1, 2, \ldots, n$, and $T = S_1 \times S_2 \times \cdots \times S_n$. Operations on T are defined as follows:
$$(a_1, a_2, \ldots, a_n) + (b_1, b_2, \ldots, b_n) = (a_1 + b_1, \ldots, a_n + b_n),$$
$$(a_1, a_2, \ldots, a_n) \cdot (b_1, b_2, \ldots, b_n) = (a_1 \cdot b_1, \ldots, a_n \cdot b_n).$$
   (a) Prove that $(T, +, \cdot)$ is a ring.
   (b) If $N = \{(s_1, z, z, \ldots, z) : s_1 \in S_1\}$, prove that N is an ideal of T.
   (c) Can you find other ideals of T?

6. Let integral domain $(\mathbf{Z}, +, \cdot)$ be given.
   (a) Verify that $(\mathbf{Z} \times \mathbf{Z}, +, \cdot)$ is a commutative ring with unity, but is not an integral domain.
   (b) Give three ideals of $\mathbf{Z} \times \mathbf{Z}$.

7. (a) Construct the addition and multiplication tables of ring $\mathbf{Z}_2 \times \mathbf{Z}_2$.
   (b) Give all subrings of ring $\mathbf{Z}_2 \times \mathbf{Z}_2$, and indicate which subrings are ideals.

8. (a) List the eight elements of $\mathbf{Z}_2 \times \mathbf{Z}_2 \times \mathbf{Z}_2$.
   (b) Find all proper subrings and ideals of $\mathbf{Z}_2 \times \mathbf{Z}_2 \times \mathbf{Z}_2$.

9. Find an element, $a + b\sqrt{6}$, in $\mathbf{Z}(\sqrt{6})$ such that ideal
$$\langle 15 + 7\sqrt{6}, -3 - 2\sqrt{6} \rangle = \langle a + b\sqrt{6} \rangle.$$

10. Determine whether ideal $\langle 1 + \sqrt{-5}, 2 - \sqrt{-5} \rangle$ is principal in $\mathbf{Z}(\sqrt{-5})$.

11. If a is a fixed element of a commutative ring S, and if
$$N = \{s : sa = z, s \in S\},$$
verify that N is an ideal of S.

12. Prove Theorem 6.2.

13. Prove Theorem 6.4.

14. Let A and B be subrings of ring S. Prove:
    (a) $A \cap B$ is a subring of S.
    (b) If A and B are ideals of S, then $A \cap B$ is an ideal of S.

15. Was Dedekind considered to be a second-rate mathematician because he taught in a high school? (See Appendix E-17.)

16. A ring S is said to satisfy *the ascending chain condition for ideals* (ACC) if every ascending chain of ideals N_i of S:
$$N_1 \subseteq N_2 \subseteq N_3 \subseteq \cdots,$$
consists of a finite number of ideals. (That is, there exists a k such that $N_r = N_k$ for all $r \geq k$.) A commutative ring with unity that satisfies ACC is called a *Noetherian ring*, so named in honor of an outstanding algebraist, Emmy Noether, who first observed the important role that ACC plays in the study of rings. Verify that $(\mathbf{Z}, +, \cdot)$ is a Noetherian ring.

17 What were two "strikes" against Emmy Noether over which she had no control? (See Appendix E–21.)

6.2 Factor Rings

Let $(N, +, \cdot)$ be an ideal of a ring $(S, +, \cdot)$. This makes $(N, +)$ a normal subgroup of the abelian group $(S, +)$, and so $(S/N, +)$ is a factor group, by Theorem 5.18, with $S/N = \{a + N : a \in S\}$, and

$$(a + N) + (b + N) = (a + b) + N.$$

We want to define an operation of multiplication on the factor set S/N in such a manner that $(S/N, +, \cdot)$ will be a ring. This we now do.

THEOREM 6.8

Let $(N, +, \cdot)$ be an ideal of a ring $(S, +, \cdot)$. If $(S/N, +)$ is the additive factor group of S modulo N, then multiplication of cosets may be defined as follows:

$$(a + N) \cdot (b + N) = (a \cdot b) + N.$$

Further, $(S/N, +, \cdot)$ is a ring. If S is a commutative ring, then S/N is a commutative ring; if S has a unity u, then S/N has unity $u + N$.

Proof

First we must verify that the multiplication defined is a well-defined operation on the additive cosets. To this end we must show that

$$a + N = x + N, \quad b + N = y + N \Rightarrow (ab) + N = (xy) + N.$$

Since the cosets are equivalence classes,

$$a + N = x + N, \quad b + N = y + N \Rightarrow x \in a + N, \quad y \in b + N$$
$$\Rightarrow x = a + n, \quad y = b + m, \quad n, m \in N$$
$$\Rightarrow xy = (a + n)(b + m)$$
$$\Rightarrow xy = ab + (nb + am + nm).$$

It is apparent that $nb, am, nm \in N$, since N is an ideal. Hence $nb + am + nm \in N$, and so $xy \in ab + N$. Therefore, $xy + N = ab + N$.

We know, by Theorem 5.19, that $(S/N, +)$ is an abelian group. We need verify only the associative law of multiplication and the distributive laws. The left distributive law is proved as follows:

$$(a + N)[(b + N) + (c + N)] = (a + N)[(b + c) + N]$$
$$= a(b + c) + N = (ab + ac) + N = (ab + N) + (ac + N)$$
$$= (a + N)(b + N) + (a + N)(c + N).$$

Note that, in the above string of equalities, we used the definition of addition and multiplication of cosets several times, and the fact that the left distributive law holds in ring S. The other properties of S/N are verified in similar fashion. ☐

Ring $(S/N, +, \cdot)$ is called the *factor ring of S modulo N*. A factor ring also goes by the name of *quotient ring*, or *residue class ring*. But since S/N is a factor set, and $(S/N, +)$ is a factor group, it is consistent (and less confusing) to call $(S/N, +, \cdot)$ a *factor ring*. We used this terminology in Section 3.6. Our next example shows that there is no conflict between our use there and our use in the present section.

EXAMPLE 6.6

Since $(n\mathbf{Z}, +, \cdot)$ is an ideal of ring $(\mathbf{Z}, +, \cdot)$, where n is any integer, then $(\mathbf{Z}/n\mathbf{Z}, +, \cdot)$ is a factor ring of $\mathbf{Z}$ modulo $n\mathbf{Z}$.

But
$$\mathbf{Z}/n\mathbf{Z} = \{n\mathbf{Z}, 1 + n\mathbf{Z}, 2 + n\mathbf{Z}, \ldots, (n-1) + n\mathbf{Z}\}$$
$$= \{\bar{0}, \bar{1}, \bar{2}, \ldots, \overline{n-1}\} = \mathbf{Z}_n.$$

Hence the ring of integers modulo n is a factor ring, in the sense of Theorem 6.8. ☐

EXAMPLE 6.7

Let $S = \mathbf{Z}_2 \times \mathbf{Z}_2 \times \mathbf{Z}_4$, with component-wise addition and multiplication of the ordered triples of S. Then $(S, +, \cdot)$ is a commutative ring with unity u, where $u = (\bar{1}, \bar{1}, \bar{1})$. (See Problem 6.1–5.) It can be verified that N is an ideal of S if

$$N = \{(\bar{0}, \bar{0}, \bar{0}), (\bar{0}, \bar{1}, \bar{0}), (\bar{0}, \bar{0}, \bar{2}), (\bar{0}, \bar{1}, \bar{2})\} = z.$$

The additive cosets of N are:

$$(\bar{1}, \bar{1}, \bar{1}) + N = \{(\bar{1}, \bar{1}, \bar{1}), (\bar{1}, \bar{0}, \bar{1}), (\bar{1}, \bar{1}, \bar{3}), (\bar{1}, \bar{0}, \bar{3})\} = a,$$
$$(\bar{1}, \bar{0}, \bar{0}) + N = \{(\bar{1}, \bar{0}, \bar{0}), (\bar{1}, \bar{1}, \bar{0}), (\bar{1}, \bar{0}, \bar{2}), (\bar{1}, \bar{1}, \bar{2})\} = b,$$
$$(\bar{0}, \bar{0}, \bar{1}) + N = \{(\bar{0}, \bar{0}, \bar{1}), (\bar{0}, \bar{1}, \bar{1}), (\bar{0}, \bar{0}, \bar{3}), (\bar{0}, \bar{1}, \bar{3})\} = c.$$

The operation tables of factor ring S/N are as follows:

+	z	a	b	c		·	z	a	b	c
z	z	a	b	c		z	z	z	z	z
a	a	z	c	b		a	z	a	b	c
b	b	c	z	a		b	z	b	b	z
c	c	b	a	z		c	z	c	z	c

Observe that both S and S/N have zero divisors. ☐

An ideal M of a ring S is called a *maximal ideal* of S if $M \subset S$ and there is no ideal N of S such that $M \subset N \subset S$. In other words, if N is an ideal such that $M \subseteq N \subseteq S$, then either $N = M$ or $N = S$.

EXAMPLE 6.8

In ring $\mathbf{Z}$ there exist these chains of ideals:

$$24\mathbf{Z} \subset 8\mathbf{Z} \subset 4\mathbf{Z} \subset 2\mathbf{Z} \subset \mathbf{Z}, \quad 24\mathbf{Z} \subset 12\mathbf{Z} \subset 6\mathbf{Z} \subset 3\mathbf{Z} \subset \mathbf{Z}.$$

Ideals $3\mathbf{Z}$ and $2\mathbf{Z}$ are maximal ideals, while the other ideals listed are not maximal in $\mathbf{Z}$. Note, however, that $24\mathbf{Z}$ is a maximal ideal of $8\mathbf{Z}$.

Also observe that the factor rings of $\mathbf{Z}$ modulo the maximal ideals are fields: $\mathbf{Z}/2\mathbf{Z} = \mathbf{Z}_2$, and $\mathbf{Z}/3\mathbf{Z} = \mathbf{Z}_3$. The factor rings modulo the ideals that are not maximal are not fields. For instance, $\mathbf{Z}/8\mathbf{Z} = \mathbf{Z}_8$. □

The next theorem turns out to be very useful. It gives a method for obtaining a field by starting with a ring.

THEOREM 6.9

Let S be a commutative ring with unity. If M is a maximal ideal of S, then S/M is a field.

Proof

By Theorem 6.8 we know that S/M is a commutative ring with unity. We must prove that each nonzero element of S/M has a multiplicative inverse. Let $a + M$ be any nonzero element of S/M. If $a \in M$, then $a + M = M$. But M is the zero of S/M. So $a \notin M$. Define N to be the sum of two ideals:

$$N = \langle a \rangle + M = \{ra + m : r \in S, m \in M\}.$$

By Theorem 6.3, $\langle a \rangle$ is an ideal, and so N is an ideal of S, by Theorem 6.2. Also, $M \subseteq N \subseteq S$. But M is a maximal ideal of S. Hence either $N = M$ or $N = S$. But $a = ua + z \in N$, and $a \notin M$. We conclude that $N = S$. Hence $u \in N$, and so there is an $r \in S$ and an $m \in M$ such that $u = ra + m$. Then

$$u + M = (ra + m) + M = (ra + M) + (m + M) = ra + M = (r + M)(a + M).$$

Therefore, $(a + M)^{-1} = r + M$, and so S/M is a field. □

We call the field S/M of Theorem 6.9 the *factor field* of S modulo M. The converse of Theorem 6.9 will be proved in Section 6.3. Observe that the theorem and its converse are the analogue of Theorem 5.38. We can make these parallel statements: If N is a maximal normal subgroup of a group G, then G/N has no proper normal subgroups. If M is a maximal ideal of commutative ring with unity S, then S/M has no proper ideals.

The next result is a corollary of Theorem 6.9.

THEOREM 6.10

Let M be a maximal ideal of S, where S is a commutative ring with unity. If $a \cdot b \in M$, then either $a \in M$ or $b \in M$.

6.2 Factor Rings

Proof

$a \cdot b \in M \Rightarrow ab + M = M \Rightarrow (a + M)(b + M) = M$. Now S/M is a field, and M is the zero of S/M. Since a field has no zero divisors, we conclude that either $a + M = M$, or $b + M = M$, which implies that either $a \in M$, or $b \in M$. □

EXAMPLE 6.9

$6\mathbf{Z}$ is an ideal of $\mathbf{Z}$, but is not a maximal ideal of $\mathbf{Z}$. We see that $8 \cdot 3 \in 6\mathbf{Z}$, for instance, but $8 \notin 6\mathbf{Z}$, and $3 \notin 6\mathbf{Z}$. □

EXAMPLE 6.10

In this example we look at a factor field of $\mathbf{Q}[x]$ modulo the ideal M, where $M = \langle x^2 + x + 3 \rangle = \langle m(x) \rangle$.

We know, by Theorem 6.6, that every ideal of $\mathbf{Q}[x]$ is principal. We first show that M is maximal. Assume that $M \subseteq N \subseteq \mathbf{Q}[x]$, for some ideal N; then $N = \langle g(x) \rangle$, for some $g(x)$. This means that $m(x) \in \langle g(x) \rangle$, and hence that $m(x) = g(x) \cdot h(x)$, for some $h(x) \in \mathbf{Q}[x]$. However, $m(x)$ is prime over $\mathbf{Q}$, and so $g(x)$ is either an associate of $m(x)$ or is a unit. That is, either $g(x) = a \cdot (x^2 + x + 3)$ or $g(x) = b$, where $a, b \in \mathbf{Q}^*$. Now

and
$$g(x) = a(x^2 + x + 3) \Rightarrow N = M,$$
$$g(x) = b \Rightarrow N = \mathbf{Q}[x].$$

Therefore, M is a maximal ideal, and so $\mathbf{Q}[x]/M$ is a field.

What do the elements of $\mathbf{Q}[x]/M$ look like? By the Division Theorem, if $f(x) \in \mathbf{Q}[x]$, then there exist unique $q(x)$ and $r(x) = ax + b$ such that

$$f(x) = (x^2 + x + 3)q(x) + (ax + b), \quad \text{with } a, b \in \mathbf{Q}.$$

Hence, $f(x) \in (ax + b) + M$,

and so $f(x) + M = (ax + b) + M$.

Therefore, $\mathbf{Q}[x]/M = \{(ax + b) + M : a, b \in \mathbf{Q}\} = \overline{\{ax + b : a, b \in \mathbf{Q}\}}$.

Suppose we want to find the multiplicative inverse of $(x + 2) + M$. Applying the Euclidean Algorithm to $(x + 2)$ and $m(x)$, we get that

$$(x + 2)(-\tfrac{1}{5}x + \tfrac{1}{5}) + (x^2 + x + 3)(\tfrac{1}{5}) = 1.$$

Hence, $[(x + 2) + M] \cdot [(-\tfrac{1}{5}x + \tfrac{1}{5}) + M] = 1 + M.$

Therefore, $[(x + 2) + M]^{-1} = (-\tfrac{1}{5}x + \tfrac{1}{5}) + M.$ □

6.2 Problems

1. Construct the operation tables of the factor ring $(\mathbf{Z}_{12}/\langle \overline{4} \rangle, +, \cdot)$, and find a proper ideal of the factor ring.

2   (a) Verify that subset N of Example 6.7 is an ideal of $\mathbf{Z}_2 \times \mathbf{Z}_2 \times \mathbf{Z}_4$.

(b) Note that in the factor ring S/N of Example 6.7, $ab = bb$. Explain why you cannot then conclude that $a = b$.

3   (a) Verify that $\mathbf{Z}_2 \times \mathbf{Z}_2 \times \{\bar{0}\}$ is an ideal of $\mathbf{Z}_2 \times \mathbf{Z}_2 \times \mathbf{Z}_4$.

(b) Construct the operation tables of the factor ring,

$$\frac{\mathbf{Z}_2 \times \mathbf{Z}_2 \times \mathbf{Z}_4}{\mathbf{Z}_2 \times \mathbf{Z}_2 \times \{\bar{0}\}}.$$

4   (a) Verify that $\{\bar{0}\} \times \{\bar{0}\} \times \mathbf{Z}_4$ is an ideal of $\mathbf{Z}_2 \times \mathbf{Z}_2 \times \mathbf{Z}_4$.

(b) Construct the operation tables of the factor ring,

$$\frac{\mathbf{Z}_2 \times \mathbf{Z}_2 \times \mathbf{Z}_4}{\{\bar{0}\} \times \{\bar{0}\} \times \mathbf{Z}_4}.$$

5   Complete the proof of Theorem 6.8.

6   Let N be an ideal of ring S. If S is an integral domain, is S/N an integral domain? (Either prove that it is, or give a counterexample.)

7   If S is a ring, describe the factor rings: S/S and $S/\{z\}$.

8   List all the maximal ideals of ring $\mathbf{Z}$ which contain $\langle n \rangle$, if $n = 6, 9, 30, 210, 1024$.

9   Prove: $\langle n \rangle$ is a maximal ideal of $\mathbf{Z}$ iff n is prime.

10   If $S = \mathbf{Z}_4 \times \mathbf{Z}_5 \times \mathbf{Z}_7$, give a maximal ideal M of ring $(S, +, \cdot)$, and identify the field S/M.

11   Let $N = \langle g(x) \rangle$, with $g(x) = x^4 - 2x^2 - 3 \in \mathbf{Q}[x]$.

(a) Show that N is not a *maximal* ideal of $\mathbf{Q}[x]$, and that ring $\mathbf{Q}[x]/N$ has zero divisors.

(b) Generalize your findings of part (a).

12   Let integral domain $\mathbf{Q}[x]$ and ideal $\langle x^3 - x + 2 \rangle$ be given.

(a) Verify that if $m(x) = x^3 - x + 2$, then $m(x)$ is irreducible in $\mathbf{Q}[x]$.

(b) If $M = \langle x^3 - x + 2 \rangle$, explain why M is a *maximal* ideal of $\mathbf{Q}[x]$.

(c) Show that if $f(x)$ is any polynomial in $\mathbf{Q}[x]$, then there is exactly one coset, $r(x) + M$, with $r(x) = a + bx + cx^2$, where $a, b, c \in \mathbf{Q}$, such that $f(x) \in r(x) + M$.

(d) Show that, in factor field $\mathbf{Q}[x]/M$, if $s(x) \in r(x) + M$, where deg $s(x) < 3$ and deg $r(x) < 3$, then $s(x) = r(x)$. Therefore,

$$\mathbf{Q}[x]/M = \{(a + bx + cx^2) + M : a, b, c \in \mathbf{Q}\}.$$

(e) Find the multiplicative inverse of coset, $(3 + 2x + x^2) + M$.

13   Suppose that $m(x) = x^2 + \bar{2}x + \bar{2} \in \mathbf{Z}_3[x]$, and $M = \langle m(x) \rangle$.

(a) Explain why M is a maximal ideal of $\mathbf{Z}_3[x]$.

(b) Exhibit the nine elements of the field, $\mathbf{Z}_3[x]/M$.

(c) Find the multiplicative inverse of each nonzero element of $\mathbf{Z}_3[x]/M$.

6.3 Ring Morphisms

Every theorem in this section should look familiar, since each one has its analogue in Chapter 5. We shall use our knowledge of *group* morphisms to develop analogous results concerning *ring* morphisms.

According to an earlier definition, a *morphism* from a ring $(S, +, \cdot)$ to a ring $(T, \oplus, \otimes)$ is a map, $\theta: S \to T$, such that for all $a, b \in S$:

$$(a + b)\theta = a\theta \oplus b\theta, \quad \text{and} \quad (a \cdot b)\theta = a\theta \otimes b\theta.$$

An *isomorphism* is, of course, a bijective morphism. Rings S and T are said to be *isomorphic rings* if an isomorphism exists from S onto T. An *endomorphism* of ring $(S, +, \cdot)$ is a morphism from S to S, an an *automorphism* of ring S is an isomorphism from S to S.

We saw that under a group morphism, a subgroup maps onto a subgroup, and a normal subgroup maps onto a normal subgroup. We should not be surprised to find out that under a ring morphism, a subring maps onto a subring, and an ideal maps onto an ideal. The next theorem gives some basic results that first appeared in Problem 2.10–19.

THEOREM 6.11

Let $\theta: (S, +, \cdot) \to (T, \oplus, \otimes)$ be a morphism.
(a) If S is a ring, then $S\theta$ is a ring.
   If ring S is commutative, then ring $S\theta$ is commutative.
   If ring S has unity u, then ring $S\theta$ has unity $u\theta$.
(b) If S is an integral domain, and θ is one–one, then $S\theta$ is an integral domain.
(c) If S is a field, and if θ is not the zero map, then $S\theta$ is a field.

Proof

(a) By applying Theorem 2.30 to the morphisms,

$$\theta: (S, +) \to (T, \oplus), \quad \text{and} \quad \theta: (S, \cdot) \to (T, \otimes),$$

it follows that $S\theta$ satisfies all the ring properties except possibly the distributive laws, which must be verified. Let $a\theta, b\theta, c\theta \in S\theta$.

Then
$$\begin{aligned}
a\theta \otimes (b\theta \oplus c\theta) &= a\theta \otimes (b + c)\theta \\
&= [a \cdot (b + c)]\theta = [(a \cdot b) + (a \cdot c)]\theta \\
&= (a \cdot b)\theta \oplus (a \cdot c)\theta \\
&= (a\theta \otimes b\theta) \oplus (a\theta \otimes c\theta).
\end{aligned}$$

Similarly, the right distributive law holds in $S\theta$.

(b) Let z and z' be zeros of S and T, respectively. Then

$$\begin{aligned}
(a\theta)(b\theta) = z' &\Rightarrow (ab)\theta = z' \Rightarrow ab = z \\
&\Rightarrow a = z, \quad \text{or} \quad b = z \\
&\Rightarrow a\theta = z', \quad \text{or} \quad b\theta = z'.
\end{aligned}$$

(c) $a\theta \neq z' \Rightarrow a \neq z \Rightarrow a^{-1}$ exists, and $(a\theta)^{-1} = (a^{-1})\theta$. We specify here that θ is not the zero map, since a field must contain at least two elements. □

EXAMPLE 6.11

The map, $\theta:(\mathbf{Z}, +, \cdot) \to (\mathbf{Z}_6, +, \cdot)$, is a morphism if $n\theta = \bar{n}$, for all $n \in \mathbf{Z}$. (Verify this.) Note that ideals map to ideals. For instance,

$(2\mathbf{Z})\theta = (4\mathbf{Z})\theta = (8\mathbf{Z})\theta = \{\bar{0}, \bar{2}, \bar{4}\}$, $(3\mathbf{Z})\theta = (9\mathbf{Z})\theta = (15\mathbf{Z})\theta = \{\bar{0}, \bar{3}\}$,
$(6\mathbf{Z})\theta = (12\mathbf{Z})\theta = (18\mathbf{Z})\theta = \{\bar{0}\}$, $(5\mathbf{Z})\theta = (7\mathbf{Z})\theta = (11\mathbf{Z})\theta = \mathbf{Z}_6$. □

THEOREM 6.12

Let $\theta:(S, +, \cdot) \to (T, \oplus, \otimes)$ be a morphism from ring S to ring T.
(a) If M is a subring of S, then $M\theta$ is a subring of T.
If M is an ideal of S, then $M\theta$ is an ideal of $S\theta$.
(b) If N is a subring of T, then $N\theta^{-1}$ is a subring of S.
If N is an ideal of T, then $N\theta^{-1}$ is an ideal of S.

Proof

(a) The first statement is true, by Theorem 6.11(a). Now suppose that M is an ideal of S. Let $m\theta \in M\theta$, and $s\theta \in S\theta$. But $(m\theta) \otimes (s\theta) = (m \cdot s)\theta$, and $m \cdot s \in M$. So $(m \cdot s)\theta \in M\theta$. Similarly, $(s\theta) \otimes (m\theta) \in M\theta$. Therefore, $M\theta$ is an ideal of $S\theta$.

(b) Since $(N, \oplus)$ is an abelian subgroup of $(T, \oplus)$, then $(N\theta^{-1}, +)$ is an abelian subgroup of $(S, +)$, by Theorem 5.23(b). Let $a, b \in N\theta^{-1}$. Then $a\theta, b\theta \in N$, and so $(a\theta) \otimes (b\theta) \in N$. Hence $(a \cdot b)\theta \in N$, and so $a \cdot b \in N\theta^{-1}$. Therefore, $N\theta^{-1}$ is a subring of S.

Given that N is an ideal of T. Let $a \in N\theta^{-1}$, and $s \in S$. Then $a\theta \in N$ and $s\theta \in S\theta$, and so $(a \cdot s)\theta \in N$ and $(s \cdot a)\theta \in N$. Hence, $a \cdot s, s \cdot a \in N\theta^{-1}$. Therefore, $N\theta^{-1}$ is an ideal of S. □

If z' is the zero of ring T, then the *kernel* K of a ring morphism, $\theta: S \to T$, is defined as follows:

$$K = \{x : x \in S, \text{ and } x\theta = z'\}.$$

Just as the kernel of a group morphism is the set of elements that map to the identity of the image group, the kernel of a ring morphism is the set of elements that map to the additive identity (the zero) of the image ring. We continue the analogy in the theorems that follow.

THEOREM 6.13

Let $\theta: S \to T$ be a morphism from ring S to ring T, with kernel K. Then:
(a) K is an ideal of S, and
(b) θ is 1-1 if and only if $K = \{z\}$.

6.3 Ring Morphisms

Proof

(a) By Theorem 5.36(a) we know that $(K, +)$ is a subgroup of $(S, +)$. We need only to prove that K is closed under multiplication by any element of S. Suppose that $k \in K$ and $r \in S$. Then

$$(kr)\theta = (k\theta)(r\theta) = z'(r\theta) = z', \quad \text{and} \quad (rk)\theta = (r\theta)(k\theta) = (r\theta)z' = z'.$$

Therefore, $kr, rk \in K$, and so K is an ideal of S.

(b) This is true, by Theorem 5.36(b). □

If N is an ideal of a ring S, we define the *natural map* of S onto factor ring S/N to be the map $\lambda: r \to r + N$, for all $r \in S$.

THEOREM 6.14

If N is an ideal of ring S, then the natural map, $\lambda: S \to S/N$, is a morphism with kernel N.

Proof

By Theorem 5.37, $\lambda:(S, +) \to (S/N, +)$ is a group morphism with kernel N. We need only verify that multiplication is preserved. If $r, s \in S$, then

$$(r \cdot s)\lambda = (r \cdot s) + N = (r + N)(s + N) = (r\lambda)(s\lambda). \quad \square$$

The converse of Theorem 6.9 is easily proved by using the morphism of Theorem 6.14.

THEOREM 6.15

Let M be an ideal of S, where S is a commutative ring with unity. If S/M is a field, then M is a maximal ideal of S.

Proof

We shall prove that if M is not maximal, then S/M is not a field. Let A be an ideal of S such that $M \subset A \subset S$. Under the natural map λ we have:

$$\lambda: \quad S \to S/M = \{s + M : s \in S\},$$
$$A \to \{a + M : a \in A\} = A\lambda,$$
$$M \to \{M\} = \text{the zero ideal of } S/M.$$

The image of an ideal is an ideal, and so $A\lambda$ is an ideal. We must show that $A\lambda$ is a *proper* ideal of S/M. There is an $r \in S$ such that $r \notin A$. So $r + M \notin A\lambda$, and thus $A\lambda \subset S/M$. Also, there is a $b \in A$ such that $b \notin M$. So $b + M \in A\lambda$, but $b + M \neq M$. Hence, $\{M\} \subset A\lambda \subset S/M$. However, a field has no proper ideals (Theorem 6.7(b)), and so S/M is not a field. □

THEOREM 6.16 *(Fundamental Theorem on Ring Morphisms)*

If $\theta: S \to T$ is a morphism of ring S onto ring T with kernel K, then $S/K \cong T$.

Proof

We define map ψ as follows:

$$\psi: S/K \to T,$$

where
$$r + K \to r\theta.$$

We proved in Theorem 5.39 that $\psi: (S/K, +) \to (T, +)$ is an isomorphism. We need to verify only that multiplication is preserved. If $a, b \in S$, then

$$[(a + K)(b + K)]\psi = [ab + K]\psi = (ab)\theta = (a\theta)(b\theta) = [(a + K)\psi][(b + K)\psi].$$

Therefore, ring S/K is isomorphic to ring T. □

Using the notation of Theorem 6.16: If $\lambda: S \to S/K$ is the natural map, then maps λ, θ, ψ, and ψ^{-1} may be depicted as in Figure 6–2.

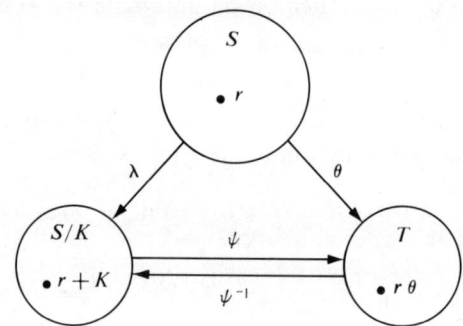

Figure 6–2

EXAMPLE 6.12

We verify that the map, $\theta: (\mathbf{Z}_6, +, \cdot) \to (\mathbf{Z}_{12}, +, \cdot)$, is a morphism, if $\overline{m}\theta = \overline{4m}$ for all $m \in \mathbf{Z}_6$. First, for all $\overline{m}, \overline{n} \in \mathbf{Z}_6$:

$$\overline{m} = \overline{n} \Rightarrow m \equiv n \pmod{6} \Rightarrow 4m \equiv 4n \pmod{12} \Rightarrow \overline{m}\theta = \overline{n}\theta.$$

Hence θ is indeed a map. Also, $\overline{m}\theta + \overline{n}\theta = \overline{4m} + \overline{4n} = \overline{4(m + n)} = \overline{m + n}\theta$, and $(\overline{m}\theta)(\overline{n}\theta) = (\overline{4m})(\overline{4n}) = \overline{16mn} = \overline{4mn} = \overline{mn}\theta$. Under this map:

$$\overline{0} \to \overline{0}, \quad \overline{3} \to \overline{0},$$
$$\overline{1} \to \overline{4}, \quad \overline{4} \to \overline{4},$$
$$\overline{2} \to \overline{8}, \quad \overline{5} \to \overline{8}.$$

So θ maps $\mathbf{Z}_6$ onto T, where $T = \{\overline{0}, \overline{4}, \overline{8}\}$. Kernel $K = \{\overline{0}, \overline{3}\}$, and factor ring $\mathbf{Z}_6/K = \{K, \overline{1} + K, \overline{2} + K\}$, where $\overline{1} + K = \{\overline{1}, \overline{4}\}$, and $\overline{2} + K = \{\overline{2}, \overline{5}\}$. Hence, $\mathbf{Z}_6/K \cong T \cong \mathbf{Z}_3$.

Note that if $\lambda: (\mathbf{Z}_6, +) \to (\mathbf{Z}_{12}, +)$ is a map such that $\overline{m}\lambda = \overline{2m}$, for all $\overline{m} \in \mathbf{Z}_6$, then λ is a *group* morphism. However,

$$(\overline{m}\lambda)(\overline{n}\lambda) = (\overline{2m})(\overline{2n}) = \overline{4mn}, \quad \text{but} \quad \overline{mn}\lambda = \overline{2mn} \neq \overline{4mn},$$

so $\lambda: (\mathbf{Z}_6, +, \cdot) \to (\mathbf{Z}_{12}, +, \cdot)$ is not a *ring* morphism. □

6.3 Problems

1. (a) Verify that if $m\theta = \overline{5m}$, for $m \in \mathbf{Z}$, then $\theta:(\mathbf{Z}, +) \to (\mathbf{Z}_6, +)$ is a group morphism, but $\theta:(\mathbf{Z}, +, \cdot) \to (\mathbf{Z}_6, +, \cdot)$ is not a ring morphism.
   (b) If $m\lambda = \overline{3m}$, for $m \in \mathbf{Z}$, is $\lambda:(\mathbf{Z}, +, \cdot) \to (\mathbf{Z}_6, +, \cdot)$ a morphism?

2. Find all endomorphisms of $(\mathbf{Z}_{12}, +, \cdot)$. Which endomorphisms are automorphisms?

3. Find all morphisms of:
   (a) Group $(\mathbf{Z}_8, +)$ to group $(\mathbf{Z}_{12}, +)$; ring $(\mathbf{Z}_8, +, \cdot)$ to ring $(\mathbf{Z}_{12}, +, \cdot)$,
   (b) $(\mathbf{Z}_{12}, +)$ to $(\mathbf{Z}_8, +)$, and $(\mathbf{Z}_{12}, +, \cdot)$ to $(\mathbf{Z}_8, +, \cdot)$.

4. Given the group morphism, $\theta:(\mathbf{Z}_n, +) \to (\mathbf{Z}_k, +)$, with $\overline{1}\theta = \bar{a}$; prove that $\theta:(\mathbf{Z}_n, +, \cdot) \to (\mathbf{Z}_k, +, \cdot)$ is a ring morphism iff $a^2 \equiv a \pmod{k}$.

5. Find all endomorphisms of ring $(\mathbf{Z}_n, +, \cdot)$ if:
   (a) $n = 6$,   (b) $n = 10$,   (c) $n = 11$.

6. Prove:
   (a) The only automorphism of $(\mathbf{Z}, +, \cdot)$ is the identity map.
   (b) If p is a prime, the only endomorphisms of $(\mathbf{Z}_p, +, \cdot)$ are the zero map and the identity map.

7. (a) Referring to Theorem 6.11(a), verify that the right distributive law holds in $S\theta$. Give a reason for each step.
   (b) Why, in Theorem 6.11(b), was it necessary to state that θ is 1–1?

8. (a) Find all ring morphisms of $\mathbf{Z}$ into $\mathbf{Z}$.
   (b) Is there a ring morphism of $\mathbf{Z}$ onto $2\mathbf{Z}$?
   (c) Find three ring morphisms of $\mathbf{Z} \times \mathbf{Z}$ into $\mathbf{Z}$, where the operations on $\mathbf{Z} \times \mathbf{Z}$ are componentwise addition and multiplication.

9. Verify that ring T of Problem 6.1–3(a) is isomorphic to ring S of Problem 2.7–5(b).

10. Find:
    (a) Each ring morphism, $\theta: \mathbf{Z}_{12} \to \mathbf{Z}_6$.
    (b) The kernel K of each θ.
    (c) The factor ring, $\mathbf{Z}_{12}/K$.
    (d) The image of each element of $\mathbf{Z}_{12}$ under the natural map,
    $$\psi: \mathbf{Z}_{12} \to \mathbf{Z}_{12}/K.$$
    (e) The image of each element of $\mathbf{Z}_{12}/K$ under the isomorphism,
    $$\lambda: \mathbf{Z}_{12}/K \to \mathbf{Z}_{12}\theta.$$

11. Prove: There is only one field of order 3.

12. Prove:
    (a) If $(F, +, \cdot)$ is a field of order 4, then the additive group cannot be cyclic. (Show that the distributive property fails to hold.)
    (b) There is exactly one field of order 4.

13 Prove: The only automorphism of the rational field $\mathbf{Q}$ is the identity map.

14 Assume $(S_i, \oplus_i, \otimes_i)$ is a ring, for $i = 1, 2, \ldots, n$, and $S = \{(r_1, r_2, \ldots, r_n) : r_i \in S_i\}$. Operations on S are defined as follows:
$$(r_1, \ldots, r_n) + (t_1, \ldots, t_n) = (r_1 \oplus_1 t_1, \ldots, t_n \oplus_n t_n),$$
$$(r_1, \ldots, r_n) \cdot (t_1, \ldots, t_n) = (r_1 \otimes_1 t_1, \ldots, t_n \otimes_n t_n).$$

Prove:

(a) $(S, +, \cdot)$ is a ring.

(b) There is an endomorphism of S onto a subring S_i', where S_i is isomorphic to S_i', for each i.

15 Prove: There are only two automorphisms of the quadratic field, $\mathbf{Q}(\sqrt{n})$:
$$\varepsilon: a + b\sqrt{n} \to a + b\sqrt{n}, \quad \text{and}$$
$$\theta: a + b\sqrt{n} \to a - b\sqrt{n}.$$

16 Let $\theta: F \to F$ be a ring morphism of a field F into F. Verify that if θ is not the zero map, then θ is injective.

17 Let $(S, \oplus, \otimes)$ be a ring with unity. If a is any element of S, define a map $\psi_a: S \to S$ such that $s\psi_a = s \otimes a$, for all $s \in S$. Let $T = \{\psi_a : a \in S\}$. Prove:

(a) ψ_a is an endomorphism of the additive group $(S, \oplus)$.

(b) If the operations on T are $+$ and $\circ$, where $+$ is addition of maps and $\circ$ is composition of maps, then $\psi_a + \psi_b = \psi_{a \oplus b}$, and $\psi_a \circ \psi_b = \psi_{a \otimes b}$, for all $a, b \in S$.

(c) $(S, \oplus, \otimes)$ is isomorphic to $(T, +, \circ)$, and so $(T, +, \circ)$ is a ring with unity.

[Note: This is the *Representation Theorem for Rings*. Compare with Theorem 5.35, the Representation Theorem for Groups.]

18 Let $A(E)$ be the set of all automorphisms of a field E, and let $\circ$ be the operation of composition of maps.

(a) Prove that $(A(E), \circ)$ is a group.

(b) Let F be a subfield of E. We say that an automorphism ψ of E *fixes* F if $a\psi = a$, for all $a \in F$. If $A(E/F)$ is the set of all automorphisms of E that fix F, prove that $(A(E/F), \circ)$ is a subgroup of $(A(E), \circ)$.

6.4 The Characteristic of a Ring

We prove a lemma before introducing the main topic of this section. Recall that if a is in a ring S, and if $k \in \mathbf{Z}^+$, then, by definition, $1a = a$, $(k+1)a = ka + a$, and $(-k)a = -(ka)$. Also, $0a = z$, the zero of S.

THEOREM 6.17

If a and b are in a ring $(S, +, \cdot)$, and $m, n \in \mathbf{Z}$, then

(a) $n(a \cdot b) = (na) \cdot b = a \cdot (nb)$, and

(b) $(ma) \cdot (nb) = (mn)(a \cdot b)$.

6.4 The Characteristic of a Ring

Proof

(a) The theorem is obviously true if $n = 0$. If $n > 0$, then

$$n(ab) = \sum_{i=1}^{n} (ab) = \left(\sum_{i=1}^{n} a\right)b = (na)b, \quad \text{and} \quad \sum_{i=1}^{n} (ab) = a\left(\sum_{i=1}^{n} b\right) = a(nb).$$

If $n < 0$, let $n = -k$, so that $k > 0$. Then

$$n(ab) = (-k)(ab) = -[k(ab)] = -[(ka)b] = [-(ka)]b = [(-k)a]b = (na)b.$$

Similarly, $n(ab) = a(nb)$.

(b) This is readily proved by using part (a). □

The *characteristic* of a ring S is defined to be the smallest positive integer k such that $ka = z$, for all $a \in S$. If no such k exists, then the characteristic is defined to be 0. We write: $\text{ch}(s) = k$, or $\text{ch}(s) = 0$.

EXAMPLE 6.13

(a) In ring $\mathbf{Z}_6$:

$$1(\bar{0}) = \bar{0}, \quad\quad 3(\bar{2}) = \bar{6} = \bar{0}, \quad\quad 3(\bar{4}) = \overline{12} = \bar{0},$$
$$6(\bar{1}) = \bar{6} = \bar{0}, \quad\quad 2(\bar{3}) = \bar{6} = \bar{0}, \quad\quad 6(\bar{5}) = \overline{30} = \bar{0}.$$

If $\bar{m}$ is any element in $\mathbf{Z}_6$, then $6(\bar{m}) = \overline{6m} = \bar{0}$, and 6 is the smallest positive integer that works for all $\bar{m}$. Therefore, $\text{ch}(\mathbf{Z}_6) = 6$.

(b) S is a subring of $\mathbf{Z}_6$, where $S = \{\bar{0}, \bar{2}, \bar{4}\}$. The characteristic of S is 3, illustrating the fact that the characteristic of a subring does not have to be the same as the characteristic of the ring. □

If S is a ring with unity u, we can determine the characteristic of S simply by finding the smallest positive integer k such that $ku = z$. We verify this next.

THEOREM 6.18

Let S be a ring with unity u. If k is the smallest positive integer such that $ku = z$, then $\text{ch}(S) = k$. If no such positive integer k exists, then $\text{ch}(S) = 0$.

Proof

Because k is the smallest integer in $\mathbf{Z}^+$ such that $ku = z$, we know that $\text{ch}(S) \geq k$. Now let a be any element of S. Using Theorem 6.17(a): $ka = k(ua) = (ku)a = za = z$. So k works for all a, and therefore, $\text{ch}(S) = k$.

If no k exists such that $ku = z$, then $\text{ch}(S) = 0$, by definition. □

According to Theorem 6.18, if $k > 0$, the characteristic k of a ring with unity is the additive order of the unity.

EXAMPLE 6.14

(a) In $\mathbf{Z}_n$, we see that $n(\bar{1}) = \bar{n} = \bar{0}$, and if $1 \leq m < n$, then $m(\bar{1}) = \bar{m} \neq \bar{0}$. Hence, $\mathrm{ch}(\mathbf{Z}_n) = n$.

(b) $(\mathbf{Z}_2 \times \mathbf{Z}_2, +, \cdot)$ is a commutative ring with unity u, where $u = (\bar{1}, \bar{1})$. Since $2u = (\bar{0}, \bar{0})$, this ring is of characteristic 2. Thus, the characteristic of a finite ring can be less than the order of the ring. We leave it as an exercise for you to show that the characteristic is never greater than the order of the ring.

(c) In ring $\mathbf{Z}$, $k(1) = k \neq 0$, for each $k \in \mathbf{Z}^+$. So $\mathrm{ch}(\mathbf{Z}) = 0$. □

THEOREM 6.19

(a) The characteristic of an integral domain is either a prime or 0.
(b) The characteristic of a finite field is a prime.

Proof

(a) Given that D is an integral domain, and that $\mathrm{ch}(D) = k > 0$, then k is the smallest positive integer such that $ku = z$. Assume that k is composite. Then $k = mn$, with $1 < m < k$, and $1 < n < k$. So $z = ku = (mn)(uu) = (mu)(nu)$, by Theorem 6.17(b). But D has no zero divisors, and so either $mu = z$ or $nu = z$, which is a contradiction of the minimality of k. Therefore, k is a prime.

(b) $\mathrm{ch}(F) = k$, and $|F| = n \Rightarrow |u| = k \Rightarrow k \mid n \Rightarrow k \neq 0$. (We used the Lagrange Theorem on the finite group $(F, +)$.) Since field F is an integral domain, then k is a prime, by part (a). □

It is of interest to learn that *every* integral domain contains a subdomain structurally like either $\mathbf{Z}$ or $\mathbf{Z}_p$. We now prove this.

THEOREM 6.20

Let D be an integral domain.
(a) If $\mathrm{ch}(D) = 0$, then D contains a subdomain isomorphic to domain $\mathbf{Z}$.
(b) If $\mathrm{ch}(D) = p$, where p is a prime, then D contains a subdomain isomorphic to field $\mathbf{Z}_p$.

Proof

If u is the unity of D, then $\mathbf{Z}u \subseteq D$, where $\mathbf{Z}u = \{nu : n \in \mathbf{Z}\}$.

(a) Given that $\mathrm{ch}(D) = 0$, let θ be this map:

$$\theta: \mathbf{Z} \to \mathbf{Z}u,$$

where
$$n \to nu = n\theta.$$

It is readily shown that this is a bijective map that preserves addition and multiplication. Therefore, $\mathbf{Z}u$ is an integral domain, by Theorem 6.11(b), that is isomorphic to $\mathbf{Z}$.

(b) Given that ch(D) = p, let ψ be this map:

$$\psi: \mathbf{Z}_p \to \mathbf{Z}u,$$

where
$$\bar{n} \to nu = \bar{n}\psi.$$

We note that p is the order of u in the additive group, (D, +). Now,

$$\bar{n}\psi = \bar{m}\psi \Rightarrow nu = mu \Rightarrow (n - m)u = z$$
$$\Rightarrow p \mid (n - m) \Rightarrow n \equiv m \pmod{p} \Rightarrow \bar{n} = \bar{m}.$$

Hence, ψ is 1–1. It is a simple matter to show that $\mathbf{Z}u$ is isomorphic to $\mathbf{Z}_p$. □

We know that, given any positive integer m, there is at least one group of order m. Is a similar statement true of fields? Our next theorem makes use of the characteristic to answer this question.

THEOREM 6.21

If F is a field of order m and of characteristic p, then $m = p^n$, for some $n \geq 1$.

Proof

The additive order of unity u is p, so $p \mid m$. Let q be any positive prime such that $q \mid m$. By Theorem 5.34 (or 5.49) there is a subgroup of (F, +) of order q. Hence there is an element x whose additive order is q, so $qx = z$. But ch(F) = $p \Rightarrow px = z \Rightarrow q \mid p \Rightarrow q = p$. Therefore, p is the only prime dividing m, and so $m = p^n$, for some $n \in \mathbf{Z}^+$. □

We have verified the significant result that the order of a finite field must be a power of a prime. Hence there is no field of order 6, 10, 12, 14, or 15, for instance.

We remark that although a *finite* field must have a prime as its characteristic, an *infinite* integral domain or field may have either a prime or zero as its characteristic.

6.4 Problems

1. Write out the equations of Theorem 6.17 if $n = 3$ and $m = 2$, and illustrate the fact that these are generalized distributive laws if n and m are positive integers.

2. Complete the proof of Theorem 6.17 by proving that:
   (a) $n(ab) = a(nb)$ if $n < 0$,   (b) $(ma)(nb) = (mn)(ab)$.

3. Give the characteristic of each of these integral domains:
   (a) $\mathbf{Q}[x]$,   (b) $\mathbf{Z}(\sqrt{6})$,   (c) $\mathbf{Z}_5[x]$.

4   Give the characteristic of these rings:

$$\mathbf{Z} \times \mathbf{Z}, \quad \mathbf{Z} \times \mathbf{Z}_3, \quad \mathbf{Z}_3 \times \mathbf{Z}_5, \quad \mathbf{Z}_4 \times \mathbf{Z}_6, \quad \mathbf{Z}_m \times \mathbf{Z}_n.$$

In each ring, the operations are componentwise addition and multiplication.

5   Let S be a commutative ring with unity. Under what conditions does $(a + b)^2 = a^2 + b^2$, for all $a, b \in S$? Is it ever possible that $(a + b)^3 = a^3 + b^3$, for all a and b in some ring?

6   Prove: If S is a ring of order m, and $\text{ch}(S) > 0$, then $\text{ch}(S) | m$.

7   Complete the proof of Theorem 6.20.

8   If a finite field is of order m, with $m < 100$ and m composite, list all possible numbers that m might be.

9   Let p and q be distinct positive primes. Verify that if $\text{ch}(F) = p$, then field F has no subfield isomorphic to $\mathbf{Z}_q$.

10 Let $D = (\mathbf{Z} \times \mathbf{Z}, +, \otimes)$ be an algebra with these operations:

$$(a, b) + (c, d) = (a + c, b + d),$$
$$(a, b) \otimes (c, d) = (ac - bd, ad + bc).$$

(a) Verify that D is an integral domain, and that $\text{ch}(D) = 0$.
(b) Find a subdomain of D to which $\mathbf{Z}$ is isomorphic.

11 Let $F = (\mathbf{Z}_3 \times \mathbf{Z}_3, +, \otimes)$ be an algebra with these operations:

$$(a, b) + (c, d) = (a + c, b + d),$$
$$(a, b) \otimes (c, d) = (ac + \bar{2}bd, ad + bc).$$

(a) List the nine elements of F. Identify z, u, and the multiplicative inverse of each nonzero element.
(b) Verify that F is a field, and that $\text{ch}(F) = 3$.
(c) Find a subfield of F to which $(\mathbf{Z}_3, +, \cdot)$ is isomorphic.

12 Prove: If $(F, +, \cdot)$ is a field of order 8, then

$$(F, +) \cong (\mathbf{Z}_2 \times \mathbf{Z}_2 \times \mathbf{Z}_2, +), \quad \text{and} \quad (F^*, \cdot) \cong (\mathbf{Z}_7, +).$$

13 Define an algebra $(S, +, \cdot)$ as follows:

$$S = \{(a_1, a_2, a_3, \ldots) : a_i \in \mathbf{Z}_3, \text{ all but a finite number of } a_i \text{ are } \bar{0}\}.$$

That is, each element of S is an infinite sequence of elements of $\mathbf{Z}_3$, with only a finite number of the components being $\bar{1}$ or $\bar{2}$. To add two sequences, we add the corresponding components, and to multiply two sequences, we multiply corresponding components. Verify:

(a) S is an infinite set, and is closed under the two operations.
(b) S has zero divisors.
(c) S is a commutative ring without a unity.
(d) $\text{ch}(S) = 3$.

14 Let $(D, \oplus, \otimes)$ be any ordered integral domain with the property that the set D^+ of positive elements is well-ordered, and let $(\mathbf{Z}, +, \cdot)$ be the domain of rational integers. Prove that $\mathbf{Z}$ is isomorphic to D. (Hint: If u is the unity of D, first show that u is the smallest element in D^+, and that every element of D^+ is of the form mu, where $m \in \mathbf{Z}^+$. Next show that every element of D is uniquely expressible in the form nu, with $n \in \mathbf{Z}$. Finally, define map θ from $\mathbf{Z}$ to D so that $n\theta = nu$, and show that θ is an isomorphism.)

6.5 Quotient Fields and Other Ring Extensions

In Chapter 5 we discussed the concept of embedding a group in a larger group. Similarly, we can embed a ring in a larger ring. We say that ring $(S, +, \cdot)$ is *embedded* in ring $(T, \oplus, \otimes)$, or that T is an *extension* of S, if T contains a subring S' that is isomorphic to S. This is depicted in Figure 6–3. In symbols:

$$(S, +, \cdot) \cong (S', \oplus, \otimes) \leq (T, \oplus, \otimes).$$

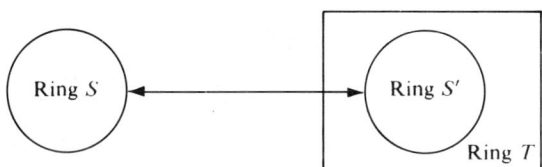

Figure 6–3

EXAMPLE 6.15

We illustrate the embedding of field $\mathbf{Z}_2$ in ring $\mathbf{Z}_2 \times \mathbf{Z}_2$. We know that $(\mathbf{Z}, +, \cdot)$ is a field of order 2. Let $S = \mathbf{Z}_2 \times \mathbf{Z}_2$, and denote the elements of S as follows:

$$z = (\bar{0}, \bar{0}), \quad u = (\bar{1}, \bar{1}), \quad a = (\bar{1}, \bar{0}), \quad b = (\bar{0}, \bar{1}).$$

If operations, $\oplus$ and $\otimes$, on S are componentwise addition and multiplication, then $(S, \oplus, \otimes)$ is a ring with the following operation tables.

$\oplus$	z	u	a	b
z	z	u	a	b
u	u	z	b	a
a	a	b	z	u
b	b	a	u	z

$\otimes$	z	u	a	b
z	z	z	z	z
u	z	u	a	b
a	z	a	a	z
b	z	b	z	b

We see that a and b are zero divisors of S, so S is not a field.

If $T = \{z, a\}$, then $(T, \oplus, \otimes)$ is a subring of S, with the tables that follow.

$\oplus$	z	a
z	z	a
a	a	z

$\otimes$	z	a
z	z	z
a	z	a

The map, $\theta: \mathbf{Z}_2 \to T$, is an isomorphism, if $\overline{0} \to z$, and $\overline{1} \to a$. Therefore,

$$(\mathbf{Z}_2, +, \cdot) \cong (T, \oplus, \otimes) < (S, \oplus, \otimes). \quad \square$$

We may be given a ring S that does not possess a particular property that we would like it to have, and so we seek a larger ring T that has the desired property. At the same time, we don't want to lose S and its properties. Hence, T must be an extension of S. This means that we must devise a technique for embedding S in a ring T that does what we want it to do. What kind of properties are we talking about? For instance, if S has no unity, we may want to embed S in a ring T that has a unity. Or perhaps some elements of S do not have multiplicative inverses, and we want to find an extension T of S in which every nonzero element has a multiplicative inverse. Or we may be given a polynomial,

$$f(x) = a_3 x^3 + a_2 x^2 + a_1 x + a_0, \quad \text{with } a_i \in S,$$

and we find that no element in S is a root of $f(x)$. So we look for an extension T of S such that T contains a root of $f(x)$.

EXAMPLE 6.16

The equation $ax = b$ with $a, b \in \mathbf{Z}, a \neq 0$, might not have a solution in $\mathbf{Z}$, but it has the unique solution, b/a, in $\mathbf{Q}$. We see that the rational field $\mathbf{Q}$ is an extension of the integral domain $\mathbf{Z}$, since $\mathbf{Q} = \{m/n : m, n \in \mathbf{Z}, n \neq 0\}$, and if $\mathbf{Z}' = \{m/1 : m \in \mathbf{Z}\}$, then $\theta: \mathbf{Z} \to \mathbf{Z}'$ is an isomorphism under the map: $m\theta = m/1$. Up to this point we have identified these isomorphic sets so completely that we have blithely written $m = m/1$, not bothering to distinguish between the integer m and the rational number $m/1$. $\quad \square$

We now give a technique for embedding any integral domain in a field. The method we use is such that if the domain is $\mathbf{Z}$, then the extension field will turn out to be $\mathbf{Q}$. You might keep that particular case in mind as we proceed.

THEOREM 6.22

Let $(D, +, \cdot)$ be an integral domain.

(a) If $\sim$ is defined on $D \times D^*$ as follows: $(a, b) \sim (c, d)$ if and only if $a \cdot d = b \cdot c$; then $\sim$ is an equivalence relation on $D \times D^*$.

(b) If $\overline{(a, b)} = \{(x, y) : (x, y) \sim (a, b)\}$ and $F = \{\overline{(a, b)}\}$, and if $\oplus$ and $\otimes$ are defined as follows:

$$\overline{(a, b)} \oplus \overline{(c, d)} = \overline{(ad + bc, bd)}, \qquad \overline{(a, b)} \otimes \overline{(c, d)} = \overline{(ac, bd)};$$

then $\oplus$ and $\otimes$ are well-defined operations on F.

(c) $(F, \oplus, \otimes)$ is a field.

Proof

(a) (i) $ab = ba$, so $(a, b) \sim (a, b)$, for all $(a, b) \in D \times D^*$.

(ii) $(a, b) \sim (c, d) \Rightarrow ad = bc \Rightarrow cb = da \Rightarrow (c, d) = (a, b)$.

6.5 Quotient Fields and Other Ring Extensions

(iii) $(a, b) \sim (c, d), (c, d) \sim (e, f) \Rightarrow ad = bc, cf = de \Rightarrow a(de) = bce,$
$de = cf \Rightarrow a(cf) = bce \Rightarrow af = be \Rightarrow (a, b) = (e, f)$.

(b) We must show that if $\overline{(a, b)} = \overline{(q, r)}$ and $\overline{(c, d)} = \overline{(s, t)}$, then $\overline{(a, b)} \oplus \overline{(c, d)} = \overline{(q, r)} \oplus \overline{(s, t)}$ and $\overline{(a, b)} \otimes \overline{(c, d)} = \overline{(q, r)} \otimes \overline{(s, t)}$. Now $\overline{(a, b)} = \overline{(q, r)}, \overline{(c, d)} = \overline{(s, t)} \Rightarrow ar = bq, ct = ds \Rightarrow (ar)dt + (ct)br = (bq)dt + (ds)br \Rightarrow (ad + bc)rt = bd(qt + rs) \Rightarrow \overline{(ad + bc, bd)} = \overline{(qt + rs, rt)} \Rightarrow \overline{(a, b)} \oplus \overline{(c, d)} = \overline{(q, r)} \oplus \overline{(s, t)}$. The proof that $\otimes$ is well-defined is left as an exercise.

(c) The proof is relatively simple, but it is time-consuming, since each of the field postulates must be verified. We leave this as an exercise. You must show, among other things, that (1) for all $t \neq z$, $\overline{(at, bt)} = \overline{(a, b)}$; (2) zero is $\overline{(z, u)}$, unity is $\overline{(u, u)}$; (3) $-\overline{(a, b)} = \overline{(-a, b)}$, and if $a \neq z$, then $\overline{(a, b)}^{-1} = \overline{(b, a)}$. (Hint: Mentally write element $\overline{(a, b)}$ as a/b in thinking about the proof, because that is the form in which we shall eventually express the elements of F.) □

Field F, of Theorem 6.22, is called the *quotient field* of D. Our next result shows that we have, indeed, embedded D in F.

THEOREM 6.23

Let F be the quotient field of integral domain D, and define D' to be the following subset of F: $D' = \{\overline{(a, u)}: a \in D\}$. Then (a) $D \cong D'$; and (b) if $\overline{(a, u)}$ is denoted by a, then $F = \{a/b: a, b \in D, b \neq z\}$.

Proof

(a) Define $\theta: D \to D'$ to be the map such that $a\theta = \overline{(a, u)}$. It is readily verified that θ is an isomorphism.

(b) It is at this point that our identification of isomorphic rings becomes so complete that we even agree to use the same symbols to write the elements of D and of D'. We replace the statement, $a\theta = \overline{(a, u)}$, by the statement, $a = \overline{(a, u)}$. Then $\overline{(u, b)} = \overline{(b, u)}^{-1} = b^{-1}$. So $\overline{(a, b)} = \overline{(a, u)} \cdot \overline{(u, b)} = a \cdot b^{-1} = a/b$. Therefore,

$$F = \{a/b: a, b \in D, b \neq z\}. \quad \square$$

Appropriately, each element in the quotient field F consists of the *quotient* of two elements of D. If $D = \mathbf{Z}$, it is evident that $F = \mathbf{Q}$, thus substantiating our earlier statement that $\mathbf{Q}$ is the quotient field of $\mathbf{Z}$.

Suppose we agree to let $+$ and $\cdot$ represent the operations in both D and F. Now that we have $\overline{(a, b)} = a/b$, with $a, b \in D$, we can rewrite some of the expressions given in Theorem 6.22 as follows:

$$\frac{a}{b} = \frac{c}{d} \Leftrightarrow ad = bc, \quad \frac{a}{b} + \frac{c}{d} = \frac{ad + bc}{bd}, \quad \frac{a}{b} \cdot \frac{c}{d} = \frac{ac}{bd}.$$

These equations should come as no surprise, in view of Theorem 2.20 and the fact that a, b, c, d are now considered to be members of field F.

If integral domain D is not a field, then there is some element a of D that has no multiplicative inverse in D. Since a has an inverse in quotient field F, we conclude that D' is a *proper* subset of F. What happens if integral domain D is actually a field? We leave it as an exercise for you to show that if F is the quotient field of a field E, then $E \cong F$. In this case the F we construct simply gives us the set with which we started.

EXAMPLE 6.17

Consider the integral domain D, where

$$D = \{a + b\sqrt{6} : a, b \in \mathbf{Z}\} = \mathbf{Q}[\sqrt{6}].$$

An element of D such as $2 + 3\sqrt{6}$ has no multiplicative inverse in D, but it *will* have an inverse in the quotient field F of D. Let's find out what the elements of the quotient field look like. If $a, b, c, d \in \mathbf{Z}$, then

$$\frac{a + b\sqrt{6}}{c + d\sqrt{6}} = \frac{(a + b\sqrt{6})(c - d\sqrt{6})}{(c + d\sqrt{6})(c - d\sqrt{6})} = \frac{ac - 6bd}{c^2 - 6d^2} + \frac{(bc - ad)}{c^2 - 6d^2}\sqrt{6} = r + s\sqrt{6},$$

where $r, s \in \mathbf{Q}$.

Conversely, every real number of the form, $r + s\sqrt{6}$, with $r, s \in \mathbf{Q}$, is in the quotient field, since if $r = u/v$ and $s = x/y$, where $u, v, x, y \in \mathbf{Z}$, then

$$(u/v) + (x/y)\sqrt{6} = \frac{uy + vx\sqrt{6}}{vy + 0\sqrt{6}},$$

which is the quotient of two elements of D. Thus,

$$F = \{r + s\sqrt{6} : r, s \in \mathbf{Q}\} = \mathbf{Q}(\sqrt{6}),$$

and so the quadratic field $\mathbf{Q}(\sqrt{6})$ is the quotient field of the quadratic domain $\mathbf{Q}[\sqrt{6}]$. □

EXAMPLE 6.18

The quotient field of the polynomial domain $\mathbf{R}[x]$ is $\mathbf{R}(x)$, where

$$\mathbf{R}(x) = \{f(x)/g(x) : f(x), g(x) \in \mathbf{R}[x], g(x) \neq 0\}.$$

For instance, $(x^2 - 3)/(x^3 + 5x + 1)$ is an element of $\mathbf{R}(x)$. □

The next result assures us that isomorphic integral domains give rise to isomorphic quotient fields.

THEOREM 6.24

Let F be the quotient field of integral domain D, and let F' be the quotient field of integral domain D'. If $D \cong D'$, then $F \cong F'$.

6.5 Quotient Fields and Other Ring Extensions

Proof

Let $\theta: D \to D'$ be an isomorphism. Since θ is a bijective map, we can express the elements of D' as follows:

$$D' = \{a\theta : a \in D\}.$$

So $\quad F = \{a/b : a, b \in D, b \neq z\}, \quad F' = \{a\theta/b\theta : a\theta, b\theta \in D', b \neq z\}.$

Define ψ to be the following map:

$$\psi: F \to F',$$
$$a/b \to (a/b)\psi = a\theta/b\theta,$$
$$c/d \to (c/d)\psi = c\theta/d\theta.$$

Map ψ is obviously well-defined, and surjective. We'll let you do the rather mechanical manipulations that verify these statements:

(1) $(a/b)\psi = (c/d)\psi \Rightarrow a/b = c/d,$
(2) $(a/b + c/d)\psi = (a/b)\psi + (c/d)\psi,$
(3) $[(a/b) \cdot (c/d)]\psi = (a/b)\psi \cdot (c/d)\psi.$

Therefore, ψ is an isomorphism. □

In Theorem 6.20 we proved that every field of characteristic p is an extension of field $\mathbf{Z}_p$. We shall now verify that every field of characteristic zero is an extension of the rational field $\mathbf{Q}$.

THEOREM 6.25

The rational field $\mathbf{Q}$ is embedded in every field of characteristic zero.

Proof

If $ch(F) = 0$, then F contains a subdomain $\mathbf{Z}'$ such that $\mathbf{Z} \cong \mathbf{Z}'$, by Theorem 6.20. If $E = \{a/b : a, b \in \mathbf{Z}'; b \neq z\}$, then E is a subfield of F. But E is the quotient field of $\mathbf{Z}'$, and $\mathbf{Q}$ is the quotient field of $\mathbf{Z}$; hence $\mathbf{Q} \cong E$, by Theorem 6.24. Therefore, $\mathbf{Q}$ is embedded in F. □

There are two popular methods for embedding the rational field in the real field (and, by so doing, actually developing the real numbers and their properties), but these methods ((1) sequences, and (2) Dedekind cuts) would get us into the realm of advanced calculus, for which we have no time in this course.

We now show how it is possible to "invent" the complex field if we are given the real field as a starting point.

THEOREM 6.26

If $(\mathbf{R}, +, \cdot)$ is the real field and

$$\mathbf{C} = \mathbf{R} \times \mathbf{R} = \{(a, b) : a, b \in \mathbf{R}\},$$

with addition and multiplication of ordered pairs defined as follows:

$$(a, b) + (c, d) = (a + c, b + d), \qquad (a, b) \odot (c, d) = (ac - bd, ad + bc);$$

then $(\mathbf{C}, +, \odot)$ is a field.

We leave the proof to you. In verifying that the field postulates hold, show that: $z = (0, 0)$, $u = (1, 0)$, $-(a, b) = (-a, -b)$, and that if a and b are not both zero, then

$$(a, b)^{-1} = \left(\frac{a}{a^2 + b^2}, \frac{-b}{a^2 + b^2} \right).$$

Field $\mathbf{C}$ is called the *complex field*, and each element of $\mathbf{C}$ is called a *complex number*. We now show that we have embedded $\mathbf{R}$ in $\mathbf{C}$.

THEOREM 6.27

If $\mathbf{R}' = \{(a, 0) : a \in \mathbf{R}\}$, then $\mathbf{R}'$ is a subfield of $\mathbf{C}$, and $\mathbf{R}$ is isomorphic to $\mathbf{R}'$.

Proof

It is a simple matter to show that the map, $\theta : \mathbf{R} \to \mathbf{R}'$, given by $x\theta = (x, 0)$, is an isomorphism. Since $\mathbf{R}$ is a field, then $\mathbf{R}'$ is a field. □

We now identify the real number x with the complex number $(x, 0)$ by agreeing to let x represent $(x, 0)$. That is, "$x\theta = (x, 0)$" becomes: "$x = (x, 0)$". We also define $(0, 1) = i$. Thus i is nothing more than the name we have assigned to the ordered pair $(0, 1)$. We observe that $i^2 = (0, 1)^2 = (-1, 0) = -1$, which tells us that i is not an element in any set that is isomorphic to a set of real numbers. Now

$$a, b \in \mathbf{R} \Rightarrow a + bi = (a, 0) + (b, 0)(0, 1) = (a, 0) + (0, b) = (a, b),$$

thus giving us two ways to represent a complex number. The remarks of this paragraph are summarized in the next theorem.

THEOREM 6.28

Let (a, b) be an element of field $\mathbf{C}$. If $(x, 0) = x$ and $(0, 1) = i$, then $i^2 = -1$ and $(a, b) = a + bi$.

Since the set, $\mathbf{R} \times \mathbf{R}$, under appropriate operations, extended the real field to the complex field, it is reasonable to ask what would happen if we defined similar operations on the Cartesian product, $\mathbf{C} \times \mathbf{C}$. It turns out that we can embed the complex field in a ring that is isomorphic to the quaternion ring of Problem 2.7-16. The quaternion ring is the most well-known example of that class of algebras known as *division rings*, a *division ring* being a ring with unity in which each nonzero element has a multiplicative inverse. A *commutative* division ring is, of course, a field. Quaternions were discovered in 1843 by the Irish mathematician, Sir William Rowan Hamilton.

6.5 Problems

1. Describe the quotient field F of D if:
   (a) $D = \{a + bi : a, b \in \mathbf{Z}\} = \mathbf{Q}[i]$.
   (b) $D = \{a + bi : a, b \in \mathbf{Q}\} = \mathbf{Q}(i)$.
   (c) $D = \{a + b\sqrt{2} : a, b \in \mathbf{Z}\} = \mathbf{Q}[\sqrt{2}]$.

2. Prove: If n is any square-free rational integer, then the quotient field of the quadratic domain, $\mathbf{Q}[\sqrt{n}]$, is the quadratic field, $\mathbf{Q}(\sqrt{n})$.

3. (a) Describe the quotient field of the polynomial domain $\mathbf{Z}[x]$, and show that $\mathbf{Q}[x]$ is embedded in the quotient field of $\mathbf{Z}[x]$.
   (b) Is the quotient field of $\mathbf{Z}[x]$ isomorphic to the quotient field of $\mathbf{Q}[x]$?

4. Complete the proof of Theorem 6.22.

5. Prove Theorem 6.23.

6. Let F be the quotient field of an integral domain D. Prove: If D is a field, then $F = D$.

7. Verify statements (1), (2), and (3) in the proof of Theorem 6.24.

8. Prove Theorem 6.26.

9. Show that the real field $\mathbf{R}$ is embedded in the ring of matrices, $M_2(\mathbf{R})$.

10. (a) Prove: (1) Ring $2\mathbf{Z}$ is embedded in ring $\mathbf{Z}$.
    (2) Ring $2\mathbf{Z}$ is embedded in ring $2\mathbf{Z} \times \mathbf{Z}$.
    (b) Since $2\mathbf{Z}$ is embedded in $\mathbf{Z}$, and $2\mathbf{Z}$ is embedded in $2\mathbf{Z} \times \mathbf{Z}$, does this mean that $\mathbf{Z}$ and $2\mathbf{Z} \times \mathbf{Z}$ are isomorphic, or are these two structurally distinct extensions of $2\mathbf{Z}$?

11. Prove: If $\theta : F \to F'$ is an isomorphic map of field F onto field F', where each contains $\mathbf{Q}$, then $a\theta = a$, for all $a \in \mathbf{Q}$.

12. Show that any ring $(S, +, \cdot)$ *without* a unity can be embedded in a ring *with* unity, as follows: Define an algebra $(S \times \mathbf{Z}, \oplus, \otimes)$, where

    $(x, m) \oplus (y, n) = (x + y, m + n),$
    $(x, m) \otimes (y, n) = (xy + nx + my, mn).$

    (a) Prove that $(S \times \mathbf{Z}, \oplus, \otimes)$ is a ring with unity $(z, 1)$.
    (b) Verify that $S \times \{0\}$ is a subring of $S \times \mathbf{Z}$.
    (c) Show that S is isomorphic to $S \times \{0\}$.

13. The polynomial, $x^2 - 2$, has no root in the field $\mathbf{Q}$, and so we want to embed $\mathbf{Q}$ in a field in which $x^2 - 2$ has a root. Define an algebra, $(F, +, \cdot)$, as follows:

    $F = \{(a, b) : a, b \in \mathbf{Q}\},$
    $(a, b) + (c, d) = (a + c, b + d),$
    $(a, b) \cdot (c, d) = (ac + 2bd, ad + bc).$

Prove:

(a) $(F, +, \cdot)$ is a field.

(b) If $\mathbf{Q}' = \{(a, 0) : a \in \mathbf{Q}\}$, then $\mathbf{Q} \cong \mathbf{Q}'$.
Hence, we let $(a, 0) = a$.

(c) $(0, 1)$ is a root in F of the polynomial, $x^2 - 2$.

(d) If $(0, 1)$ is denoted by the symbol, $\sqrt{2}$, then $F = \{a + b\sqrt{2} : a, b \in \mathbf{Q}\}$.
(Note: We have embedded $\mathbf{Q}$ in the quadratic field $\mathbf{Q}(\sqrt{2})$.)

14 We shall construct an extension K of the complex field $\mathbf{C}$ as follows:
Let $K = \mathbf{C} \times \mathbf{C} = \{(\alpha, \beta) : \alpha, \beta \in \mathbf{C}\}$, and let the operations on K be as follows: For all $\alpha, \beta, \mu, \sigma$ in $\mathbf{C}$,

$$(\alpha, \beta) + (\mu, \sigma) = (\alpha + \mu, \beta + \sigma),$$
$$(\alpha, \beta) \otimes (\mu, \sigma) = (\alpha\mu - \beta\sigma^*, \alpha\sigma + \beta\mu^*),$$

where σ^* and μ^* are the conjugates of σ and μ, respectively.

(a) Prove that $(K, +, \otimes)$ is isomorphic to the quaternion ring of Problem 2.7-16.

(b) Show that if $(\alpha, \beta) \neq (0, 0)$, then

$$(\alpha, \beta)^{-1} = \left(\frac{\alpha^*}{N(\alpha) + N(\beta)}, \frac{-\beta}{N(\alpha) + N(\beta)} \right).$$

(Recall that $N(x + yi) = x^2 + y^2$.)

(c) Prove: If $\mathbf{C}' = \{(\alpha, 0) : \alpha \in \mathbf{C}\}$, then $\mathbf{C} \cong \mathbf{C}'$.

(d) Verify: If we define $(0, 1) = j$; $(a + bi, 0) = a + bi$, where $a, b \in \mathbf{R}$; and $ij = k$, then: $(a + bi, c + di) = a + bi + cj + dk$, where $a, b, c, d \in \mathbf{R}$. Also, $i^2 = j^2 = k^2 = -1$, $jk = i$, $ki = j$, $ji = -k$, $kj = -i$, $ik = -j$.
(Note: In Problem 2.4-4 you discovered that the subset $\mathbf{Q}_8$ of K is a group under multiplication, where $\mathbf{Q}_8 = \{\pm 1, \pm i, \pm j, \pm k\}$.)

15 Let $\alpha = 2 - 3i + k$ and $\beta = 2j + 5k$, where α and β are quaternions. (See Problem 14.)

(a) As elements of $\mathbf{C} \times \mathbf{C}$, $\alpha = \underline{\quad}$ and $\beta = \underline{\quad}$.
Compute the following:

(b) $\alpha + \beta, -\beta, \alpha - \beta$.   (c) α^{-1}, β^{-1}.   (d) $\alpha\beta, \beta\alpha$.

16 (a) Hamilton's discovery of quaternions brought him fame, but probably ruined him mathematically. Why? (See Appendix E-11.)

(b) What did Hamilton have to do with converting *algebra* to *algebras*?

6.6 Algebraic Extension Fields

Chapter 4 introduced some useful ideas about polynomials. We now return to polynomials, approaching the subject by means of the mathematics developed in Chapter 6.

For centuries mathematicians have been attempting to "solve" polynomial equations, or to assure themselves that solutions exist. In the language of today,

6.6 Algebraic Extension Fields

they were trying to find roots of polynomials over various fields. We know, for instance, that $x^2 + \bar{1}$ has no root in Z_3. The question arises: If $f(x)$ has no root in a field F, is there an extension field of F in which $f(x)$ *does* have a root? We shall answer the question in the affirmative in this section. This quite remarkable and important result, attributed to the mathematician Kronecker (1823–1891), assures us that, given *any* polynomial over *any* field, there must exist an extension field in which the polynomial has a root.

We now have at our disposal the mathematical machinery that is necessary in order to arrive at the main result of this section. We know that if S is a commutative ring with unity, and if N is an ideal of S, then S/N is a commutative ring with unity (Theorem 6.8). Furthermore, factor ring S/N is a field if and only if N is a maximal ideal of S, according to Theorems 6.9 and 6.15. Suppose $p(x) \in F[x]$, where $F[x]$ is, of course, the integral domain consisting of all polynomials with coefficients in the field F. If we let $S = F[x]$, $N = \langle p(x) \rangle$, and $E = S/N = F[x]/\langle p(x)\rangle$, then E is a field if and only if $\langle p(x)\rangle$ is a maximal ideal of $F[x]$. (We proved, in Theorem 6.6, that every ideal of $F[x]$ is a principal ideal.)

If $E = F[x]/\langle p(x)\rangle$, then our immediate goal is to establish that:

(1) E is a field iff $p(x)$ is a prime polynomial in $F[x]$ (Theorems 6.29, 6.30).

(2) E is an extension field of F (Theorem 6.31).

(3) $p(x)$ has a root in E (Theorem 6.32).

THEOREM 6.29

Let $p(x) \in F[x]$. Ideal $\langle p(x) \rangle$ is a maximal ideal of $F[x]$ if and only if $p(x)$ is irreducible over F.

Proof

(1) Assume that $p(x)$ is not prime over F. Then there exist polynomials $f(x), g(x) \in F[x]$, each of degree at least one, such that $p(x) = f(x)g(x)$. Hence,

$$\langle p(x) \rangle \subseteq \langle f(x) \rangle \subseteq F[x].$$

To verify that the *proper* inclusions, $\subset$, hold in the above, we note that $f(x) \in \langle f(x) \rangle$, but $f(x) \notin \langle p(x) \rangle$, and that if 1 is the unity of F, then $1 \in F[x]$, but $1 \notin \langle f(x) \rangle$. Therefore, $\langle p(x) \rangle$ is not a maximal ideal of $F[x]$.

(2) Assume that $p(x)$ is prime over F, and there is an $h(x)$ in $F[x]$ such that

$$\langle p(x) \rangle \subseteq \langle h(x) \rangle \subseteq F[x].$$

This implies that $p(x) = h(x) \cdot k(x)$, for some $k(x) \in F[x]$. But $p(x)$ is prime, so either $\deg h(x) = 0$ or $\deg k(x) = 0$. Now

$$\deg h(x) = 0 \Rightarrow \langle h(x)\rangle = F[x], \quad \deg k(x) = 0 \Rightarrow \langle h(x)\rangle = \langle p(x)\rangle.$$

Therefore, $\langle p(x) \rangle$ is a maximal ideal of $F[x]$. ☐

The following result is a corollary to Theorem 6.29.

THEOREM 6.30

Factor ring $F[x]/\langle p(x) \rangle$ is a field if and only if $p(x)$ is irreducible over F.

We know that $E = \{f(x) + \langle p(x) \rangle : f(x) \in F[x]\}$. Employing our usual notation for an equivalence class, we write $f(x) + \langle p(x) \rangle = \overline{f(x)}$. We now verify that each coset $\overline{f(x)}$ contains either the zero polynomial or exactly one polynomial whose degree is less than the degree of $p(x)$.

THEOREM 6.31

Let $E = F[x]/\langle p(x) \rangle$, with $p(x)$ prime over F, and deg $p(x) = n$.
(a) Then $E = \{\overline{s(x)} : s(x) \in F[x]; s(x) = 0 \text{ or deg } s(x) < n\}$; if $t(x) = 0$ or deg $t(x) < n$, then $\overline{t(x)} = \overline{s(x)}$ only if $t(x) = s(x)$.
(b) If $F' = \{\bar{a} : a \in F\}$, then $F \cong F' < E$, and so E is an extension field of F.

Proof

(a) Let $f(x)$ be any element of $F[x]$. By the Division Theorem, there exist unique $q(x)$ and $s(x)$ in $F[x]$ such that

$$f(x) = p(x) \cdot q(x) + s(x), \quad \text{with } s(x) = 0 \quad \text{or} \quad \deg s(x) < n.$$

Hence $f(x) \in s(x) + \langle p(x) \rangle$, so $f(x) + \langle p(x) \rangle = s(x) + \langle p(x) \rangle$, and $\overline{f(x)} = \overline{s(x)}$. Therefore, $E = \{\overline{s(x)} : s(x) = 0 \text{ or deg } s(x) < n\}$.

Now suppose that $\overline{t(x)} = \overline{s(x)}$, where $t(x) = 0$ or deg $t(x) < n$. Then

$$t(x) = s(x) + p(x) \cdot k(x), \quad \text{for some } k(x) \in F[x],$$

and so $\quad t(x) - s(x) = p(x) \cdot k(x)$.

But if $t(x) - s(x) \neq 0$, then

$$n > \deg[t(x) - s(x)] = \deg[p(x)k(x)] = \deg p(x) + \deg k(x) \geq n,$$

and so $n > n$, an impossibility. Thus $t(x) - s(x) = 0$, and so $t(x) = s(x)$.
(b) Define $\theta : F \to F'$ to be the map such that $a\theta = \bar{a}$, with $a \in F$. This map is obviously surjective. Also, $\bar{a} = \bar{b} \Rightarrow a = b$, by part (a). Thus, θ is injective. The operations are preserved, for

$$(a + b)\theta = \overline{a + b} = \bar{a} + \bar{b} = a\theta + b\theta, \quad (ab)\theta = \overline{ab} = \bar{a} \cdot \bar{b} = (a\theta)(b\theta).$$

Therefore, θ is an isomorphism. □

Field $F[x]/\langle p(x) \rangle$ is called an *algebraic extension* of F.

6.6 Algebraic Extension Fields

EXAMPLE 6.19

Let $p(x) = x^2 + \bar{1}$, where $p(x) \in \mathbf{Z}_3[x]$. Then $p(x)$ is prime over $\mathbf{Z}_3$. According to Theorem 6.31,

$$E = \frac{\mathbf{Z}_3[x]}{\langle x^2 + \bar{1} \rangle} = \{\overline{a + bx} : a, b \in \mathbf{Z}_3\}.$$

So $\quad E = \{\bar{\bar{0}}, \bar{\bar{1}}, \bar{\bar{2}}, \bar{x}, \overline{\bar{1} + x}, \overline{\bar{2} + x}, \overline{\bar{2}x}, \overline{\bar{1} + \bar{2}x}, \overline{\bar{2} + \bar{2}x}\}.$

(Unfortunately, we must use double bars in this example. Be sure you understand why this is necessary.)

Now $F = \mathbf{Z}_3 = \{\bar{0}, \bar{1}, \bar{2}\}$, and $F' = \{\bar{\bar{0}}, \bar{\bar{1}}, \bar{\bar{2}}\}$, so field F is embedded in field E. Note that we have constructed algebraic extension field E of order 9 by starting with a field F of order 3.

Recall that set E is a partition of the set of all polynomials over $\mathbf{Z}_3$. For instance,

$$\bar{\bar{0}} = \langle x^2 + \bar{1} \rangle = \{(x^2 + \bar{1}) \cdot f(x) : f(x) \in \mathbf{Z}_3[x]\},$$
$$\overline{\bar{2} + x} = (\bar{2} + x) + \langle x^2 + \bar{1} \rangle = \{(\bar{2} + x) + (x^2 + \bar{1}) \cdot f(x) : f(x) \in \mathbf{Z}_3[x]\}.$$

By definition of addition and multiplication of cosets and of polynomials, we add and multiply elements of field E as follows, for example:

$$\overline{\bar{1} + x} + \overline{\bar{2} + x} = \overline{(\bar{1} + x) + (\bar{2} + x)} = \overline{\bar{2}x},$$
$$\overline{(\bar{1} + x)} \cdot \overline{(\bar{2} + x)} = \overline{(\bar{1} + x)(\bar{2} + x)} = \overline{\bar{2} + x^2} = \bar{\bar{1}} + \overline{(\bar{1} + x^2)} = \bar{\bar{1}} + \bar{\bar{0}} = \bar{\bar{1}}.$$

Thus, $\overline{\bar{2} + x}$ is the multiplicative inverse of $\overline{\bar{1} + x}$. □

We now perform our usual stunt of simplifying the notation in the extension field E. Referring to Theorem 6.31: Since $\theta: F \to F'$ is an isomorphism, with $a\theta = \bar{a}$, we shall agree to let $\bar{a} = a$. The next theorem makes clear the form that E now takes, and shows that r is a root of $p(x)$.

THEOREM 6.32

Let $E = F[x]/\langle p(x) \rangle$, where $p(x)$ is prime over F, and $\deg p(x) = n$. If $\bar{x} = r$, and $\bar{a} = a$, for $a \in F$, then:

(a) $\overline{f(x)} = f(r)$, for all $f(x) \in F[x]$,
(b) $E = \{a_0 + a_1 r + \cdots + a_{n-1} r^{n-1} : a_i \in F\}$, and
(c) $p(r) = 0$, so $p(x)$ has a root in E.

Proof

(a) Let $f(x) = b_0 + b_1 x + \cdots + b_k x^k$, with $b_i \in F$. Then

$$\overline{f(x)} = \overline{b_0 + b_1 x + \cdots + b_k x^k} = \overline{b_0} + \overline{b_1 x} + \cdots + \overline{b_k x^k}$$
$$= \overline{b_0} + \overline{b_1} \bar{x} + \cdots + \overline{b_k}(\bar{x})^k = b_0 + b_1 r + \cdots + b_k r^k = f(r).$$

(b)
$$E = \{\overline{s(x)} : s(x) \in F[x], \deg s(x) < n\}$$
$$= \{\overline{a_0 + a_1 x + \cdots + a_{n-1} x^{n-1}} : a_i \in F\}$$
$$= \{a_0 + a_1 r + \cdots + a_{n-1} r^{n-1} : a_i \in F\}.$$

(c) $0 = \overline{0} = \langle p(x) \rangle = \overline{p(x)} = p(r)$. □

EXAMPLE 6.20

Continuing with Example 6.19, we now can write E as follows:

$$E = \mathbf{Z}[x]/\langle x^2 + \overline{1} \rangle = \{\overline{0}, \overline{1}, \overline{2}, r, \overline{1} + r, \overline{2} + r, \overline{2}r, \overline{1} + \overline{2}r, \overline{2} + \overline{2}r\}.$$

We proved that $r^2 + \overline{1} = \overline{0}$. Also, $(\overline{2}r)^2 + \overline{1} = r^2 + \overline{1} = \overline{0}$. So r and $\overline{2}r$ are both roots of $x^2 + \overline{1}$ in E. Since a polynomial of degree n over a field cannot have more than n roots, we know that these are the only roots $x^2 + \overline{1}$ has in E, and $x^2 + \overline{1} = (x - r)(x - \overline{2}r)$.

Since $r^2 + \overline{1} = \overline{0}$, we have that $r^2 = -\overline{1} = \overline{2}$. Hence, when we multiply elements of E, and obtain r^2 in the product, we can always replace r^2 by $\overline{2}$. For instance,

$$(\overline{2} + r)(\overline{1} + \overline{2}r) = \overline{2} + \overline{5}r + \overline{2}r^2 = \overline{2} + \overline{2}r + \overline{2}(\overline{2}) = \overline{6} + \overline{2}r = \overline{2}r.$$ □

EXAMPLE 6.21

Let $p(x) = x^3 - 3x + 1 \in \mathbf{Q}[x]$. We apply Theorems 4.21 and 4.13 to verify that $p(x)$ is prime over $\mathbf{Q}$. By Theorem 6.32, $p(x)$ has a root r in E, with

$$E = \mathbf{Q}[x]/\langle x^3 - 3x + 1 \rangle = \{a_0 + a_1 r + a_2 r^2 : a_i \in \mathbf{Q}, r^3 = 3r - 1\}.$$

In E, $p(x)$ factors as follows:

$$p(x) = x^3 - 3x + 1 = (x - r)(x^2 + rx + (r^2 - 3)).$$

Comparing this example with Example 4.21, we note that E is structurally the algebraic number field, $\mathbf{Q}(r)$. □

You might well ask: What is the difference between what we are doing in the present section, and what we did in Section 4.7? In Section 4.7, we *assumed* that we had an extension field E of a field F, and *assumed* that there was an element s in E and a prime polynomial $p(x)$ in $F[x]$ such that $p(s) = 0$. We then obtained the extension field $F(s)$, which contains s, where $F \leq F(s) \leq E$. In the present section we start with a prime polynomial $p(x)$ in $F[x]$, and from this we obtain the extension field $F[x]/\langle p(x) \rangle$, in which there is a number r such that $p(r) = 0$. It is satisfying to know that the two routes (no pun intended) come together, by virtue of the fact that $F(r) = F[x]/\langle p(x) \rangle$; and if r and s are roots of $p(x)$, then $F(r) \cong F(s)$, and so $F(s) \cong F[x]/\langle p(x) \rangle$.

The main result, given in Theorem 6.33, is essentially a corollary to Theorems 6.31 and 6.32.

THEOREM 6.33 *(Kronecker)*

Let F be a field and let $f(x)$ be any polynomial of degree ≥ 1 in $F[x]$. Then there exists an algebraic extension field E of F and an element $r \in E$ such that $f(r) = 0$.

Proof

By Theorem 4.9, there exist prime polynomials $p_i(x)$ in $F[x]$ such that

$$f(x) = p_1(x)p_2(x) \cdots p_k(x), \quad \text{with } k \geq 1.$$

Let $E_i = F[x]/\langle p_i(x)\rangle$, for $i = 1, 2, \ldots, k$. By Theorem 6.32, there is an element r_i in E_i such that $p_i(r_i) = 0$, and hence, $f(r_i) = 0$. Therefore, $f(x)$ has a root in each one of the k factor fields that can be constructed. □

EXAMPLE 6.22

Let $f(x) = x^5 - 6x^3 + x^2 + 9x - 3 \in \mathbf{Q}[x]$. If $f(x)$ has a root in $\mathbf{Q}$, it would be one of the integers: $1, -1, 3, -3$. None of these is a root, so we seek an extension field of $\mathbf{Q}$ in which $f(x)$ has a root. If $f(x)$ is factorable, it must factor into the product of two polynomials, one of degree 2 and the other of degree 3. We find that

$$f(x) = (x^3 - 3x + 1)(x^2 - 3),$$

where each of the factors is prime over $\mathbf{Q}$. We can construct two factor fields that are extensions of $\mathbf{Q}$:

$$E_1 = \mathbf{Q}[x]/\langle x^3 - 3x + 1\rangle = \{a_0 + a_1 r + a_2 r^2 : a \in \mathbf{Q}, r^3 = 3r - 1\},$$
$$E_2 = \mathbf{Q}[x]/\langle x^2 - 3\rangle = \{b_0 + b_1 t : b_i \in \mathbf{Q}, t^2 = 3\}.$$

So r is a root of $f(x)$ over field E_1, and t is a root of $f(x)$ over field E_2. Field E_1 is a cubic number field, while E_2 is a quadratic number field. □

The method of extending a field that we have described was first used by Cauchy when he defined the complex field $\mathbf{C}$ as an extension of the real field $\mathbf{R}$ by letting $\mathbf{C} = \mathbf{R}[x]/\langle x^2 + 1\rangle$. Then Kronecker generalized the method so that it would apply to any field F and to any polynomial in $F[x]$.

We saw, by Theorem 6.21, that the order of any finite field of characteristic p must be p^n, for some $n \geq 1$. As a result of Theorem 6.32 we know that if $m(x)$ is a prime polynomial of degree n over $\mathbf{Z}_p$, then $\mathbf{Z}_p[x]/\langle m(x)\rangle$ is a field of order p^n. Conversely, it can be proved that, given *any* prime p and *any* integer n, there is a prime polynomial of degree n in $\mathbf{Z}_p[x]$, and hence there exists a field of order p^n. Furthermore, this field is structurally unique. We summarize these remarks in the next theorem, the proof of which will not be given.

THEOREM 6.34

Let $n, p \in \mathbf{Z}^+$, where p is a prime.

(a) There is a prime polynomial in $\mathbf{Z}_p[x]$ of degree n, and hence there is a field of order p^n.

(b) Any two fields of order p^n are isomorphic.

Once we have found a prime polynomial $m(x)$ of degree n in $\mathbf{Z}_p[x]$, we can construct the field E of order p^n, where

$$E = \frac{\mathbf{Z}_p[x]}{\langle m(x) \rangle} = \{a_0 + a_1 r + \cdots + a_{n-1} r^{n-1} : a_i \in \mathbf{Z}_p, \ m(r) = 0\}.$$

This unique field of order p^n is called a *Galois Field*, and is denoted by $GF(p^n)$. It can also be verified that the multiplicative group of $GF(p^n)$ is the cyclic group of order $p^n - 1$.

EXAMPLE 6.23

We want to construct $GF(2^3)$, so we must find a prime polynomial over $\mathbf{Z}_2$ of degree 3. There are two prime polynomials in $\mathbf{Z}_2[x]$:

$$x^3 + x^2 + \bar{1}, \qquad x^3 + x + \bar{1}.$$

Then

$$E_1 = \mathbf{Z}_2[x]/\langle x^3 + x^2 + \bar{1} \rangle = \{a + br + cr^2 : a, b, c \in \mathbf{Z}_2, r^3 = \bar{1} + r^2\},$$
$$E_2 = \mathbf{Z}_2[x]/\langle x^3 + x + \bar{1} \rangle = \{a + bs + cs^2 : a, b, c \in \mathbf{Z}_2, s^3 = \bar{1} + s\}.$$

According to Theorem 6.34(b), fields E_1 and E_2 are isomorphic. To find an isomorphism, $\theta: E_1 \to E_2$, let $r\theta = a + bs + cs^2$. If we can find a, b, and c, then θ is completely determined. Since $r^3 = \bar{1} + r^2$, then

$$r^3 \theta = (\bar{1} + r^2)\theta = \bar{1}\theta + (r\theta)^2 = \bar{1} + (r\theta)^2.$$
So
$$(a + bs + cs^2)^3 = \bar{1} + (a + bs + cs^2)^2.$$

If we simplify each side of the preceding equation, we end up with three equations in the three unknowns. It turns out that there are three solutions. Hence, there are three possibilities for the image of r:

$$r \to \bar{1} + s, \qquad r \to \bar{1} + s^2, \qquad r \to \bar{1} + s + s^2.$$

If we choose to let $r\theta = \bar{1} + s$, we then have that

$$\theta: a + br + cr^2 \to (a + b + c) + bs + cs^2.$$

We leave it to you to prove that θ is an isomorphism. □

6.6 Problems

1. Let $E = \mathbf{Z}_3[x]/\langle x^2 + \bar{1} \rangle$. (See Examples 6.19 and 6.20.)
   (a) Find the additive inverse of each element of E.
   (b) Find the multiplicative inverse of each nonzero element of E.
   (c) To which group is $(E, +)$ isomorphic?
   (d) Find the multiplicative orders of r and $\bar{1} + \bar{2}r$.

(e) To which group is $(E^*, \cdot)$ isomorphic?

(f) What is the characteristic of E?

(g) List all monic polynomials in $\mathbf{Z}_3[x]$ of degree 1 or 2 that are prime over $\mathbf{Z}_3$, and show that each element of E is a root of exactly one of these polynomials.

2. These questions refer to Example 6.21, where $E = \mathbf{Q}[x]/\langle x^3 - 3x + 1\rangle$.

(a) What is the characteristic of E?

(b) Find $-(5 - 4r + 2r^2)$ and $(5 - 4r + 2r^2)^{-1}$.

(c) Show that $(E, +) \cong (\mathbf{Q} \times \mathbf{Q} \times \mathbf{Q}, +)$.

(d) Is $(E^*, \cdot)$ cyclic?

3. Referring to Example 6.22, show that there is no map $\theta: E_1 \to E_2$ that is an isomorphism. (Hint: Assume that an isomorphic map θ exists. Then $r \to c + dt$, for some $c, d \in \mathbf{Q}$. Find r^{-1} and $(c + dt)^{-1}$, and show that $(r^{-1})\theta \neq (r\theta)^{-1}$.)

4. Referring to Example 6.23:

(a) List the elements in E_1, and factor $x^3 + x^2 + \bar{1}$ in $E_1[x]$, if $E_1 = \mathbf{Z}_2[x]/\langle x^3 + x^2 + \bar{1}\rangle$.

(b) Verify that $(E_1^*, \cdot)$ is the cyclic group of order 7, by showing that $E_1^* = \langle r \rangle$.

(c) Compute: $(\bar{1} + r + r^2)(r + r^2)$, $-(\bar{1} + r^2)$, $(\bar{1} + r + r^2)^{-1}$.

(d) Find the additive order of each element of E_1, and identify $(E_1, +)$.

(e) Find the minimal polynomial over $\mathbf{Z}_2$ of each element of E_1.

5. (a) Explain why $GF(2^3) = \mathbf{Z}_2[x]/\langle x^3 + x^2 + \bar{1}\rangle$.

(b) Find all automorphisms of $GF(2^3)$, and identify the automorphism group. (See Problem 4.)

6. Prove that the map θ in Example 6.23, defined by
$$(a + br + cr^2)\theta = (a + b + c) + bs + cs^2,$$
is an isomorphism.

7. In Example 6.23, if ψ is the map, $r\psi = \bar{1} + s$, give the image of each element of E_1 under ψ.

8. Give a formula for the product of any two elements in each of these fields:

(a) $\mathbf{Q}[x]/\langle x^2 + x + 1\rangle$. (b) $\mathbf{Z}_2[x]/\langle x^2 + x + \bar{1}\rangle$.

9. If $f(x) \in \mathbf{Q}[x]$, where $f(x) = (x^3 + 5x - 2)(x^2 - 4x + 1)$, find two extension fields of $\mathbf{Q}$ in which $f(x)$ has a root.

10. Verify that $\mathbf{R}[x]/\langle x^2 + 1\rangle$ is the complex field $\mathbf{C}$.

11. Prove: If $E_1 = \mathbf{Q}[x]/\langle x^2 - 3x + 1\rangle$, and $E_2 = \mathbf{Q}[x]/\langle x^2 - 5\rangle$, then $E_1 = E_2$.

12. If $\deg p(x) = 1$, describe $F[x]/\langle p(x)\rangle$.

13. Show that if $g(x)$ is not prime over F, then ring $F[x]/\langle g(x)\rangle$ has zero divisors.

14 Verify: If $p(x)$ is in $F[x]$, and if $q(x)$ is an associate of $p(x)$, then $\langle p(x) \rangle = \langle q(x) \rangle$, and hence, $F[x]/\langle p(x) \rangle = F[x]/\langle q(x) \rangle$.

15 Given that $p(x), h(x), k(x) \in F[x]$, with deg $h(x) = 0$, prove:
(a) $\langle h(x) \rangle = F[x]$.
(b) If $p(x) = h(x) \cdot k(x)$, then $\langle k(x) \rangle = \langle p(x) \rangle$.

16 Give the addition and multiplication tables of $GF(2^2)$. Is $(GF(2^2), +)$ isomorphic to $\mathbf{Z}_4$ or K_4?

17 Galois Field $GF(p^n)$ can be described as follows:
$$GF(p^n) = \{a_0 + a_1 r + \cdots + a_{n-1} r^{n-1} : a_i \in \mathbf{Z}_p,$$
$$r^n = b_0 + b_1 r + \cdots + b_{n-1} r^{n-1}\},$$
where $m(x)$ is prime over $\mathbf{Z}_p$, and $m(x) = x^n - b_{n-1} x^{n-1} - \cdots - b_1 x - b_0$. Describe $GF(p^n)$ by finding an appropriate prime polynomial $m(x)$, if
(a) $p = 5, n = 2$.
(b) $p = 5, n = 3$.
(c) $p = 2, n = 5$.
(d) $p = 3, n = 5$.

18 (a) List all abelian groups of order 3^5.
(b) What is the characteristic of $GF(3^5)$?
(c) Use (a) and (b) to identify the additive group $(GF(3^5), +)$.
(d) Identify the group $(GF(p^n), +)$.

19 Let $E = GF(p^n)$, and let $p^n = k$. Assuming that $(E^*, \cdot)$ is cyclic, verify that every element of $GF(p^n)$ is a root of $f(x) \in \mathbf{Z}_p[x]$, where $f(x) = x^k - x$.

20 (a) Cauchy got into trouble over two non-mathematical subjects. What were they? (See Appendix E–9.)

(b) What was an important contribution that Cauchy made to abstract algebra?

21 (a) What was unusual about Kronecker's position at the University of Berlin? (See Appendix E–16.)

(b) What was Kronecker's attitude toward mathematics?

6.7 Splitting Field and Galois Group of a Polynomial

In this section we describe some really beautiful mathematics that relates *polynomials*, the *field* in which the roots of a polynomial are found, and an important *group* of automorphisms of this field. Then in Section 6.8 we shall discuss some famous old algebraic and geometric problems that were finally solved, once and for all, through the use of the concepts we develop in the present section. These two sections are giving you a brief glimpse of what is known as "Galois theory". This theory is attributed chiefly to two men: Galois and Abel.

Suppose that $f(x) \in F[x]$. This means, of course, that the coefficients of the polynomial $f(x)$ are in the field F. Now if F has no proper subfield that contains the coefficients, then F is called the *coefficient field* of $f(x)$. Thus, there may be many fields that contain the coefficients, but only one coefficient field.

6.7 Splitting Field and Galois Group of a Polynomial

EXAMPLE 6.24

Let $f(x) = 4x^3 - 5x + 7$, and $g(x) = 4x^3 - \sqrt{5}x + 7$. The coefficients of both $f(x)$ and $g(x)$ lie in **R** and in **C**. However, the coefficient field of $f(x)$ is **Q**, while the coefficient field of $g(x)$ is the quadratic field $\mathbf{Q}(\sqrt{5})$. □

Let $f(x)$ be a polynomial of degree n in $F[x]$. It may happen that every root of $f(x)$ is in F. Suppose $f(x)$ has no roots in F. By Kronecker's Theorem there is an algebraic extension field E_1 of F in which $f(x)$ has a root r_1, and by the Factor Theorem, $f(x) = (x - r_1)f_1(x)$, where $f_1(x) \in E_1[x]$. Now if every root of $f(x)$ is in field E_1, then $f_1(x)$ factors into a product of linear factors over E. However, if $f_1(x)$ has a factor, $g_1(x)$, of degree at least two that is prime in $E_1[x]$, then, by the Kronecker Theorem, there is an extension field E_2 of E_1 in which $g_1(x)$ has a root r_2, and so $(x - r_2)$ is a factor of $g_1(x)$. Thus,

$$f(x) = (x - r_1)(x - r_2)f_2(x), \quad \text{with } f_2(x) \in E_2[x].$$

After k extensions, where $0 \leq k < n$, we arrive at an extension field E_k of F that contains *all* the roots of $f(x)$. That is,

$$F = E_0 < E_1 < E_2 < \cdots < E_k = E, \quad \text{and if } f(r_i) = 0, \quad \text{then } r_i \in E.$$

A field E which contains all the roots of a polynomial $f(x)$ is called a *root field* of $f(x)$. If E is a root field of $f(x)$, and if E contains no proper subfield that is a root field of $f(x)$, then E is called the *splitting field* of $f(x)$. Thus, the splitting field E is the "smallest" field in which $f(x)$ "splits," or factors, into a product of first-degree polynomials. A polynomial can have several root fields, but exactly one splitting field. Except for the "uniqueness" part, the following theorem is a corollary to the Kronecker Theorem.

THEOREM 6.35

Let $f(x)$ be a polynomial over a field F. Then $f(x)$ has a unique splitting field E, where E is an algebraic extension field of F.

Let it be noted that this is an "appreciation of Galois Theory" section, and we shall do very little in the way of proving theorems in it. If we did otherwise, the section would expand into one or two more chapters!

EXAMPLE 6.25

Let $f(x) = x^2 - 3$. Then $f(x) \in \mathbf{Q}[x]$, and **Q** is the coefficient field. The roots being $\sqrt{3}$ and $-\sqrt{3}$, the real field **R** and the complex field **C** are both root fields of $f(x)$. But a smaller root field of $f(x)$ is:

$$\mathbf{Q}[x]/\langle x^2 - 3 \rangle = \{a + b\sqrt{3} : a, b \in \mathbf{Q}\} = \mathbf{Q}(\sqrt{3}).$$

Of course, $\quad \mathbf{Q} < \mathbf{Q}(\sqrt{3}) < \mathbf{R} < \mathbf{C},$

and the quadratic field $\mathbf{Q}(\sqrt{3})$ is the splitting field of $x^2 - 3$.

If $g(x) = x^2 - 4$, then **Q** is both the coefficient field and the splitting field of $g(x)$.

In general, suppose $f(x) = ax^2 + bx + c \in \mathbf{Z}[x]$. If r is a root, then $r = s + t\sqrt{n}$, with $s, t \in \mathbf{Q}$, and $n \in \mathbf{Z}$, where n is square-free. If $n = 0$, then $f(x)$ splits in **Q**, and if $n \neq 0$, then $f(x)$ splits in the quadratic field, $\mathbf{Q}(\sqrt{n})$. □

EXAMPLE 6.26

Let $f(x) = x^4 - x^2 - 2 \in \mathbf{Q}[x]$. We wish to find the splitting field of $f(x)$. Now $f(x) = (x^2 + 1)(x^2 - 2)$, so define

$$E_1 = \mathbf{Q}[x]/\langle x^2 + 1 \rangle = \{a + br : a, b \in \mathbf{Q}, r^2 = -1\} = \mathbf{Q}(r).$$

Then r is a root of $x^2 + 1$, r is in the quadratic field $\mathbf{Q}(r)$, and $f(x)$ factors as follows in the integral domain, $E_1[x]$:

$$f(x) = (x - r)(x + r)(x^2 - 2).$$

Now $x^2 - 2$ is prime over E_1. Define E as follows:

$$\begin{aligned} E &= E_1[x]/\langle x^2 - 2 \rangle = \{c + ds : c, d \in E_1, s^2 = 2\} \\ &= E_1(s) = (\mathbf{Q}(r))(s) = \{(a_1 + a_2 r) + (a_3 + a_4 r)s : a_i \in \mathbf{Q}\} \\ &= \{(a_1 + a_3 s) + (a_2 + a_4 s)r\} = \{h + kr : h, k \in E_2; E_2 = \mathbf{Q}(s)\} \\ &= (\mathbf{Q}(s))(r) = E_2(r). \end{aligned}$$

Since $E = (\mathbf{Q}(r))(s) = (\mathbf{Q}(s))(r)$, we write: $E = \mathbf{Q}(r, s)$. Thus

$$\mathbf{Q} < \mathbf{Q}(r) < \mathbf{Q}(r, s) = E,$$
and
$$\mathbf{Q} < \mathbf{Q}(s) < \mathbf{Q}(r, s) = E.$$

In this example we can let $r = i$, and $s = \sqrt{2}$. Then

$$E = \{a_1 + a_2 i + a_3 \sqrt{2} + a_4 i\sqrt{2} : a_i \in \mathbf{Q}\} = \mathbf{Q}(i, \sqrt{2}),$$

and E is the splitting field of $f(x)$. To obtain E we made two successive algebraic extensions. It is readily checked that $E = \mathbf{Q}(t)$, where $t = r + s$. Hence, one properly chosen extension can suffice! □

In Problem 6.3–18 you were asked to prove that, under the operation of composition of maps, the set $A(E)$ of automorphisms of a field E is a group, and that $A(E/F)$ is a subgroup of $A(E)$, if $A(E/F)$ is the set of automorphisms of E that fix a subfield F of E. (Recall that automorphism, $\theta : E \to E$, fixes F if $a\theta = a$, for all $a \in F$.) We state these facts formally now.

THEOREM 6.36

(a) If F is a subfield of E, and if $A(E/F)$ is the set of automorphisms of E that fix F, then $(A(E), \circ)$ is a group, and

$$(A(E/F), \circ) \leq (A(E), \circ).$$

(b) If E is a field of characteristic 0, then every automorphism of E fixes the rational field $\mathbf{Q}$. That is, $A(E/\mathbf{Q}) = A(E)$.

Proof

(a) Do Problem 6.3–18, if you have not already done so.

(b) By Theorem 6.25, $\mathbf{Q}$ is a subfield of every field of characteristic 0, so $\mathbf{Q} \le E$. If $\theta: E \to E$ is an automorphism, then $1\theta = 1$, since 1 is the unity of E. From this (as in Problem 6.3–14), we arrive at the conclusion that $(a/b)\theta = a/b$, for all $(a/b) \in \mathbf{Q}$. ☐

If F is the coefficient field of a polynomial $f(x)$, and if E is the splitting field of $f(x)$, then the group $(A(E/F), \circ)$ is called the *Galois group* of $f(x)$. The next theorem (which is more general than our needs demand) reveals important information about the elements of the Galois group.

THEOREM 6.37

Let $f(x) \in F[x]$ and let E be a root field of $f(x)$. If r is a root of $f(x)$ and $\theta \in A(E/F)$, then $r\theta$ is a root of $f(x)$.

Proof

Let $f(x) = a_n x^n + \cdots + a_1 x + a_0 \in F[x]$. If r is a root of $f(x)$ and $\theta \in A(E/F)$, then $f(r) = 0$ and $a_i \theta = a_i$. Thus,

$$0 = 0\theta = [f(r)]\theta = (a_n r^n + \cdots + a_1 r + a_0)\theta$$
$$= a_n(r\theta)^n + \cdots + a_1(r\theta) + a_0 = f(r\theta). \quad \square$$

According to Theorem 6.37, if F is the coefficient field of $f(x)$ and E is the splitting field of $f(x)$, then each root of $f(x)$ must map to a root under any automorphism that belongs to the Galois group of the polynomial. If E is the splitting field of $f(x)$, it can be shown that

$$E = F(r_1, r_2, \ldots, r_m),$$

where the r_i are roots of $f(x)$, and if the degree of r_i is $n_i + 1$, then

$$E = \{a_0 + a_1 r_1 + \cdots + a_m r_m + a_{m+1} r_1 r_2 + \cdots + a_s r_1^{n_1} r_2^{n_2} \cdots r_m^{n_m} : a_i \in F\}.$$

(This is a generalization of the E of Example 6.26.) Since each element of E is expressed in terms of elements of F and the roots of $f(x)$, then any automorphism of E that fixes F is completely determined, once the images of the roots are given. So if $\theta \in A(E/F)$, then θ is determined by a permutation of the roots of $f(x)$. Since the maximum number of distinct roots of $f(x)$ is n, and the maximum number of permutations of n elements is $n!$, it is clear that the order of the Galois group of a polynomial of degree n has to be at most $n!$. By Cayley's Theorem, every finite group is isomorphic to a subgroup of S_n. Thus we have the following result.

THEOREM 6.38

If $\deg f(x) = n$, then the Galois group of $f(x)$ is isomorphic to a subgroup of the symmetric group S_n.

EXAMPLE 6.27

Let $f(x) = ax^2 + bx + c \in \mathbf{Q}[x]$, and suppose $f(x)$ is prime over $\mathbf{Q}$. Then the splitting field of $f(x)$ is the quadratic field $\mathbf{Q}(\sqrt{n})$, for some square-free n. According to Problem 6.3-15, the only automorphisms of $\mathbf{Q}(\sqrt{n})$ are:

$$\varepsilon: \ a + b\sqrt{n} \to a + b\sqrt{n},$$
$$\theta: \ a + b\sqrt{n} \to a - b\sqrt{n}.$$

So the Galois group of $f(x)$ is S_2, since

$$\{\varepsilon, \theta\} = A(\mathbf{Q}(\sqrt{n})/\mathbf{Q}) = A(\mathbf{Q}(\sqrt{n})) \cong \mathbf{Z}_2 \cong S_2. \ \square$$

EXAMPLE 6.28

We shall determine the Galois group of $x^4 - x^2 - 2 \in \mathbf{Q}[x]$. According to Example 6.26, the splitting field is

$$E = \mathbf{Q}(i, \sqrt{2}) = \{a + bi + c\sqrt{2} + di\sqrt{2} : a, b, c, d \in \mathbf{Q}\}.$$

Every automorphism of E is completely determined by the images of the two elements: i and $\sqrt{2}$. Since a root must map to a root, the only possible images of i are: $i, -i, \sqrt{2},$ and $-\sqrt{2}$. But

$$i \to \pm\sqrt{2} \Rightarrow i^2 \to (\pm\sqrt{2})^2 \Rightarrow -1 \to 2,$$

a contradiction of the fact that $-1 \to -1$. There are but four automorphisms, each determined by the images of i and $\sqrt{2}$:

$$\varepsilon: \ i \to i \qquad \theta: \ i \to i \qquad \lambda: \ i \to -i \qquad \eta: \ i \to -i$$
$$\sqrt{2} \to \sqrt{2} \qquad \sqrt{2} \to -\sqrt{2} \qquad \sqrt{2} \to \sqrt{2} \qquad \sqrt{2} \to -\sqrt{2}$$

For instance, θ is the automorphism of E such that

$$(a + bi + c\sqrt{2} + di\sqrt{2})\theta = a + bi - c\sqrt{2} - di\sqrt{2}.$$

The group table of $A(E/\mathbf{Q})$ is given as follows:

$\circ$	ε	θ	λ	η
ε	ε	θ	λ	η
θ	θ	ε	η	λ
λ	λ	η	ε	θ
η	η	λ	θ	ε

6.7 Splitting Field and Galois Group of a Polynomial

Therefore, $A(E/\mathbf{Q}) = A(E) \cong K_4 \cong \mathbf{Z}_2 \times \mathbf{Z}_2$, and so the Galois group is isomorphic to a proper subgroup of S_4. $\square$

EXAMPLE 6.29

We now illustrate the fact that we can find the Galois group of a polynomial over a field whose characteristic is not 0. Let

$$m(x) = x^3 + x + \bar{1} \in \mathbf{Z}_2[x].$$

The coefficient field of $m(x)$ is $\mathbf{Z}_2$, and $m(x)$ is prime over $\mathbf{Z}_2$. Then

$$E = \mathbf{Z}_2[x]/\langle m(x) \rangle = \{a + br + cr^2 : a, b, c \in \mathbf{Z}_2, r^3 = \bar{1} + r\} = \mathbf{Z}_2(r).$$

So

$$E = \{\bar{0}, \bar{1}, r, \bar{1} + r, r^2, \bar{1} + r^2, r + r^2, \bar{1} + r + r^2\}.$$

Then
$$m(x) = (x - r)(x - r^2)(x - (r + r^2)).$$

Hence, E is the splitting field of $m(x)$.

The three distinct roots of $m(x)$ are: r, r^2, and $r + r^2$, and each automorphism in the Galois group of $m(x)$ is determined by the image of root r. According to Theorem 4.28(b) each of the three possible maps is an automorphism. Therefore, $A(E/\mathbf{Z}_2) = \{\varepsilon, \alpha, \beta\} \cong \mathbf{Z}_3 \cong A_3 < S_3$, where ε is the identity map, and α and β are the following maps:

$$(a + br + cr^2)\alpha = a + b(r^2) + c(r^2)^2 = a + cr + (b + c)r^2,$$
$$(a + br + cr^2)\beta = a + b(r + r^2) + c(r + r^2)^2 = a + (b + c)r + br^2,$$

for all $a, b, c \in \mathbf{Z}_2$.

Bear in mind that each automorphism is a permutation of the entire set E, even though our interest is in the possible permutations of the *roots* of $m(x)$. One can, for instance, write the automorphic map α as follows:

$$\alpha = \begin{pmatrix} \bar{0} & \bar{1} & r & \bar{1} + r & r^2 & \bar{1} + r^2 & r + r^2 & \bar{1} + r + r^2 \\ \bar{0} & \bar{1} & r^2 & \bar{1} + r^2 & r + r^2 & \bar{1} + r + r^2 & r & \bar{1} + r \end{pmatrix}$$
$$= (r, r^2, r + r^2) \circ (\bar{1} + r, \bar{1} + r^2, \bar{1} + r + r^2).$$

Since $\mathbf{Z}_2$ consists of just the zero and the unity of E, $\mathbf{Z}_2$ is fixed under every automorphism of E, and so $A(E/\mathbf{Z}_2) = A(E)$.

Incidentally, note that splitting field E is the Galois field $GF(2^3)$. Also, the additive order of each nonzero element is 2, so $(E, +) \cong (\mathbf{Z}_2 \times \mathbf{Z}_2 \times \mathbf{Z}_2, +)$. Further, $(E^*, \cdot) \cong (\mathbf{Z}_7, +)$. $\square$

6.7 Problems

1 Give two root fields (that are not the splitting field) and the splitting field of each polynomial $f(x)$, where the coefficient field of $f(x)$ is $\mathbf{Q}$. Also, find the Galois group of $f(x)$.
(a) $f(x) = x^2 - 7$
(b) $f(x) = 3x^2 - 2x - 2$
(c) $f(x) = x^2 - 5x + 6$
(d) $f(x) = x^3 + 2x^2 - 3x - 6$

2 Let $f(x) = x^3 + \bar{3} \in \mathbf{Z}_7[x]$. Factor $f(x)$ over its splitting field E, and find the automorphisms of E. What is the Galois group of $f(x)$?

3 If $f(x) = x^4 - 9 \in \mathbf{Q}[x]$, find the splitting field E of $f(x)$, factor $f(x)$ over E, and find the Galois group of $f(x)$.

4 If $f(x) = x^4 + 4x^2 + 2 \in \mathbf{Q}[x]$, factor $f(x)$ over its splitting field E, and find the Galois group of $f(x)$. (Hint: $E = \mathbf{Q}(r)$, r is a root.)

5 Let E be the splitting field of $p(x)$, where $p(x) = x^3 + 3x^2 + 4 \in \mathbf{Q}[x]$. Show that $A(E/\mathbf{Q}) \cong S_3$.

6 Show that the Galois group of $x^4 - 3$ over $\mathbf{Q}$ is the octic group, D_4.

7 Give three second-degree polynomials over $\mathbf{Q}$ that have the same splitting field E, with $\mathbf{Q} < E$.

8 Discuss the truth or falsity of this statement: Every polynomial in $\mathbf{Z}[x]$ splits over a subfield of the complex field.

9 Suppose that F is the coefficient field of $f(x)$, $\deg f(x) = n$, r is a root of $f(x)$, and the splitting field E is $F(r)$. Explain why $|A(E/F)| \leq n$.

10 Prove: If $f(x)$ and $g(x)$ are in $F[x]$, $f(x)|g(x)$, and E is the splitting field of $g(x)$, then E contains a subfield K that is the splitting field of $f(x)$.

11 By definition, an extension field E of F is called an *extension by radicals* if there exists a finite chain of fields,
$$F = E_0 < E_1 < \cdots < E_m = E,$$
such that $E_{i+1} = E_i(\sqrt[n_i]{s_i})$, for $s_i \in E_i$, $i = 0, 1, \ldots, m - 1$,
and $\sqrt[n_i]{s_i}$ is a root of a prime polynomial, $x^{n_i} - s_i$ in $E_i[x]$, $n_i \geq 2$. If $t = 3\sqrt{5} + \sqrt[3]{2} - \sqrt{7} - \sqrt[4]{6} \in E$, describe the chain of subfields E_i of E such that E is an extension by radicals of $\mathbf{Q}$. (Find the smallest E that contains t.)

6.8 Two Applications of Galois Theory

The solutions to certain problems have been sought for decades, and even centuries, by many eminent mathematicians. Sometimes a solution is not found, simply because the problem is an *impossible* one. Before the impossible problem can be given a decent burial, the mathematician must prove that it, indeed, *is* impossible. A proof that something *cannot* be done is just as important as a proof

6.8 Two Applications of Galois Theory

that something else *can* be accomplished. In this section we discuss two problems that were finally laid to rest through the use of Galois theory: (1) Polynomials whose roots cannot be expressed by means of a formula involving radicals (such as the quadratic formula for solving second-degree polynomials). (2) Geometric constructions that cannot be made with straightedge and compass.

(1) SOLUTION BY RADICALS

A person who is asked to "solve" the equation $x^2 - 5x + 3 = 0$ would say that the solution is: $x = (5 \pm \sqrt{13})/2$, thus expressing the roots of the polynomial in terms of the coefficients. By the Kronecker theorem, we know that *every* nth degree polynomial over a field F has a root r in some extension field of F. The question still to be answered is: Can r be written explicitly in terms of the coefficients of the polynomial? We first clarify what this means, and then answer the question.

Let s be in a field F, and let $n \in \mathbf{Z}^+$. We say we are *extracting an nth root of s* if we find an element t such that $t^n = s$. (We usually represent a particular nth root by the symbol, $\sqrt[n]{s}$. For instance, $\sqrt{9} = +3, \sqrt[3]{-8} = -2$.) Such an nth root always exists, since the polynomial $x^n - s$ has a root r in the algebraic extension field E, where $E = F[x]/\langle x^n - s \rangle$. Of course, all the nth roots of s lie in E, if E is the splitting field of $x^n - s$.

Let $f(x)$ be a polynomial with coefficient field F, where $\text{ch}(F) = 0$. The polynomial $f(x)$ is said to be *solvable by radicals* if each of its roots can be calculated from its coefficients in a finite number of steps by using the field operations (addition, multiplication, subtraction, division) and root extractions.

We see that every second-degree polynomial, $ax^2 + bx + c$, is solvable by radicals, since a root $r = (-b + \sqrt{b^2 - 4ac})/2a$. Note that, in calculating r, we perform one addition, four multiplications, one subtraction, one division, and one root extraction, for a total of eight steps.

In 1545 the French mathematician Cardan published a solution by radicals of the general cubic polynomial. It may be of interest to see what the solution looks like, in our present-day notation. Since a polynomial has the same roots as each of its associates, we need only consider the problem of solving *monic* polynomials. We first observe that the three roots of $x^3 - 1$ are $1, \omega,$ and ω^2, where $\omega = (-1 + \sqrt{-3})/2$, and $\omega^2 = (-1 - \sqrt{-3})/2$. Then if a is any complex number, and if $\sqrt[3]{a}$ represents one of the cube roots of a, the three roots of $x^3 - a$ are:

$$\sqrt[3]{a}, \quad \omega\sqrt[3]{a}, \quad \text{and} \quad \omega^2\sqrt[3]{a}.$$

Assume that $f(x) = x^3 + bx^2 + cx + d$, with $b, c,$ and d in a field F of characteristic zero. (Recall that F contains the rational field.) If we make the substitution, $x = y - b/3$, then the equation becomes: $y^3 + py + q = 0$, where $p = c - b^2/3, q = d - bc/3 + 2b^3/27$. Next, let $y = t - p/3t$, and obtain the equation, $t^6 + qt^3 - p^3/27 = 0$. Solving this equation for one value of t^3, we get:

$$t^3 = \frac{-q + \sqrt{q^2 + 4p^3/27}}{2}, \quad \text{and so } t = \sqrt[3]{\frac{-q + \sqrt{q^2 + 4p^3/27}}{2}}.$$

Putting these equations together, we get the following result.

THEOREM 6.39 *(Cardan's cubic formula)*

Let $f(x) = x^3 + bx^2 + cx + d$, where b, c, d are in a field F of characteristic zero. If

$$p = \frac{3c - b^2}{3}, \quad q = \frac{2b^3 - 9bc + 27d}{27}, \quad t = \sqrt[3]{\frac{-9q + \sqrt{81q^2 + 12p^3}}{18}},$$

$$\omega = \frac{-1 + \sqrt{-3}}{2}, \quad v_1 = t, \quad v_2 = \omega t, \quad v_3 = \omega^2 t,$$

then the roots of $f(x)$ are r_1, r_2, and r_3, where

$$r_j = v_j + \frac{b^2 - 3c}{9v_j} - \frac{b}{3}, \quad j = 1, 2, 3.$$

EXAMPLE 6.30

We shall use Theorem 6.39 to obtain the roots of $f(x)$, where $f(x) = x^3 + 3x^2 + 4$. Since $b = 3$, $c = 0$, and $d = 4$, we get that $p = -3$, $q = 6$, $t = \sqrt[3]{-3 + 2\sqrt{2}}$, and the three roots are:

$$r_1 = t + 1/t - 1, \quad r_2 = \omega t + 1/\omega t - 1, \quad r_3 = \omega^2 t + 1/\omega^2 t - 1.$$

Hence, $\quad r_1 = \sqrt[3]{-3 + 2\sqrt{2}} + \dfrac{1}{\sqrt[3]{-3 + 2\sqrt{2}}} - 1.$

The formula gives us the *exact* value of r_1. From this expression we can, of course, obtain a rational approximation of r_1, which happens to be -3.35. The approximate values of r_2 and r_3 are, respectively, $.18 + 1.1i$, and $.18 - 1.1 i$. ☐

There are several methods for *approximating* a real root of a third degree polynomial over the real field, so that one does not need to use the rather messy formula of Theorem 6.39 for that purpose. But it is mathematically satisfying to know that such a formula exists!

The Italian mathematician, Ferrari, who was Cardan's pupil, was the first person to obtain a solution by radicals of the general *fourth* degree polynomial. It was natural for those mathematicians of the 16th century to look for a solution by radicals of the general fifth degree polynomial, $x^5 + bx^4 + cx^3 + dx^2 + ex + f$, and of polynomials of higher degrees. It was not until 1825 that the question was finally resolved by Abel, who proved that it is *impossible* for a solution by radicals to exist for the general polynomial of degree n, if $n \geq 5$. This remarkable result is based on the following theorem, whose proof is left to a later course.

THEOREM 6.40

If $f(x) \in F[x]$, where $\text{ch}(F) = 0$, then $f(x)$ is solvable by radicals if and only if the Galois group of $f(x)$ is a solvable group.

6.8 Two Applications of Galois Theory

Recall that in Section 5.4 we defined a finite group to be a *solvable group* if its composition factors are all groups of prime orders, and in Section 6.7 we defined the *Galois group* of a polynomial $f(x)$ to be the group of automorphisms of the splitting field that leave the coefficient field fixed.

EXAMPLE 6.31

Given that $f(x) = x^5 - 4x^4 + 2x + 2 \in \mathbf{Q}[x]$; then, by the Fundamental Theorem of Algebra, we know that $f(x)$ has five complex roots. Using elementary calculus to obtain information about maximum and minimum points and concavity, we can graph the equation $y = f(x)$, and thus ascertain that $f(x)$ has three distinct real roots: one negative and two positive. So $f(x)$ must have two conjugate complex roots that are not real. Since there are five distinct roots, and each element of the Galois group of $f(x)$ is determined by a permutation of those roots, we know that the Galois group is a subgroup, proper or improper, of S_5. For this particular polynomial, it can be shown that its Galois group is S_5. But, by Theorem 5.20, S_5 is not a solvable group. Hence, $f(x)$ is not solvable by radicals. □

Now let $f(x) = x^5 + a_4 x^4 + a_3 x^3 + a_2 x^2 + a_1 x + a_0$, where the coefficient field is of characteristic zero. If $f(x)$ were solvable by radicals, then the general formula so obtained, expressed in terms of the a_i, could be used to find a root of the polynomial of Example 6.31. But this is impossible. Hence, no such formula can exist.

Let's look at the general situation. It can be proved that A_n is the only proper *normal* subgroup of S_n, for $n \geq 5$, and that A_n has no proper normal subgroups. Hence, the orders of the composition factors of S_n are 2 and $n!/2$, which means that S_n is not solvable, for $n \geq 5$. It can also be shown that, for each n, there is *some* polynomial $f(x)$ of degree n whose Galois group is S_n. These results give us the famous and significant theorem of Abel.

THEOREM 6.41 (Abel)

Given the general polynomial of degree n whose coefficient field is of characteristic zero:

$$f(x) = a_n x^n + a_{n-1} x^{n-1} + \cdots + a_1 x + a_0, \quad n \geq 5;$$

then $f(x)$ is not solvable by radicals.

(2) GEOMETRIC CONSTRUCTIONS

Changing the subject, we turn now to some famous construction problems of Euclidean geometry. Every high school geometry student is familiar with some of the constructions that can be made with the aid of a straightedge and compass. One can, for instance, construct a line that is perpendicular to a given line, bisect a given angle, and determine a line segment whose length is $\sqrt{3}$. One starts all this by drawing a line segment with the straightedge, and then arbitrarily marking off a unit length on the segment. (See Figure 6–4.) The unit length is called 1.

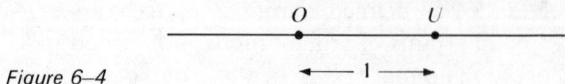

Figure 6–4

The question arises as to just which lengths can be constructed, and which points in the plane can be determined by means of straightedge and compass. We let $\overline{AB}$ represent the line segment joining points A and B, and let AB represent the length of $\overline{AB}$.

EXAMPLE 6.32

If $a, b \in \mathbf{Z}^+$, one can construct rational number a/b as follows: Use compass and unit length OU to determine segment $\overline{OB}$ of length b (Figure 6–5). Construct a line perpendicular to $\overline{OB}$ at B, and on this perpendicular determine A so that $BA = a$. Construct line k through U perpendicular to $\overline{OB}$. Then the intersection of k and $\overline{OA}$ is a point D such that $UD = a/b$.

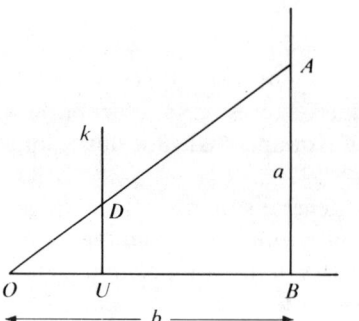

Figure 6–5 □

If we construct a perpendicular to $\overline{OU}$ at O, we can set up a coordinate system, and construct any point (s, t) in $\mathbf{Q} \times \mathbf{Q}$, and any length $r \in \mathbf{Q}$. We shall call (s, t) a *rational point*. If a straightedge passes through two rational points, (s, t) and (v, w), the equation of the line drawn through the two points will be either $y = mx + b$, or $x = a$, where $m, b, a \in \mathbf{Q}$. It is easy to verify that two lines, each of which has a rational slope and a rational y–intercept, will intersect in a rational point. So to construct *irrational* lengths, we need to use the compass and draw arcs of circles. If a circle has a rational radius and has its center at a rational point, then its equation is of the form,

$$x^2 + y^2 + dx + ey + f = 0, \quad \text{with } d, e, f \in \mathbf{Q}.$$

The geometry of constructing points with straightedge and compass has its algebraic counterpart in the simultaneous solution of two equations, where each equation is that of either a line or a circle. The next theorem indicates that these constructions produce points with coordinates that are in either $\mathbf{Q}$ or E, where E is obtained from $\mathbf{Q}$ by a succession of quadratic field extensions. The proof involves high school algebra.

THEOREM 6.42

Let F be a subfield of the real field, and let

$$L_1 = \{(x, y) : y = m_1 x + b_1\},$$
$$L_2 = \{(x, y) : y = m_2 x + b_2\},$$
$$C_1 = \{(x, y) : x^2 + y^2 + d_1 x + e_1 y + f_1 = 0\},$$
$$C_2 = \{(x, y) : x^2 + y^2 + d_2 x + e_2 y + f_2 = 0\},$$

with $m_i, b_i, d_i, e_i, f_i \in F$. If $(s_1, s_2) \in L_1 \cap L_2$, then $s_i \in F$. If $(s_1, s_2) \in L_1 \cap C_1$, or $(s_1, s_2) \in C_1 \cap C_2$, then $s_i \in F(\sqrt{n})$, for an $n \in F$.

According to Theorem 6.42, a sequence of constructions with straightedge and compass enables one to obtain a line segment of length s, where s is in an extension field E of $\mathbf{Q}$, and E is built up from $\mathbf{Q}$ by the following sequence of subfields:

$$\mathbf{Q} = E_0 < E_1 < E_2 < \cdots < E_k = E,$$

where $E_{i+1} = E_i(\sqrt{n_i})$, $n_i \in E_i$, and $x^2 - n_i$ is prime over E_i. Thus, E_{i+1} is a quadratic extension of E_i, and

$$E_1 = \mathbf{Q}(\sqrt{n_0}),$$
$$E_2 = E_1(\sqrt{n_1}) = \mathbf{Q}(\sqrt{n_0}, \sqrt{n_1}) = \mathbf{Q}(\sqrt{n_0} + \sqrt{n_1}),$$
$$E_3 = E_2(\sqrt{n_2}) = \mathbf{Q}(\sqrt{n_0}, \sqrt{n_1}, \sqrt{n_2}) = \mathbf{Q}(\sqrt{n_0} + \sqrt{n_1} + \sqrt{n_2}),$$
$$\cdots\cdots\cdots\cdots\cdots\cdots$$
$$E = E_k = E_{k-1}(\sqrt{n_{k-1}}) = \mathbf{Q}(\sqrt{n_0}, \sqrt{n_1}, \ldots, \sqrt{n_{k-1}})$$
$$= \mathbf{Q}(\sqrt{n_0} + \sqrt{n_1} + \cdots + \sqrt{n_{k-1}}),$$

so the degree of E_i over $\mathbf{Q}$ is 2^i, for $i = 1, 2, \ldots, k$. (This is a generalization of Problem 4.8–11.)

THEOREM 6.43

If a length s is constructible by straightedge and compass, then s is in an algebraic extension field E of $\mathbf{Q}$ of degree 2^k, where k is some nonnegative rational integer.

EXAMPLE 6.33

One can construct a segment $\overline{OS}$ of length $\sqrt{3}$, as follows (Figure 6–6): Point P is the intersection of two circles, each with radius 2, one having its center at $(1, 0)$, and the other having its center at $(-1, 0)$. The coordinates (x, y) of P are a solution of the pair of equations:

$$(x - 1)^2 + y^2 = 4, \quad (x + 1)^2 + y^2 = 4.$$

This solution is $(0, \sqrt{3})$. So $OP = \sqrt{3}$, which can be marked off as the length OS on the x–axis. We have thus constructed a length that is in the

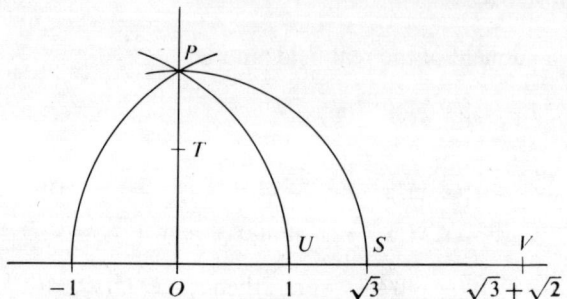

Figure 6-6

quadratic extension field $\mathbf{Q}(\sqrt{3})$ of $\mathbf{Q}$. Further, if $OT = 1$, where T is on $\overline{OP}$, then $TU = \sqrt{2}$, and so we can construct a segment OV of length $\sqrt{3} + \sqrt{2}$. This number is in E_2, where $E_1 = \mathbf{Q}(\sqrt{3})$, and

$$E_2 = E_1(\sqrt{2}) = \mathbf{Q}(\sqrt{3}, \sqrt{2}) = \{a_1 + a_2\sqrt{3} + a_3\sqrt{2} + a_4\sqrt{6} : a \in \mathbf{Q}\}.$$

We note that $\mathbf{Q} < E_1 < E_2$, and that E_1 is of degree 2 over $\mathbf{Q}$, and E_2 is of degree 4 over $\mathbf{Q}$. □

As far back as the fifth century B.C., Greek geometers were attempting to solve these problems:

(a) *Doubling a cube*: Given a cube whose edge is of unit length 1, and whose volume, of course, is 1 cubic unit; using straightedge and compass, construct length r such that a cube whose edge is of length r has a volume of 2 cubic units.

(b) *Trisecting an angle*: Given *any* angle θ; using straightedge and compass, construct the angle $\theta/3$.

It was not until 1837 that these constructions were proved to be impossible. The impossibility follows rather easily from the next theorem.

THEOREM 6.44

If $f(x) = x^3 + px + q$, where $f(x) \in \mathbf{Q}[x]$, $f(x)$ is prime over $\mathbf{Q}$, and r is a real root of $f(x)$; then it is impossible to construct length r with straightedge and compass.

Proof

Assume that a root r is constructible. Then, by Theorem 6.43, r lies in an extension field E_k of $\mathbf{Q}$, where

$$E_k = \mathbf{Q}(\sqrt{n_0}, \sqrt{n_1}, \ldots, \sqrt{n_{k-1}}), \quad \text{and} \quad \deg E_k = 2^k.$$

Further, $f(x) = (x - r)(x^2 + cx + d)$, with $c, d \in E_k$.

So the other two roots of $f(x)$ lie in either E_k or $E_{k+1} = E_k(\sqrt{c^2 - 4d})$.

Let 2^m be the smallest integer such that field E_m of degree 2^m contains a root of $f(x)$. If r_1 is the root in E_m, then

$$r_1 = a + b\sqrt{n_{m-1}}, \quad \text{where } b \neq 0; \quad ab, n_{m-1} \in E_{m-1}; \quad \text{and}$$
$$E_{m-1} = \mathbf{Q}(\sqrt{n_0}, \sqrt{n_1}, \ldots, \sqrt{n_{m-2}}).$$

(If $b = 0$, then $r_1 \in E_{m-1}$, contradicting the minimality of 2^m.)
If $r_2 = a - b\sqrt{n_{m-1}}$, then r_2 is also a root of $f(x)$. (This follows from the fact that the map $\theta : a + b\sqrt{n_{m-1}} \to a - b\sqrt{n_{m-1}}$, is an automorphism of field E_m that fixes field E_{m-1}. Proof is analogous to that of Theorem 4.18.) If r_3 is the third root of $f(x)$, then $r_1 + r_2 + r_3 = 0$, since the coefficient of x^2 in $f(x)$ is 0. (See Problem 4.4-13.) Hence, $r_3 = -(r_1 + r_2) = -2a$. But $a \in E_{m-1}$, and so $r_3 \in E_{m-1}$, a contradiction, since there is no root of $f(x)$ in any field of degree less than 2^m. Therefore, there is no root of $f(x)$ that is constructible with straightedge and compass. □

THEOREM 6.45

The following constructions are *impossible* by means of straightedge and compass:
(a) Construction of the edge of a cube whose volume is double that of a cube with a given edge.
(b) Trisection of an arbitrary angle.

Proof

(a) If the volume of the cube is to be doubled, this means that we must construct a number r such that $r^3 = 2$. But then r is a root of $x^3 - 2$, and no root of $x^3 - 2$ can be constructed with straightedge and compass, according to Theorem 6.44.

(b) Although it is possible to trisect special angles, such as 45° and 90°, we shall verify that it is impossible to trisect 60°. If the special angle of 20° is not constructible, then there is no possible method for trisecting an *arbitrary* angle.

We first note that 20° is constructible if and only if length cos 20° is constructible (see Figure 6–7).

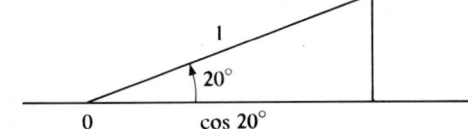

Figure 6–7

We shall make use of the trigonometric identity:

$$\cos 3\theta = 4\cos^3\theta - 3\cos\theta.$$

(This can be derived from the formulas for $\cos(A + B)$ and for $\cos 2A$ by letting $A = \theta, B = 2\theta$.) If $\theta = 20°$, we have:

$$\cos 60° = 4\cos^3(20°) - 3\cos 20°.$$

Since $\cos 60° = 1/2$, this equation reduces to:

$$(\cos 20°)^3 - (3/4)(\cos 20°) - 1/8 = 0.$$

So $\cos 20°$ is a root of $x^3 - (\tfrac{3}{4})x - (\tfrac{1}{8})$, which is prime over $\mathbf{Q}$. Therefore, by Theorem 6.44, $\cos 20°$ is not constructible. Hence, $20°$ is not constructible. This means that $60°$ cannot be trisected. So, given an arbitrary angle θ, is is impossible to construct $\theta/3$. □

6.8 Problems

1. Use Cardan's cubic formula to obtain the roots of $x^3 + 3x^2 + 9x + 5$.

2. Note that if $f(x) = x^3 - x^2 + x - 1$, then $f(x) = (x - 1)(x - i)(x + i)$. Use Cardan's formula to show that r_1 is a root, where

$$r_1 = \frac{\sqrt[3]{10 + 6\sqrt{3}}}{3} - \frac{2}{3\sqrt[3]{10 + 6\sqrt{3}}} + \frac{1}{3}.$$

   Also, verify that $r_1 = 1$. (This illustrates the fact that a simple answer may appear in a complicated form when the cubic formula is used.)

3. Fill in the details of the proof of Theorem 6.39.

4. Referring to Example 6.30: The root r_1 of $f(x)$ is in an algebraic extension field E of $\mathbf{Q}$. Describe E, in terms of an extension by radicals of $\mathbf{Q}$. (See Problem 6.7–11.)

5. If $f(x) = x^5 - 4x^4 + 2x + 2 \in \mathbf{Q}[x]$, show that
   (a) $f(x)$ is irreducible over $\mathbf{Q}$,
   (b) $f(x)$ has one negative, two positive, and two imaginary roots.

6. (a) Look in a theory of equations book for Ferrari's solution by radicals of the general fourth degree polynomial, and give the equations needed, as was done for the cubic in Theorem 6.39.
   (b) Find the roots of $f(x)$, if $f(x) = x^4 + 2\sqrt{5}x^3 + 5x^2 - 1$.

7. Let F be a subfield of the real field. Prove:
   (a) If $(a, b), (c, d) \in F \times F$, then the equation of the line through (a, b) and (c, d) has coefficients in F.
   (b) If $(a, b) \in F \times F$, and $r \in F$, then the equation of the circle with radius r and center at (a, b) has coefficients in F.

8. Prove Theorem 6.42.

9. Referring to the proof of Theorem 6.44, prove that if $a + b\sqrt{n_k}$ is a root of $f(x)$, then $a - b\sqrt{n_k}$ is a root, where $a, b, n_k \in E_k$, and $a + b\sqrt{n_k} \in E_{k+1}$.

10. Show how to trisect angles of $90°$ and $45°$ with straightedge and compass.

11. Verify: Angle θ is constructible $\Leftrightarrow \cos \theta$ is constructible.

12 Some of his contemporaries probably called Cardan a sneaky, underhanded scoundrel. Why? (See Appendix E–3.)

6.9 Epilogue

Many mathematics textbooks contain a prologue, or introduction. Practically none contains an epilogue. Perhaps this is because the author feels that there is nothing more to say, once he has written down the last proof of the last theorem in the last section of the last chapter. Or perhaps he is just too exhausted to attempt even one more coherent paragraph. This I can well understand! (I am now dropping the editorial "we".) However, I should like to conclude this book on a more personal note than was practical up to this point, and to have you end your hard work in a rather relaxed atmosphere.

First, I wish to express the hope that you have found your brief trip through abstract algebra useful, satisfying, and even pleasurable, and that you will return, via independent reading or another course, to explore in greater depth some of the topics to which you have now been exposed.

Second, I want to share with you some rather philosophical thoughts about the very nature of mathematics itself. Do you have a "philosophy of mathematics"? To many mathematicians, mathematics is a *creative art*, and the creator regards his creation with the same kind of aesthetic appreciation as does the musician who creates a symphony. In their book, *Mathematics and the Imagination* (New York: Simon & Schuster, Inc., 1940), Kasner and Newman wrote: "Mathematics is man's own handiwork, subject only to the limitations imposed by the laws of thought." The British mathematician–philosopher, Whitehead, said that in mathematics relationships "are exhibited which, apart from the agency of human reason, are extremely unobvious," and that because of this, mathematics "may claim to be the most original creation of the human spirit."

Then there are others who take the point of view that man *discovers* mathematics, rather than *creates* it. Didn't the quadratic formula hold true, even before man's little mind stumbled upon it? He didn't *make* it true by his act of recording it on paper! It could be that all mathematics has always been in existence, and that it has been patiently waiting for someone to discover it. The French mathematician, Philip Jourdain, wrote: "Mathematics is eternal and unchanging, and therefore has no history, but is something which is discovered, in the course of time, by human minds. Mathematics is independent of us personally and of the world outside, and our own discoveries and views do not affect the Truth itself, but only the extent to which we or others see it. Some of us discover things in science, but we do not really create anything in science any more than Columbus created America." An interesting thought!

Finally, let me ask you this: Has a little of the *beauty* of abstract algebra come through to you? "But, as for everything else, so for a mathematical theory— beauty can be perceived but not explained." This was said by Cayley in 1883 in his presidential address to the British Association for the Advancement of Science. This theme was elaborated upon by the eminent English mathematician, G. H. Hardy, in his little book, *A Mathematician's Apology* (New York: Cambridge University Press, 1940). Hardy felt that mathematics should be *serious* as well as *beautiful*, but not serious in the sense of being useful; he was not interested in the practical applications of mathematics. Rather than paraphrasing Hardy, let me quote him:

"A mathematician, like a painter or a poet, is a maker of patterns. If his patterns are more permanent than theirs, it is because they are made with ideas. A painter makes patterns with shapes and colours, a poet with words. A mathematician, on the other hand, has no material to work with but ideas, and so his patterns are likely to last longer, since ideas wear less with time than words. The mathematician's patterns, like the painter's or the poet's must be *beautiful*; the ideas, like the colours or the words, must fit together in a harmonious way. Beauty is the first test; there is no permanent place in the world for ugly mathematics. It may be very hard to *define* mathematical beauty, but that is just as true of beauty of any kind—we may not know quite what we mean by a beautiful poem, but that does not prevent us from recognizing one when we read it . . . .

"The best mathematics is serious as well as beautiful . . . . The 'seriousness' of a mathematical theorem lies, not in its practical consequences, which are usually negligible, but in the *significance* of the mathematical ideas which it connects. We may say, roughly, that a mathematical idea is 'significant' if it can be connected, in a natural and illuminating way, with a large complex of other mathematical ideas. Thus a serious mathematical theorem, a theorem which connects significant ideas, is likely to lead to important advances in mathematics itself and even in other sciences . . . . The beauty of a mathematical theorem depends a great deal on its seriousness, as even in poetry the beauty of a line may depend to some extent on the significance of the ideas which it contains."

They probably wouldn't sell very well, but, thanks to G. H. Hardy, here is a new slogan for bumper stickers:

MATHEMATICS IS BEAUTIFUL!

6.9 Problems

1. Browse through the six chapters of this book, and choose from each chapter a theorem which, in your estimation, would be considered particularly significant by G. H. Hardy, in that it can be connected "with a large complex of other mathematical ideas."

2. From each chapter select an idea [which might be a theorem, an example, a definition] that was aesthetically pleasing to you—that you found to contain some beauty.

3. Choose a topic that particularly interested you in this course, use it as a starting point in looking up material in the library, and write a short paper that introduces some mathematics that was not covered in this book.

4. Why, in your opinion, did Hardy dislike any mathematics that could be applied? (See Appendix E–20.)

Appendix A

Mathematical Logic, Its Use in Proving Theorems

A system of logic is the basic tool that the mathematician needs if he is to give a valid proof of anything. He must be able to decide if one mathematical statement he makes is a logical consequence of another mathematical statement. In this appendix we introduce a few of the basic notions of logic that are of use to a mathematics student.

We denote *statements*, or *propositions*, by $p, q, r, \ldots$. Table A–1 defines and illustrates some fundamental *connectives*, or *operations*, on statements. A statement p is called a *prime statement*, while a statement containing one or more connectives is called a *composite statement*.

Table A–1 The Basic Connectives

Symbol for Connective	Word for Connective	Name of Composite Statement	Symbol for Composite Statement	Example, where p is: "It is raining," q is: "I am wet."
$\sim$	Not	Negation	$\sim p$	It is *not* raining.
$\wedge$	And	Conjunction	$p \wedge q$	It is raining *and* I am wet.
$\vee$	Or	Disjunction	$p \vee q$	It is raining *or* I am wet.
$\rightarrow$	Implies	Conditional (or Implication)	$p \rightarrow q$	*If* it is raining, *then* I am wet.
$\leftrightarrow$	Is equivalent to	Bi-conditional	$p \leftrightarrow q$	It is raining *if and only if* I am wet.

We postulate that a statement p is either true, T, or false, F, and call T or F the *truth value* of p. The truth or falsity of $\sim p$ is postulated in Table A–2. Such a table of truth values is called a *truth table*.

Table A–2 Truth Values of Negation

p	$\sim p$
T	F
F	T

A composite statement that contains any one of the other four connectives will contain two prime statements, p and q, and each of these can be either true or false, giving us four combinations of truth values: TT, TF, FT, and FF. We postulate the truth or falsity of the composite statements $p \wedge q, p \vee q, p \to q$, and $p \leftrightarrow q$, as shown in Table A-3.

Table A-3 Truth Values of Conjunction, Disjunction, Conditional, Biconditional

p	q	$p \wedge q$	$p \vee q$	$p \to q$	$p \leftrightarrow q$
T	T	T	T	T	T
T	F	F	T	F	F
F	T	F	T	T	F
F	F	F	F	T	T

We can use these truth values to determine the truth or falsity of more complicated statements. Suppose we want the truth values of the statement:

$$[(p \to q \vee r) \wedge \sim q] \to (p \to r).$$

By agreement, if parentheses or other signs of grouping are omitted, one performs the operations in this order: $\sim, \wedge, \vee, \to, \leftrightarrow$. If $s = [(p \to q \vee r) \wedge \sim q]$, and $t = p \to r$, then the statement $s \to t$ can be built up as shown in Table A-4.

Table A-4 Truth Values of $[(p \to q \vee r) \wedge \sim q] \to (p \to r)$

p	q	r	$q \vee r$	$p \to q \vee r$	$\sim q$	$(p \to q \vee r) \wedge \sim q$ s	$p \to r$ t	$s \to t$
T	T	T	T	T	F	F	T	T
T	T	F	T	T	F	F	F	T
T	F	T	T	T	T	T	T	T
T	F	F	F	F	T	F	F	T
F	T	T	T	T	F	F	T	T
F	T	F	T	T	F	F	T	T
F	F	T	T	T	T	T	T	T
F	F	F	F	T	T	T	T	T

Note that the statement, $s \to t$, is *always* true, regardless of the truth values of the prime statements, p, q, and r.

A statement that is always true is called a *valid statement*, or a *tautology*. If a conditional statement $s \to t$ is a valid statement, we say that s *logically implies* t, and write $s \Rightarrow t$. Mathematicians prefer to replace $\to$ by $\Rightarrow$, thus reserving the symbol $\to$ for other uses. When we have a logical implication, we know that the conclusion t is true whenever the premise s is true, and that is exactly the situation that a mathematician wants! He wants the *implication* to be true, regardless of the truth or falsity of the premise. For instance, the implication,

$$(x + 2 = 9) \to (x = 7),$$

is always true, even if x should happen to be 4. So we write:

$$(x + 2 = 9) \Rightarrow (x = 7).$$

Two statements, s and t, are said to be *logically equivalent* if they have the same truth values. This means that one statement is true whenever the other statement is true, and false whenever the other is false. To indicate that s and t are logically equivalent, we write $s \Leftrightarrow t$. We illustrate logical equivalence in Table A–5.

Table A–5 Verification that $\sim(p \wedge q) \Leftrightarrow \sim p \vee \sim q$

p	q	$\sim p$	$\sim q$	$\sim p \vee \sim q$	$p \wedge q$	$\sim(p \wedge q)$
T	T	F	F	F	T	F
T	F	F	T	T	F	T
F	T	T	F	T	F	T
F	F	T	T	T	F	T

We see that $\sim(p \wedge q)$ has the same truth values as $\sim p \vee \sim q$, and so we can write

$$\sim(p \wedge q) \Leftrightarrow \sim p \vee \sim q.$$

Similarly, it can be shown that $\sim(p \vee q) \Leftrightarrow \sim p \wedge \sim q$.

An important set of truth values for people trying to prove theorems is given in Table A–6.

Table A–6 Truth Values of Implication, Contrapositive, Converse, and Negated Implications

p	q	$p \to q$	$\sim q \to \sim p$	$q \to p$	$\sim(p \to q)$	$p \wedge \sim q$
T	T	T	T	T	F	F
T	F	F	F	T	T	T
F	T	T	T	F	F	F
F	F	T	T	T	F	F

According to the table we see that

$$p \to q \Leftrightarrow \sim q \to \sim p.$$

The statement, $\sim q \to \sim p$, is called the *contrapositive* of the implication $p \to q$. Hence an implication and its contrapositive are logically equivalent. Statement $q \to p$ is called the *converse* of $p \to q$. The converse is *not* equivalent to the implication. Note also that

$$\sim(p \to q) \Leftrightarrow p \wedge \sim q.$$

Each of these last two equivalent expressions represents the *negation* of the given implication.

We now discuss eight tautologies that are commonly used in proofs, and illustrate their use. Each tautology can be verified by means of a truth table.

(1) DIRECT INFERENCE: $(p \to r) \wedge (r \to s) \Rightarrow (p \to s)$.

Some of the steps in almost any proof involve this tautology. For example, to solve the equation, $2x + 3 = 11$, a ninth-grade algebra student would probably give these steps:

1. $2x + 3 = 11$ 2. $2x = 8$ 3. $x = 4$

The student is, in effect, verifying that the following is a logical implication:

$$\text{If } 2x + 3 = 11, \text{ then } x = 4. \qquad (p \to s)$$

He does so by asserting that these are logical implications:

$$\text{If } 2x + 3 = 11, \text{ then } 2x = 8. \qquad (p \to r)$$
$$\text{If } \quad 2x = 8, \text{ then } x = 4. \qquad (r \to s)$$

Hence, he subconsciously uses direct inference to arrive at his conclusion.

(2) RULE OF DETACHMENT: $(p \to r) \wedge p \Rightarrow r$.

We use the rule of detachment in proofs when we reason that *if* (a) p logically implies r, and (b) p is true, *then* r must be true. The following steps in a proof illustrate the use of this rule, assuming that Step 1 has been proved earlier, and that Step 2 is given as a premise.

(1) If a is a real number, then $a^2 \geq 0$. $\qquad (p \to r)$
(2) a is a real number. $\qquad (p)$
(3) Therefore, $a^2 \geq 0$. $\qquad (r)$

(3) PROOF OF EQUIVALENCE OF TWO STATEMENTS: $p \leftrightarrow r \Leftrightarrow (p \to r) \wedge (r \to p)$.

We frequently prove that a statement p is true if and only if a statement q is true by breaking the proof into two parts: We prove that p logically implies r, and then we prove that r logically implies p. We are, of course, proving an implication and its converse. □

(4) CONTRAPOSITIVE PROOF: $(p \to r) \Leftrightarrow (\sim r \to \sim p)$.

We have already verified that an implication and its contrapositive have the same truth values. There are times when it is easier to prove the contrapositive of a theorem rather than the theorem itself.

(5) PROOF BY CONTRADICTION: $s \Leftrightarrow \sim s \to (r \wedge \sim r)$.

We note first that the statement $r \wedge \sim r$ is always false, no matter what statement is represented by r, and no matter what the truth value of r is. We call $r \wedge \sim r$ a *contradiction*. Now $\sim s \to (r \wedge \sim r)$ is true if and only if $\sim s$ is false, which means that s is true. Thus, s is true if and only if the negation of s logically implies a false statement, or contradiction.

As a corollary to this tautology, suppose that the theorem s to be proved takes the form $p \to q$. Then we must show that the statement $\sim(p \to q) \to (r \wedge \sim r)$ is true. We saw earlier that $\sim(p \to q) \Leftrightarrow p \wedge \sim q$. Hence, the tautology which we often use to justify a proof by contradiction takes this form:

$$p \to q \Leftrightarrow (p \wedge \sim q) \to (r \wedge \sim r).$$

For example, let

$$p = (n^2 \text{ is an odd integer.}), \qquad q = (n \text{ is an odd integer.})$$

We want to prove: If n^2 is odd, then n is odd. We can show that $p \to q$ is valid as follows:

$$\sim(p \to q) \Leftrightarrow p \wedge \sim q \Leftrightarrow (n^2 \text{ is odd}) \quad \text{and} \quad (n \text{ is even})$$
$$\Rightarrow (n^2 \text{ is odd}) \quad \text{and} \quad (n = 2k, \text{ for some } k)$$
$$\Rightarrow (n^2 \text{ is odd}) \quad \text{and} \quad (n^2 = 4k^2)$$
$$\Rightarrow (n^2 \text{ is odd}) \quad \text{and} \quad (n^2 \text{ is even})$$
$$\Rightarrow p \wedge \sim p.$$

Therefore, $p \to q$ is a valid statement.

(6) PROOF BY CASES: $(p \to r) \wedge (q \to r) \Rightarrow (p \vee q) \to r$.

Suppose we want to prove the following: If two integers are either both even or both odd, then their sum is even. We prove this by considering the two cases: (i) m and n are both even [p]. (ii) m and n are both odd [q]. We must show that each case implies that $m + n$ is even [r]. Thus,

(i) $p \Rightarrow m = 2a, n = 2b \Rightarrow m + n = 2(a + b) \Rightarrow r$,
(ii) $q \Rightarrow m = 2a + 1, n = 2b + 1 \Rightarrow m + n = 2(a + b + 1) \Rightarrow r$.

Therefore, $p \vee q \Rightarrow r$.

(7) PROOF BY ELIMINATION OF CASES: $(p \vee r) \wedge \sim p \Rightarrow r$.

In a proof we are sometimes confronted with two alternatives: Either p has to be true or r has to be true. If we go on to verify that p is false, then we *must* conclude that r is true. This tautology can be extended, of course, to any finite number of cases. That is,

$$(p_1 \vee p_2 \vee \cdots \vee p_n \vee r) \wedge \sim p_1 \wedge \sim p_2 \wedge \cdots \wedge \sim p_n \Rightarrow r.$$

Another tautology that is used in a proof by elimination of cases is:

$$(s \to p \vee r) \wedge \sim p \Rightarrow s \to r.$$

(8) CONDITIONAL PROOF: $p \to (q \to r) \Leftrightarrow (p \wedge q) \to r$.

According to this tautology these two statements, for instance, are equivalent:

(i) If $a > 0$, then $a < b$ implies that $a^2 < b^2$. $[p \to (q \to r)]$
(ii) If $a > 0$ and $a < b$, then $a^2 < b^2$. $[(p \wedge q) \to r]$

It is rather intriguing to note that the converses of these equivalent statements are not equivalent. That is, $(q \to r) \to p \not\Leftrightarrow r \to (p \wedge q)$. □

Many postulates and theorems of mathematics involve *quantified statements*. The *universal quantifier* is denoted by the symbol $(\forall x)$ and is read in any one of these ways:

For all x.   For every x.   For each x.

If $g(x)$ is a statement that involves an element x from a specified set S under consideration, then the symbol $(\forall x)g(x)$ means that $g(x)$ is a true statement, no matter which element of S is substituted for the x. For instance, if the set is the set $\mathbf{R}$ of real numbers, we could write:

$$(\forall x)(x^2 \geq 0).$$

Specifying the set under consideration can be part of the statement $g(x)$. Thus,

$$(\forall x)(x \in \mathbf{R} \Rightarrow x^2 \geq 0).$$

The *existential quantifier* is denoted by the symbol $(\exists x)$ and is read in any one of these ways:

For some x.   For at least one x.   There exists an x such that.

The symbol $(\exists x)g(x)$ means that there exists at least one element x for which $g(x)$ is a true statement. One can, for instance, express symbolically the fact that the polynomial $x^2 + 4x - 3$ has a real root by writing:

$$(\exists x)[(x \in \mathbf{R}) \wedge (x^2 + 4x - 3 = 0)].$$

A mathematician frequently lets the symbol $\ni$ stand for "such that", and then writes the last statement as follows:

$$\exists x \in \mathbf{R} \ni x^2 + 4x - 3 = 0.$$

The basic relationship between the universal quantifier and the existential quantifier is postulated to be the following:

$$\sim [(\forall x)g(x)] \Leftrightarrow (\exists x)[\sim g(x)].$$

That statement certainly makes sense, and is not difficult to accept. From it, we can easily deduce:

$$\sim [(\exists x)g(x)] \Leftrightarrow (\forall x)[\sim g(x)].$$

A statement can involve more than one quantifier. The fact that the product of two real numbers is a real number can be written:

$$(\forall x)(\forall y)[x \in \mathbf{R}, y \in \mathbf{R} \Rightarrow xy \in \mathbf{R}].$$

A mathematician frequently uses a comma instead of the conjunctive connective. Note that we wrote $x \in \mathbf{R}, y \in \mathbf{R}$ instead of $(x \in \mathbf{R}) \wedge (y \in \mathbf{R})$. This is often abbreviated still more as $x, y \in \mathbf{R}$. Thus:

$$(\forall x)(\forall y)(x, y \in \mathbf{R} \Rightarrow xy \in \mathbf{R}).$$

If a statement involves both the universal and the existential quantifiers, one must be careful about the order in which they are written. (One always works from left to right.) For instance, of these two statements concerning real numbers:

$$(\forall x)(\exists y)(x + y = 6), \quad (\exists y)(\forall x)(x + y = 6),$$

the first is true, while the second is false. The first statement says that if x is any real number, then there exists a real y such that the sum of x and y is 6. Of course, y is $6 - x$. The second statement says that *every* real number x is equal to the same number, $6 - y$, where y is some (fixed) real number. The second statement says, in effect, that all real numbers are equal!

It is good to acquire the habit of translating the statement of a theorem into the symbols of logic, because this forces one to think very clearly and precisely about the meaning of the statement, and it helps one to select correctly the method of proof to be used. Also, one is less likely to err in writing either the negation, contrapositive, or the converse of the given statement. For example, suppose one is asked to prove a statement which, written in the language of logic, is:

$$(\forall x)(\exists y)[f(x, y) \to g(x, y) \vee h(x, y)].$$

If one decides to prove this by contradiction, one must show that the negation of this statement logically implies some false statement, $r \wedge \sim r$. Letting $f = f(x, y)$, etc., one finds the negation as follows:

$$\sim[(\forall x)(\exists y)(f \to g \vee h)] \Leftrightarrow (\exists x)[(\sim(\exists y)(f \to g \vee h)]$$
$$\Leftrightarrow (\exists x)(\forall y)[\sim(f \to g \vee h)]$$
$$\Leftrightarrow (\exists x)(\forall y)[f \wedge \sim(g \vee h)]$$
$$\Leftrightarrow (\exists x)(\forall y)[f \wedge \sim g \wedge \sim h].$$

Thus, with the knowledge of a few tautologies and rules of logic, the work of negating a complicated statement becomes almost mechanical.

We have introduced you to a few symbols and tautologies that are useful to anyone who is studying mathematics. One would examine these concepts in greater detail in a course in mathematical logic, or symbolic logic, as it is also called.

Appendix B

Permutations: Cycles, Transpositions, Parity

In this appendix we prove the theorems that were stated without proof in Section 2.5. The terms we use here are defined and illustrated in that section. It is also convenient to use the Induction Principle (Theorem 3.11) in some of the proofs.

THEOREM B–1

If λ and μ are disjoint cycles of a finite set S, then $\lambda \circ \mu = \mu \circ \lambda$.

Proof

If $\lambda = (a_1, a_2, \ldots, a_r)$, $\mu = (b_1, b_2, \ldots, b_t)$, and S is of order n, then

$$S = \{a_1, a_2, \ldots, a_r, b_1, b_2, \ldots, b_t, c_1, c_2, \ldots, c_v\},$$

where $r + t + v = n$, and $v \geq 0$. Maps λ and μ behave as follows:

$$\begin{array}{llll} \lambda: & a_i \to a_{i+1}, & \mu: \ a_i \to a_i, & i = 1, 2, \ldots, r, \\ & b_j \to b_j, & b_j \to b_{j+1}, & j = 1, 2, \ldots, t, \\ & c_k \to c_k, & c_k \to c_k, & k = 1, 2, \ldots, v. \end{array}$$

(We agree to let $a_{r+1} = a_1$, and $b_{t+1} = b_1$.) Hence,

$$\begin{aligned} a_i(\lambda \circ \mu) &= (a_i\lambda)\mu = a_{i+1}\mu = a_{i+1} = a_i\lambda = (a_i\mu)\lambda = a_i(\mu \circ \lambda), \\ b_j(\lambda \circ \mu) &= (b_j\lambda)\mu = b_j\mu = b_{j+1} = b_{j+1}\lambda = (b_j\mu)\lambda = b_j(\mu \circ \lambda), \\ c_k(\lambda \circ \mu) &= (c_k\lambda)\mu = c_k\mu = c_k = c_k\lambda = (c_k\mu)\lambda = c_k(\mu \circ \lambda). \end{aligned}$$

Therefore, $s(\lambda \circ \mu) = s(\mu \circ \lambda)$, for all $s \in S$, and so $\lambda \circ \mu = \mu \circ \lambda$. □

Next we show that every permutation on n symbols can be expressed as a product of a finite set of disjoint cycles.

THEOREM B–2

If μ is a permutation of a set S of order n, then $\mu = \sigma_1 \circ \sigma_2 \circ \cdots \circ \sigma_m$, where the σ_i are disjoint cycles, and $m \geq 1$.

Proof

We agree to write $\varepsilon = (1)$. Suppose that $\mu \neq \varepsilon$. To obtain the cycle σ_1, we let a_1 be an element such that $a_1\mu \neq a_1$. So

$$a_1\mu = a_2, \quad a_2\mu = a_3, \ldots, \quad a_{r-1}\mu = a_r, \quad a_r\mu = a_{r+1},$$

where a_{r+1} is the *first* element that equals an element already in the cycle. (This must happen, since there are but a finite number n of distinct elements.) Now

$$a_{r+1} = a_i, \quad \text{with } i > 1 \Rightarrow a_r\mu = a_{i-1}\mu \Rightarrow a_r = a_{i-1},$$

since μ is 1-1. But this contradicts the fact that a_{r+1} is the first repeater. Hence, $a_{r+1} = a_1$, and so $\sigma_1 = (a_1, a_2, \ldots, a_r)$.

If every other element maps to itself, we are done. If there is a $b_1 \neq a_i$, for any i, and such that $b_1\mu = b_2$, where $b_1 \neq b_2$, then we proceed as we just did, and get that $\sigma_2 = (b_1, b_2, \ldots, b_t)$.

We must verify that σ_1 and σ_2 are disjoint. Assume that $b_i = a_j$, for some i, j. Now $b_i\sigma_2{}^k = b_i\mu^k = b_1$, where $k = t + 1 - i$, and $a_j\sigma_1{}^k = a_j\mu^k = a_v$, for some v. So if $b_i = a_j$, then $b_1 = a_v$, a contradiction of the choice of b_1. Therefore, σ_1 and σ_2 are disjoint.

Continuing in this manner, we obtain a finite number m of disjoint cycles such that $\mu = \sigma_1 \circ \sigma_2 \circ \cdots \circ \sigma_m$. $\square$

THEOREM B–3

A cycle of length n, with $n \geq 2$, can be expressed as a product of $n - 1$ transpositions, as follows:

$$(a_1, a_2, \ldots, a_{n-1}, a_n) = (a_1, a_2) \circ (a_1, a_3) \circ \cdots \circ (a_1, a_{n-1}) \circ (a_1, a_n).$$

Proof

Proof will be by induction on n. Let P_n be the statement:

$$(a_1 a_2 \cdots a_n) = (a_1 a_2) \circ (a_1 a_3) \circ \cdots \circ (a_1 a_n).$$

(i) P_2 is trivially true, since $(a_1 a_2)$ is already a product of $2 - 1$, or 1 transposition of the required form.

(ii) We show that $P_k \Rightarrow P_{k+1}$ as follows: By inspection, we see that $(a_1 a_2 \cdots a_k a_{k+1}) = (a_1 a_2 \cdots a_k) \circ (a_1 a_{k+1})$. But, by P_k, $(a_1 a_2 \cdots a_k) = (a_1 a_2) \circ (a_1 a_3) \circ \cdots \circ (a_1 a_k)$. So by substitution, we get the statement P_{k+1}. Therefore, by the Induction Principle, P_n is true for every $n \geq 2$. $\square$

As an immediate consequence of Theorems B–2 and B–3, we obtain this corollary:

THEOREM B–4

Every permutation on n symbols can be expressed as a product of a finite number of transpositions.

The next theorem is a lemma needed in order to prove Theorem B-6.

THEOREM B-5

Let $S = \{1, 2, \ldots, n\}$, with $n \geq 3$.

(a) If (a, b) is a transposition of S, with $a \neq 1$, $b \neq 1$, then

$$(a, b) = (1, a) \circ (1, b) \circ (1, a).$$

(b) Every permutation of S can be written as a product of transpositions of the form $(1, c)$, where c is one of the integers: $2, 3, \ldots, n$.

Proof

(a) Under the map (ab), $a \to b$ and $b \to a$. Under the map $(1a)(1b)(1a)$, $1 \to a \to 1$, $a \to 1 \to b$, and $b \to 1 \to a$. Hence, the two maps are equal.

(b) This is immediate, from Theorems B-4 and B-5(a). □

THEOREM B-6

Let $S = \{1, 2, \ldots, n\}$, and let ε be the identity permutation of S. If $\varepsilon = \sigma_1 \sigma_2 \cdots \sigma_r$, where each σ_i is a transposition, then r is an even integer.

Proof

Let $\varepsilon = \sigma_1 \sigma_2 \cdots \sigma_r$. If $\sigma_i = (ab)$, where $a \neq 1$, $b \neq 1$, for some i, we replace σ_i by $(1a)(1b)(1a)$. Then ε is a product of $r + 2$ transpositions. If we must make m such replacements, we have then expressed ε as a product of $r + 2m$ transpositions. Let $s = r + 2m$. Then r and s have the same parity, and

$$\varepsilon = (1c_1)(1c_2)\cdots(1c_s),$$

where the c_i are not all distinct. (If they were, we'd have $1 \to c_1$, instead of $1 \to 1$.)

If $n = 2$, then $\varepsilon = (12)(12)\cdots(12)$. Since $(12)(12) = \varepsilon$, there would obviously have to be an even number of transpositions in the factoring, or we'd have the contradiction $\varepsilon = (12)$.

Let P_n be the statement: If the order of S is n, then s is even. We just proved this to be true for $n = 2$, and we now verify that $P_{n-1} \Rightarrow P_n$.

Let ε be the identity permutation of a set of n symbols, where

$$\varepsilon = (1c_1)(1c_2)\cdots(1c_s).$$

Now $(1c_1)$ must occur again in the product, or otherwise we'd have $1 \to c_1$. Suppose that $c_k = c_1$, and $c_i \neq c_1$, for $1 < i < k$. Then ε can be written as follows:

$$\varepsilon = [(1c_1)(1c_2)\cdots(1c_{k-1})(1c_1)](1c_{k+1})\cdots(1c_s).$$

Permutations: Cycles, Transpositions, Parity

The product of the first k transpositions gives the map $1 \to c_1 \to 1$. Hence the permutation, $(1c_{k+1})$, must occur again, or else we'd have the entire product resulting in 1 mapping to c_{k+1}. So we bracket off the product that begins with $(1c_{k+1})$ and ends with the next occurrence of $(1c_{k+1})$. Suppose that $c_t = c_{k+1}$, with $t > k+1$. Continuing this reasoning, we bracket off the transpositions so that each bracketed expression begins and ends with the same transposition. That is,

$$\varepsilon = [(1c_1)(1c_2)\cdots(1c_{k-1})(1c_1)][(1c_{k+1})(1c_{k+2})\cdots(1c_{t-1})(1c_{k+1})]\cdots [(1c_s)\cdots(1c_{s-1})(1c_s)].$$

The first bracketed expression can be modified so as to eliminate 1, since

$$(1c_1)(1c_2)\cdots(1c_{k-1})(1c_1) = (c_1c_2)(c_1c_3)\cdots(c_1c_{k-1}).$$

(To check this, we observe that the above equation can be written:

$$(1c_1c_2\cdots c_{k-1}c_1) = (c_1c_2\cdots c_{k-1}).)$$

Each bracketed expression can be modified similarly. Hence

$$\varepsilon = [(c_1c_2)(c_1c_3)\cdots(c_1c_{k-1})][(c_{k+1}c_{k+2})\cdots(c_{k+1}c_{t-1})]\cdots[\cdots(c_sc_{s-1})],$$

and ε is a product of an even number of transpositions iff s is even. (Note that when 1 was eliminated, the number of transpositions in each bracketed expression was reduced by two.) Since we have eliminated 1, this last factoring is a representation of ε as the identity permutation of a set of $n-1$ symbols. By the induction hypothesis, there are an even number of transpositions in this last factoring. Therefore, s is even, and so r is even. □

Now that we have shown that the identity permutation always factors into a product of an even number of transpositions, it is a relatively simple matter to obtain the main result that we're after.

THEOREM B–7

Let μ be a permutation of a finite set. If μ can be expressed as a product of s transpositions and also as a product of t transpositions, then s and t are of the same parity.

Proof

Assume that s and t are of opposite parity, and let $s = 2k$, and $t = 2m + 1$. Then

$$\mu = \sigma_1\sigma_2\cdots\sigma_{2k}, \quad \text{and} \quad \mu = \eta_1\eta_2\cdots\eta_{2m+1},$$

where the σ_i and η_j are transpositions. Now μ^{-1} is a permutation, and so can be written as a product of transpositions. Let $\mu^{-1} = \lambda_1\lambda_2\cdots\lambda_r$. We can then express ε as a product of transpositions in two ways:

$$\varepsilon = \mu^{-1}\mu = (\lambda_1\lambda_2\cdots\lambda_r)(\sigma_1\sigma_2\cdots\sigma_{2k}),$$

a product of $r + 2k$ transpositions, and

$$\varepsilon = \mu^{-1}\mu = (\lambda_1\lambda_2\cdots\lambda_r)(\eta_1\eta_2\cdots\eta_{2m+1}),$$

a product of $r + 2m + 1$ transpositions. But $r + 2k$ and $r + 2m + 1$ are of opposite parity, a contradiction of Theorem B–6. So we must abandon our original assumption, and conclude that s and t are of the same parity. □

Appendix C

From Peano's Postulates to the Rational Integers

In Example 2.31 we *postulated* that $(\mathbf{Z}, +, \cdot)$ is an integral domain. However, it is possible to *prove* that $\mathbf{Z}$ is an integral domain, deriving the properties of $\mathbf{Z}$ from a more basic set of postulates. We shall outline the procedure by which this can be done.

The postulates with which we begin are attributed to the Italian mathematician, Peano. They have been given in various forms throughout the past several decades. The form we give is in terms of a mapping, and is in agreement with current mathematical terminology. We shall use the postulates to verify some basic properties of the natural numbers.

Peano's Postulates

Let N be a set of elements, called *natural numbers*.

PP–1 There is an element in N, which shall be denoted as "1".
PP–2 There is an injective map, $\sigma: N \to N$, such that 1 has no antecedent.
PP–3 If S is any subset of N such that

$$1 \in S, \quad \text{and} \quad n\sigma \in S, \quad \text{for all } n \in S,$$

then $S = N$. □

It is evident from the postulates that $N = \{1\} \cup N\sigma$. That is, 1 is the only element of N that is not in the image set $N\sigma$. From PP–3 we obtain the Induction Principle, of which we shall make great use in proving the theorems in this appendix. (Compare Theorem C–1 with Theorem 3.11.)

THEOREM C–1 *(Induction Principle)*

Let a statement P_m be associated with each natural number m. If:

(i) P_1 is true, and
(ii) $P_{m\sigma}$ is true whenever P_m is true, where m is any natural number;
then P_m is true, for every $m \in N$.

Proof

Let S be a subset of N such that $m \in S$ iff P_m is true. Then $1 \in S$, by premise (i). Also, if $m \in S$, then $m\sigma \in S$, by premise (ii). Hence $S = N$, by PP–3. Therefore, P_m is true, for every $m \in N$. □

We now define a binary operation of *addition*, $+$, on N recursively as follows:

(i) For all $a \in N$, $a + 1 = a\sigma$;

(ii) if $a + b$ is defined, where $a, b \in N$, then $a + b\sigma = (a + b)\sigma$.

Note that this definition tells us how to add to a any element of N (since every element of N except 1 is in $N\sigma$).

It would probably be helpful at this point to indicate the base-10 names that we give to the elements of N. By *definition*, we denote the elements of N as follows: $N = \{1, 2, 3, 4, \ldots\}$, where these natural numbers appear as images in this way, under the map σ:

$$1 \to 1\sigma = 1 + 1 = 2, \qquad 2 \to 2\sigma = 2 + 1 = 3,$$
$$3 \to 3\sigma = 3 + 1 = 4, \qquad \ldots, n \to n\sigma = n + 1, \qquad \ldots .$$

Thus, 2 is the image of 1, 3 is the image of 2, etc.

We are now in a position to prove the associative and commutative properties of addition of the natural numbers.

THEOREM C–2

If a, b, and c are any elements of N, then

$$(a + b) + c = a + (b + c).$$

Proof

Let a and b be any elements of N. Define P_c as follows:

$$(a + b) + c = a + (b + c), \qquad c \in N.$$

(i) P_1 is true, since

$$\begin{aligned}(a + b) + 1 &= (a + b)\sigma && [\text{Def. of } +] \\ &= a + b\sigma && [\text{Def. of } +] \\ &= a + (b + 1). && [\text{Def. of } +]\end{aligned}$$

(ii) We show that $P_m \Rightarrow P_{m\sigma}$ as follows:

$$\begin{aligned}(a + b) + m\sigma &= [(a + b) + m]\sigma && [\text{Def. of } +] \\ &= [a + (b + m)]\sigma && [P_m \text{ is true.}] \\ &= a + (b + m)\sigma && [\text{Def. of } +] \\ &= a + (b + m\sigma). && [\text{Def. of } +]\end{aligned}$$

So by Theorem C–1, $(a + b) + c = a + (b + c)$, for all $a, b, c \in N$. □

Before proving that addition is commutative we verify a special case.

THEOREM C–3

$a + 1 = 1 + a$, for all $a \in N$.

Proof

Let P_a be the statement: $a + 1 = 1 + a$.

(i) P_1 is trivially true, since $1 + 1 = 1 + 1$.

(ii) $P_m \Rightarrow m\sigma + 1 = (m + 1) + 1 = (1 + m) + 1$
$$= 1 + (m + 1) = 1 + m\sigma \Rightarrow P_{m\sigma}.$$

Therefore, by Theorem C–1, $a + 1 = 1 + a$, for all $a \in N$. □

THEOREM C–4

$a + b = b + a$, for all $a, b \in N$.

Proof

Let a be any element of N, and define P_b to be the statement: $a + b = b + a$. Now P_1 is true, by Theorem C–3. We show that $P_m \Rightarrow P_{m\sigma}$ as follows:

$a + m\sigma = (a + m)\sigma = (m + a)\sigma$ [Def. of $+$, and premise P_m]
$ = m + a\sigma = m + (a + 1)$ [Def. of $+$]
$ = m + (1 + a) = (m + 1) + a$ [Theorems C–3, C–2]
$ = m\sigma + a.$ [Def. of $+$]

Therefore, $a + b = b + a$, for all $a, b \in N$, by C–1. □

The cancellation property of addition can also be proved by using Theorem C–1. In the proof we use the fact that σ is an injective map.

THEOREM C–5

For all $a, b, c \in N$, if $a + c = b + c$, then $a = b$.

The next result will be used in a later proof.

THEOREM C–6

Let $a, b \in N$, with $a \neq b$. Then there exists either an $x \in N$ such that $x + a = b$, or a $y \in N$ such that $y + b = a$.

Proof

We let P_a be the above statement of the theorem, and use the Induction Principle. If $a = 1$, then $b \neq 1$, so $b = s\sigma$, for some $s \in N$. Then $s + 1 = b$, and P_1 is true.

Next, we verify that $P_m \Rightarrow P_{m\sigma}$. We assume that $m\sigma \neq b$.

Case 1: $m = b$. Then $m\sigma = b\sigma$, so $b + 1 = m\sigma$. Hence, $P_{m\sigma}$ is true.

Case 2: $m \neq b$. Under the assumption that P_m is true, there is either an s or a t in N such that $s + m = b$ or $t + b = m$. If $s + m = b$, then $s \neq 1$ (for then $m\sigma = b$). So $s = x\sigma$ for some $x \in N$, and

$$b = x\sigma + m = (x + 1) + m = x + (m + 1) = x + m\sigma.$$

If $t + b = m$, then $(b + t)\sigma = m\sigma$. So $b + t\sigma = m\sigma$, and $t\sigma + b = m\sigma$. Therefore, $P_m \Rightarrow P_{m\sigma}$; and P_a is true for all a, by Theorem C–1. □

Next we define a binary operation of multiplication, $\cdot$, on N:

(i) For all $a \in N, a \cdot 1 = a$

(ii) if $a \cdot b$ is defined, where $a, b \in N$, then $a \cdot (b\sigma) = a \cdot b + a$.

We leave it to the reader to use the Induction Principle to prove the following theorems involving multiplication.

THEOREM C–7

$a \cdot (b + c) = a \cdot b + a \cdot c,$ for all $a, b, c \in N$.

THEOREM C–8

$1 \cdot a = a, \quad (b\sigma) \cdot a = b \cdot a + a,$ for all $a, b \in N$.

THEOREM C–9

$(a \cdot b) \cdot c = a \cdot (b \cdot c),$ for all $a, b, c \in N$.

THEOREM C–10

$a \cdot b = b \cdot a,$ for all $a, b \in N$.

THEOREM C–11

$a \cdot c = b \cdot c \Rightarrow a = b,$ for all $a, b, c \in N$.

We have now verified that the set N of natural numbers is associative and commutative with respect to both addition and multiplication, the cancellation laws of both addition and multiplication are satisfied, there is an identity of multiplication, and the distributive properties hold. The algebra $(N, +, \cdot)$ satisfies all the properties of a commutative ring with unity except for two: (1) There is no identity of addition, and (2) additive inverses do not exist. The next step in our program is to "invent" zero and the negative integers. To do this we need to use a certain equivalence relation that can be defined on the Cartesian product $N \times N$.

THEOREM C-12

Let $\sim$ be a relation on $N \times N$ such that $(a, b) \sim (c, d)$ if and only if $a + d = b + c$.

(a) Relation $\sim$ is an equivalence relation on $N \times N$.
(b) If $\overline{(s, t)} = \{(x, y) : (x, y) \sim (s, t)\}$, and if a, b and m are any elements of N, then
  (i) $\overline{(a, a)} = \overline{(b, b)}$,
  (ii) $\overline{(m + a, a)} = \overline{(m + b, b)}$,
  (iii) $\overline{(a, m + a)} = \overline{(b, m + b)}$, and
  (iv) $\overline{(a + m, b + m)} = \overline{(a, b)}$.

Proof

(a) The equivalence relation properties follow directly from the definition of $\sim$. The transitive property is verified as follows:

$(a, b) \sim (c, d), (c, d) \sim (e, f) \Rightarrow a + d = b + c, c + f = d + e$
$\Rightarrow (a + d) + (c + f) = (b + c) + (d + e)$
$\Rightarrow (a + f) + (c + d) = (b + e) + (c + d)$
$\Rightarrow a + f = b + e \Rightarrow (a, b) \sim (e, f)$.

(b) These four statements are equivalent, respectively, to:

(i) $a + b = a + b$, (ii) $(m + a) + b = a + (m + b)$,
(iii) $a + (m + b) = (m + a) + b$, (iv) $(a + m) + b = (b + m) + a$.

The last three statements are true by Theorems C-2 and C-4. □

The equivalence classes of Theorem C-12 will be the elements in the integral domain that we now introduce.

THEOREM C-13

Let $\mathbf{Z} = \{\overline{(a, b)} : a, b \in N\}$, where $\mathbf{Z}$ is the set of equivalence classes of Theorem C-12, and let addition, $+$, and multiplication, $\cdot$, on $\mathbf{Z}$ be defined as follows:

$$\overline{(a, b)} + \overline{(c, d)} = \overline{(a + c, b + d)}, \qquad \overline{(a, b)} \cdot \overline{(c, d)} = \overline{(ac + bd, ad + bc)}.$$

(a) Addition and multiplication are well-defined.
(b) $(\mathbf{Z}, +, \cdot)$ is an ordered integral domain.

Proof

(a) $\overline{(a, b)} = \overline{(s, t)}, \overline{(c, d)} = \overline{(x, y)} \Rightarrow a + t = b + s, \quad c + y = d + x$
$\Rightarrow (a + c) + (t + y) = (b + d) + (s + x)$
$\Rightarrow \overline{(a + c, b + d)} = \overline{(s + x, t + y)} \Rightarrow \overline{(a, b)} + \overline{(c, d)} = \overline{(s, t)} + \overline{(x, y)}$.

So addition is well-defined.

Also,

$a + t = b + s, \quad c + y = d + x$

$\Rightarrow (a + t)c + (b + s)d = (b + s)c + (a + t)d,$
$(c + y)s + (d + x)t = (d + x)s + (c + y)t$
$\Rightarrow [(a + t)c + (b + s)d] + [(c + y)s + (d + x)t]$
$= [(b + s)c + (a + t)d] + [(d + x)s + (c + y)t]$
$\Rightarrow (ac + bd + sy + tx) + (ct + ds + cs + dt)$
$= (bc + ad + sx + ty) + (cs + dt + ds + ct)$
$\Rightarrow (ac + bd) + (sy + tx) = (ad + bc) + (sx + ty)$
$\Rightarrow \overline{(ac + bd, ad + bc)} = \overline{(sx + ty, sy + tx)}$
$\Rightarrow \overline{(a, b)} \cdot \overline{(c, d)} = \overline{(s, t)} \cdot \overline{(x, y)}.$

Therefore, multiplication is well-defined.

(b) The associative, commutative, and distributive properties follow readily from the definitions of $+$ and $\cdot$ and the fact that those properties hold in $(N, +, \cdot)$. Also, one can show that $\overline{(1, 1)}$ is the zero, $\overline{(b, a)}$ is the additive inverse of $\overline{(a, b)}$, and $\overline{(1 + c, c)}$ is the unity. Hence $(\mathbf{Z}, +, \cdot)$ is a commutative ring with unity.

We next show that $\mathbf{Z}$ has no zero divisors. Now $\overline{(a, b)} \cdot \overline{(c, d)} = \overline{(1, 1)} \Rightarrow \overline{(ac + bd, ad + bc)} = \overline{(1, 1)} \Rightarrow ac + bd + 1 = ad + bc + 1 \Rightarrow ac + bd = ad + bc$. If $c = d$, then $\overline{(c, d)} = \overline{(d, d)} = \overline{(1, 1)}$. We must show that if $c \neq d$, then $a = b$. If $c \neq d$, then either $c = d + r$ or $d = c + s$, for some $r, s \in N$, by Theorem C-6. We see that

$c = d + r, \quad ac + bd = ad + bc \Rightarrow$
$a(d + r) + bd = ad + b(d + r) \Rightarrow$
$ad + ar + bd = ad + bd + br \Rightarrow$
$ar = br \Rightarrow a = b.$

Similarly, if $d = c + s$, then $a = b$. So $\mathbf{Z}$ has no zero divisors, and hence is an integral domain.

It remains to be shown that $\mathbf{Z}$ can be ordered. Define

$\mathbf{Z}^+ = \{\overline{(m + 1, 1)} : m \in N\}.$

It is a simple matter to verify that for all $m, n \in N$,

$\overline{(m + 1, 1)} + \overline{(n + 1, 1)} = \overline{(m + n + 1, 1)}, \quad \text{and}$
$\overline{(m + 1, 1)} \cdot \overline{(n + 1, 1)} = \overline{(mn + 1, 1)}.$

Hence $\mathbf{Z}^+$ is closed under the two operations. If $\mathbf{Z}^- = \{\overline{(1, m + 1)} : m \in N\}$, then a nonzero element $\overline{(a,b)}$ of $\mathbf{Z}$ is either in $\mathbf{Z}^-$ or in $\mathbf{Z}^+$, and $\overline{(a,b)} \in \mathbf{Z}^+$ if and only if $-\overline{(a,b)} \in \mathbf{Z}^-$. This follows from the fact that either $a = b + r$ or $b = a + s$, for some $r, s \in N$, and so one of these holds:

$$\overline{(b+r, b)} = \overline{(r+1, 1)} \in \mathbf{Z}^+, \qquad \overline{(a, a+s)} = \overline{(1, s+1)} \in \mathbf{Z}^-.$$

Therefore, $\mathbf{Z}^+$ is the set of positive elements of $\mathbf{Z}$, and so $(\mathbf{Z}, +, \cdot)$ is ordered. □

We are now in a position to verify that $\mathbf{Z}$ is an extension of N, in the sense that N is isomorphic to the subset $\mathbf{Z}^+$ of $\mathbf{Z}$.

THEOREM C–14

$(N, +, \cdot)$ is isomorphic to $(\mathbf{Z}^+, +, \cdot)$.

Proof

Define $\theta: N \to \mathbf{Z}^+$ to be the map such that $m\theta = \overline{(m+1, 1)}$, for all $m \in N$. Map θ is obviously surjective. It is injective, since

$$m\theta = n\theta \Rightarrow \overline{(m+1, 1)} = \overline{(n+1, 1)} \Rightarrow (m+1) + 1 = 1 + (n+1) \Rightarrow$$
$$(m+1) + 1 = (n+1) + 1 \Rightarrow m = n.$$

Also, $\quad m\theta + n\theta = \overline{(m+1, 1)} + \overline{(n+1, 1)} = \overline{(m+n+1, 1)} = (m+n)\theta,$
$$(m\theta) \cdot (n\theta) = \overline{(m+1, 1)} \cdot \overline{(n+1, 1)} = \overline{(mn+1, 1)} = (mn)\theta.$$

Therefore θ is an isomorphism. □

We have just proved that the set of positive rational integers is structurally the same as the set of natural numbers. Because of this we agree to simplify our notation for elements of $\mathbf{Z}^+$, *defining* $\overline{(m+1, 1)} = m$. Then $\overline{(1, m+1)} = -m$. If we let $\overline{(1, 1)} = 0$, then

$$\mathbf{Z} = \{0\} \cup \{m : m \in N\} \cup \{-m : m \in N\}.$$

Furthermore, any element $\overline{(a, b)}$ of $\mathbf{Z}$ can now be written as follows:

$$\overline{(a, b)} = \overline{(a+1, 1)} + \overline{(1, b+1)} = a + (-b) = a - b.$$

This completes the story.

Appendix D

Reference List of Groups of Order n, for $n \leq 12$

Each group whose order is not more than 12 is defined in this appendix in terms of one, two, or three generators. The group table is given for each noncyclic group. These tables are included so that the student will have quick and easy access to several examples of subgroups, factor groups, automorphism groups, etc. The groups are numbered so that the digit to the left of the decimal point is the order n of the group, and the digit to the right of the point distinguishes the various groups of that order.

1.1 $Z_1 \cong S_1 \cong \langle e \rangle = \{e\}$.

2.1 $Z_2 \cong S_2 \cong \langle a \rangle = \{a^i : |a| = 2\} = \{e, a\}$.

3.1 $Z_3 \cong \langle a \rangle = \{a^i : |a| = 3\} = \{e, a, a^2\}$.

4.1 $Z_4 \cong \langle a \rangle = \{a^i : |a| = 4\} = \{e, a, a^2, a^3\}$.

4.2 $Z_2 \times Z_2 \cong K_4 \cong \langle a, b \rangle = \{a^i b^j : |a| = |b| = 2, ba = ab\}$.

	e	a	b	ab
e	e	a	b	ab
a	a	e	ab	b
b	b	ab	e	a
ab	ab	b	a	e

5.1 $Z_5 \cong \langle a \rangle = \{a^i : |a| = 5\} = \{e, a, a^2, a^3, a^4\}$.

6.1 $Z_6 \cong \langle a \rangle = \{a^i : |a| = 6\} = \{e, a, a^2, a^3, a^4, a^5\}$.

6.2 $D_3 \cong S_3 \cong \langle a, b \rangle = \{a^i b^j : |a| = 3, |b| = 2, ba = a^2 b\}$.

	e	a	a^2	b	ab	$a^2 b$
e	e	a	a^2	b	ab	$a^2 b$
a	a	a^2	e	ab	$a^2 b$	b
a^2	a^2	e	a	$a^2 b$	b	ab
b	b	$a^2 b$	ab	e	a^2	a
ab	ab	b	$a^2 b$	a	e	a^2
$a^2 b$	$a^2 b$	ab	b	a^2	a	e

7.1 $Z_7 \cong \langle a \rangle = \{a^i : |a| = 7\}$.

8.1 $Z_8 \cong \langle a \rangle = \{a^i : |a| = 8\}$.

Reference List of Groups of Order n, for n ≤ 12

8.2 $\mathbf{Z}_4 \times \mathbf{Z}_2 \cong \langle a, b \rangle = \{a^i b^j : |a| = 4, |b| = 2, ba = ab\}$.

	e	a	a^2	a^3	b	ab	a^2b	a^3b
e	e	a	a^2	a^3	b	ab	a^2b	a^3b
a	a	a^2	a^3	e	ab	a^2b	a^3b	b
a^2	a^2	a^3	e	a	a^2b	a^3b	b	ab
a^3	a^3	e	a	a^2	a^3b	b	ab	a^2b
b	b	ab	a^2b	a^3b	e	a	a^2	a^3
ab	ab	a^2b	a^3b	b	a	a^2	a^3	e
a^2b	a^2b	a^3b	b	ab	a^2	a^3	e	a
a^3b	a^3b	b	ab	a^2b	a^3	e	a	a^2

8.3 $\mathbf{Z}_2 \times \mathbf{Z}_2 \times \mathbf{Z}_2 \cong \langle a, b, c \rangle = \{a^i b^j c^k : |a| = |b| = |c| = 2, ba = ab, ca = ac, cb = bc\}$.

	e	a	b	ab	c	ac	bc	abc
e	e	a	b	ab	c	ac	bc	abc
a	a	e	ab	b	ac	c	abc	bc
b	b	ab	e	a	bc	abc	c	ac
ab	ab	b	a	e	abc	bc	ac	c
c	c	ac	bc	abc	e	a	b	ab
ac	ac	c	abc	bc	a	e	ab	b
bc	bc	abc	c	ac	b	ab	e	a
abc	abc	bc	ac	c	ab	b	a	e

8.4 $D_4 \cong \langle a, b \rangle = \{a^i b^j : |a| = 4, |b| = 2, ba = a^3 b\}$.

	e	a	a^2	a^3	b	ab	a^2b	a^3b
e	e	a	a^2	a^3	b	ab	a^2b	a^3b
a	a	a^2	a^3	e	ab	a^2b	a^3b	b
a^2	a^2	a^3	e	a	a^2b	a^3b	b	ab
a^3	a^3	e	a	a^2	a^3b	b	ab	a^2b
b	b	a^3b	a^2b	ab	e	a^3	a^2	a
ab	ab	b	a^3b	a^2b	a	e	a^3	a^2
a^2b	a^2b	ab	b	a^3b	a^2	a	e	a^3
a^3b	a^3b	a^2b	ab	b	a^3	a^2	a	e

8.5 $C_8 \cong Q_8 \cong \langle a, b \rangle = \{a^i b^j : |a| = |b| = 4, b^2 = a^2, ba = a^3 b\}$.

	e	a	a^2	a^3	b	ab	a^2b	a^3b
e	e	a	a^2	a^3	b	ab	a^2b	a^3b
a	a	a^2	a^3	e	ab	a^2b	a^3b	b
a^2	a^2	a^3	e	a	a^2b	a^3b	b	ab
a^3	a^3	e	a	a^2	a^3b	b	ab	a^2b
b	b	a^3b	a^2b	ab	a^2	a	e	a^3
ab	ab	b	a^3b	a^2b	a^3	a^2	a	e
a^2b	a^2b	ab	b	a^3b	e	a^3	a^2	a
a^3b	a^3b	a^2b	ab	b	a	e	a^3	a^2

9.1 $\mathbf{Z}_9 \cong \langle a \rangle = \{a^i : |a| = 9\}$.

9.2 $\mathbf{Z}_3 \times \mathbf{Z}_3 \cong \langle a, b \rangle = \{a^i b^j : |a| = |b| = 3, ba = ab\}$.

	e	a	a^2	b	ab	a^2b	b^2	ab^2	a^2b^2
e	e	a	a^2	b	ab	a^2b	b^2	ab^2	a^2b^2
a	a	a^2	e	ab	a^2b	b	ab^2	a^2b^2	b^2
a^2	a^2	e	a	a^2b	b	ab	a^2b^2	b^2	ab^2
b	b	ab	a^2b	b^2	ab^2	a^2b^2	e	a	a^2
ab	ab	a^2b	b	ab^2	a^2b^2	b^2	a	a^2	e
a^2b	a^2b	b	ab	a^2b^2	b^2	ab^2	a^2	e	a
b^2	b^2	ab^2	a^2b^2	e	a	a^2	b	ab	a^2b
ab^2	ab^2	a^2b^2	b^2	a	a^2	e	ab	a^2b	b
a^2b^2	a^2b^2	b^2	ab^2	a^2	e	a	a^2b	b	ab

10.1 $\mathbf{Z}_{10} \cong \langle a \rangle = \{a^i : |a| = 10\}$.

10.2 $D_5 \cong \langle a, b \rangle = \{a^i b^j : |a| = 5, |b| = 2, ba = a^4 b\}$.

	e	a	a^2	a^3	a^4	b	ab	a^2b	a^3b	a^4b
e	e	a	a^2	a^3	a^4	b	ab	a^2b	a^3b	a^4b
a	a	a^2	a^3	a^4	e	ab	a^2b	a^3b	a^4b	b
a^2	a^2	a^3	a^4	e	a	a^2b	a^3b	a^4b	b	ab
a^3	a^3	a^4	e	a	a^2	a^3b	a^4b	b	ab	a^2b
a^4	a^4	e	a	a^2	a^3	a^4b	b	ab	a^2b	a^3b
b	b	a^4b	a^3b	a^2b	ab	e	a^4	a^3	a^2	a
ab	ab	b	a^4b	a^3b	a^2b	a	e	a^4	a^3	a^2
a^2b	a^2b	ab	b	a^4b	a^3b	a^2	a	e	a^4	a^3
a^3b	a^3b	a^2b	ab	b	a^4b	a^3	a^2	a	e	a^4
a^4b	a^4b	a^3b	a^2b	ab	b	a^4	a^3	a^2	a	e

11.1 $\mathbf{Z}_{11} \cong \langle a \rangle = \{a^i : |a| = 11\}$.

12.1 $\mathbf{Z}_{12} \cong \langle a \rangle = \{a^i : |a| = 12\}$.

12.2 $\mathbf{Z}_6 \times \mathbf{Z}_2 \cong \langle a, b \rangle = \{a^i b^j : |a| = 6, |b| = 2, ba = ab\}$.

	e	a	a^2	a^3	a^4	a^5	b	ab	a^2b	a^3b	a^4b	a^5b
e	e	a	a^2	a^3	a^4	a^5	b	ab	a^2b	a^3b	a^4b	a^5b
a	a	a^2	a^3	a^4	a^5	e	ab	a^2b	a^3b	a^4b	a^5b	b
a^2	a^2	a^3	a^4	a^5	e	a	a^2b	a^3b	a^4b	a^5b	b	ab
a^3	a^3	a^4	a^5	e	a	a^2	a^3b	a^4b	a^5b	b	ab	a^2b
a^4	a^4	a^5	e	a	a^2	a^3	a^4b	a^5b	b	ab	a^2b	a^3b
a^5	a^5	e	a	a^2	a^3	a^4	a^5b	b	ab	a^2b	a^3b	a^4b
b	b	ab	a^2b	a^3b	a^4b	a^5b	e	a	a^2	a^3	a^4	a^5
ab	ab	a^2b	a^3b	a^4b	a^5b	b	a	a^2	a^3	a^4	a^5	e
a^2b	a^2b	a^3b	a^4b	a^5b	b	ab	a^2	a^3	a^4	a^5	e	a
a^3b	a^3b	a^4b	a^5b	b	ab	a^2b	a^3	a^4	a^5	e	a	a^2
a^4b	a^4b	a^5b	b	ab	a^2b	a^3b	a^4	a^5	e	a	a^2	a^3
a^5b	a^5b	b	ab	a^2b	a^3b	a^4b	a^5	e	a	a^2	a^3	a^4

Reference List of Groups of Order n, for n ≤ 12

12.3 $D_6 \cong \langle a, b \rangle = \{a^i b^j : |a| = 6, |b| = 2, ba = a^5 b\}$.

	e	a	a^2	a^3	a^4	a^5	b	ab	a^2b	a^3b	a^4b	a^5b
e	e	a	a^2	a^3	a^4	a^5	b	ab	a^2b	a^3b	a^4b	a^5b
a	a	a^2	a^3	a^4	a^5	e	ab	a^2b	a^3b	a^4b	a^5b	b
a^2	a^2	a^3	a^4	a^5	e	a	a^2b	a^3b	a^4b	a^5b	b	ab
a^3	a^3	a^4	a^5	e	a	a^2	a^3b	a^4b	a^5b	b	ab	a^2b
a^4	a^4	a^5	e	a	a^2	a^3	a^4b	a^5b	b	ab	a^2b	a^3b
a^5	a^5	e	a	a^2	a^3	a^4	a^5b	b	ab	a^2b	a^3b	a^4b
b	b	a^5b	a^4b	a^3b	a^2b	ab	e	a^5	a^4	a^3	a^2	a
ab	ab	b	a^5b	a^4b	a^3b	a^2b	a	e	a^5	a^4	a^3	a^2
a^2b	a^2b	ab	b	a^5b	a^4b	a^3b	a^2	a	e	a^5	a^4	a^3
a^3b	a^3b	a^2b	ab	b	a^5b	a^4b	a^3	a^2	a	e	a^5	a^4
a^4b	a^4b	a^3b	a^2b	ab	b	a^5b	a^4	a^3	a^2	a	e	a^5
a^5b	a^5b	a^4b	a^3b	a^2b	ab	b	a^5	a^4	a^3	a^2	a	e

12.4 $C_{12} \cong \langle a, b \rangle = \{a^i b^j : |a| = 6, |b| = 4, b^2 = a^3, ba = a^5 b\}$.

	e	a	a^2	a^3	a^4	a^5	b	ab	a^2b	a^3b	a^4b	a^5b
e	e	a	a^2	a^3	a^4	a^5	b	ab	a^2b	a^3b	a^4b	a^5b
a	a	a^2	a^3	a^4	a^5	e	ab	a^2b	a^3b	a^4b	a^5b	b
a^2	a^2	a^3	a^4	a^5	e	a	a^2b	a^3b	a^4b	a^5b	b	ab
a^3	a^3	a^4	a^5	e	a	a^2	a^3b	a^4b	a^5b	b	ab	a^2b
a^4	a^4	a^5	e	a	a^2	a^3	a^4b	a^5b	b	ab	a^2b	a^3b
a^5	a^5	e	a	a^2	a^3	a^4	a^5b	b	ab	a^2b	a^3b	a^4b
b	b	a^5b	a^4b	a^3b	a^2b	ab	a^3	a^2	a	e	a^5	a^4
ab	ab	b	a^5b	a^4b	a^3b	a^2b	a^4	a^3	a^2	a	e	a^5
a^2b	a^2b	ab	b	a^5b	a^4b	a^3b	a^5	a^4	a^3	a^2	a	e
a^3b	a^3b	a^2b	ab	b	a^5b	a^4b	e	a^5	a^4	a^3	a^2	a
a^4b	a^4b	a^3b	a^2b	ab	b	a^5b	a	e	a^5	a^4	a^3	a^2
a^5b	a^5b	a^4b	a^3b	a^2b	ab	b	a^2	a	e	a^5	a^4	a^3

12.5 $A_4 \cong \langle a, b, c \rangle = \{a^i b^j c^k : |a| = |b| = 2, |c| = 3, ba = ab, ca = bc, cb = abc\}$.

	e	a	b	ab	c	ac	bc	abc	c^2	ac^2	bc^2	abc^2
e	e	a	b	ab	c	ac	bc	abc	c^2	ac^2	bc^2	abc^2
a	a	e	ab	b	ac	c	abc	bc	ac^2	c^2	abc^2	bc^2
b	b	ab	e	a	bc	abc	c	ac	bc^2	abc^2	c^2	ac^2
ab	ab	b	a	e	abc	bc	ac	c	abc^2	bc^2	ac^2	c^2
c	c	bc	abc	ac	c^2	bc^2	abc^2	ac^2	e	b	ab	a
ac	ac	abc	bc	c	ac^2	abc^2	bc^2	c^2	a	ab	b	e
bc	bc	c	ac	abc	bc^2	c^2	ac^2	abc^2	b	e	a	ab
abc	abc	ac	c	bc	abc^2	ac^2	c^2	bc^2	ab	a	e	b
c^2	c^2	abc^2	ac^2	bc^2	e	ab	a	b	c	abc	ac	bc
ac^2	ac^2	bc^2	c^2	abc^2	a	b	e	ab	ac	bc	c	abc
bc^2	bc^2	ac^2	abc^2	c^2	b	a	ab	e	bc	ac	abc	c
abc^2	abc^2	c^2	bc^2	ac^2	ab	e	b	a	abc	c	bc	ac

Appendix E

Thumbnail Sketches of 22 Mathematicians

The names of certain mathematicians always crop up in the study of abstract algebra. This is not the place to go into *great* detail about any of these persons. However, we have assembled here a collection of short short stories about some of the mathematicians whose love of mathematics and devotion to research have made this book possible.

E–1 Euclid   *(4th Century B.C.)*

Very little is known about Euclid, the man. It is assumed that he was Greek, and that he came to Alexandria from Athens at the invitation of Ptolemy, the King of Egypt. He help found the University of Alexandria, and he taught and wrote there for twenty or thirty years.

The great legacy that Euclid left the mathematical world was his *Elements*, consisting of thirteen parchment scrolls, in which he tried to organize and coordinate *all* the mathematics that was known in his day! The mathematical historian, D. E. Smith, said of Euclid: "He was the most successful textbook writer that the world has ever known." For over 2200 years the geometry taught was based on the postulates and theorems in the *Elements*. But Euclid was not concerned just with geometry. Books VII, VIII, and IX are on the theory of numbers. At the beginning of Book VII he gave a procedure for finding the greatest common divisor of two integers. This procedure is now called the *Euclidean algorithm*. It is suspected, however, that Euclid did not invent the algorithm. A very famous proof is Euclid's proof that the number of primes is infinite. (See Problem 3.4–5.) The British mathematician, G. H. Hardy, said that this proof is "as fresh and significant as when it was discovered—two thousand years has not written a wrinkle on it."

Tradition has it that after a pupil of Euclid's learned to prove his first theorem, he asked: "But what advantage shall I get by learning these things?" So Euclid called for his servant (his slave, actually), and said, "Give him threepence, since he must needs make profit out of what he learns." Students haven't changed much in 2300 years, have they?

E–2 Diophantus   *(3rd Century A.D.)*

Diophantus was a Greek algebraist who migrated to Alexandria, Egypt, and lived to be about 84 years old. His fame is indicated by the fact that, in about 280 A. D., a treatise on the Egyptian method of reckoning was dedicated to him. From his surviving works—six of the 13 books (which were actually scrolls) of his *Arithmetica*, and parts of his *Polynomial Numbers* and *Porisms*—it has been concluded that he was the earliest known master of the theory of numbers. He solved systems of equations in as many as four unknowns, and found solutions of systems in which there were more unknowns than equations. He insisted on solutions that were positive integers or positive fractions, since negative numbers and zero had not been discovered yet. He even used the quadratic formula in solving certain second-degree equations.

The success that Diophantus had in solving equations is attributed to the fact that he introduced abbreviations for words, thus developing a primitive sort of algebraic symbolism. Almost no symbols were used up to the time of Diophantus. The two Greek letters, $\Delta \gamma$, was his abbreviation for the word, *power*. He used σ for 200, and ν for 50. Thus, $\Delta \gamma \sigma \nu$ meant $250x^2$. Strangely enough, very little advancement was made in either number theory or in notation from the time of Diophantus until the time of Fermat, about 14 centuries later! It is difficult to realize that the present-day symbolism of elementary algebra is only around 300 years old.

E–3 Cardan   (1501–1576)

Like many of the early mathematicians, Girolamo Cardan did not confine his work solely to mathematics. Although he occupied the Chair of Mathematics at the University of Milan, Italy, he was a practicing physician, and also dabbled in astrology and philosophy. In addition, he was quite a gambler.

The formula for solving a cubic polynomial by radicals is known as "Cardan's solution of the cubic," but it must be said that Cardan gained this claim to fame in an unethical manner. Another Italian, known by the nickname of Tartaglia because he stammered (*tartaglione* means *stammerer* in Italian), had produced a solution of the cubic. Cardan learned of this, and begged Tartaglia to give him the solution, promising to tell no one. Then, in 1545, he proceeded to publish the solution as his own original work in his book, *Ars Magna*. He also took more credit than he should have for the solution by radicals of the general fourth degree polynomial, which solution was devised by his pupil, Ferrari. And all this underhandedness was really quite unnecessary, since Cardan was considered to be an algebraic genius, and actually did much original work in algebra in his *Ars Magna*. But some people are never satisfied!

E–4 Fermat   (1601–1665)

Pierre de Fermat has been called the "prince of amateurs," since for him mathematics was but a hobby to be indulged in during his leisure time. Nevertheless, he is also regarded as the greatest number theorist between Diophantus (of the third century) and Euler (of the eighteenth century).

Fermat was born near Toulouse, France, the son of a leather merchant. He received his early education at home, but at age 30 he began his career as a lawyer for the local parliament at Toulouse. When not engaged in legal work, he was "playing" with mathematics. He didn't publish much, but he corresponded with many of the best mathematicians, and so influenced *their* work. As a result, this humble, retiring man has been said to be the greatest French mathematician of the seventeenth century. The two volumes of his *Opera mathematica* were edited by his son, and published several years after his death.

Although he worked in many areas of mathematics, including probability and analytic geometry, and has even been given credit for inventing differential calculus, his greatest contributions were in number theory. He became interested in number theory through reading *Arithmetica*, by Diophantus. As he read the book, he jotted down statements in the margins, often without proof. Later, these statements were proved by others to be true. His most famous statement was found in the margin opposite Problem 8, Book II of *Arithmetica*. Problem 8 was: "To divide a given square number into two squares." This meant that, given a positive integer c, find positive integers a and b such that $a^2 + b^2 = c^2$. In the margin Fermat wrote: "To divide a cube into two cubes, a fourth power, or, in general, any power whatever, into two powers of the same denomination above the second is impossible, and I have assuredly found an admirable proof of this, but the margin is too narrow to contain it." In symbols, his statement declared that if $n > 2$, then there exist no positive integers, a, b, and c, such that $a^n + b^n = c^n$. This statement, called *Fermat's Last Theorem*, has never been proved, although mathematicians, both professional and amateur, have been attempting to prove or disprove it for the past 300 years. By 1955, research had progressed to the point where it was known that the statement is true for all $n < 4003$, and for many other special values of n. (It is not difficult to verify that the theorem is true for any integer that is a multiple of 4.) Fermat's Last Theorem has the dubious distinction of being the mathematical problem for which the greatest number of incorrect proofs have been published. The weary and disappointed authors of those "proofs" probably have wished, with a fervor, that Fermat had chosen some hobby other than mathematics!

E–5 Euler (1707–1783)

Leonhard Euler was truly prolific, as both man and mathematician. He had 13 children, and he did enough mathematical research to fill 80 large volumes. Evidently he budgeted his time carefully!

As a boy in Basel, Switzerland, Leonhard was taught by his father, who was a Calvinist minister and a mathematician. By the age of 17, the boy earned his master's degree at the University of Basel, and he did his first original mathematics at the age of 19, receiving honorable mention from the Paris Academy. In the eighteenth century the universities were not the centers of research. Instead, the rulers of the European countries established royal academies, and paid the most prominent research scientists to come to them. So, at the age of 20, Euler accepted a mathematics position at the Academy of St. Petersburg, which was formed by Peter the Great of Russia. He stayed in Russia for 14 years, and then went to Berlin as Chief Mathematician of the Prussian Academy, at the invitation of Frederick the Great. He worked in Berlin for 25 years, and then went back to Russia at the urging of Catherine the Great, where he remained for the rest of his life.

Euler made some great contributions to number theory, but he also did research in calculus, differential equations, differential geometry, calculus of variations, applied mathematics, mathematical theory of investment, and found time to write about mathematical puzzles and recreations. In Berlin he worked also on some practical problems relating to navigation canals and pension systems, for example, and in St. Petersburg he wrote the elementary mathematics textbooks for the Russian schools, helped reform the system of weights and measures, and supervised the government Department of Geography. Among the mathematics notation that Euler invented are the following symbols: $f(x)$ for a functional value, e for the base of natural logarithms, i for $\sqrt{-1}$, Σ for summation. Total blindness in later years failed to stop him, for he then dictated material to his sons.

If there were a Mathematical Hall of Fame, Euler's name would probably appear in a most prominent position in the hall!

E–6 Lagrange   (1736–1813)

The two greatest mathematicians of the 18th century were Euler and Lagrange. Joseph-Louis Lagrange, of a famous French family with some Italian blood in it, was born in Turin, Italy, of a rich father who didn't invest wisely, and lost everything. Lagrange later said: "If I had inherited a fortune, I should probably not have cast my lot with mathematics." But early, he *did* develop a burning passion for mathematics. One gets the picture of a modest but courtly man, with an elegant style of writing, a winning personality, and many friends.

When only 16, Lagrange became Professor of Mathematics in the Royal School of Artillery at Turin. Frederick the Great of Prussia liked his sophistication, something which he felt was lacking in Euler, and invited Lagrange, at the age of 30, to be the director of the physics-mathematics division of the Berlin Academy after Euler left that post. Lagrange worked in Berlin for twenty years, but disliked the atmosphere there after Frederick died. He went to Paris, where he became a member of the French Academy, and was received with great enthusiasm. He had quarters in the Louvre until the French Revolution, and was a favorite of Marie Antoinette. In those first years in Paris he was despondent and not very productive, feeling that mathematics was on the decline, and that he was burned out mathematically. (Also, his wife of many years had died in Berlin after a long illness, just before he came to Paris.) However, the French Revolution broke his apathy, and he once again became an active and dynamic mathematician. Napoleon made him a Senator, a Count of the Empire, and a Grand Officer of the Legion of Honor. His second marriage, at age 56 to a girl in her teens, the daughter of a French astronomer, was said to be a very happy one.

Most of the professional life of Lagrange was devoted exclusively to research, since he was 61 when he took his first teaching position, instructing at L'École Polytechnique, a school which was to train many renowned mathematicians of modern France. Like Euler, Lagrange had a wide range of mathematical interests, working in many areas of both analysis and algebra. Thanks to him, we have the derivative notation: $f'(x)$, $f''(x)$. He has been called the first real analyst. His book, *Mécanique Analytique*, has been described as a "scientific poem." He proved some of Fermat's theorems, and his discoveries in the theory of equations paved the way for Abel and Galois and their algebraic accomplishments. Like others, Lagrange attempted to solve by radicals the general fifth degree polynomial. He failed to do so, of course, but his work provided a clue that helped Abel to prove the impossibility of a solution just 20 years after the death of Lagrange.

Perhaps teachers shouldn't criticize students who do their homework while listening to their radio or record player. Lagrange liked to work to music. When asked why he liked music, Lagrange replied: "I like it because it isolates me. I hear the first three measures; at the fourth I distinguish nothing; I give myself up to my thoughts; nothing interrupts me; and it is thus that I have solved more than one difficult problem."

E–7 Wilson (1741–1793)

John Wilson's claim to mathematical fame was a single theorem that he discovered while he was an undergraduate at Cambridge University. (See Problem 3.6–16(c).) His proof was published in *Meditationes algebraicae* by Waring, his teacher, and has since been called *Wilson's Theorem*, even though Leibniz knew the theorem, and Leibniz died before Wilson was born! Wilson was senior wrangler (which means that he was the winner of the highest honors in mathematics at Cambridge), but then did not live up to the expectations that his mathematics instructors had for him. He abandoned mathematics, went into law, rose to a judgeship, and was knighted. One can hardly classify him as a failure!

E–8 Gauss (1777–1855)

Carl Friedrich Gauss was the greatest mathematician of the nineteenth century, even though his father was but a poor gardener, brick layer, and canal tender. Born in Brunswick, Germany, his genius was recognized early, for when he was three years old he found an error in his father's bookkeeping. His ability to do mental calculations was amazing. The Duke of Brunswick learned of him, paid for his education, and supported him in his research. In his doctoral dissertation, Gauss gave the first good proof of the Fundamental Theorem of Algebra. The Duke died when Gauss was 29, and so Gauss had to look for another patron. He was invited to St. Petersburg, but he went to the University of Göttingen, Germany, instead, and remained there as Director of the Observatory and Professor of Mathematics. He enjoyed family life, was married twice, and had three children by each wife.

Gauss was one of the first mathematicians to recognize mathematics as an important study in its own right, and not just a tool of physics and astronomy. He deplored the sloppy mathematics of others, and attempted to fill in all the gaps, and put the subject on a rigorous footing. He made tremendous contributions to complex analysis and to number theory. It was he who said: "Mathematics is the queen of the sciences, and the theory of numbers is the queen of mathematics." He wrote a monumental work in number theory, called *Disquisitiones Arithmeticae*. He invented the complex integers, now known as the *Gaussian integers*, and the notion of *congruence modulo n*. He discovered the quadratic integers, and his work paved the way for the development of the theory of algebraic numbers.

Gauss was a serene man, with little personal ambition. It is said that his capacity for intense and prolonged concentration was one secret of his success. He cared deeply for the cause of advancing mathematics, and if others sometimes published work which he had done, but hadn't had time to publish, this didn't bother him. In later years he liked to test his mind by learning new languages. He began studying Russian at the age of 62, and in two years he could read and speak it!

If Gauss lived today, he would probably be a champion of women's rights, for he had a very liberal attitude toward women scientists. Around 1807 he carried on a mathematical correspondence with Sophie Germain, who worked in number theory and physics. It was said that "his broadmindedness in this respect would have been remarkable for any man of his generation; for a German it was almost without precedent." This attitude alone is enough to make us women approve of the title given him some years ago: "Prince of Mathematicians."

E–9 Cauchy (1789–1857)

Baron Augustin-Louis Cauchy, born in Paris only a few weeks after the fall of the Bastille, was to live in turbulent political times for most of his life. He and his Monarchist family escaped to their country place, where the family was often short of food. However, Cauchy was well-fed intellectually and religiously by his cultured parents. He soon attracted the attention of their neighbor on the next estate, mathematician Marquis Laplace. In 1800 the family moved back to Paris, where Cauchy's lawyer father was elected Secretary of the Senate. Young Cauchy would often work at mathematics in his father's office, and there met the famous Lagrange, who predicted a great career for the boy.

Cauchy graduated from L'École Polytechnique and then from a civil engineering school. He worked as a military engineer for three years for Napoleon in Cherbourg, but still found time for mathematics. While in his twenties he attained international fame with his proof of one of Fermat's unproved theorems, and by winning the Grand Prize of the French Academy for a paper on mathematical physics. He soon attained a chair at the Academy, and was lecturing at the Polytechnique, the College de France, and the Sorbonne.

But so loyal was Cauchy to King Charles X that he went into self-imposed exile in Switzerland when Charles was forced to give up the throne in 1830. He taught at the University at Turin, Italy, for a while, until he was called to Prague to educate the exiled king's 13-year-old son. In 1838 he returned to Paris and the Academy, but could not accept a position at the College de France because he refused to take the oath of allegiance to the Government. However, the Bureau des Longitudes needed him, and kept him on in spite of his attitude toward the Government. He eventually got back to the Sorbonne.

Besides mathematics, Cauchy's other passion was religion. He tried to convert everyone with whom he came in contact to Roman Catholicism. His dogmatic views on religion and politics frequently caused him trouble. He was even accused of judging candidates for the Academy by their religious and political beliefs, instead of merely by their scientific accomplishments.

Cauchy published 789 papers, filling 24 large volumes. In his analysis lectures at the Polytechnique are found the modern rigorous definitions of limit, continuity, and convergence of infinite series. He is considered the first of the modern French mathematicians—men who insisted on a rigorous rather than an intuitive approach to mathematics. In his fifties, Cauchy wrote several papers on permutations (called *substitutions* at that time) and permutation groups, and is credited with being the founder and organizer of the theory of groups of finite order, even though some fundamental results had been reached earlier by others. In 1847 he published the intriguing and imaginative notion of congruence modulo $(x^2 + 1)$, thus enabling him to work with complex numbers without introducing $\sqrt{-1}$. This was the seed that later resulted in Kronecker's work on roots of polynomials in extension fields.

It is fitting that the last words of this man obsessed with mathematics and religion should be to the Archbishop of Paris, to whom he said, just before dying: "Men pass away but their deeds abide."

E–10 Abel   (1802–1829)

Neils Henrick Abel, son of a Protestant minister, was born to a poor but happy family in the village of Findö, Norway. At the age of 15 he discovered "who he was" under the instruction of a schoolmaster who excelled at mathematics. He soon had read the works of Euler, Lagrange, and Newton, and he began writing papers on algebra and analysis. When he was only 18 his father died, and he took on private tutoring jobs to support his mother and his six brothers and sisters. He managed to graduate from the University of Kristiania (now Oslo) at the age of 19, and was able to get a Norwegian Government grant so that he could travel for a year in France and Germany. It was his dream to talk mathematics with the great French and German mathematicians of his day. He had already succeeded in proving that the general fifth degree polynomial could not be solved by radicals, a feat that had defied many prominent mathematicians, but when he sent his proof to the great Gauss, Gauss assumed that the proof *had* to be erroneous, and wouldn't waste time looking at it!

Abel went to Germany in 1826, and made friends with August Crelle, who had just started the first journal in the world devoted exclusively to mathematical research. Crelle's first three volumes of his *Journal for Pure and Applied Mathematics* contained 22 of Abel's papers. Abel went on to Paris, but could not stir up much interest in his work. In Paris he learned that he had tuberculosis. While Crelle was trying to get Abel a position at the University of Berlin, Abel returned penniless to Norway, but could not get a professorship there either. He spent his last days in the home of an English family, doing research and part-time teaching, and falling in love with the family's governess. Unfortunately, he died before reaching even his 27th birthday—ironically, two days before Crelle wrote him a letter informing him that he was to be given a professorship of mathematics at the University of Berlin!

The mathematician Hermite said of Abel: "He has left mathematicians something to keep them busy for 500 years." Abel got little recognition while he lived, but *abelian groups* will honor his name forever.

E–11 Hamilton (1805–1865)

Sir William Rowan Hamilton, a native of Dublin, Ireland, was versatile enough to have life-long interests in languages, poetry, astronomy, and mathematics. He had mastered thirteen languages by the age of thirteen, encouraged in this pursuit by his curate uncle, who was his teacher until he entered Trinity College, Dublin. At Trinity he received the highest honors in both mathematics and the classics, and was elected to a professorship of astronomy at Trinity at age 22. When he was 30 he was knighted in Dublin at a meeting of the British Association for the Advancement of Science. Later he became a professor at the University of Dublin, and was a Fellow or member of 16 scientific societies.

Although he did important work in optics, Hamilton is usually remembered for his quaternions. Knowing that complex numbers could be used to express rotations in the plane, Hamilton searched for fifteen years for an extension of complex numbers that could express rotations in three-dimensional space. Finally, he discovered, or invented, quaternions. It is said that he was so excited and impressed with his discovery that on the day of October 16, 1843, when he was out walking with his wife, he carved the basic equations, $ij = k, jk = i$, etc., into the stone of the bridge on which he found himself at the moment.

There were three things that ruined Hamilton: marriage, alcohol, and quaternions. An unhappy marriage and a drinking problem one can understand, but why quaternions? Hamilton erroneously believed that quaternions held the key to all the mathematics of the physical universe, and he wasted his time and energy by devoting the last twenty-two years of his life almost exclusively to the study of quaternions. Because of his work in quaternions, he was elected as the first foreign member of the rather new National Academy of Sciences of the United States. The mathematical world was fascinated with quaternions, and attached far too much significance to them. Quaternions obsessed Hamilton to the point that he became almost a recluse, and the room in which he worked became full of masses of confused papers, dishes, half-eaten meals, all piled together.

Quaternions are, of course, an excellent example of a division ring—a noncommutative ring with unity in which every nonzero element has a multiplicative inverse. It can be said that Hamilton's algebra paved the way for algebraists to invent new algebras by deleting or altering one or more of the field postulates. Thus, *algebra* became *algebras*, just as, through altering the parallel postulate of Euclidean geometry, *geometry* became *geometries*.

E–12 Galois (1811–1832)

You probably would not find it too startling if you were to read the following article in a newspaper published in 1970:

"A radical student died today of injuries suffered recently in a fight over a worthless girl with whom he had had a brief affair. He was only 21 years old, but he had a police record, and had been a known troublemaker for several years, stirring up his fellow students to riot against the establishment and the school authorities, hating the Government, wanting the 'people' to have more power, treating with contempt any school courses that he did not feel were relevant to *him*. Aside from political issues, he spent all his time on mathematics, and even had the audacity to send some of his work to the prestigious Academy of Sciences. When he tried to find out what had happened to the papers he submitted, he learned that they were 'missing'. It is reported that his hatred of tyranny and autocratic authority was instilled in him at an early age by his liberal father. Incidentally, during the time that his father held an elective public office, underhanded smear tactics were used against the father by the reactionary right in order to discredit him. In despair and desperation, the father committed suicide."

Such a news item could have been dated: Paris, June, 1832. The young man was Évariste Galois, a mathematical genius, who was a pioneer in the development of abstract algebra. Born in Bourg-la-Reine, just outside Paris, Évariste was taught by his mother through age 12, when he was sent to the Lycée Louis-le-Grand in Paris. He hated the discipline of the place (more like a jail, he thought), and his own brilliance made him view most of his teachers as being rather stupid. His mathematics texts were too simple for him, so he began reading, on his own, the works of the great mathematicians, and was "turned on" by Legendre, Lagrange, and Abel. He was inspired by but one teacher—a mathematics teacher by the name of Louis Richard, who sat in on lectures at the Sorbonne in order to keep abreast of modern mathematics. Galois took the entrance examinations for the Polytechnique at age 16, and again at age 18, and failed both times. It is said that Galois did so many steps in his head that the examiners for the oral examination couldn't follow him, and hence thought he didn't know much. Instructor Richard was so impressed with the brilliance of Galois that he recommended that Galois be admitted to the university without examination.

During his brief life, Galois became bitter because everything seemed to go against him. When he was 17, Cauchy promised to present some work of Galois to the Academy, but lost the manuscript. When he was 19, Galois sent in a remarkable paper in the theory of equations to the Secretary of the Academy of Sciences, in competition for the Grand Prize in Mathematics. The Secretary of the Academy died before he looked at it, and the manuscript was never found. Galois remarked, "Genius is condemned by a malicious social organization to an eternal denial of justice in favor of fawning mediocrity." He suspected that the two losses of his work were not accidental.

During the French uprising of 1830 he was expelled from school for his championing of the masses, and, not being able to support himself with his mathematics, he joined the National Guard. Again he sent a paper to the Academy, work on polynomials that is now known as *Galois theory*, and the referee, Poisson, said the mathematics was incomprehensible. Galois turned more of his attention to politics, was arrested twice for his political activities, was acquitted the first time, and given a jail sentence of six months the second time.

Soon after being released from jail, Galois was challenged to a duel over some girl who wasn't of any great importance to him. (Some felt that the challenge was a malicious plot of the right-wing to exterminate him.) But, being a Frenchman, his honor made him accept the challenge. He was sure he was going to die, and so he spent the night before the duel writing down as much mathematics as he could—abstract algebra theory that he had previously worked out mentally, but had not recorded. Several times in the margin of his notes he wrote: "I have not time; I have not time." But he had enough time to give posterity some exciting algebra, including conditions under which a polynomial equation can be solved by radicals. The duel was fought with pistols, at 25 paces. Galois was shot, left to die where he fell, was taken to a hospital by someone passing by, and died shortly thereafter.

E–13 Boole   (1815–1864)

To George Boole goes the credit for developing mathematical logic (also called symbolic logic) as another branch of pure mathematics. He was born in Lincoln, England, the son of a keeper of a small shop, and so was considered to be "lower class." Since the upper classes were exposed to Latin and Greek, he taught himself those languages, and his father helped him to learn some mathematics. To help support his parents he became an assistant teacher at two elementary schools at the early age of 16. He had intended to become a clergyman, but at the age of 20 he opened up his own school, and became intensely interested in mathematics, the logic of mathematics, and the mathematics of logic. He studied on his own, and corresponded with well-known mathematicians of his day. In 1848 he published a pamphlet on *The Mathematical Analysis of Logic*, and was so well thought of that in 1849 he was appointed Professor of Mathematics at Queen's College at Cork, Ireland.

Boole was one of the British mathematicians who developed the modern concept of abstract algebra, and investigated abstract algebraic systems, apart from concrete examples of those systems. His big contribution was his creation of a system of mathematical logic. This was contained in his monumental work of 1854: *An Investigation of the Laws of Thought, on which are founded the Mathematical Theories of Logic and Probabilities*. In this work he formulated logic in terms of an abstract algebra, now known as *Boolean algebra*. It has been said that *pure mathematics* was really invented by him. After his death, of pneumonia at age 49, not much attention was paid to mathematical logic for the next half-century, but all that changed with the publication of Whitehead and Russell's *Principia Mathematica* (1910–1913). The groundwork that Boole had laid was finally appreciated.

E–14 Cayley   (1821–1895)

Arthur Cayley presents the picture of a calm, reasonable person whose mathematical talent was recognized early, and whose life was happy and fulfilled. Born in Surrey, England, he lived his first few years on the family estate in Normandy, but spent most of his life in England. He outshone every other student while at Trinity College, Cambridge, published his first paper while an undergraduate, was a Fellow of Trinity and assistant tutor for three years, during which time he published 25 papers on algebra and related subjects. The mathematics that he began to investigate then was to occupy him for the next fifty years. However, he recognized the need to earn a living, and left Trinity to study law for three years, which he then practiced for the next fourteen.

Cayley never abandoned his mathematical research while practicing law, and in 1863 he left his law practice to accept the Sadlerian Professorship of Mathematics at Cambridge University, even though this meant a cut in income. His life was then devoted to teaching, research, and university administration, at which he was adept. Like Gauss, Cayley displayed a liberal attitude toward women. A big issue at Cambridge University was the admission of women. It was mainly through Cayley's efforts that Cambridge opened its doors to female students for the first time. He relaxed by enjoying good literature, travel on the Continent, mountaineering, natural beauty, painting, and architecture. He saw something of the New World also, as he went to Baltimore, Maryland, in 1881 to lecture on algebra at Johns Hopkins University for a half-year. That institution was only six years old at the time.

Cayley might be called the founding father of the subject now known as *linear algebra.* He, together with his friend Sylvester, did much work on the theory of algebraic invariants, and he wrote some important papers on the subject of linear transformations. He introduced the idea of an n-dimensional space, and he invented matrices and the algebra of matrices, these being a natural outgrowth of his work on linear transformations. His work with groups shows up in his theorem that states that every group is isomorphic to a group of permutations of its elements. The operation table of a finite group is sometimes called a *Cayley table.* He also worked in geometry, endeavoring to unify the notions of Euclidean geometry and projective geometry. The *Collected Mathematical Papers* of Cayley consists of 966 papers bound in 13 large volumes, comprising a total of approximately 7800 pages!

E–15 Eisenstein (1823–1852)

The brilliant young German, Ferdinand Eisenstein, studied mathematics at the University of Göttingen under the master, Gauss, and was said to be the favorite pupil of Gauss. He obtained a teaching position at the University of Berlin, and did research in number theory and the theory of equations. When he was only 24, his collected papers were published. Unfortunately, he was only 29 when he died, and so he did not have the opportunity to fulfill the expectations of Gauss, who once said: "There have been only three epoch-making mathematicians: Archimedes, Newton, and Eisenstein."

E–16 Kronecker (1823–1891)

Versatile Leopold Kronecker was an expert gymnast, swimmer, and mountain climber, as well as an accomplished pianist and vocalist. Furthermore, he was rich! Kronecker was born in Leibnitz, Prussia, the son of wealthy parents. In school he excelled in languages, philosophy, and mathematics. He earned his Ph.D. from the University of Berlin, his thesis being on the units in the integral domains of certain algebraic number fields. After college he married a rich cousin, went into business for eight years, and by age thirty he was affluent enough to retire from business and devote himself exclusively to mathematics. Musicians as well as mathematicians would visit his elegant home in Berlin. He counted Felix Mendelssohn among his friends.

Kronecker lectured at the University of Berlin without pay from 1861 to 1883. Only at the age of 60 did he become a regular member of the University faculty. His research was in the theory of numbers, the theory of equations, and elliptic functions. It is said that he was about the only mathematician of his time to really master Galois theory, as evidenced by his 1853 paper on the solvability of equations. He was more interested in the pure mathematics of abstract algebra than in the applied mathematics arising from geometry and calculus, and for this reason he carried on a mathematical feud with analyst Weierstrass. Because very few students were interested in mathematics for its own sake in those days, Kronecker had a *small* but enthusiastic group of students, some of whom later became prominent mathematicians.

When he was an old man, Kronecker remarked that music is the finest of all the fine arts, with the possible exception of mathematics. He thought mathematics was like poetry—further evidence that he was a pure rather than an applied mathematician!

E–17 Dedekind (1831–1916)

The son of a lawyer in Brunswick, Germany, Richard Dedekind turned from his earlier interest in physics and chemistry, and received his doctorate in mathematics under the supervision of Gauss at the University of Göttingen. He was a lecturer at Göttingen for four years, professor at Zurich Polytechnique for five, and then spent the next fifty years as professor at the technical high school in his home town of Brunswick. One wonders why he did not obtain a more prestigious position.

Dedekind taught the first-recorded formal course on Galois theory in 1857. In his early forties he published papers that made two outstanding contributions to the mathematical world: (1) He extended the rational field to the real field by a scheme now called a *Dedekind cut*. (2) He invented *ideals*. The notion of an ideal arose out of his discovery that some algebraic domains are not unique factorization domains. He developed an algebra of ideals in which every ideal *is* uniquely factorable into a product of prime ideals. His work did much to advance the state of algebraic number theory.

The death of Dedekind, during World War I, marked the end of an era in mathematics. He was the last of the nineteenth-century pioneers—those great mathematicians who planted the seeds that produced such an abundance of blooms in the twentieth century. In an address before the Royal Society of Göttingen in 1917, the mathematician Landau said: "Richard Dedekind was not only a great mathematician, but one of the wholly great in the history of mathematics, now and in the past, the last hero of a great epoch, the last pupil of Gauss, for four decades himself a classic, from whose works not only we, but our teachers and the teachers of our teachers, have drawn."

E–18 Klein (1849–1925)

Felix Klein taught, wrote, and did research in Germany for over 50 years, editing a mathematics journal in his spare time. In addition to his abiding interest in groups and their use in studying geometries, he was concerned with the history of mathematics, and with the improvement of mathematics teaching at all levels. He gave inspiring lectures, and was so enthusiastic that he often made the mathematics he was discussing appear simpler than it actually was. His influence among both mathematicians and mathematics educators was great during his lifetime.

Klein was born in Dusseldorf, studied at the Universities of Bonn, Göttingen, and Berlin, and was a professor at the Universities of Erlangen, Munich, Leipzig, and Göttingen, remaining at Göttingen for 27 years. When he arrived at Erlangen, at age 23, he gave an inaugural lecture in which he showed how groups could be used to characterize the various geometries, defining a *geometry* as the study of those properties of figures that remain unchanged under a particular *group* of mappings. This lecture has since come to be known as the *Erlangen Programm*. He published, among other things, a book on the icosahedron (regular solid with 20 faces), in which he showed that the group of rigid motions of an icosahedron is the symmetric group S_5.

As a young man, Klein did mathematics day and night, and it is said that he took some kind of drug to stop his mind from working so that he could get a little sleep. As a result, he had a tragic nervous breakdown at one point in his life. Fortunately, he recovered, and continued his fine work. He wrote much, and travelled a great deal, lecturing in the United States at one time. Students flocked to Göttingen from all over the world to study under him. "Functional thinking" was his slogan, in the movement he led to reform the teaching of mathematics in Germany. Topology students know of Klein as the discoverer of the *Klein bottle*, and abstract algebra students become acquainted early with the *Klein four-group*.

E–19 Peano (1858–1932)

Giuseppe Peano was an Italian mathematician and logician who formulated a set of postulates from which one can derive the properties of the natural numbers. He hoped that his postulational method would not only revolutionize symbolic logic, but would encompass most branches of mathematics. His main results are found in *Formulario Mathematico*, published from 1894 to 1908. In the original formulation of the postulates, Peano reduced undefined mathematical terms to three: *zero*, *number*, and *successor*. He listed these five postulates:

(1) Zero is a number.

(2) If a is a number, the successor of a is a number.

(3) Zero is not the successor of a number.

(4) Two numbers of which the successors are equal are themselves equal.

(5) If a set S of numbers contains zero and also the successor of every number in S, then every number is in S.

Peano was a professor at the University of Turin from 1890 until his retirement. In addition to his work in logic, foundations of mathematics, topology, etc., Peano created an international language, called the *latino sine flexione*, which was based on the words most commonly used in western Europe. In 1915 he published a vocabulary for this language. This endeavor evidently was not received with great enthusiasm.

Bertrand Russell, who was greatly influenced by Peano's work in mathematical logic, said: "In all discussions, Peano and his pupils had a precision which was not possessed by others. As soon as I had mastered his notation, I saw that it extended the region of mathematical precision backwards towards regions which had been given over to philosophical vagueness." If there is any one thing that a mathematician abhors, it's *vagueness*!

E–20 Hardy   (1877–1947)

"I cannot remember ever having wanted to be anything but a mathematician. . . . I thought of mathematics in terms of examinations and scholarships: I wanted to beat other boys, and this seemed to be the way in which I could do so most decisively. . . . I found at once, when I came to Cambridge, that a Fellowship implied 'original work', but it was a long time before I formed any definite idea of research. . . . I shall never forget the astonishment with which I read that remarkable work (Jordan's *Cours d'analyse*, 1893–1896), the first inspiration for so many mathematicians of my generation, and learnt for the first time as I read it what mathematics really meant. From that time onwards I was in my way a real mathematician, with sound mathematical ambitions and a genuine passion for mathematics."

So wrote the brilliant, complex Godfrey Harold Hardy, at age 63, in his little confessional, *A Mathematician's Apology* (New York and London: Cambridge University Press, 1940). Britisher Hardy was educated at Cambridge, and held a professorship there for most of his life, except for the twelve years he spent at Oxford. One gets the impression that teaching bored him, while research excited him. He published over 300 research papers, and his 1908 book, *A Course of Pure Mathematics*, influenced the work of many students in the decades to follow.

Hardy was a man of intense likes and dislikes. Besides mathematics, his other passions were cricket and baseball. It was said that when he was in Boston in 1936 to give a series of lectures at Harvard, he could hardly spare the time to deliver the lectures, because the Red Sox were playing some home games.

His violent dislikes included God and war. He said that God was his personal enemy, and he refused to enter any church or chapel, even for the purpose of attending a nonreligious meeting!

If he had been alive and in the United States in the late sixties, he would have most certainly been a leader in the war protest movement. His hatred of war led him to categorize mathematics into two types: *real* and *trivial*. He said that *real mathematics* was any mathematics that had no applications to the world outside of mathematics, and that *trivial mathematics* was any mathematics that could be applied in the mundane world around us. He abhorred trivial mathematics, particularly that which helped to produce weapons of destruction. To him, real mathematics was beautiful, and trivial mathematics was exceedingly ugly.

But perhaps Hardy really *was* apologizing for his life as a mathematician, and was attempting to rationalize his contributions to the world. He said: "I have never done anything 'useful'. No discovery of mine has made, or is likely to make, directly or indirectly, for good or ill, the least difference to the amenity of the world.... The case for my life, then, or for that of any one else who has been a mathematician in the same sense in which I have been one, is this: that I have added something to knowledge, and helped others to add more, and that these somethings have a value which differs in degree only, and not in kind, from that of the creations of the great mathematicians, or of any of the other artists, great or small, who have left some kind of memorial behind them."

E–21 Noether (1882–1935)

On May 3, 1935, the *New York Times* carried a three-paragraph obituary written by Albert Einstein. These are his last two paragraphs:

"Within the past few days a distinguished mathematician, Professor Emmy Noether, formerly connected with the University of Göttingen and for the past two years at Bryn Mawr College, died in her fifty-third year. In the judgment of the most competent living mathematicians, Fräulein Noether was the most significant creative mathematical genius thus far produced since the higher education of women began. In the realm of algebra, in which the most gifted mathematicians have been busy for centuries, she discovered methods which have proved of enormous importance in the development of the present-day younger generation of mathematicians. Pure mathematics is, in its way, the poetry of logical ideas. One seeks the most general ideas of operation which will bring together in simple logical and unified form the largest possible circle of formal relationships. In this effort toward logical beauty spiritual formulae are discovered necessary for the deeper penetration into the laws of nature."

"Born in a Jewish family distinguished for the love of learning, Emmy Noether, who, in spite of the efforts of the great Göttingen mathematician, Hilbert, never reached the academic standing due her in her own country, none the less surrounded herself with a group of students and investigators at Göttingen, who have already become distinguished as teachers and investigators. Her unselfish, significant work over a period of many years was rewarded by new rulers of Germany with a dismissal, which cost her the means of maintaining her simple life and the opportunity to carry on her mathematical studies. Farsighted friends of science in this country were fortunately able to make such arrangements at Bryn Mawr College and at Princeton that she found in America up to the day of her death not only colleagues who esteemed her friendship but grateful pupils whose enthusiasm made her last years the happiest and perhaps the most fruitful of her entire career."

What manner of woman was this, to be eulogized by the great Einstein? Amalie Emmy Noether began life as a plain, acutely nearsighted little girl who lived with her parents and three younger brothers in a flat in Erlangen, Germany. After Emmy became qualified as a teacher of French and English at age 18, she decided to attend classes at the University of Erlangen, where she was one of two females in a student population of almost a thousand. She had to get special consent of her professors to obtain course credits, since credits were normally awarded only to male students. Here she switched her interest from languages to mathematics, and was allowed to graduate. After attending some classes conducted by eminent mathematicians at the University of Göttingen, Emmy returned to Erlangen to do graduate work, and received her doctorate at age 26. Her dissertation was filled with formal computations; the elegant axiomatic approach to ring theory which made her famous was to come later.

There is no doubt that Emmy was influenced by her father, who was a highly respected professor of mathematics for 46 years at the University of Erlangen. Of the relationship between Emmy and Max Noether, Hermann Weyl said: "This scientific kinship of father and daughter—who became in a certain sense his successor in algebra, but stands beside him independent in her fundamental attitude and in her problems—is something extremely beautiful and gratifying."

In 1916 (at the age of 34) Emmy Noether left Erlangen for the University of Göttingen, where she was welcomed by the mathematicians, but had trouble getting the administration to pay her for her efforts. At a University Senate meeting Hilbert said in her defense: "I do not see that the sex of the candidate is an argument against her admission as *Privatdozent*. After all, we are a university and not a bathing establishment." In 1922 she was finally given a salary and the title of Extraordinary Professor. She founded the Göttingen School of Algebra, which soon attracted the attention of mathematicians from the Soviet Union when they visited Göttingen. Thus began her close ties with the University of Moscow, where she taught a course in abstract algebra in 1928–29. This is how the Russian P. S. Alexandroff viewed her in his 1935 address to the Moscow Mathematical Society:

"Emmy Noether was interested ... in what had been taking place in the whole area of mathematics (and not only in the area of mathematics) in Soviet Russia; she did not hide her sympathies with our country and our social and governmental system, in spite of the fact that the manifestation of these sympathies seemed outrageous and unseemly to the majority of representatives of Western European academic circles. The matter had reached the point where Emmy Noether was literally banished from one of the Göttingen boarding houses (where she had settled and lived) at the demand of the student corporation, resident in the same house, who did not want to live under the same roof with a 'Marxist-inclined Jewess'. ... A spokesman of the most abstract areas of mathematical science, she distinguished herself at the same time by a surprising sensitivity in understanding the great historical movements of our epoch; always vitally interested in politics, hating war with her whole being, and hating chauvinism in all its manifestations, she never in this area knew any vacillation; her sympathies always and unchangingly belonged to the Soviet Union, in which she saw the beginning of a new era in the history of mankind and firm support for everything progressive for which human thought has lived and lives still."

Noether reached the high point of her career in 1932, when she presented her work to the International Congress of Mathematicians in Zurich, and was acclaimed highly. Her world was to collapse in 1933 when Hitler's Nazis came into power, and she was ousted from the university. She was able to migrate to the

United States, where she spent the last two years of her life lecturing at both Bryn Mawr College and the Institute for Advanced Study at Princeton. As Alexandroff said: "Such was Emmy Noether, the greatest of women mathematicians, a great scientist, an amazing teacher, and an unforgettable person.... She loved people, science, life, with all the warmth, all the cheerfulness, all the unselfishness, and all the tenderness of which a deeply sensitive—and feminine—soul is capable."

E–22 Stark, Harold M. (1939–)

It has been said that the number of mathematicians alive right now is greater than the number of mathematicians who lived and died in all the hundreds of years before this. Because of their numbers, and because they have not yet completed their work, it would be virtually impossible for anyone to select the hundred greatest living mathematicians. However, I should like to conclude these sketches with the story of *one* very-much-alive mathematician. I have selected Harold Stark for this purpose because the nature of the problem he solved can be understood by beginning abstract algebra students, and because reference was made to his work in Section 4.10. We shall never know what mathematical historians of the year 2100 will have to say about him, or *if* they will say anything!

Born in Los Angeles at the beginning of World War II, Harold Stark did his undergraduate work at the California Institute of Technology (B.S. 1961), received his Ph.D. in mathematics from the University of California, Berkeley, in 1964, and only eight years later became a full Professor of Mathematics at the prestigious Massachusetts Institute of Technology. How did he do it? His success hinges on the fact that he was able to solve a problem that had been bugging mathematicians for over a century, a problem that had been unsuccessfully tackled by mathematical giants such as Gauss.

During his graduate years at Berkeley, Stark became interested in the unsolved problem of determining all imaginary quadratic domains in which factorization is unique. Back in 1934, it was proved that unique factorization holds in $Q[\sqrt{n}]$, for nine negative values of n: $-1, -2, -3, -7, -11, -19, -43, -67, -163$, and for *at most one more n*. The object was either to find that other n, or to show that no such n exists. Stark doggedly kept at this problem while he held two one-year appointments as Instructor at the University of Michigan, 1964–66. (He was not re-appointed for even a third year at Ann Arbor.) By 1966 it had been established that if a tenth value of n exists, that $n < -10^{9,000,000}$. Then, in the summer of 1966, Stark proved that there was no other n. As he tells it himself (in a letter sent to me in April, 1974): "During the summer of 1966, at the end of three months of traveling in Europe, I proved my result on unique factorization (beginning in Paris and continuing in New York, Washington D.C., and the last piece in Ann Arbor a couple of days after returning). At this point, I became employable again."

Stark became famous almost overnight (to coin a phrase), for as early as December, 1966, M.I.T. made him a job offer, but he stayed on at Michigan (at which he was now welcome) so that his wife Betty could complete her graduate work in mathematics at Ann Arbor. In 1969 he accepted an Associate

Professorship at M.I.T., and in 1972 he was promoted to Professor. He said: "As much as anything, we are in Boston now because of the large number of good universities in the area. . . . Betty is now an Assistant Professor at Northeastern University, and doing quite well."

Harold Stark has written an undergraduate textbook on number theory (published in 1970), is the author of more than 25 research papers, and has produced some of his own Ph.D. students. His story illustrates the fact that there is always room at the top for the person who has, not only ability, but also the will to pursue a goal until it is reached!

A Short Reading List

It is hoped that you will do much browsing and become acquainted with the books in your school library. Here are a few books and journals with which you might start, but don't limit yourself to these.

Abstract Algebra Books

Birkhoff, Garrett, and Saunders MacLane, *A Survey of Modern Algebra*, 3rd Edition. New York: The Macmillan Co., 1965.

Dean, Richard A., *Elements of Abstract Algebra*. New York: John Wiley & Sons, Inc., 1966.

Deskins, W. E., *Abstract Algebra*. New York: The Macmillan Co., 1964.

Fraleigh, John B., *A First Course in Abstract Algebra*. Reading, Massachusetts: Addison-Wesley Publishing Co., Inc., 1967.

Herstein, I. N., *Topics in Algebra*, 2nd Edition. Lexington, Massachusetts: Xerox College Publishing, 1975.

McCoy, Neal H., *Fundamentals of Abstract Algebra*. Boston: Allyn and Bacon, Inc., 1972.

Books on Special Subjects

Borofsky, Samuel, *Elementary Theory of Equations*. New York: The Macmillan Co., 1950.

Burton, David A., *A First Course in Rings and Ideals*. Reading, Massachusetts: Addison-Wesley Publishing Co., Inc., 1970.

Grossman, Israel, and Wilhelm Magnus, *Groups and Their Graphs*. New York: Random House, Inc., 1964.

Kazarinoff, Nicholas D., *Ruler and the Round*. Boston: Prindle, Weber & Schmidt, Inc., 1970.

Lieber, L. R. and H. G., *Galois and The Theory of Groups*. New York: Science Press, 1956.

Niven, Ivan, and Herbert S. Zuckerman, *An Introduction to the Theory of Numbers*, 2nd Edition. New York: John Wiley & Sons, Inc., 1966.

Pollard, Harry, *The Theory of Algebraic Numbers*, Carus Monograph 9. New York: John Wiley & Sons, Inc., 1950.

Scott, W. R., *Group Theory*. Englewood Cliffs, N.J.: Prentice-Hall, Inc., 1964.

Stark, Harold M., *An Introduction to Number Theory*. Chicago: Markham Publishing Co., 1970.

Stoll, Robert R., *Set Theory and Logic*. San Francisco: W. H. Freeman and Co., Publishers, 1963.

General References

American Mathematical Monthly (Official Journal of The Mathematical Association of America). January, 1894, to the current issue.

Mathematics Magazine (a Mathematical Association of America publication with articles written especially for undergraduates). October, 1926, to the current issue.

World of Mathematics (a four-volume collection of the literature of mathematics), edited by James R. Newman. New York: Simon and Schuster, 1956.

Answers
to Selected Problems

SECTION 1.1 (page 3)

1. (a) $p \Rightarrow q$. (b) $q \Rightarrow p$. (c) $p \Leftrightarrow q$. (d) $p \Leftrightarrow q$. (e) $q \Rightarrow p$. (f) None. (g) $q \Rightarrow p$.
2. (a) F (converse), T (contrapositive). (b) T, F. (c) T, T. (d) F, F. (e) F, T.

SECTION 1.2 (page 9)

1. $[0, \infty), [1, 2), [0, 1), [2, \infty), [0, 1) \cup [2, \infty)$.
3. (a) (3), (6) are true. (b) (1), (3), (4), (6), (7) are true.
4. (a) $\{\emptyset\}$. (b) $\emptyset$. (c) $\{\emptyset, \{\emptyset\}\}$. (d) $\{\{\emptyset\}\}$. (e) $\{\emptyset\}$. (f) $\{\emptyset\}$. (g) $\{\emptyset, \{\emptyset\}\}$.
6. (a)

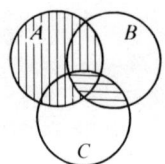 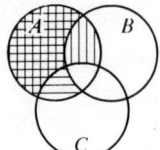

$B \cap C$: ≡
$A - (B \cap C)$: ||||||

$A - B$: ≡, $A - C$: ||||||
$(A - B) \cup (A - C)$: Shaded region

7.

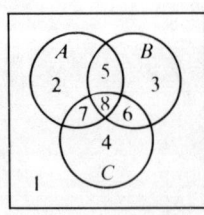

$1 = U - (A \cup B \cup C)$ $3 = B - (A \cup C)$ $5 = A \cap B - C$ $7 = A \cap C - B$
$2 = A - (B \cup C)$ $4 = C - (A \cup B)$ $6 = B \cap C - A$ $8 = A \cap B \cap C$

8. (a) $A = B$. (b) $A \cap B = \emptyset$.
9. (a) T. (b) $A \subseteq A \cup B$. (c) $A - B \subseteq A + B$. (d) T. (e) T. (f) T. (g) T. (h) T. (i) $\emptyset - A = \emptyset$. (j) T. (k) T. (l) $A + A = \emptyset$. (m) T.
10. Prove: $A \cap B = A \Rightarrow A \cup B = B$. Obviously, $B \subseteq A \cup B$. If $x \in A \cup B$, then $x \in A$ or $x \in B$. But $x \in B \Rightarrow A \cup B \subseteq B$, and $x \in A \Rightarrow x \in A \cap B$ (since $A \cap B = A$) $\Rightarrow x \in B \Rightarrow A \cup B \subseteq B$. Therefore, $A \cup B = B$.
12. (a) 16 subsets, including $\emptyset$ and A.
14. (a) $[-1, +1], [-2, +2], [-3, +3]$. (b) $[-3, -1) \cup (+1, +3], \emptyset, A_3 - A_1$. (c) $A_5, A_1, \mathbf{R}, A_1$.
15. (a) $[0, 3], [0, \frac{3}{2}], [0, \frac{4}{3}], [0, \infty), \{0\}$. (b) $\mathbf{R}^+, \emptyset$.

336

Answers to Selected Problems 337

SECTION 1.3 (page 15)

1 (a) 256.   (b) 16 relations. The two equivalence relations are:
$\{(5, 5), (6, 6)\}$ and $\{(5, 5), (6, 6), (5, 6), (6, 5)\}$.   (c) 15.
2 (a) 2.   (b) 1, 2, 3.   (c) 2, 3.   (d) 3.   (e) 1, 2.   (f) 1, 3.   (g) 1.   (h) none.
3 (a) $(x, x) \in \rho$, for all $x \in \mathbf{R}$.
4 $(x, y) \in (A \cap B) \times (C \cap D) \Leftrightarrow x \in A, x \in B, y \in C, y \in D \Leftrightarrow (x, y) \in A \times C, (x, y) \in B \times D \Leftrightarrow (x, y) \in (A \times C) \cap (B \times D)$.
7 The alternate definition enables us to *prove* that $(a, b) = (c, d)$ iff $a = c$ and $b = d$. Proof asked for in Problem 1.2–13.
8 (a) $\{(1, 1), (2, 2), (3, 3), (1, 2), (2, 1), (1, 3), (3, 1), (2, 3), (3, 2), (4, 4), (5, 5), (4, 5), (5, 4), (6, 6)\}$.
   (b) (i) $(a, a) \in \rho$, since a and a are in the same subset, for all $a \in S$.
   (ii) $(a, b) \in \rho \Rightarrow a$ and b are in same subset $\Rightarrow b$ and a are in same subset $\Rightarrow (b, a) \in \rho$.
   (iii) $(a, b) \in \rho, (b, c) \in \rho \Rightarrow a$ and b are in a subset T, b and c are in a subset $V \Rightarrow T = V$ (since b is not in two distinct subsets) $\Rightarrow a, c \in T \Rightarrow (a, c) \in \rho$.
9 (a) Not transitive, since $(3, 5), (5, 4) \in \rho$, but $(3, 4) \notin \rho$.   (b) $\{(1, 1), (2, 2), (3, 3), (4, 4), (5, 5), (2, 3), (3, 2), (2, 5), (5, 2), (3, 5), (5, 3), (4, 5), (5, 4)\}$.
10 (b)

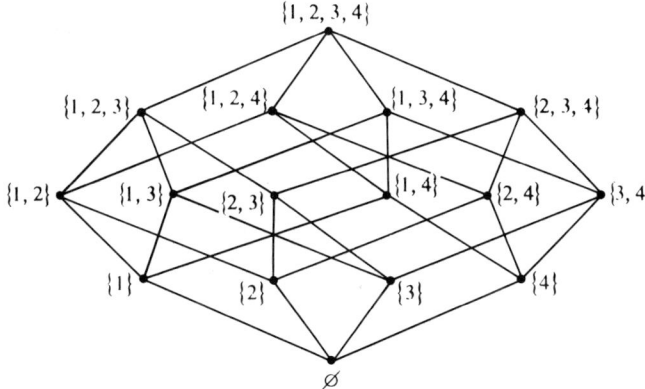

SECTION 1.4 (page 21)

1 (a) $P_1 = \{\{a\}, \{b\}, \{c\}\}, P_2 = \{\{a, b\}, \{c\}\}, P_3 = \{\{a, c\}, \{b\}\}, P_4 = \{\{b, c\}, \{a\}\}, P_5 = \{\{a, b, c\}\}$.
   (b) $\rho_1 = \{(a, a), (b, b), (c, c)\}, \rho_2 = \{(a, a), (b, b), (c, c), (a, b), (b, a)\}$,
   $\rho_3 = \{(a, a), (b, b), (c, c), (a, c), (c, a)\}, \rho_4 = \{(a, a), (b, b), (c, c), (b, c), (c, b)\}$,
   $\rho_5 = \{(a, a), (b, b), (c, c), (a, b), (b, a), (a, c), (c, a), (b, c), (c, b)\}$.
   (c) $P_i \to \rho_i$, for $i = 1, 2, 3, 4, 5$.
2 52.
5 (a) Let P_n be the set of all polygons having n sides, so that P_3 is the set of all triangles, P_4 is the set of all quadrilaterals, etc. Then $P = \{P_3, P_4, P_5, \ldots\}$.
   (b) $\bar{1} = \{1, 6, 11, \ldots\}, \bar{2} = \{2, 7, 12, \ldots\}, \bar{3} = \{3, 8, 13, \ldots\}, \bar{4} = \{4, 9, 14, \ldots\}, \bar{5} = \{5, 10, 15, \ldots\}$.
   $P = \{\bar{1}, \bar{2}, \bar{3}, \bar{4}, \bar{5}\}. |P| = 5$.
   (c) $P = \{\bar{0}, \bar{1}, \bar{2}, \bar{3}\}. |P| = 4$.
   (d) $|P| = 366$. Each equivalence class contains all the people that were born on the same day and month of the year. If S_j is the set of all people born on the jth day (of a leap year), then $P = \{S_1, S_2, \ldots, S_{366}\}$.
   (e) Each class, C_r, consists of the set of points on a circle of radius r with center at c, where $0 \le r < 1$. The given set is partitioned into a collection of infinitely many concentric circles. $P = \{C_r : 0 \le r < 1\}$.

SECTION 1.5 (page 27)

1 (a) $\begin{pmatrix} 1 & 2 & 3 \\ 4 & 4 & 4 \end{pmatrix}, \begin{pmatrix} 1 & 2 & 3 \\ 5 & 5 & 5 \end{pmatrix}, \begin{pmatrix} 1 & 2 & 3 \\ 4 & 4 & 5 \end{pmatrix}, \begin{pmatrix} 1 & 2 & 3 \\ 4 & 5 & 4 \end{pmatrix}, \begin{pmatrix} 1 & 2 & 3 \\ 5 & 4 & 4 \end{pmatrix}, \begin{pmatrix} 1 & 2 & 3 \\ 4 & 5 & 5 \end{pmatrix}, \begin{pmatrix} 1 & 2 & 3 \\ 5 & 4 & 5 \end{pmatrix}, \begin{pmatrix} 1 & 2 & 3 \\ 5 & 5 & 4 \end{pmatrix}$. All maps except the first two are surjective.

(b) No 1–1 maps from A to B. Six 1–1 maps from B to A.
2 (a) 512 relations, 27 maps.   (b) Six permutations of S.
3 Order of S:

	n	2	3	4
No. of relations on S:	$2^{(n^2)}$	16	512	65,536
No. of maps:	n^n	4	27	256
No. of bijective maps:	$n!$	2	6	24

5 (a) $n \to 2n$.   (b) $n \to n/2$ if n is even, $n \to n$ if n is odd.   (c) $n \to n + 1$ if n is even, $n \to n$ if n is odd.   (d) $n \to n + 3$.
8 (a) 1–1, onto, $\mathbf{R}\mu = \mathbf{R}$.   (b) $\mathbf{R}\mu = [1, \infty)$.   (c) 1–1, $\mathbf{R}\mu = \mathbf{R}^+$.   (d) 1–1, onto, $\mathbf{R}\mu = \mathbf{R}$.
  (e) 1–1, onto, $\mathbf{R}\mu = \mathbf{R}$.   (f) $\mathbf{R}\mu = [-4, +4]$.   (g) Onto, $\mathbf{R}\mu = \mathbf{R}$.   (h) 1–1, onto, $\mathbf{R}\mu = \mathbf{R}$.
9 Range: (a) $\mathbf{R} \times \mathbf{R}$.   (b) $\mathbf{R} \times \mathbf{R}$.   (c) Line, $y = 1$.   (d) $\{(x, y) : x \geq 0, y \leq 1\}$.   (e) $\mathbf{R} \times \mathbf{R}$.
  (f) $\mathbf{R} \times \mathbf{R}$.   (g) Line, $2x - y = 2$.   (h) $\mathbf{R} \times \mathbf{R}$.
  The 1–1 maps are: a, b, e, f.
10 (a) $v : \mathbf{R} \to S$, with $S = [1, \infty)$, and $xv = x^2 + 1$. $\mu \neq v$.   (b) $S = \{-1, +1\}$. T can be any set containing 3. If $T = \{3\}$, then μ and v are surjective maps.

SECTION 1.6   (page 32)

1 (a) $\begin{pmatrix} 1 & 2 & 3 \\ y & x & z \end{pmatrix}$.   (b) $\begin{pmatrix} 1 & 2 & 3 \\ 4 & 6 & 5 \end{pmatrix}$.   (c) $\begin{pmatrix} n & p & q & r \\ 4 & 6 & 6 & 5 \end{pmatrix}$.   (d) $\begin{pmatrix} 1 & 2 & 3 \\ 4 & 6 & 5 \end{pmatrix}$.

2 (a) $\lambda \circ \mu = \begin{pmatrix} 1 & 2 & 3 & 4 \\ 1 & 2 & 4 & 3 \end{pmatrix}$, $\mu \circ \lambda = \begin{pmatrix} 1 & 2 & 3 & 4 \\ 2 & 1 & 3 & 4 \end{pmatrix}$.   (b) $\eta = \begin{pmatrix} 3 & 4 & 1 & 2 \\ 1 & 2 & 3 & 4 \end{pmatrix} = \begin{pmatrix} 1 & 2 & 3 & 4 \\ 3 & 4 & 1 & 2 \end{pmatrix}$,
  $v = \begin{pmatrix} 4 & 3 & 1 & 2 \\ 1 & 2 & 3 & 4 \end{pmatrix} = \begin{pmatrix} 1 & 2 & 3 & 4 \\ 3 & 4 & 2 & 1 \end{pmatrix}$.

3 (a) $n(\lambda \circ \mu) = n(\mu \circ \lambda) = n - 4$.   (b) $n\eta = n - 5, nv = n + 9$.
4 (a) $\sin x^2, [-1, +1]$. $\sin^2(x - 1) + 1, [1, 2]$.   (b) $3 - x^2, (-\infty, 3]$. $(4 - x)^2 + 1, [1, \infty)$.
  (c) $4 - \sin(x - 1), [3, 5]$. $\sin(3 - x), [-1, +1]$.   (d) $\sin[\sin(x - 1) - 1], [-1, 0]$. $x, (-\infty, +\infty)$.
  (e) $4 - \sin(x^2), [3, 5]$. $[4 - \sin(x - 1)]^2, [10, 26]$.
5 $x(\eta \circ \mu) = x(\mu \circ \eta) = x$, so $\eta \circ \mu = \mu \circ \eta = \varepsilon$.

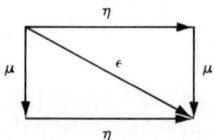

6

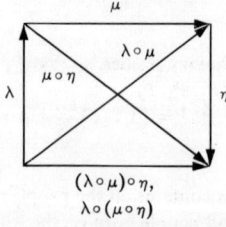

SECTION 1.7   (page 36)

1 Binary operation: (a), (f), (g).
2 (a) Two of the 16 possible operation tables are the following:

	a	b
a	a	a
b	a	a

	a	b
a	a	b
b	b	a

(b) 16, 19683, $n^{(n^2)}$.

Answers to Selected Problems   339

3

	1	i	-1	$-i$
1	1	i	-1	$-i$
i	i	-1	$-i$	1
-1	-1	$-i$	1	i
$-i$	$-i$	1	i	-1

4 (a) t, r.   (b) t, r, r, s.   (c) s, r, s, t.

5 $r^n = \begin{cases} r \text{ if } n \text{ is odd,} \\ s \text{ if } n \text{ is even.} \end{cases}$ $s^n = s$, for all $s \in \mathbb{Z}^+$. $t^n = \begin{cases} t \text{ if } n \text{ is odd,} \\ r \text{ if } n \text{ is even.} \end{cases}$

7

	ε	α	β	η	ρ	ν
ε	ε	α	β	η	ρ	ν
α	α	β	ε	ρ	ν	η
β	β	ε	α	ν	η	ρ
η	η	ν	ρ	ε	β	α
ρ	ρ	η	ν	α	ε	β
ν	ν	ρ	η	β	α	ε

8 $n = 1, 3, 3, 2, 2, 2$.

9 (a) $-4, -13, -40. -10, -34, -106$.   (b) $1, 1, 1. 4, 16, 256$.   (c) $5 + 5i, 5\sqrt{2} + 5i, 5\sqrt{3} + 5i$
                                                                        $2 + 2i, 2\sqrt{2} + 2i, 2\sqrt{3} + 2i$.

SECTION 2.1   (page 40)

1 (a) , (d).
2 (a) c, c, b, a. No.   (b) c, c, d, d. No. $[(ab)c \neq a(bc).]$ (c) $e, e' = e$.
3 (a) Eight cases.   (b) 729. $n^{(n/2)(n+1)}$
4

	Commut.	Assoc.	Identity	Inverses
(a)	Yes	Yes	0	$a' = (-a)/(a + 1)$, if $a \neq -1$
(b)	Yes	Yes	0	$0' = 0$
(c)	Yes	Yes	1/2	$a' = 1/4a$, if $a \neq 0$
(d)	No	No	—	—
(e)	Yes	No	—	—
(f)	Yes	Yes	—	—
(g)	Yes	Yes	2	$a' = 4 - a$.

SECTION 2.2   (page 44)

3

$\cap$	$\emptyset$	S	A	B
$\emptyset$	$\emptyset$	$\emptyset$	$\emptyset$	$\emptyset$
S	$\emptyset$	S	A	B
A	$\emptyset$	A	A	$\emptyset$
B	$\emptyset$	B	$\emptyset$	B

$\cup$	$\emptyset$	S	A	B
$\emptyset$	$\emptyset$	S	A	B
S	S	S	S	S
A	A	S	A	S
B	B	S	S	B

$+$	$\emptyset$	S	A	B
$\emptyset$	$\emptyset$	S	A	B
S	S	$\emptyset$	B	A
A	A	B	$\emptyset$	S
B	B	A	S	$\emptyset$.

Each algebra is commutative and associative. Identities are $S, \emptyset, \emptyset$, respectively. Each element of $(P(S), +)$ is its own inverse.

4 (a) Left, neither, right.   (b) An element appears at least twice in one of the rows.
   (c) An element appears at least twice in one of the columns.

8

(a)

	a	b	c	d
a	a	c	d	b
b	b	d	a	c
c	c	a	b	d
d	d	b	c	a

(b)

	a	b	c	d
a	a	b	c	d
b	b	d	a	c
c	c	a	d	b
d	d	c	b	a.

SECTION 2.3 (page 49)

1 (a) Semigroup. (b) Monoid. (c) Abelian group. (d) Quasigroup. (e) Groupoid. (f) Loop. (g) Monoid. (h) Group.

5 $\{r_0, r_2\}$ is a subgroup.

	r_0	r_1	r_2	r_3
r_0	r_0	r_1	r_2	r_3
r_1	r_1	r_2	r_3	r_0
r_2	r_2	r_3	r_0	r_1
r_3	r_3	r_0	r_1	r_2

6 (a) Abelian group. (b) Abelian group. (c) Groupoid.

8 This is an abelian group.

$\otimes$	1	2	3	4
1	1	2	3	4
2	2	4	1	3
3	3	1	4	2
4	4	3	2	1

10 (a) $rstuv = (rstu)v = [(rst)u]v = \{[(rs)t]u\}v = \{(rs)(tu)\}v = (rs)[(tu)v] = (rs)(tuv)$.

SECTION 2.4 (page 54)

1 Subgroups: (a), (b), (d), (e), (g). **3** Yes, by Prob. 2.2–2.

4 (b) $Q_8, A = \{1, -1, i, -i\}, B = \{1, -1, j, -j\}, C = \{1, -1, k, -k\}, D = \{1, -1\}, E = \{1\}$.

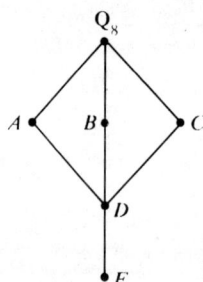

9 Hint: $a^2 = e, b^2 = e, (a \otimes b)^2 = e \Rightarrow (a \otimes b)^2 = a^2 \otimes b^2$.

11 The products in the table below are given, and since the Cayley table must be a Latin square, there is only one way to complete the table.

	e	a	b	c
e	e	a	b	c
a	a	e		
b	b		e	
c	c			e.

$A = \{e, a\}, B = \{e, b\}, C = \{e, c\}, E = \{e\}$.

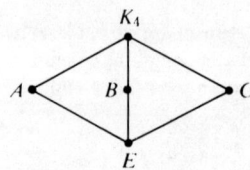

13 $a \in H \Rightarrow a' \in H \Rightarrow a \otimes a' \in H \Rightarrow e \in H \Rightarrow H \leq G$.

SECTION 2.5 (page 64)

1 (a) $\lambda^2 = (12)(34)$, $\lambda^3 = (1423)$, $\lambda^4 = \varepsilon$, $\lambda' = \lambda^3$, $(\lambda')^2 = \lambda^2 = (\lambda^2)'$.
   (b) $\lambda^2 = \varepsilon$, $\lambda^3 = \lambda$, $\lambda^4 = \varepsilon$, $\lambda' = \lambda$, $(\lambda')^2 = \varepsilon = (\lambda^2)'$.
2 (a) $\mu^2 = \lambda^3 = \eta^4 = \varepsilon$.   (b) $\mu' = \mu$, $\lambda' = \lambda^2$, $\eta' = \eta^3$.
3 (a) $\mu = (125)(364) = (12)(15)(36)(34)$.   (b) $\mu = (153462) = (15)(13)(14)(16)(12)$.
   (c) $\mu = (24)(35)$.   (d) $\mu = (125)(34) = (12)(15)(34)$.
4 Even, even, odd.
5 (1234), (1243), (1324), (1342), (1423), (1432), (12), (13), (14), (23), (24), (34).
6 (a) (35).   (b) (35)(14).   (c) (34)(12)(35).   (d) (1432).   (e) (153)(264).   (f) (26543)(1432).
7 (a) (12)(35).   (b) (246)(35).   (c) (145362).   (d) (1).   (e) (1453).   (f) (13562).
   (g) (13542).
8

Type:	(abcde)	(abc)	(ab)	(a)	(ab)(cd)
No.:	6	8	6	1	3

9

Type:	(abcde)	(abcd)	(abc)	(ab)	(a)	(abc)(de)	(ab)(cd)
No.:	24	30	20	10	1	20	15

11 (a) $\mu^2 = (13524)$, $\lambda^2 = (135)(246)$.   (b) $(1, 3, 5, \ldots, n, 2, 4, \ldots, n-1)$,
    $(1, 3, 5, \ldots, n-1)(2, 4, \ldots, n)$.   (c) If η is a cycle of length $3n$, then η^3 is a product of three
    disjoint cycles. If η is of length $3n + 1$ or $3n + 2$, then η^3 is one cycle.
12 (a) 3.   (b) 4.   (c) 100.
17 (a) $D_5 = \{(1), (12345), (13524), (14253), (15432), (25)(43), (13)(45), (15)(24), (12)(35), (14)(23)\}$.

SECTION 2.6 (page 71)

1 (a) Commutative ring.   (b) Integral domain.   (c) Field.   (d) Not a ring (not closed under $\cdot$).
   (e) Field.
3 $\cup$ and $\cap$ are associative and commutative. Zero is $\emptyset$, unity is I. $\cup$ is distributive with respect to $\cap$,
   and $\cap$ is distributive with respect to $\cup$. Not a ring, because nonzero elements do not possess
   additive inverses.
4 (a) c, b, a.   (b) z, b, z, b.   (c) $(T, +, \otimes)$ and $(T, \oplus, \cdot)$ are structurally like $(T, \oplus, \odot)$.
5 $\begin{bmatrix} 1 & 1 \\ 3 & 5 \end{bmatrix} \otimes \begin{bmatrix} x & u \\ y & v \end{bmatrix} = \begin{bmatrix} 1 & 0 \\ 0 & 1 \end{bmatrix} \Rightarrow \begin{matrix} x + y = 1, & u + v = 0, \\ 3x + 5y = 0, & 3u + 5v = 1. \end{matrix} \Rightarrow A^{-1} = \begin{bmatrix} \frac{5}{2} & -\frac{1}{2} \\ -\frac{3}{2} & \frac{1}{2} \end{bmatrix}$.
   B^{-1} fails to exist because these equations have no solution in common: $3x - 2y = 1$,
   $-6x + 4y = 0$.
7 $M_2(E)$ is a ring without unity.
9 $A^{-1} = \begin{bmatrix} \dfrac{d}{ad-bc} & \dfrac{-b}{ad-bc} \\ \dfrac{-c}{ad-bc} & \dfrac{a}{ad-bc} \end{bmatrix}$.
10 (a) $\begin{bmatrix} 3 & 4 \\ 2 & 3 \end{bmatrix}, \begin{bmatrix} -5 & 3 \\ 7 & -4 \end{bmatrix}$.   (b) $\begin{bmatrix} a & b \\ c & d \end{bmatrix}$ has a multiplicative inverse in $M_2(\mathbf{Z})$ iff $ad - bc = \pm 1$.
    (c) $\begin{bmatrix} 5 & 2 \\ 0 & 0 \end{bmatrix}, \begin{bmatrix} -4 & 0 \\ 10 & 0 \end{bmatrix}$ are zero divisors, since their product is the zero matrix.
11 Each algebra is associative and commutative with respect to both operations. The zero is 1, and the
    unity is 0, in each case. The distributive property holds in (b) and (c), but not in (a).
    (a) Additive inverses do not exist. $a^{-1} = -a$.   (b) $-a = 2 - a$. a^{-1} does not exist. (Integral
    domain).   (c) $-a = 1/a$. $a^{-1} = a/(a - 1)$. (Field).
15 (a) Hint: $bc = (d + c)c = dc + cc = c + c = a$.
    (b) No unity. b, c, d are zero divisors. $\odot$ is not commutative.

SECTION 2.7 (page 76)

1 $a(b - c) = a, b, c$. $(b - c)c = c, b, a$.
3 (a) $(a \oplus b) \oplus (-a - b) = (b \oplus a) \oplus [(-a) \oplus (-b)] = [(b \oplus a) \oplus (-a)] \oplus (-b) =$
   $[b \oplus (a \oplus -a)] \oplus -b = (b \oplus z) \oplus -b = b \oplus -b = z$. But $(a \oplus b) \oplus [-(a \oplus b)] = z$. Therefore,
   $-(a \oplus b) = -a - b$.

5 (a) Commutative ring with unity. $u = (1, 1, 1)$. There are 8 units: $(\pm 1, \pm 1, \pm 1)$. (There are zero divisors.)   (b) Not commutative. Unity = $(1, 0, 1)$. Set of units = $\{(\pm 1, n, \pm 1): n \in \mathbf{Z}\}$.
7 (a) b, b.   (b) c, z.   (c) u, u.   (d) u, b, c.   (e) a, a, z, a.
9 (b) Units are $1, -1, i, -i$.
10 $\mathbf{Q}(i)$ is closed under addition and multiplication. Also, $0, 1 \in \mathbf{Q}(i)$, $-(a + bi) = (-a) + (-b)i$, $(a + bi)^{-1} = \dfrac{a}{a^2 + b^2} + \dfrac{-b}{a^2 + b^2} i \in \mathbf{Q}(i)$. $\mathbf{Q}(i)$ is a subset of a field, and so the associative, commutative, and distributive properties hold in $\mathbf{Q}(i)$ because they hold in $\mathbf{C}$.
14 $I = \{a, b\}$, and $P(S) = \{\emptyset, I, A, B\}$, with $A = \{a\}, B = \{b\}$.

+	$\emptyset$	I	A	B
$\emptyset$	$\emptyset$	I	A	B
I	I	$\emptyset$	B	A
A	A	B	$\emptyset$	I
B	B	A	I	$\emptyset$

$\cap$	$\emptyset$	I	A	B
$\emptyset$	$\emptyset$	$\emptyset$	$\emptyset$	$\emptyset$
I	$\emptyset$	I	A	B
A	$\emptyset$	A	A	$\emptyset$
B	$\emptyset$	B	$\emptyset$	B

SECTION 2.8 (page 81)

1 $9 < 11$ and $2 < 7$, but $9 - 2 \not< 11 - 7$.
2 (a) $r \in D^+, s \in D^- \Rightarrow r, -s \in D^+ \Rightarrow r(-s) \in D^+ \Rightarrow -(rs) \in D^+ \Rightarrow rs \in D^-$.
5 (a) $5 > -2$ and $-2 > -4$, but $-10 \not> 8$.   (b) $a > b, c > d \Rightarrow ac > bc, bc \geq bd \Rightarrow ac > bd$.
8 $a \neq b \Rightarrow a < b, b < a \Rightarrow a < a$, a contradiction.
13 (a) No. If $D = \mathbf{Z}$, then $-3 < 2$, but $9 > 4$.

SECTION 2.9 (page 86)

1 Let u be unity, and v be a unit. Then $v = ab \Rightarrow v^{-1}v = v^{-1}ab \Rightarrow b^{-1} = v^{-1}a, a^{-1} = v^{-1}b \Rightarrow a$ and b are units $\Rightarrow v$ has no proper divisors.
3 (a) $a \mid b, b \mid c \Rightarrow b = as, c = bt$, for some $s, t \in D \Rightarrow c = (as)t = a(st) \Rightarrow a \mid c$.
5 (a) $m = a^2 n$, where n is square-free, $a > 1$. So $\mathbf{Z}(\sqrt{m}) = \{x + y\sqrt{m} : x, y \in \mathbf{Z}\} = \{x + ay\sqrt{n}\} \subset \mathbf{Z}(\sqrt{n})$. (Note that $1 + \sqrt{n} \in \mathbf{Z}(\sqrt{n})$, but $1 + \sqrt{n} \notin \mathbf{Z}(\sqrt{m})$.) The proof that $\mathbf{Z}(\sqrt{m})$ is an integral domain does not depend upon whether m is square-free or not, and so is the same as the proof of Theorem 2.27.
   (b) $\mathbf{Z}(\sqrt{3}) = \{x + y\sqrt{3} : x, y \in \mathbf{Z}\}$, $\mathbf{Z}(\sqrt{12}) = \{x + 2y\sqrt{3} : x, y \in \mathbf{Z}\}$. So $\mathbf{Z}(\sqrt{12}) \subset \mathbf{Z}(\sqrt{3})$.
8 (a) $2 + \sqrt{3}$, $7 + 4\sqrt{3}$, $26 + 15\sqrt{3}$.
9 (a) $4 - 5i, -4 + 5i, 5 + 4i, -5 - 4i$.   (b) $3, -3, 3i, -3i$.
10 (a) $(2 - i)(1 + 2i)$.   (b) 7.   (c) $(1 + 2i)(2 + i)$.   (d) $(2 + i)(2 - i)$.   (e) $(1 + 4i)(-1 - i)$.   (f) $4 - 5i$.
12 Hint: Show that if $(1 + \sqrt{2})^n = a_n + b_n \sqrt{2}$, for $n = 1, 2, 3, \ldots$, then $a_n < a_{n+1}$. Hence, units so obtained are all distinct.
14 (a) $\bar{n} = \{n, -n\}$. Each equivalence class, except $\bar{0}$, contains two integers.
   (b) $\overline{a + bi} = \{a + bi, -a - bi, -b + ai, b - ai\}$. Each class, except $\bar{0}$, contains 4 complex numbers.

SECTION 2.10 (page 93)

1 Hint: Construct the operation table of H, writing the elements in the headings in the new order: r_0, r_3, r_2, r_1. Compare this table with the table for G.
2 (a) If $\theta: (G, \cdot) \to (G, \otimes)$ is the map such that $a\theta = a'$, then θ is bijective, and $(a \cdot b)\theta = (a \cdot b)' = b' \cdot a' = a' \otimes b' = (a\theta) \otimes (b\theta)$.   (b) Yes, by Theorem 2.30.
3 Hint: Compare the operation tables, and set up the 1–1 map that is an isomorphism.
4 (a) No. In $(P(S), +)$, each element is its own additive inverse, while in $(S, \oplus)$, $a \oplus a \neq z$. Since inverses of corresponding elements must also correspond (by Theorem 2.30(c)), an isomorphic map is impossible to set up.   (b) Yes. Map: $\emptyset \to a, I \to b, A \to c, B \to d$, is an isomorphism.
7 (a) $a\theta = e$, for all $a \in G$, where e is the identity of H.
   (b) $a\theta = z$, for all $a \in S$, where z is the zero of ring T.
11 (a) One of order 2, one of order 3. (Construct their Cayley tables.)

Answers to Selected Problems

12 There are two rings of order 2, one of which is a field.
15 (a) $(a\psi) \otimes (b\psi) = (\cos a, \sin a) \otimes (\cos b, \sin b) = (\cos(a + b), \sin(a + b)) = (a + b)\psi$.
(b) R is the real line (or x-axis). G is the set of points on the circle $x^2 + y^2 = 1$. The map "winds" the x-axis around the circle, mapping each point of the x-axis to a point on the circle. Each point on the circle has infinitely many antecedents.

SECTION 3.1 (page 99)

1 (a) 17 distinct factorings, apart from order of factors.
  (b) There are 2 units, 4 primes, 10 composites.
5 (a)

$x =$	8	7	6	5	4	3
$38 - 7x =$	-18	-11	-4	3	10	17

  (b) If $a = 0$, let $x = -1$. If $a > 0$, let $x = 0$. If $a < 0$, let $x = 2a$.
7 Hint: If n is any integer, there exist q and r such that $n = 3q + r$, where r is 0, 1, or 2. So $n^2 = \_\_\_$.
8 (b) No, since $a \mid b$ and $b \mid a$ does not imply that $a = b$ (if $a, b \in \mathbf{Z}$).
9 (a) 21.   (b) 1029.   (c) 63.   10 (a) $(1214)_6$.   (b) $(24040)_6$.   (c) $(20130035113)_6$.

SECTION 3.2 (page 106)

1 (a) $x = 1 + 2n, y = -1 - 3n$, for any $n \in \mathbf{Z}$. $x = 2 + 15n, y = -1 - 8n$, for any $n \in \mathbf{Z}$.
  (b) No, since $3 \mid 6x + 15y$, but $3 \nmid 4$.
3 If $g = (a, b)$, then: (i) $g \in \{1, 2, 3, 6\}$, (ii) $g = 1$, (iii) $g \in \{1, 5\}$.
4 (a) $3 = 867(-71) + 810(76)$.   (b) $1 = 1111(-9) + 500(20)$.
  (c) $14 = 397{,}670(-323) + 12{,}054(10{,}656)$.
5 234,090,   555,500,   342,393,870.   7 180,   210,   2310.

(b) 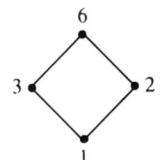   (d)    (f)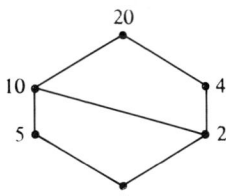

9 $\{\{2, 3, 5, 7, 11, 13, 17, 19, 23, 29\}, \{4, 9, 25\}, \{8, 27\}, \{16\}, \{6, 10, 14, 15, 21, 22, 26\}, \{12, 18, 20, 28\},$
  $\{24\}\}$.

$$n \sim m \Leftrightarrow n = p_1^{n_1} p_2^{n_2} \cdots p_k^{n_k}, m = q_1^{n_1} q_2^{n_2} \cdots q_k^{n_k},$$

where n and m each has k distinct primes as factors.
10 10 and 300, 20 and 150, 30 and 100, 50 and 60.   11 $1, n(n + 1)$.
13 $4 \mid n^2 + 2 \Rightarrow n^2 + 2 = 4k \Rightarrow n^2 = 4k - 2 \Rightarrow 2 \mid n^2 \Rightarrow 2 \mid n \Rightarrow n = 2m \Rightarrow 4m^2 = 4k - 2 \Rightarrow$
  $2(m^2 - 2k) = 1 \Rightarrow 2 \mid 1$, a contradiction.
15 (b) $g \mid h \Rightarrow (g, h) = g, [g, h] = h$.
19 $x_0 = 69, y_0 = -115$. Solution set $= \{(69 + 10n, -115 - 17n) : n \in \mathbf{Z}\}$.

SECTION 3.3 (page 111)

1 (a) $P_1 \Leftrightarrow 1^2 = (\frac{1}{6})(1 + 1)(2 + 1)$. $P_k \Rightarrow 1^2 + 2^2 + \cdots + k^2$
  $= (k/6)(k + 1)(2k + 1) \Rightarrow (1^2 + \cdots + k^2) + (k + 1)^2$
  $= (k/6)(k + 1)(2k + 1) + (k + 1)^2 = [(k + 1)/6][k(2k + 1) + 6(k + 1)]$
  $= [(k + 1)/6](2k^2 + 7k + 6) = [(k + 1)/6](k + 2)(2k + 3) \Rightarrow P_{k+1}$.
  So P_n is true for all n, by Theorem 3.11.
3 (a) $P_n \Leftrightarrow 2 \mid n^2 + n$. (i) P_1 is true, since $2 \mid 1^2 + 1$. (ii) Prove: $2 \mid k^2 + k \Rightarrow 2 \mid (k + 1)^2 + (k + 1)$. Now
  $(k + 1)^2 + (k + 1) = k^2 + 3k + 2 = (k^2 + k) + 2(k + 1)$. But $2 \mid k^2 + k$, and $2 \mid 2(k + 1)$, so
  $P_k \Rightarrow P_{k+1}$. Therefore, $2 \mid n^2 + n$, for all $n \in \mathbf{Z}$, by Theorem 3.11.

4 No. P_n is true if $1 \leq n \leq 39$, but P_{40} is false. **5** $P_k \Rightarrow P_{k+1}$, but P_n is false for *every* $n \in \mathbf{Z}^+$.
(Note: In Prob. 4, hyp. (i) of Theorem 3.11 is satisfied, but (ii) is not. In Prob. 5, hyp. (ii) is satisfied, but (i) is not.)

6 Hint: $a < 0 \Rightarrow -a > 0 \Rightarrow$ There exist q_1, r_1 such that $-a = bq_1 + r_1$, with $0 \leq r_1 < b$. (Multiply each side by -1, and find q and r.)

8 Hint: (a) Use induction to prove that $n \in H$, for all $n \in \mathbf{Z}^+$. From this, show that $-n, 0 \in H$.
(b) $4s + 7t = 1$, for some $s, t \in H$.

11 $P_3 : a_1 * a_2 * a_3 = (a_1 * a_2) * a_3$ (by def.) $= a_1 * (a_2 * a_3)$ (by assoc. prop.)
Let $1 \leq r < k + 1$, and let P_k be given. Now $a_1 * a_2 * \cdots * a_k * a_{k+1}$
$= (a_1 * a_2 * \cdots * a_k) * a_{k+1}$ by def. (making P_{k+1} true if $r = k$)
$= [(a_1 * \cdots * a_r) * (a_{r+1} * \cdots * a_k)] * a_{k+1}$ by P_k, if $r < k$
$= (a_1 * \cdots * a_r) * [(a_{r+1} * \cdots * a_k) * a_{k+1}]$ by assoc. prop.
$= (a_1 * \cdots * a_r) * (a_{r+1} * \cdots * a_k * a_{k+1})$ by def.

SECTION 3.4   (page 116)

1 (a) 12, 7560. (b) 200, 858,000.
3 $m = p_1^{m_1} \cdot p_2^{m_2} \cdots \cdot p_s^{m_s} \Rightarrow m^2 = p_1^{2m_1} p_2^{2m_2} \cdots p_s^{2m_s} \Rightarrow n_i = 2m_i$.
5 (a) $q \leq p_n \Rightarrow q \mid p_1 p_2 \cdots p_n \Rightarrow q \mid (p_1 p_2 \cdots p_n) - t \Rightarrow q \mid 1$.
(b) Assume there are but a finite number of primes p_i, $i = 1, 2, \ldots, n$. If $t = 1 + (p_1 p_2 \cdots p_n)$, then, by part (a), there is a prime $q > p_i$ such that $q \mid t$, a contradiction of the assumption that there are only n primes.
6 (a) If $(n + 1)! = k$, then S is a set of n consecutive integers, where
$S = \{k + 2, k + 3, \ldots, k + (n + 1)\}$.   (b) $2 + 1{,}000{,}001$, and $1{,}000{,}001 + (1{,}000{,}001)!$.
The last integer factors into $1{,}000{,}001[1 + (1{,}000{,}000)!]$, which is a product of two odd integers.

SECTION 3.5   (page 121)

1 (a) $\bar{0} = \{0, \pm 2, \pm 4, \ldots\}$, $\bar{1} = \{\pm 1, \pm 3, 5, \ldots\}$.
(b) $\{0, \pm 3, \pm 6\} \subset \bar{0}$, $\{1, 4, 7, -2, -5\} \subset \bar{1}$, $\{2, 5, 8, -1, -4\} \subset \bar{2}$.
2 (a) 1, 2, 3, 4, 6, 12. (b) 1, 7.
5 Every integer is congruent to one of these modulo 4: $-1, 0, 1, 2$. $p \equiv 0$ or $2 \pmod 4 \Rightarrow 2 \mid p$.
6 $s \in \bar{c} \Rightarrow s \equiv c \Rightarrow as \equiv ac \pmod n \Rightarrow as \equiv b \pmod n$.
7 Solution sets: (a) $\bar{3}$. (b) $\bar{4} \cup \overline{11}$. (c) $\bar{1} \cup \bar{4} \cup \bar{7} \cup \overline{10} \cup \overline{13}$. (d) $\varnothing$. (e) $\bar{1} \cup \bar{2} \cup \bar{4} \cup \bar{5}$.
8 (a) $as \equiv b \pmod n \Leftrightarrow as = b + nt \Leftrightarrow as - nt = b \Leftrightarrow (a, n) \mid b$.
(b) $28x \equiv 35 \pmod{1200}$, $28x \equiv 44 \pmod{1200}$.
9 (a) $(a, n) = 1 \Rightarrow as + nt = 1 \Rightarrow a(bs) + n(bt) = b \Rightarrow a(bs) \equiv b \pmod n \Rightarrow bs$ is a sol. $\Rightarrow$ each element in $\overline{bs}$ is a sol. Further, $av \equiv b \Rightarrow av \equiv a(bs) \Rightarrow v \equiv bs \Rightarrow v \in \overline{bs}$.
10 (a) $\bar{8} \cup \overline{25} \cup \overline{42}$. (b) $\bar{2} \cup \bar{7} \cup \overline{12} \cup \overline{17} \cup \overline{22} \cup \overline{27} \cup \overline{32} \cup \overline{37}$. (c) $\overline{61}$.
11 (a) $x = 30 + 147n$, $y = 50 + 250n$, $n \in \mathbf{Z}$. (b) $x = 84 + 89n$, $y = 4 - 71n$, $n \in \mathbf{Z}$.
(c) $x = 12 + 30n$, $y = 602 - 1001n$, $n \in \mathbf{Z}$.
13 (a) $\overline{23} \pmod{28}$. (b) $\overline{107} \pmod{252}$.   **14** 333 coins.   **15** $\overline{34} \cup \overline{69} \cup \overline{104} \pmod{105}$.
16 (a) $1 + 4s = 2 + 6t \Rightarrow 4s - 6t = 1 \Rightarrow 2 \mid 1$, an impossibility. (b) $\bar{5} \pmod{12}$.

SECTION 3.6   (page 126)

1 (c) $\mathbf{Z}_6$:

+	$\bar{0}$	$\bar{1}$	$\bar{2}$	$\bar{3}$	$\bar{4}$	$\bar{5}$
$\bar{0}$	$\bar{0}$	$\bar{1}$	$\bar{2}$	$\bar{3}$	$\bar{4}$	$\bar{5}$
$\bar{1}$	$\bar{1}$	$\bar{2}$	$\bar{3}$	$\bar{4}$	$\bar{5}$	$\bar{0}$
$\bar{2}$	$\bar{2}$	$\bar{3}$	$\bar{4}$	$\bar{5}$	$\bar{0}$	$\bar{1}$
$\bar{3}$	$\bar{3}$	$\bar{4}$	$\bar{5}$	$\bar{0}$	$\bar{1}$	$\bar{2}$
$\bar{4}$	$\bar{4}$	$\bar{5}$	$\bar{0}$	$\bar{1}$	$\bar{2}$	$\bar{3}$
$\bar{5}$	$\bar{5}$	$\bar{0}$	$\bar{1}$	$\bar{2}$	$\bar{3}$	$\bar{4}$

$\cdot$	$\bar{0}$	$\bar{1}$	$\bar{2}$	$\bar{3}$	$\bar{4}$	$\bar{5}$
$\bar{0}$	$\bar{0}$	$\bar{0}$	$\bar{0}$	$\bar{0}$	$\bar{0}$	$\bar{0}$
$\bar{1}$	$\bar{0}$	$\bar{1}$	$\bar{2}$	$\bar{3}$	$\bar{4}$	$\bar{5}$
$\bar{2}$	$\bar{0}$	$\bar{2}$	$\bar{4}$	$\bar{0}$	$\bar{2}$	$\bar{4}$
$\bar{3}$	$\bar{0}$	$\bar{3}$	$\bar{0}$	$\bar{3}$	$\bar{0}$	$\bar{3}$
$\bar{4}$	$\bar{0}$	$\bar{4}$	$\bar{2}$	$\bar{0}$	$\bar{4}$	$\bar{2}$
$\bar{5}$	$\bar{0}$	$\bar{5}$	$\bar{4}$	$\bar{3}$	$\bar{2}$	$\bar{1}$

2 (f) $n = 9$: $\bar{1}^{-1} = \bar{1}, \bar{2}^{-1} = \bar{5}, \bar{4}^{-1} = \bar{7}, \bar{8}^{-1} = \bar{8}$
$n = 11$: $\bar{1}^{-1} = \bar{1}, \bar{2}^{-1} = \bar{6}, \bar{3}^{-1} = \bar{4}, \bar{5}^{-1} = \bar{9}, \bar{7}^{-1} = \bar{8}, \overline{10}^{-1} = \overline{10}$.

Answers to Selected Problems   345

3 $Z_1 = \{0\} = \{Z\}$. **5** No, since in Z_3 we are adding *sets* of integers, while in Z we are adding integers.

7 (a) 200. (b) 20,000,000. (c) 480. (d) 162.

9

·	$\bar{1}$	$\bar{5}$	$\bar{7}$	$\overline{11}$	$\overline{13}$	$\overline{17}$
$\bar{1}$	$\bar{1}$	$\bar{5}$	$\bar{7}$	$\overline{11}$	$\overline{13}$	$\overline{17}$
$\bar{5}$	$\bar{5}$	$\bar{7}$	$\overline{17}$	$\bar{1}$	$\overline{11}$	$\overline{13}$
$\bar{7}$	$\bar{7}$	$\overline{17}$	$\overline{13}$	$\bar{5}$	$\bar{1}$	$\overline{11}$
$\overline{11}$	$\overline{11}$	$\bar{1}$	$\bar{5}$	$\overline{13}$	$\overline{17}$	$\bar{7}$
$\overline{13}$	$\overline{13}$	$\overline{11}$	$\bar{1}$	$\overline{17}$	$\bar{7}$	$\bar{5}$
$\overline{17}$	$\overline{17}$	$\overline{13}$	$\overline{11}$	$\bar{7}$	$\bar{5}$	$\bar{1}$

10 (b) U_{10} is isomorphic to U_5. **11** (a) $Z_8, A = \{\bar{0}, \bar{2}, \bar{4}, \bar{8}\}, B = \{\bar{0}, \bar{4}\}, E = \{\bar{0}\}$.

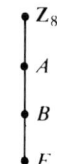

12 (a) A bijective map is impossible, since the groups are of different orders.

13 (a) $\{\bar{0}, \bar{4}, \bar{8}, \overline{12}, \overline{16}, \overline{20}\}$. (b) $\{\bar{0}, \overline{12}\}$. (c) $\{\bar{0}\}$. (d) Z_{24}. (e) Z_{24}.
  (f) $\{\bar{0}, \bar{3}, \bar{6}, \bar{9}, \overline{12}, \overline{15}, \overline{18}, \overline{21}\}$.

14 (a) $\bar{3}$. (b) $\bar{4}, \overline{11}$. (c) None. (d) $\bar{1}, \bar{4}, \bar{7}, \overline{10}, \overline{13}$. (e) $\bar{1}, \bar{2}, \bar{4}, \bar{5}$. (f) None.
  (g) $\bar{1}, \bar{3}, \bar{7}, \bar{9}$.

15 (a) For each $n \in Z: 3n \to \bar{0}, 3n+1 \to \bar{1}, 3n+2 \to \bar{2}. \bar{0}\theta^{-1} = \{0, \pm 3, \pm 6, \ldots\}$.

SECTION 4.1   (page 134)

1 $f + g = (7, -1, 3, 12, -1, 0, 0, \ldots) = 7 - x + 3x^2 + 12x^3 - x^4$.
  $fg = (1^0, 1, 13, 30, -5, 26, 32, -8, 0, 0, \ldots)$.

2 (a) $(\bar{2}, \bar{0}, \bar{2}, \bar{2}, \bar{1}, \bar{0}, \bar{0}, \bar{0}, \bar{0}, \bar{2}, \bar{2}, \bar{1}, \bar{0}, \bar{0}, \ldots)$, deg $fg = 11$. (b) $(\bar{1}, \bar{0}, \bar{0}, \ldots), (\bar{2}, \bar{0}, \bar{0}, \ldots)$.
  (c) 18 of degree 2, 54 of degree 3, $2 \cdot 3^n$ of degree n.

6 (a)
$$-f = \left(\begin{bmatrix} -2 & -1 \\ 0 & 13 \end{bmatrix}, \begin{bmatrix} 0 & 0 \\ 0 & 0 \end{bmatrix}, \begin{bmatrix} -1 & 1 \\ 0 & 0 \end{bmatrix}, 0, 0, \ldots\right)$$
$$f + g = \left(\begin{bmatrix} 3 & 1 \\ 0 & 4 \end{bmatrix}, \begin{bmatrix} 4 & 0 \\ 0 & 4 \end{bmatrix}, \begin{bmatrix} 1 & -1 \\ 0 & 0 \end{bmatrix}, \begin{bmatrix} 1 & 3 \\ 1 & 3 \end{bmatrix}, 0, 0, \ldots\right)$$
$$fg = \left(\begin{bmatrix} 2 & 1 \\ 0 & 3 \end{bmatrix}, \begin{bmatrix} 8 & 4 \\ 0 & 12 \end{bmatrix}, \begin{bmatrix} 1 & -1 \\ 0 & 0 \end{bmatrix}, \begin{bmatrix} 7 & 5 \\ 3 & 9 \end{bmatrix}, 0, 0, \ldots\right)$$
$$gf = \left(\begin{bmatrix} 2 & 1 \\ 0 & 3 \end{bmatrix}, \begin{bmatrix} 8 & 4 \\ 0 & 12 \end{bmatrix}, \begin{bmatrix} 1 & -1 \\ 0 & 0 \end{bmatrix}, \begin{bmatrix} 6 & 6 \\ 2 & 10 \end{bmatrix}, 0, \begin{bmatrix} 1 & -1 \\ 1 & -1 \end{bmatrix}, \ldots\right).$$

(b) $\deg(-f) = 2, \deg(f+g) = 3, \deg(fg) = 3, \deg(gf) = 5$. (c) $\begin{bmatrix} 1 & 0 \\ 0 & 1 \end{bmatrix}$.

8 Hint: If $\theta: S \to T$ is an isomorphism, with $a_i \to a_i\theta$, define $\lambda: S[x] \to T[x]$ to be the map such that $(a_1, a_2, a_3, \ldots) \to (a_1, a_2, a_3, \ldots)\lambda = (a_1\theta, a_2\theta, \ldots)$, and show that λ is an isomorphism.

10 Hint: Use Theorem 4.3 in your proof.

SECTION 4.2   (page 139)

1 (a) $q(x) = (\tfrac{1}{2})x - \tfrac{3}{4}, r(x) = (-\tfrac{9}{4})x^2 + (\tfrac{13}{4})x + \tfrac{1}{4}$. (b) $q(x) = 0, r(x) = 2x + 7$.
  (c) $q(x) = 0, r(x) = 0$. (d) $q(x) = 0, r(x) = 5$.

3 $q(x) = x^4 + x^3 + \bar{2}x + \bar{3}, r(x) = \bar{4}x + \bar{1}$.

7 (a) $2x^2 - x + 4, 5$. (b) $3x^3 - 3x^2 + 10x - 2, 0$. (c) $2x^2 + (\tfrac{5}{2})x + \tfrac{3}{4}, \tfrac{37}{4}$.

8 $x^5 + \bar{2}x^4 + \bar{3}x^3 + \bar{4}x + \bar{3}, r = \bar{2}$.

SECTION 4.3 (page 142)

1. $\bar{2}x^2 + \bar{4}x + \bar{3}, \bar{4}x^2 + \bar{3}x + \bar{1}, x^2 + \bar{2}x + \bar{4}, \bar{3}x^2 + x + \bar{2}$.
2. If a_n is the leading coefficient of $f(x)$, then $a_n^{-1} \cdot f(x)$ is the only monic polynomial that is an associate of $f(x)$ (since the inverse of an element is unique).
5. (a) $d(x) = x^2 + x + 2, s(x) = (\frac{2}{3})x - \frac{2}{3}, t(x) = \frac{-2}{3}x^2 + \frac{2}{3}x + \frac{1}{3}$.
   (b) $d(x) = x - 1, s(x) = \frac{1}{17}, t(x) = (\frac{-1}{17})(x^3 - 2x^2 + 5x - 8)$.
   (c) $d(x) = x^2 + \bar{2}x + \bar{1}, s(x) = \bar{1}, t(x) = x + \bar{2}$.
8. $f(x) = \bar{4}(x^2 + \bar{5})(x^3 + \bar{6}x + \bar{2})(x + \bar{1})$.
11. $n = mk$, where $1 < m < n, 1 < k < n$. $(x - \bar{m})(x - \bar{k}) = x(x - \overline{m + k})$.
12. Hint: Use the Euclidean Algorithm. Then $a = 0, b = 0 \Rightarrow (x^3 + x^2, x^3 + 2x^2) = x^2$; $a = 1, b = 0 \Rightarrow (x^3 + x^2 - x, x^3 + 2x^2 - 1) = x^2 + x - 1$.

SECTION 4.4 (page 148)

1. (a) $(x - \bar{5})(x^3 - x^2 + x - \bar{2})$. (b) $\bar{2}(x - \bar{1})(x - \bar{3})(x - \bar{5})$.
   (c) $x(x - \bar{1})(x - \bar{2})(x - \bar{3})(x - \bar{4})(x - \bar{5})(x - \bar{6})$. (d) $(x^2 + \bar{1})(x - \bar{4})(x - \bar{4})$.
2. (a) Since no element of $\mathbf{Z}_5$ is a root, the polynomial is prime (Theorem 4.13).
   (b) Let $f(x) = x^2 + \bar{3}x + \bar{2}$. Roots are $\bar{1}, \bar{2}, \bar{4}, \bar{5}$. $x^2 + \bar{3}x + \bar{2} = (x - \bar{4})(x - \bar{5}) = (x - \bar{1})(x - \bar{2})$.
3. $x^3 + x^2 + \bar{1}, \quad x^3 + x + \bar{1}$.
5. (a) $x^2 - 2$. (b) $x^3 - 2x, \quad x^4 + x^2 - 6, \quad x^5 - x^3 + x^2 - 2x - 2$.
8. $x^4 + 4x^2 + 3$. 9. Roots are $1 + i$ and $-3 + 2i$.

SECTION 4.5 (page 152)

1. (a) $x^3 - 3\sqrt{2}x^2 + 15x - 11\sqrt{2}, x^6 + 12x^4 + 93x^2 - 242$. (b) $x^4 - 4x^3 + 26x^2 - 100x + 25$.
2. (a) $(x - 3)(x^2 - 2) = (x - 3)(x - \sqrt{2})(x + \sqrt{2})$. (b) $(x - 1)(3x + 2)(2x - 1)$.
   (c) $x(x + 4)(x^2 + 1) = x(x + 4)(x + i)(x - i)$.
   (d) $(x - 2)(x + 2)(x^2 + 1) = (x - 2)(x + 2)(x + i)(x - i)$.
3. (a) $x^3 - (\sqrt{2} + 3i)x^2 + 3\sqrt{2}ix$. (b) $x^4 - \sqrt{2}x^3 + 9x^2 - 9\sqrt{2}x$. (c) $x^5 + 7x^3 - 18x$.
7. (a) $3 - 2i, -3 + 2i$. (b) $3i, -i$. (c) $i, 2 - i$.
10. (b) $x^2 - 3$ and $x^4 - 2x^2 - 3$ have the same real roots, but these polynomials are not associates.
11. (a) $3x^5 - x^4 - 2x^3 + 7x^2 - x - 4$. (b) $4x^5 - x^4 - 7x^3 - 2x^2 + x + 3$.
12. (a) $r = -2 \Rightarrow a = 4 \Rightarrow r_i = -2, 2, -1$. (b) $r = 0$, a root of multiplicity 4. $r = 1$ or $r = -1$, roots each of multiplicity 2. (c) $a = b = -1$.

SECTION 4.6 (page 157)

1. (a) $r = -2, \frac{1}{3}, 1$. $f(x) = (x + 2)(x - \frac{1}{3})(x - 1)(x^2 + x + 1)$.
   (b) No roots in $\mathbf{Q}$, but $f(x) = (x^2 + x + 1)(x^2 - 2)$. (c) $r = \frac{3}{2}, f(x) = (x - \frac{3}{2})(x^4 + 1)$.
   (d) $r = -1, 2$. $f(x) = (x + 1)^3(x - 2)$.
2. (a) Any monic polynomial in $\mathbf{Z}[x]$ is primitive.
   (b) $f(x) = x^n + pb_{n-1}x^{n-1} + \cdots + pb_2x^2 + pb_1x + p$, where the b_i can be any integers, and p is a prime.
3. (a) and (c) are prime, by Theorem 4.25. (b) is prime, by Theorem 4.13. (d) has no roots in $\mathbf{Q}$, but $x^4 + x^2 - 12 = (x^2 - 3)(x^2 + 4)$.
4. $f(x) = x^2 + x + 1$. 6. $f(x)$ is composite iff p and q are primes such that either $pq = 10$ or $pq = 4$.

SECTION 4.7 (page 165)

1. (a) $x - \frac{2}{3}$. (b) $x^2 - 4x + 7$. (c) $x^3 - \frac{7}{8}$. (d) $x^4 - 10x^2 + 1$. (e) $x^6 - 4x^3 - 1$.
   (f) $(\frac{1}{8})(8x^6 - 12x^4 - 96x^3 + 6x^2 - 144x + 287)$. (g) $x^2 - 5x + 7$. The algebraic integers are the numbers given in parts, b, d, e, and g.
2. (a) Any integer in $\mathbf{Z}^+$. (b) 1 or 2. (c) 1.

(a) $m(x) = x^2 + 1$, $\mathbf{Q}(i) = \{a + bi : a, b \in \mathbf{Q}\}$.
(b) $m(x) = x^2 + 1$, $\mathbf{R}(i) = \{a + bi : a, b \subset \mathbf{R}\} = \mathbf{C}$.
(c) $m(x) = x^3 - 7$, $\mathbf{Q}(\sqrt[3]{7}) = \{a + b\sqrt[3]{7} + c\sqrt[3]{49} : a, b, c \in \mathbf{Q}\}$
(d) $m(x) = x^2 + 2$, $\mathbf{R}(\sqrt{-2}) = \{a + b\sqrt{-2} : a, b \in \mathbf{R}\}$
$= \{a + bi\sqrt{2} : a, b \in \mathbf{R}\} = \mathbf{C}$
(e) $m(x) = x^2 - 5$, $F(\sqrt{5}) = \{x + y\sqrt{5} : x, y \in F\}$
$= \{a + b\sqrt{6} + (c + d\sqrt{6})\sqrt{5} : a, b, c, d \in \mathbf{Q}\}$
$= \{a + b\sqrt{6} + c\sqrt{5} + d\sqrt{30} : a, b, c, d \in \mathbf{Q}\}$.
(f) $m(x) = x^4 - 22x^2 + 25$.
$\mathbf{Q}(\sqrt{3} + 2\sqrt{2}) = \{a + b(\sqrt{3} + 2\sqrt{2}) + c(\sqrt{3} + 2\sqrt{2})^2 + d(\sqrt{3} - 2\sqrt{2})^3 : a, b, c, d \in \mathbf{Q}\} = \{x_1 + x_2\sqrt{2} + x_3\sqrt{3} + x_4\sqrt{6} : x_i \in \mathbf{Q}\} = F(\sqrt{3})$, where $F = \mathbf{Q}(\sqrt{2})$.

4 (a) $x^4 - 3 = (x^2 - \sqrt{3})(x^2 + \sqrt{3}) = (x - \sqrt[4]{3})(x + \sqrt[4]{3})(x^2 + \sqrt{3})$
$= (x - \sqrt[4]{3})(x + \sqrt[4]{3})(x - i\sqrt[4]{3})(x + i\sqrt[4]{3})$
(b) Let $r = \sqrt[4]{3}, s = -\sqrt[4]{3}, t = i\sqrt[4]{3}$. Then $\mathbf{Q}(r) = \mathbf{Q}(s)$, but $\mathbf{Q}(s) \neq \mathbf{Q}(t)$, since $\mathbf{Q}(t)$ contains imaginary numbers, while $\mathbf{Q}(s)$ does not. Of course, $\mathbf{Q}(s) \cong \mathbf{Q}(t)$.

5 (a) $(a + b\sqrt{10})^{-1} = a/(a^2 - 10b^2) - b\sqrt{10}/(a^2 - 10b^2)$.
(b) $(a + b\sqrt[3]{2} + c\sqrt[3]{4})^{-1} = (1/d)[(a^2 - 2bc) + (2c^2 - ab)\sqrt[3]{2} + (b^2 - ac)\sqrt[3]{4}]$, where $d = a^3 + 2b^3 + 4c^3 - 6abc$.   (c) $(a + b\sqrt{-2})^{-1} = [1/(a^2 + 2b^2)](a - b\sqrt{-2})$.

6 $\mathbf{Q}(r) = \{a + b\sqrt{6} : a, b \in \mathbf{Q}\}$, $\mathbf{Q}(s) = \{(x + 2y/3) + (y/2)\sqrt{6}\}$. Let $a = x + 2y/3, b = y/2$. Then $a, b \in \mathbf{Q} \Leftrightarrow x, y \in \mathbf{Q}$. So $\mathbf{Q}(r) = \mathbf{Q}(s)$.

9 $F(r) = \{a_0 + a_1 r + \cdots + a_{n-1} r^{n-1} : a_i \in F\}$. Since there are k choices for each of the n elements, a_i, there are k^n elements in $F(r)$.   10 $\mathbf{Q}(r) = \mathbf{Q}$.

14 (a) $(ad - bf - ce) + (ae + bd + 3bf + 3ce - cf)r + (af + be + cd + 3cf)r^2$.
(b) $\frac{16}{3} - (\frac{1}{3})r - (\frac{5}{3})r^2$.

15 (a) $\mathbf{Q}(r) = \{a + br : a, b \in \mathbf{Q}, r^2 = -1 - r\}$, so $(a + br)(c + dr) = (ac - bd) + (ad + bc - bd)r$.
(b) $(a + br)^{-1} = \dfrac{a - b}{a^2 - ab + b^2} + \dfrac{-b}{a^2 - ab + b^2} \cdot r$.

16 $\mathbf{Q}(r) = \{a + br + cr^2 : a, b, c \in \mathbf{Q}, r^3 = 2 - 5r^2\}$
   $\mathbf{Q}(s) = \{a + bs : a, b \in \mathbf{Q}, r^2 = -1 + 4r\}$.

17 (a) $(x - r)(x^2 + rx + r^2)$.   (b) $(x - r)(x^2 + rx + (r^2 - 3))$.   (c) $(x - r)(x + (r + b))$.
(d) $(x - r)(x^3 + (r + 3)x^2 + (r^2 + 3r)x + (r^2 + 3r^2 - 9))$.

SECTION 4.8   (page 171)

1 (a) $-\frac{20}{9}, x^2 - \frac{20}{9}$.   (b) $\frac{81}{49}, x - \frac{9}{7}$.   (c) $1, x^2 - 7x + 1$.   (d) $11, x^2 - 8x + 11$.
(e) $-\frac{4}{9}, x^2 - (\frac{2}{3})x - \frac{4}{9}$.

3 Every polynomial, $x^2 - 2ax + (a^2 - nb^2)$, where $a, b \in \mathbf{Q}, b \neq 0, n \in \mathbf{Z}, n$ is square-free, has a root r such that $\mathbf{Q}(r) = \mathbf{Q}(\sqrt{n})$.   4 $(\frac{1}{3}) + (\frac{2}{3})\sqrt{-2}$.

7 (a) $f(x) \in F[x]$, where $F = \mathbf{Q}(\sqrt{7})$.   (b) Composite, since $3 + 2\sqrt{7}$ is a root.
(c) $x^2 - 6x - 19$.

SECTION 4.9   (page 175)

1 (a) $(-\sqrt{-3})(4 + \sqrt{-3})$.   (b) $(2 + \sqrt{-3})(2 - \sqrt{-3})$.   (c) $1 + \sqrt{-3}$.
(d) $(\sqrt{-3})(8 - 5\sqrt{-3})$.   (e) Prime.   (f) Unit.

2 (a) $(1 + \sqrt{-5})(2 - 3\sqrt{-5})$.   (b) $1 + \sqrt{-5}$.   (c) $\sqrt{-5}(-\sqrt{-5})$.
(d) $(6 + \sqrt{-5})(6 - \sqrt{-5})$.   (e) 43.   (f) $(-\sqrt{-5})(\sqrt{-5})^2(1 - \sqrt{-5})(2 + \sqrt{-5})$.

3 (a) $n > 0 \Rightarrow N(s) \in \{\pm 1, \pm 2, \pm 3, \pm 6\}, n < 0 \Rightarrow N(s) \in \{1, 2, 3, 6\}$.   (b) $N(r) = 7 \Rightarrow r$ is a prime, either s or t is a unit. The one that is not a unit is an associate of r.

4 (a) $4 + 5i, -4 - 5i, -5 + 4i, 5 - 4i$.   (b) $4 + 5\sqrt{-2}, -4 - 5\sqrt{-2}$.
(c) $4 + 5\sqrt{-3}, -4 - 5\sqrt{-3}, (\frac{1}{2})(-11 + 9\sqrt{-3}), (\frac{1}{2})(11 - 9\sqrt{-3}), (\frac{1}{2})(19 + \sqrt{-3}),$
$(\frac{1}{2})(-19 - \sqrt{-3})$.

7 (a) $(3, 2)$.   (b) $(2, 1)$.   (c) $(9, 4)$.   (d) $(5, 2)$.   (e) $(8, 3)$.

8 (a) $3 + 2\sqrt{2}, 17 + 12\sqrt{2}, 99 + 70\sqrt{2}$.   (b) $2 + \sqrt{3}, 7 + 4\sqrt{3}, 26 + 15\sqrt{3}$.
(c) $9 + 4\sqrt{5}, 161 + 72\sqrt{5}, 2889 + 1292\sqrt{5}$.   (d) $5 + 2\sqrt{6}, 49 + 20\sqrt{6}, 485 + 198\sqrt{6}$.
(e) $8 + 3\sqrt{7}, 127 + 48\sqrt{7}, 2024 + 765\sqrt{7}$.

SECTION 4.10 (page 181)

1 $14 = (2)(7) = (3 + \sqrt{-5})(3 - \sqrt{-5})$.
2 $q = 1 + 4i, r = -1$. $q = 4i, r = 1 - i$. $q = 1 + 3i, r = 2i$.   3 (b) 45.
4 (a) $3 = (3 + \sqrt{6})(3 - \sqrt{6})$, $5 = (1 + \sqrt{6})(-1 + \sqrt{6})$, $19 = (5 + \sqrt{6})(5 - \sqrt{6})$, 23 is prime.
   (b) $(22 + 9\sqrt{6})/(2 + \sqrt{6}) = 5 + 2\sqrt{6}$, and $N(5 + 2\sqrt{6}) = 1$. So $22 + 9\sqrt{6}$ is an associate of $2 + \sqrt{6}$. Also, $-22 + 9\sqrt{6}$ is an associate of $-2 + \sqrt{6}$.
5 $N(1 + \sqrt{-7}) = 8$, $N(2) = 4$, and $x^2 + 7y^2 = 2$ has no solution. So $1 + \sqrt{-7}$ and 2 are prime in $\mathbf{Z}(\sqrt{-7})$. However, in $\mathbf{Q}[\sqrt{-7}]$, $2 = \left(\frac{1 + \sqrt{-7}}{2}\right)\left(\frac{1 - \sqrt{-7}}{2}\right)$,
   and $1 + \sqrt{-7} = \left(\frac{1 + \sqrt{-7}}{2}\right)^2\left(\frac{1 - \sqrt{-7}}{2}\right)$.
7 $\sqrt{-10}$ is such a prime, as is $2 + \sqrt{-10}$.
9 (a) 71% (or $\frac{22}{31}$) of the real domains are UFD.   (b) 22% (or $\frac{7}{32}$) of the imaginary domains are UFD.
10 In $\mathbf{Q}[\sqrt{-5}]$, $9 = 3 \cdot 3 = (2 + \sqrt{-5})(2 - \sqrt{-5})$, so $3 | (2 + \sqrt{-5})(2 - \sqrt{-5})$, but $3 \nmid 2 + \sqrt{-5}$ and $3 \nmid 2 - \sqrt{-5}$.
13 (a) $\pi^* = \mu\sigma = [(\mu\sigma)^*]^* = (\mu^*\sigma^*)^* \Rightarrow \pi = \mu^* \cdot \sigma^*$.   (b) Theorem 4.42 $\Rightarrow$ There is a unique prime $p \in \mathbf{Z}^+$ such that $\pi | p \Rightarrow \pi\mu = p \Rightarrow N(\pi) \cdot N(\mu) = p^2 \Rightarrow |N(\pi)| = p$ or p^2.

SECTION 5.1 (page 186)

1 (a) 1, 4, 2, 4.   (b) 1, 2, 2, 2.   (c) 1, 3, 3, 2, 2, 2.   (d) 1, 4, 2, 4, 2, 2, 2, 2.
2 The orders of $\bar{0}, \bar{1}, \bar{2}, \ldots, \overline{23}$, respectively, are: 1, 24, 12, 8, 6, 24, 4, 24, 3, 8, 12, 24, 2, 24, 12, 8, 6, 24, 4, 24, 6, 8, 12, 24.
3 (a) 3.   (b) 3.   (c) 2.   (d) 6.   (e) 2.   (f) 4.   (g) 5.   (h) 6.
4 (a) $\bar{1}, \bar{3}, \bar{9}, \bar{4}, \bar{5}$.   (b) $\bar{0}, \bar{3}, \bar{6}, \bar{8}, \bar{5}$.
5 $na = a + a + \cdots + a$, $(-n)a = n(-a) = -(na)$, $(ma) + (na) = (m + n)a$, $(m_1 a) + (m_2 a) + \cdots + (m_n a) = (m_1 + m_2 + \cdots + m_n)a$, $n(ma) = (mn)a$, $n(a + b) = na + nb$.
6 (a) $1, 3, 9, \frac{1}{3}, \frac{1}{9}$.   (b) $1 = 3^0 \in H$, $3^n \cdot 3^m = 3^{n+m} \in H$, $(3^n)^{-1} = 3^{-n} \in H$.
7 $\{0, \pm 3, \pm 6, \ldots\}$.
12 (a) 6, 4, 3, 12, 2.   (b) $\{\bar{0}, \bar{2}, \bar{4}, \bar{6}, \bar{8}, \overline{10}\}$, $\{\bar{0}, \bar{3}, \bar{6}, \bar{9}\}$, $\{\bar{0}, \bar{4}, \bar{8}\}$, $\mathbf{Z}_{12}$, $\{\bar{0}, \bar{6}\}$.
14 Let $|a^{-1}| = m$. Now $e = a^k \Rightarrow e = e^{-1} = (a^k)^{-1} = (a^{-1})^k \Rightarrow m | k$, and $e = (a^{-1})^m = (a^m)^{-1} \Rightarrow e = a^m \Rightarrow k | m$.
15 Hint: Show by induction on t that $a_i \theta^t = a_{i+t}$, and hence that $a_i \theta^k = a_{i+k} = a_i$, for $i = 1, 2, \ldots, k$.
16 Let $|a| = m$, $|a\theta| = k$. Now $(a\theta)^m = a^m\theta = e\theta = e' \Rightarrow k | m$. Also, $a^k\theta = (a\theta)^k = e'$, $e\theta = e' \Rightarrow a^k = e \Rightarrow m | k$. Therefore, $m = k$.

SECTION 5.2 (page 193)

1 (a) $\{1, \sqrt{2}, 1/\sqrt{2}, 2, \frac{1}{2}, 2\sqrt{2}, \ldots\}$.   (b) $\{1, -1\}$.   (c) $\{1, \frac{4}{5}, \frac{5}{4}, \frac{16}{25}, \frac{25}{16}, \ldots\}$.
   (d) $\{1, \pi, 1/\pi, \pi^2, 1/\pi^2, \ldots\}$.
2 (a) $\langle \bar{3} \rangle = \langle \bar{6} \rangle = \langle \bar{9} \rangle = \langle \overline{12} \rangle = \{\bar{0}, \bar{3}, \bar{6}, \bar{9}, \overline{12}\}$, $\langle \bar{5} \rangle = \langle \overline{10} \rangle = \{\bar{0}, \bar{5}, \overline{10}\}$.
   (b) $\langle \bar{2} \rangle = \langle \bar{6} \rangle = \langle \overline{10} \rangle = \langle \overline{14} \rangle = \{\bar{0}, \bar{2}, \bar{4}, \bar{6}, \bar{8}, \overline{10}, \overline{12}, \overline{14}\}$, $\langle \bar{4} \rangle = \langle \overline{12} \rangle = \{\bar{0}, \bar{4}, \bar{8}, \overline{12}\}$, $\langle \bar{8} \rangle = \{\bar{0}, \bar{8}\}$.
   (c) No proper subgroups.
3 (a)   (b)   (c)

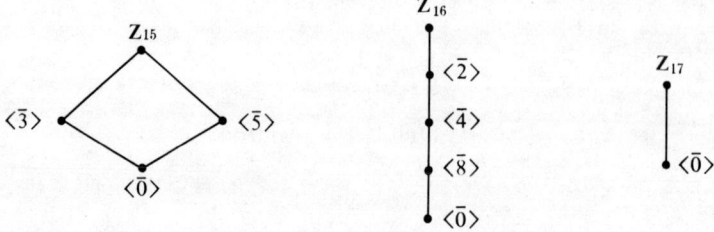

Answers to Selected Problems   349

4 (a) 80.   (b) Hint: If $d\,|\,300$, then $|a^{300/d}| = d$. There are 18 divisors of 300, and hence, 18 subgroups.
7 (a) $\{e, a, a^2\}, \{e, b\}, \{e, ab\}, \{e, a^2b\}$.   (b) $\{e, a, a^2, a^3\}, \{e, a^2\}, \{e, b\}, \{e, ab\}, \{e, a^2b\}, \{e, a^3b\}$.
8 (a) None.   (b) $\{e, a^2, b, a^2b\}, \{e, a^2, ab, a^3b\}$.
10 $\begin{bmatrix} a & b \\ c & d \end{bmatrix} + \begin{bmatrix} a & b \\ c & d \end{bmatrix} = \begin{bmatrix} 0 & 0 \\ 0 & 0 \end{bmatrix}$, for all $a, b, c, d \in \mathbf{Z}_2$. There are 15 subgroups of order 2.
11 (a) $C_4 = \{a^ib^j : |a| = 2, |b| = 4, b^2 = a, ba = ab\}$
      $= \{e, a, b, ab, b^2, ab^2, b^3, ab^3\} = \{e, b, b^2, b^3\}$.
   (b) $C_8 = \{a^ib^j : |a| = 4, |b| = 4, b^2 = a^2, ba = a^3b\}$
      $= \{e, a, a^2, a^3, b, ab, a^2b, a^3b\}$. Now $Q_8 = \langle i, j \rangle$.
   The map, $\theta: Q_8 \to C_8$, defined by: $i\theta = a, j\theta = b$, is an isomorphism, since we can define Q_8 as follows: $Q_8 = \{i^sj^t : |i| = 4, |j| = 4, j^2 = i^2, ji = -ij = i^3j\}$.
15 (a) Let p be a prime divisor of c. Now $c = (a/b)^n$, $n \in \mathbf{Z}^+ \Rightarrow cb^n = a^n \Rightarrow p\,|\,a$. Also, $c = (a/b)^{-n} \Rightarrow p\,|\,b$.   (b) No, by part (a).
16 (a) $w^k = \cos(2\pi k/m) + i\sin(2\pi k/m)$, and $w^m = 1$. So $\langle w \rangle = \{1, w, w^2, \ldots, w^{m-1}\}$.
   (b) $(w^k)^m = (w^m)^k = 1$, for all k.
17 (b)   (c)

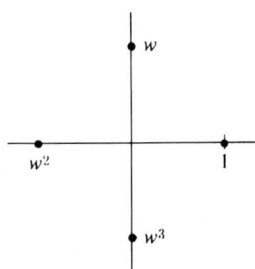

SECTION 5.3 (page 200)

1 (a) $K = \{e, a\}, bK = \{b, c\}, gK = \{g, i\}, jK = \{j, h\}, rK = \{r, u\}, tK = \{t, s\}, K = \{e, a\}$,
   $Kb = \{b, c\}, Kg = \{g, h\}, Kj = \{j, i\}, Kr = \{r, s\}, Kt = \{t, u\}$.
   (d) $N = \{e, a, b, c\}, gN = \{g, i, j, h\} = Ng, rN = \{r, u, s, t\}\} = Nr$.
2

```
                    A_4
                     |
                     L
K_1 ← K_2 ← H_1 ← H_2 → H_3 → K_3 → K_4
                    {e}
```

   $|L| = 4, |H_i| = 2, |K_j| = 3$.
3 Cosets are $\bar{0}, \bar{1}, \bar{2}, \bar{3}, \bar{4}$. Factor set is $\mathbf{Z}_5$.
6 Factor set is $\{\bar{\bar{0}}, \bar{\bar{1}}, \bar{\bar{2}}, \bar{\bar{3}}\}$, where $\bar{\bar{0}} = \{\bar{0}, \bar{4}, \bar{8}, \overline{12}, \overline{16}, \overline{20}\}, \bar{\bar{1}} = \{\bar{1}, \bar{5}, \bar{9}, \overline{13}, \overline{17}, \overline{21}\}$,
   $\bar{\bar{2}} = \{\bar{2}, \bar{6}, \overline{10}, \overline{14}, \overline{18}, \overline{22}\}, \bar{\bar{3}} = \{\bar{3}, \bar{7}, \overline{11}, \overline{15}, \overline{19}, \overline{23}\}$.
7 (b) Coset $(a, b) = (a, b) + H = \{r, b) : r \in \mathbf{R}\} = (0, b) + H$, which is a horizontal line with y-intercept of b.
9 $(mnp)^1 = (mnp), (mnp)^2 = (mpn), (mnp)^3 = (1), [(mn)(pq)]^1 = (mn)(pq), (mnp)(mn)(pq) = (nqp)$,
   $(mnp)^2(mn)(pq) = (mqp), (mnp)(mn)(pq)(mnp) = (nqm)$. Since 7 elements have been generated, we conclude that $\langle (mnp), (mn)(pq) \rangle = A_4$.
10 (a) $\{(123), (12)\}$.   (b) $\{(1234), (12)(34)\}$.   (c) $\left\{\begin{bmatrix} \bar{1} & \bar{0} \\ \bar{0} & \bar{0} \end{bmatrix}, \begin{bmatrix} \bar{0} & \bar{1} \\ \bar{0} & \bar{0} \end{bmatrix}, \begin{bmatrix} \bar{0} & \bar{0} \\ \bar{1} & \bar{0} \end{bmatrix}, \begin{bmatrix} \bar{0} & \bar{0} \\ \bar{0} & \bar{1} \end{bmatrix}\right\}$.
   (d) $\{p_1, p_2, p_3, \ldots\}$, the p_i being the distinct positive primes.

15 (a) G has a subgroup of order k, when $k = 2, 3, 4, 5$, according to a result stated (but to be proved in Section 5.9). (b) If G is abelian, then there is a subgroup of order k, for each k dividing 60.
18 (a) $|a| = p \Rightarrow |\langle a \rangle| = p$; and $|a| = p^2 \Rightarrow |a^p| = p \Rightarrow |\langle a^p \rangle| = p$.

SECTION 5.4 (page 208)

1 (a) No, since $rK = \{r, a, j\}$ and $Kr = \{r, b, h\}$. (b) No.
2 (a) $aH = \{a, a^3, a^5\} = Ha$, $bH = \{b, a^4b, a^2b\} = Hb$, $abH = \{ab, a^5b, a^3b\} = Hab$.
(b) $D_6/H = \{H, Ha, Hb, Hab\} \cong K_4$. Group D_6/H is not simple, since $\{H, aH\} \triangleleft D_6/H$.
4 (a) $H = \{e, a^3\}$, $aH = Ha = \{a, a^4\}$, $a^2H = Ha^2 = \{a^2, a^5\}$, $bH = Hb = \{b, a^3b\}$, $abH = Hab = \{ab, a^4b\}$, $a^2bH = Ha^2b = \{a^2b, a^5b\}$.
(b) $|H| = 1, |aH| = 3, |a^2H| = 3, |bH| = 2, |abH| = 2, |a^2bH| = 2$, so $C_{12}/H \cong S_3$, which is not simple. $\{H, aH, a^2H\} \triangleleft C_{12}/H$.
6 $G = H \cup aH = H \cup Ha$, if $a \notin H$, so $aH = Ha$, and $H \triangleleft G$.
8 (a) $gc = cg$ for all $c \in C$, $g \in G \Rightarrow gC = Cg \Rightarrow C \triangleleft G$. (b) Not necessarily. If $G = D_3$, then $C = \{e\}$, so $G/C \cong G$, which is not abelian.
11 (a) $K < \mathbf{R} \times \mathbf{R}$, and so is normal, since $\mathbf{R} \times \mathbf{R}$ is abelian. (b) $(\mathbf{R} \times \mathbf{R})/K$ is the family of lines in the plane, each of which has a slope of 1.
12 (a) Let $a, b \in \mathbf{Z}$, $(a, b) = 1$, and $0 \le a/b < 1$. Then $a/b + \mathbf{Z} = \{a/b + n : n \in \mathbf{Z}\}$ = the set consisting of all rational numbers that are one unit apart on the number line. (b) If $(a, b) = 1$, then $a/b + \mathbf{Z}$ is of order b. (c) There are infinitely many rationals in $[0, 1)$, and each coset contains exactly one rational number from the interval $[0, 1)$. Hence, $|\mathbf{Q}/\mathbf{Z}| = \infty$. (Note: This is an example of a group of infinite order in which each element is of finite order.)
15 No. For instance, in D_4 (Appendix D, 8.4), $\{e, b\} \triangleleft \{e, a^2, b, a^2b\} \triangleleft D_4$, but $\{e, b\}$ is not a normal subgroup of D_4.
18 Abelian group G is simple $\Leftrightarrow |G| = p$, where p is a prime. (Every group of composite order has at least one proper subgroup.)
21 (a) $\langle \bar{2} \rangle, \langle \bar{3} \rangle, \langle \bar{5} \rangle$. (b) $\mathbf{Z}_{30}/\langle \bar{2} \rangle, \mathbf{Z}_{30}/\langle \bar{3} \rangle, \mathbf{Z}_{30}/\langle \bar{5} \rangle$, of orders 2, 3, and 5, respectively. (c) One series is: $\langle \bar{0} \rangle \triangleleft \langle \overline{10} \rangle \triangleleft \langle \bar{5} \rangle \triangleleft \mathbf{Z}_{30}$.

$$\mathbf{Z}_{30}/\langle \bar{5} \rangle \cong \mathbf{Z}_5, \langle \bar{5} \rangle/\langle \overline{10} \rangle \cong \mathbf{Z}_2, \langle \overline{10} \rangle/\langle \bar{0} \rangle \cong \mathbf{Z}_3$$

22 Composition factors are isomorphic (in some order) to $\mathbf{Z}_2, \mathbf{Z}_2, \mathbf{Z}_2$, and $\mathbf{Z}_3$.
23 $(14)D_4 = \{(14), (234), (1243), (132), (23), (1342), (143), (124)\}$. $(123) = (1234)(14) \in D_4(14)$, but $(123) \notin (14)D_4$. So D_4 is not a normal subgroup of S_4, and S_4/D_4 is not a group.

SECTION 5.5 (page 215)

1 (a) $\bar{0}, \bar{2}, \bar{4}, \bar{6}, \bar{8}, \overline{10}$. (b) $\bar{0}, \bar{1}, \bar{2}, \bar{3}$. (c) $\bar{0}$. (d) $\bar{0}, \bar{3}, \bar{6}, \bar{9}, \overline{12}, \overline{15}$. (e) $\bar{0}$.
2

$\bar{1}\theta =$	$\bar{0}$	$\bar{1}$	$\bar{2}$	$\bar{3}$	$\bar{4}$	$\bar{5}$
$\mathbf{Z}_6 \theta =$	$\{\bar{0}\}$	$\mathbf{Z}_6$	$\{\bar{0}, \bar{2}, \bar{4}\}$	$\{\bar{0}, \bar{3}\}$	$\{\bar{0}, \bar{2}, \bar{4}\}$	$\mathbf{Z}_6$

There are two automorphisms.
3 (a), (c), (e), and (f) are endomorphisms, while (b) and (d) are not.
5 $\theta : \bar{1} \to \bar{a}$, is an endomorphism, for every a, since $(\bar{k} + \bar{m})\theta = \overline{(k + m)}\theta = \overline{(k + m)a} = \overline{ka + ma} = \overline{ka} + \overline{ma} = \bar{k}\theta + \bar{m}\theta$. Then $\bar{1} \to \bar{a}$ is bijective $\Leftrightarrow |\bar{a}| = n \Leftrightarrow (a, n) = 1$, by Thm. 5.11.
8 $\bar{1} \to \bar{a}$, for $a = 0, 1, \ldots, 7$. Surjective if $a = 1, 3, 5, 7$.
10 (a) $\bar{1} \to \bar{a}$, for $a = 1, 5, 7, 11$. $A_{\mathbf{Z}_{12}} \cong K_4$. (b) $\bar{1} \to \bar{a}$, for $a = 1, 2, \ldots, 12$. $A_{\mathbf{Z}_{13}} \cong \mathbf{Z}_{12}$.
(c) $\bar{1} \to \bar{a}$, for $a = 1, 3, 5, 9, 11, 13$. $A_{\mathbf{Z}_{14}} \cong \mathbf{Z}_6$.
12 (a)

Order of automorphism:	1	2	3	4
No. of auto. of that order:	1	9	8	6

A_{Q_8} is of order 24.
(b) Yes, since the orders of elements of A_{Q_8} are the same as the orders of elements of S_4.
13 Hint: Since $S_3 = \langle (123), (12) \rangle$, compute $g(123)g^{-1}$ and $g(12)g^{-1}$, for each $g \in S_3$.
17 Using the notation of Appendix D, 8.4: If $A = \{e, b\}$, $B = \{e, a^2b\}$, $C = \{e, a^2\}$, $H = \{e, ab\}$, $K = \{e, a^3b\}$, $S = \{e, a^2, b, a^2b\}$, $T = \{e, a^2, ab, a^3b\}$, then A and B are conjugates, H and K are conjugates, while each other subgroup is normal, and so has no conjugate other than itself.

Answers to Selected Problems 351

19 (See group table of Example 5.12.) $A_4 = \langle a, g \rangle$. The inner automorphism group of A_4 is isomorphic to A_4. The three subgroups of order 2 are conjugate, and the four subgroups of order 3 are conjugate. The subgroup of order 4 is normal.

SECTION 5.6 (page 220)

1 If $D_3 = \{a^i b^j : |a| = 3, |b| = 2, ba = a^2 b\}$, then $D_3 \times \mathbf{Z}_2$ has these elements:

Element:	$(e, \bar{0})$	$(a, \bar{0})$	$(a^2, \bar{0})$	$(b, \bar{0})$	$(ab, \bar{0})$	$(a^2 b, \bar{0})$
Order:	1	3	3	2	2	2

Element:	$(e, \bar{1})$	$(a, \bar{1})$	$(a^2, \bar{1})$	$(b, \bar{1})$	$(ab, \bar{1})$	$(a^2 b, \bar{1})$
Order:	2	6	6	2	2	2

$D_3 \cong \langle (a, \bar{0}), (b, \bar{0}) \rangle$, $\mathbf{Z}_6 \cong \langle (a, \bar{1}) \rangle$, $\mathbf{Z}_3 \cong \langle (a, \bar{0}) \rangle$, $K_4 \cong \langle (e, \bar{1}), (b, \bar{0}) \rangle \cong \langle (e, \bar{1}), (ab, \bar{0}) \rangle \cong \langle (e, \bar{1}), (a^2 b, \bar{0}) \rangle$. There are 7 subgroups isomorphic to $\mathbf{Z}_2$.

2 (a) There are 7 subgroups of order 2, and 4 subgroups of order 4. (b) $\mathbf{Z}_4 \cong \langle (\bar{1}, \bar{0}) \rangle \cong \langle (\bar{1}, \bar{1}) \rangle$, $K_4 \cong \langle (\bar{2}, \bar{0}), (\bar{0}, \bar{1}) \rangle$.

5 $\mathbf{Z}_2 \times \mathbf{Z}_6 = \{(a, b) : a \in \mathbf{Z}_2, b \in \mathbf{Z}_6\}$. $\mathbf{Z}_3 \times \mathbf{Z}_2 \times \mathbf{Z}_2 = \{(a, b, c) : a \in \mathbf{Z}_3, b \in \mathbf{Z}_2, c \in \mathbf{Z}_2\}$. The orders of the elements in each group are: 1, 2, 2, 2, 3, 3, 6, 6, 6, 6, 6, 6. By Thm. 5.29 and 5.30:

$$\mathbf{Z}_3 \times \mathbf{Z}_2 \times \mathbf{Z}_2 \cong (\mathbf{Z}_3 \times \mathbf{Z}_2) \times \mathbf{Z}_2 \cong \mathbf{Z}_6 \times \mathbf{Z}_2 \cong \mathbf{Z}_2 \times \mathbf{Z}_6.$$

6 $G \times H \cong \mathbf{Z}_6$, so elements are of orders 1, 2, 3, 3, 6, 6.

8 Hint: Show that these maps are isomorphisms: (a) $(g_i, h_j) \to (h_j, g_i)$, where $g_i \in G$, $h_j \in H$. (b) $(g_i, h_j, k_u) \to ((g_i, h_j), k_u)$.

9 True, since $H \cong H < G$, where H is any proper subgroup of G.

10 $(\bar{1}, \bar{1})$ is an element of greatest order in each group. Its order is: (a) 12, (b) 24, (c) 12, (d) 6.

11 (a) $\mathbf{Z}_8, \mathbf{Z}_2 \times \mathbf{Z}_4, \mathbf{Z}_2 \times \mathbf{Z}_2 \times \mathbf{Z}_2$. (b) $\mathbf{Z}_9, \mathbf{Z}_3 \times \mathbf{Z}_3$. (c) $\mathbf{Z}_{10}$. (d) $\mathbf{Z}_{12}, \mathbf{Z}_2 \times \mathbf{Z}_6$. (e) $\mathbf{Z}_{16}, \mathbf{Z}_2 \times \mathbf{Z}_8, \mathbf{Z}_2 \times \mathbf{Z}_2 \times \mathbf{Z}_4, \mathbf{Z}_4 \times \mathbf{Z}_4, \mathbf{Z}_2 \times \mathbf{Z}_2 \times \mathbf{Z}_2 \times \mathbf{Z}_2$.

12 (a) $(\bar{1}, \bar{0}) \to a, (\bar{0}, \bar{1}) \to b$. (b) $(\bar{1}, \bar{0}, \bar{0}) \to a, (\bar{0}, \bar{1}, \bar{0}) \to b, (\bar{0}, \bar{0}, \bar{1}) \to c$.

14 $(a, \bar{0}) \to (\bar{1}, \bar{0}, \bar{0}), (b, \bar{0}) \to (\bar{0}, \bar{1}, \bar{0}), (e, \bar{1}) \to (\bar{0}, \bar{0}, \bar{1})$.

15 (b) π_1 is the projection of the Cartesian plane onto the x-axis, and π_2 is its projection onto the y-axis. Under π_1 each point on the vertical line L maps to the point of intersection of L and the x-axis.

16 Hint: If e_1 and e_2 are the identities of H and K, respectively, let $H' = \{(h, e_2) : h \in H\}$, and show that $G/H' = \{(e_1, k)H' : k \in K\}$. Also, prove that $H \cong H'$, and that the map, $k + H' \to k$, is an isomorphism.

17 (a) Hint: Show that the map, $h \odot k \to (h, k)$, is an isomorphism of G onto $H \times K$.

SECTION 5.7 (page 226)

1 (a) $a \to \bar{1}, a \to \bar{5}, a \to \bar{7}, a \to \overline{11}$. (b) By symmetry and transitivity of $\cong$ we obtain: $\bar{1} \to \bar{1}$, $\bar{1} \to \bar{5}, \bar{1} \to \bar{7}, \bar{1} \to \overline{11}$.

3 $8 : \mathbf{Z}_2 \times \mathbf{Z}_2 \times \mathbf{Z}_2, \mathbf{Z}_2 \times \mathbf{Z}_4, \mathbf{Z}_8$
   $16 : \mathbf{Z}_2 \times \mathbf{Z}_2 \times \mathbf{Z}_2 \times \mathbf{Z}_2, \mathbf{Z}_2 \times \mathbf{Z}_2 \times \mathbf{Z}_4, \mathbf{Z}_2 \times \mathbf{Z}_8, \mathbf{Z}_4 \times \mathbf{Z}_4, \mathbf{Z}_{16}$.
   $18 : \mathbf{Z}_3 \times \mathbf{Z}_6, \mathbf{Z}_{18}$.

4 There are 11 abelian groups of order 64, and 7 of order 160.

5

Subgroup Order	Subgroups of These Groups		
	$\mathbf{Z}_{40}$	$\mathbf{Z}_2 \times \mathbf{Z}_{20}$	$\mathbf{Z}_2 \times \mathbf{Z}_2 \times \mathbf{Z}_{10}$
2	$\langle \overline{20} \rangle$	$\langle (\bar{1}, \bar{0}) \rangle$	$\langle \bar{1}, \bar{0}, \bar{0}) \rangle$
4	$\langle \overline{10} \rangle$	$\langle (\bar{0}, \bar{5}) \rangle$	$\langle (\bar{1}, \bar{0}, \bar{0}), (\bar{0}, \bar{1}, \bar{0}) \rangle$
5	$\langle \bar{8} \rangle$	$\langle (\bar{0}, \bar{4}) \rangle$	$\langle (\bar{0}, \bar{0}, \bar{2}) \rangle$
8	$\langle \bar{5} \rangle$	$\langle (\bar{1}, \bar{0}), (\bar{0}, \bar{5}) \rangle$	$\langle (\bar{1}, \bar{0}, \bar{0}), (\bar{0}, \bar{1}, \bar{0}), (\bar{0}, \bar{0}, \bar{5}) \rangle$
10	$\langle \bar{4} \rangle$	$\langle (\bar{0}, \bar{2}) \rangle$	$\langle (\bar{0}, \bar{0}, \bar{1}) \rangle$
20	$\langle \bar{2} \rangle$	$\langle (\bar{0}, \bar{1}) \rangle$	$\langle (\bar{0}, \bar{1}, \bar{0}), (\bar{0}, \bar{0}, \bar{1}) \rangle$

7

$$\begin{array}{r|l}
\mathbf{Z}_8: & 1, 2, 4, 4, 8, 8, 8, 8 \\
\mathbf{Q}_8: & 1, 2, 4, 4, 4, 4, 4, 4 \\
\mathbf{Z}_2 \times \mathbf{Z}_4: & 1, 2, 2, 2, 4, 4, 4, 4 \\
D_4: & 1, 2, 2, 2, 2, 2, 4, 4 \\
\mathbf{Z}_2 \times \mathbf{Z}_2 \times \mathbf{Z}_2: & 1, 2, 2, 2, 2, 2, 2, 2
\end{array}$$

8

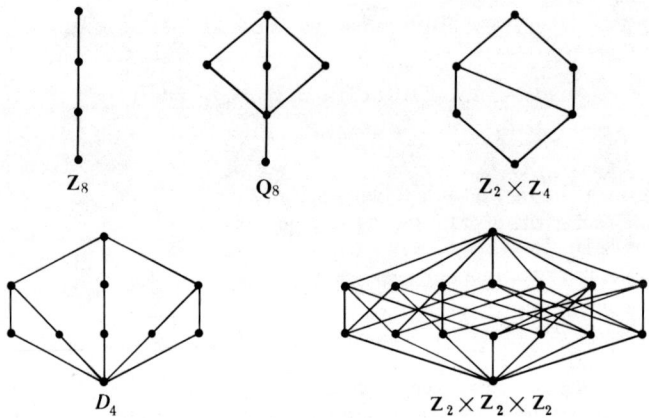

9 (a) $\mathbf{Z}_3 \times \mathbf{Z}_3$: 1 3 3 3 3 3 3 3 3, $\mathbf{Z}_9$: 1 3 3 9 9 9 9 9 9.
(b)

$\mathbf{Z}_3 \times \mathbf{Z}_3$ $\qquad$ $\mathbf{Z}_9$

10

$$\begin{array}{r|llllllllllll}
\mathbf{Z}_{12}: & 1 & 2 & 3 & 3 & 4 & 4 & 6 & 6 & 12 & 12 & 12 & 12 \\
\mathbf{Z}_2 \times \mathbf{Z}_6: & 1 & 2 & 2 & 2 & 3 & 3 & 6 & 6 & 6 & 6 & 6 & 6 \\
C_{12}: & 1 & 2 & 3 & 3 & 4 & 4 & 4 & 4 & 4 & 4 & 6 & 6 \\
D_6: & 1 & 2 & 2 & 2 & 2 & 2 & 2 & 2 & 3 & 3 & 6 & 6 \\
A_4: & 1 & 2 & 2 & 2 & 3 & 3 & 3 & 3 & 3 & 3 & 3 & 3
\end{array}$$

It is impossible to set up a bijective map from one group to another such that each element and its image are of the same order.

11 D_6. $\qquad$ **12** 53.

13 $\quad$ (a) $\mathbf{Z}_4 \to S_4$ $\qquad$ (d) $D_4 \to S_8$ (Notation of Appendix D-8.4)
$\bar{0} \to (\bar{0})$ $\qquad\qquad\qquad e \to (e)$
$\bar{1} \to (\bar{0}, \bar{1}, \bar{2}, \bar{3})$ $\qquad\quad\; a \to (e, a, a^2, a^3)$
$\bar{2} \to (\bar{0}, \bar{2})(\bar{1}, \bar{3})$ $\qquad\; a^2 \to (e, a^2)(a, a^3)(b, a^2b)(ab, a^3b)$
$\bar{3} \to (\bar{0}, \bar{3}, \bar{2}, \bar{1})$. $\qquad a^3 \to (e, a^3, a^2, a)(b, ab, a^2b, a^3b)$
$\qquad\qquad\qquad\qquad\qquad b \to (e, b)(a, ab)(a^2, a^2b)(a^3, a^3b)$
$\qquad\qquad\qquad\qquad\;\; ab \to (e, ab)(a, a^2b)(a^2, a^3b)(a^3, b)$
$\qquad\qquad\qquad\qquad a^2b \to (e, a^2b)(a, a^3b)(a^2, b)(a^3, ab)$
$\qquad\qquad\qquad\qquad a^3b \to (e, a^3b)(a, b)(a^2, ab)(a^3, a^2b).$

14 (a) $\{(n, n + 3): n \in \mathbf{Z}\}$. $\quad$ (b) $\{(x, 3x): x \in \mathbf{Q}^*\}$.

15 $\begin{bmatrix} a & b \\ c & d \end{bmatrix} \to (a, b, c, d)$ is an isomorphism under addition. $\qquad$ **18** 6, 10, 4, 6, 12, 2.

22 True. (Why?)

23 $a \to \lambda_{a^{-1}} = \lambda_{a^3} = (e, a^3, a^2, a)(b, a^3b, a^2b, ab)$
$\qquad b \to \lambda_{b^{-1}} = \lambda_b = (e, b)(a, a^3b)(a^2, a^2b)(a^3, ab)$
(The other images can be obtained from the images of a and b, with the aid of Appendix D-8.4.)

Answers to Selected Problems   353

SECTION 5.8 (page 233)

1 (a) $4 \to \bar{1}$.   (b) $\bar{1} \to \bar{1}$.

2

	Morphism, $\mathbf{Z}_n \to \mathbf{Z}_m$	Kernel K	$\mathbf{Z}_n/K$ iso. to:	$\mathbf{Z}_n/K$ simple?
(b)	$\bar{1} \to \bar{0}$	$\mathbf{Z}_8$	$\{\bar{0}\}$	Yes
	$\bar{1} \to \bar{1}$, or $\bar{1} \to \bar{3}$	$\{\bar{0}, \bar{4}\}$	$\mathbf{Z}_4$	No
	$\bar{1} \to \bar{2}$	$\{\bar{0}, \bar{2}, \bar{4}, \bar{6}\}$	$\mathbf{Z}_2$	Yes
(d)	$\bar{1} \to \bar{0}$	$\mathbf{Z}_{12}$	$\{\bar{0}\}$	Yes
	$\bar{1} \to \bar{3}$, or $\bar{1} \to \overline{15}$	$\{\bar{0}, \bar{6}\}$	$\mathbf{Z}_6$	No
	$\bar{1} \to \bar{6}$, or $\bar{1} \to \overline{12}$	$\{\bar{0}, \bar{3}, \bar{6}, \bar{9}\}$	$\mathbf{Z}_3$	Yes
	$\bar{1} \to \bar{9}$	$\{\bar{0}, \bar{2}, \bar{4}, \bar{6}, \bar{8}, \overline{10}\}$	$\mathbf{Z}_2$	Yes

3 (b) $K = A_n$. $G/K = \{A_n, B_n\}$, with A_n being the set of even permutations, and B_n the set of odd permutations.

4 (a) $[(a, b) + (c, d)]\theta = (a + c, b + d)\theta = b + d = (a, b)\theta + (c, d)\theta$.

6 (a) $G/H = \{H, (\bar{1}, \bar{0}) + H, (\bar{0}, \bar{1}) + H, (\bar{1}, \bar{1}) + H\} = \{H, A, B, C\} \cong K_4$.
    $G/L = \{L, (\bar{0}, \bar{1}) + L, (\bar{1}, \bar{2}) + L, (\bar{1}, \bar{3}) + L\} = \{L, T, T^2, T^3\} \cong \mathbf{Z}_4$.
   (b) There are 6 morphisms of G onto G/H, and 4 morphisms of G onto G/L.
   (c) No. Part (a) is a counterexample, since $H \cong L$, but G/H is not isomorphic to G/L.

7 (a) $|L| = 24$, $|G| = 168$.   (b) Group A of order 2 can always be inserted between E and K_4. If L has a normal subgroup B of order 8 that contains K_4, then B can be inserted between K_4 and L.

SECTION 5.9 (page 238)

1 (a) Let $M = \{e, a, a^2, a^3, a^4, a^5\}$, $H = \{e, a^3, b, a^3b\}$,
    $K = \{e, a^3, ab, a^4b\}$, $L = \{e, a^3, a^2b, a^5b\}$.

$g =$	e	a	a^2	a^3	a^4	a^5	b	ab	a^2b	a^3b	a^4b	a^5b
$N_g =$	D_6	M	M	D_6	M	M	H	K	L	H	K	L

   (b) $D_6 = \{e\} \cup \{a^3\} \cup \{a, a^5\} \cup \{a^2, a^4\} \cup \{b, a^2b, a^4b\} \cup \{ab, a^3b, a^5b\}$.
   (c) $12 = 1 + 1 + 2 + 2 + 3 + 3$

2 $\mathbf{Z}_{12}$: $\{e, a^4, a^8\}$, $\{e, a^3, a^6, a^9\}$.
   $\mathbf{Z}_6 \times \mathbf{Z}_2$: $\{e, a^2, a^4\}$, $\{e, a^3, b, a^3b\}$.
   D_6: $\{e, a^2, a^4\}$, $H = \{e, a^3, b, a^3b\}$, $K = \{e, a^3, ab, a^4b\}$, $L = \{e, a^3, a^2b, a^5b\}$.
   C_{12}: $\{e, a^2, a^4\}$, $H = \{e, a^3, b, a^3b\}$, $K = \{e, a^3, ab, a^4b\}$, $L = \{e, a^3, a^2b, a^5b\}$.
   A_4: $\{e, a, b, ab\}$, $H = \{e, c, c^2\}$, $K = \{e, ac, abc^2\}$, $L = \{e, bc, ac^2\}$, $M = \{e, abc, bc^2\}$.

3 Using the notation of Prob. 2:
   D_6: $a^2Ha^4 = K$, $aHa^5 = L$. C_{12}: $a^2Ha^4 = K$, $aHa^5 = L$. A_4: $bHb = K$, $abHab = L$, $aHa = M$.

SECTION 6.1 (page 244)

1
Ideal:	$\langle \bar{2} \rangle$	$\langle \bar{3} \rangle$	$\langle \bar{4} \rangle$	$\langle \bar{6} \rangle$
Unity:	None	$\bar{9}$	$\bar{4}$	None

3 (b) $N = \left\langle \begin{bmatrix} 2 & 0 \\ 0 & 2 \end{bmatrix} \right\rangle$, so N is principal.

4 Note: T is isomorphic to the complex field. (See Prob. 2.6–13 and 2.10–14.)

5 (c) There are $n - 1$ other ideals that are isomorphic to N.

6 (a) $\mathbf{Z} \times \mathbf{Z}$ is a ring, by Prob. 5. Unity is $(1, 1)$. Also, $(2, 0)$ is a zero divisor, since $(x, 0) \cdot (0, 1) = (0, 0)$.
   (b) $\langle (1, 0) \rangle$, $\langle (3, 3) \rangle$, $\langle (3, 1), (5, 2) \rangle$. (Note that the second ideal is contained in the third ideal.)

7 (a)

+	z	u	a	b		·	z	u	a	b
z	z	u	a	b		z	z	z	z	z
u	u	z	b	a		u	z	u	a	b
a	a	b	z	u		a	z	a	a	z
b	b	a	u	z		b	z	b	z	b

(b) If $S = \{z, u\}$, $T = \{z, a\}$, $V = \{z, b\}$, then T and V are ideals, but subring S is not.

8 (a) $z = (\bar{0}, \bar{0}, \bar{0})$, $u = (\bar{1}, \bar{1}, \bar{1})$, $a = (\bar{1}, \bar{0}, \bar{0})$, $b = (\bar{0}, \bar{1}, \bar{0})$, $c = (\bar{0}, \bar{0}, \bar{1})$, $d = (\bar{1}, \bar{1}, \bar{0})$, $e = (\bar{1}, \bar{0}, \bar{1})$, $f = (\bar{0}, \bar{1}, \bar{1})$.

(b) There are seven subrings of order 2. These three are ideals:

$$\{z, a\}, \{z, b\}, \{z, c\}.$$

There are seven subrings of order 4. These three are ideals:

$$\{z, a, b, d\}, \{z, a, c, e\}, \{z, b, c, f\}.$$

9 $3 + \sqrt{6}$.

SECTION 6.2 (page 249)

1

+	z	a	b	c		·	z	a	b	c
z	z	a	b	c		z	z	z	z	z
a	a	b	c	z		a	z	a	b	c
b	b	c	z	a		b	z	b	z	b
c	c	z	a	b		c	z	c	b	a

$z = \{\bar{0}, \bar{4}, \bar{8}\}$, $a = \{\bar{1}, \bar{5}, \bar{9}\}$, $b = \{\bar{2}, \bar{6}, \overline{10}\}$, $c = \{\bar{3}, \bar{7}, \overline{11}\}$. $\{z, b\}$ is an ideal of $\mathbf{Z}_{12}/\langle \bar{4}\rangle$.

3 (b) If $S = \mathbf{Z}_2 \times \mathbf{Z}_2 \times \mathbf{Z}_4$, $N = \{(\bar{0},\bar{0},\bar{0}),(\bar{1},\bar{0},\bar{0}),(\bar{0},\bar{1},\bar{0}),(\bar{1},\bar{1},\bar{0})\} = z$, $(\bar{0},\bar{0},\bar{1}) + N = a$, $(\bar{0},\bar{0},\bar{2}) + N = b$, $(\bar{0},\bar{0},\bar{3}) + N = c$, then the operation tables of S/N are the same as those in Prob. 1. That is, S/N is isomorphic to $\mathbf{Z}_{12}/\langle \bar{4}\rangle$.

4 (b) Let $z = N = \{(\bar{0},\bar{0},\bar{0}),(\bar{0},\bar{0},\bar{1}),(\bar{0},\bar{0},\bar{2}),(\bar{0},\bar{0},\bar{3})\}$, $a = (\bar{1},\bar{1},\bar{0}) + N$, $b = (\bar{0},\bar{1},\bar{0}) + N$, $c = (\bar{1},\bar{0},\bar{0}) + N$.

+	z	a	b	c		·	z	a	b	c
z	z	a	b	c		z	z	z	z	z
a	a	z	c	b		a	z	a	b	c
b	b	c	z	a		b	z	b	b	z
c	c	b	a	z		c	z	c	z	c

6 No. (Let $S = \mathbf{Z}$.)   7 $S/S = \{S\} \cong \{z\}$. $S/\{z\} = \{\{a\} : a \in S\} \cong S$.

8 $\langle 6\rangle \subset \langle 2\rangle, \langle 3\rangle$. $\langle 9\rangle \subset \langle 3\rangle$. $\langle 30\rangle \subset \langle 2\rangle, \langle 3\rangle, \langle 5\rangle$. $\langle 210\rangle \subset \langle 2\rangle, \langle 3\rangle, \langle 5\rangle, \langle 7\rangle$. $\langle 1024\rangle \subset \langle 2\rangle$.

10 $M = \mathbf{Z}_4 \times \mathbf{Z}_5 \times \{\bar{0}\} \Rightarrow S/M \cong \mathbf{Z}_7$. $M = \mathbf{Z}_4 \times \{\bar{0}\} \times \mathbf{Z}_7 \Rightarrow S/M \cong \mathbf{Z}_5$. $M = \langle \bar{2}\rangle \times \mathbf{Z}_5 \times \mathbf{Z}_7 \Rightarrow S/M \cong \mathbf{Z}_2$.

11 (a) $g(x) = (x^2 - 3)(x^2 + 1)$. N is not maximal, since $N = \langle g(x)\rangle \subset \langle x^2 + 1\rangle \subset \mathbf{Q}[x]$. $((x^2 - 3) + N)((x^2 + 1) + N) = g(x) + N = N$, and $(x^2 - 3) + N \neq N$, $(x^2 + 1) + N \neq N$.

(b) $\langle g(x)\rangle$ is a maximal ideal of $F[x]$ iff $g(x)$ is prime over F. If $g(x)$ is reducible over F, then $F[x]/\langle g(x)\rangle$ has zero divisors.

12 (a) Apply Thm. 4.13, 4.24.   (b) Show that if M is not maximal, then $m(x)$ is factorable.

(c) Use the Division Theorem on $f(x)$ and $m(x)$ to obtain $r(x)$.   (e) $(\frac{1}{4} - x/8) + M$.

13 (a) $m(x)$ is prime over $\mathbf{Z}_3$.

(b) $\mathbf{Z}_3[x]/M = \{M, \bar{1} + M, \bar{2} + M, x + M, (\bar{1} + x) + M, (\bar{2} + x) + M, \bar{2}x + M, (\bar{1} + \bar{2}x) + M, (\bar{2} + \bar{2}x) + M\}$.

(c)

Element:	$\bar{1} + M$	$\bar{2} + M$	$x + M$	$(\bar{1} + x) + M$	$\bar{2}x + M$
Inverse:	$\bar{1} + M$	$\bar{2} + M$	$(\bar{2} + x) + M$	$(\bar{2} + \bar{2}x) + M$	$(\bar{1} + \bar{2}x) + M$

SECTION 6.3 (page 255)

1 (a) $(m + n)\theta = \overline{5(m+n)} = \overline{5m} + \overline{5n} = m\theta + n\theta$. $(mn)\theta = \overline{5mn}$, but $(m\theta)(n\theta) = \overline{mn}$.

(b) Yes. $[(m\lambda)(n\lambda) = \overline{3m} \cdot \overline{3n} = \overline{9mn} = \overline{3mn} = (mn)\lambda.]$

2 $\bar{1} \to \bar{a}$ is an endomorphism iff $a^2 \equiv a \pmod{12}$. So $a = 0, 1, 4, 9$. The only automorphism is the identity automorphism.

3 (a) $\bar{1} \to \bar{a}$ is a group morphism iff $a = 0, 3, 6, 9$, and is a ring morphism iff $a = 0, 9$.

(b) $\bar{1} \to \bar{a}$ is a group morphism iff $a = 0, 2, 4, 6$, and is a ring morphism iff $a = 0$.

Answers to Selected Problems   355

5    $\bar{1} \to \bar{a}$, where a is: (a) 0, 1, 3, 4,   (b) 0, 1, 5, 6,   (c) 0, 1.
     (a) $1 \to a \Rightarrow 1^2 \to a^2 \Rightarrow a^2 = a \Rightarrow a = 0$ or $1 \Rightarrow$ identity map is the only automorphism.
     (b) $1 \to \bar{a}$ is an endomorphism iff $a^2 \equiv a \pmod{p}$, by Prob. 4. $a^2 \equiv a \pmod{p} \Leftrightarrow p \,|\, a(a-1) \Rightarrow p \,|\, a$ or $p \,|\, a - 1 \Rightarrow \bar{a} = \bar{0}$ or $\bar{a} = \bar{1}$.
8   (a) $1 \to 0$ and $1 \to 1$ are the only endomorphisms.   (b) No.   (c) $(a,b) \to 0, (a,b) \to a$, $(a,b) \to b$.
9   It is readily verified that θ is an isomorphism, if
$$\begin{bmatrix} a & 0 \\ b & c \end{bmatrix} \theta = (a, b, c).$$
10   One of the four possible answers: (a) $\bar{1} \to \bar{4}$,   (b) $K = \{\bar{0}, \bar{3}, \bar{6}, \bar{9}\}$,
     (c) $\mathbf{Z}_{12}/K = \{K, \bar{1} + K, \bar{2} + K\}$.   (d) $\bar{0}, \bar{3}, \bar{6}, \bar{9}$ map to K; $\bar{1}, \bar{4}, \bar{7}, \overline{10}$ map to $\bar{1} + K$; $\bar{2}, \bar{5}, \bar{8}, \overline{11}$ map to $\bar{2} + K$,   (e) $K\lambda = \bar{0}, (\bar{1} + K)\lambda = \bar{4}, (\bar{2} + K)\lambda = \bar{2}$.
15   Hint: Use Problem 13. Also, show that $\sqrt{n} \to c + d\sqrt{n} \Rightarrow c = 0, d = \pm 1$.
16   Hint: Use Thm. 5.36(b) on morphism $\theta: (F, +) \to (F, +)$.

SECTION 6.4   *(page 259)*

2   (a) $n(ab) = (-k)(ab) = -[k(ab)] = -[a(kb)] = a[-(kb)] = a[(-k)b] = a(nb)$.
     (b) $(ma)(nb) = m[a(nb)] = m[n(ab)] = (mn)(ab)$, the last equality being true by Thm. 5.3(a).
3   0, 0, 5.   4   0, 0, 15, 12, $[m, n]$.   5   ch$(S) = 2$. Yes, if ch$(S) = 3$.
8    $m = 4, 8, 9, 16, 25, 27, 32, 49, 64, 81$.   10   (b) $\mathbf{Z} \cong \{(a, 0) : a \in \mathbf{Z}\}$.
11   (a) $z = (\bar{0}, \bar{0}), u = (\bar{1}, \bar{0})$. $(a, b)^{-1} = (2a, b)$ if $ab \neq \bar{0}$, and $(a,b)^{-1} = (a, 2b)$ if $ab = \bar{0}$.
     (b) $\mathbf{Z}_3 \cong \{(\bar{0}, \bar{0}), (\bar{1}, \bar{0}), (\bar{2}, \bar{0})\}$.

SECTION 6.5   *(page 267)*

1   (a) $F = \{r + si : r, s \in \mathbf{Q}\} = \mathbf{Q}(i)$.   (b) $F = D$.   (c) $F = \{r + s\sqrt{2} : r, s \in \mathbf{Q}\} = \mathbf{Q}(\sqrt{2})$.
3   (a) $F = \{f(x)/g(x) : f(x), g(x) \in \mathbf{Z}[x], g(x) \neq 0\}$.
$$\mathbf{Q}[x] = \left\{\frac{a_0}{b_0} + \frac{a_1}{b_1}x + \cdots + \frac{a_n}{b_n}x^n : a_i \in \mathbf{Z}, b_i \in \mathbf{Z}^*, n = 0, 1, 2, \ldots\right\}$$
$$= \left\{\frac{c_0 + c_1 x + \cdots + c_n x^n}{b} : c_i \in \mathbf{Z}, b \in \mathbf{Z}^*\right\} \subset F.$$
9    $\mathbf{R} \cong \mathbf{R}' = \left\{\begin{bmatrix} a & 0 \\ 0 & a \end{bmatrix} : a \in \mathbf{R}\right\}$.
10   (a) $\langle 2 \rangle \subset \mathbf{Z}, 2\mathbf{Z} \cong \{(2n, 0) : n \in \mathbf{Z}\} \subset 2\mathbf{Z} \times \mathbf{Z}$.   (b) $2\mathbf{Z} \times \mathbf{Z}$ has zero divisors, so is not isomorphic to $\mathbf{Z}$.
15   (a) $(2 - 3i, i), (0, 2 + 5i)$.   (b) $2 - 3i + 2j + 6k, -2j - 5k, 2 - 3i - 2j - 4k$.
     (c) $2/\sqrt{14} + (3/\sqrt{14})i - (1/\sqrt{14})k, (-2/\sqrt{29})j - (5/\sqrt{29})k$.   (d) $-5 - 2i + 19j + 4k$, $-5 + 2i - 11j + 16k$.

SECTION 6.6   *(page 274)*

1   (a) $-\bar{1} = \bar{2}, -r = \bar{2}r, -(\bar{1} + r) = \bar{2} + \bar{2}r, -(\bar{2} + r) = \bar{1} + \bar{2}r$.
     (b) $\bar{1}^{-1} = \bar{1}, \bar{2}^{-1} = \bar{2}, r^{-1} = \bar{2}r, (\bar{1} + r)^{-1} = \bar{2} + r, (\bar{1} + \bar{2}r)^{-1} = \bar{2} + \bar{2}r$.   (c) $\mathbf{Z}_3 \times \mathbf{Z}_3$.
     (d) $|r| = 4, |\bar{1} + \bar{2}r| = 8$.   (e) $(\mathbf{Z}_8, +)$.   (f) 3.
     (g) $x, x - \bar{1}, x - \bar{2}, x^2 + \bar{1} = (x - r)(x - \bar{2}r), x^2 + x + \bar{2} = [x - (\bar{1} + r)][x - (\bar{1} + \bar{2}r)]$, $x^2 + \bar{2}x + \bar{2} = [x - (\bar{2} + r)][x - (\bar{2} + \bar{2}r)]$.
2   (a) 0.   (b) $-5 + 4r - 2r^2, (65 + 16r - 6r^2)/269$.   (c) $\theta: a_0 + a_1 r + a_2 r^2 \to (a_0, a_1, a_2)$ is an isomorphism under $+$.   (d) No.

4 (a) $E_1 = \{\bar{0}, \bar{1}, r, r^2, \bar{1} + r, r^2, \bar{1} + r^2, r + r^2, \bar{1} + r + r^2\}$. $x^3 + x^2 + \bar{1} = (x - r)(x - r^2)[x - (\bar{1} + r + r^2)]$.
   (Note that $x^3 + x^2 + \bar{1}$ has no roots in $\mathbb{Z}_2$, but has 3 roots in extension field E_1.)
   (b) $r, r^2, r^3 = \bar{1} + r^2, r^4 = \bar{1} + r + r^2, r^5 = \bar{1} + r, r^6 = r + r^2, r^7 = \bar{1}$. (c) $\bar{1} + r^2$.
   (d) Orders: 1, 2, 2, 2, 2, 2, 2, 2. $(E_1, +) \cong (\mathbb{Z}_2 \times \mathbb{Z}_2 \times \mathbb{Z}_2, +)$.
   (e)

Element:	$\bar{0}$	$\bar{1}$	$r, r^2, \bar{1} + r + r^2$	$\bar{1} + r, \bar{1} + r^2, r + r^2$
Min. Poly.:	x	$x + \bar{1}$	$x^3 + x^2 + \bar{1}$	$x^3 + x + \bar{1}$

5 (a) $E = \mathbb{Z}_2[x]/\langle x^3 + x^2 + \bar{1}\rangle$ is a field of order 8, and so must be $GF(2^3)$, since that is the only field of order 8.   (b) Automorphisms are defined by the image of r, since $E_1^* = \langle r \rangle$. The image of r must be a root of $x^3 + x^2 + 1$, so the automorphism group is $\{\theta_1, \theta_2, \theta_3\}$, with $r\theta_1 = r$, $r\theta_2 = r^2$, $r\theta_3 = 1 + r + r^2$.

7 $r \to \bar{1} + s, r^2 \to \bar{1} + s^2, \bar{1} + r^2 \to s^2, \bar{1} + r + r^2 \to \bar{1} + s + s^2, \bar{1} + r \to s, r + r^2 \to s + s^2, \bar{1} \to \bar{1}, \bar{0} \to \bar{0}$.

8 $(a + br)(c + dr) = (ac - bd) + (ad + bc - bd)r$, in each field.

9 $E_1 = \mathbb{Q}[x]/\langle x^3 + 5x - 2\rangle = \{a + br + cr^2 : a, b, c \in \mathbb{Q}, r^3 = 2 - 5r\}$,
   $E_2 = \mathbb{Q}[x]/\langle x^2 - 4x + 1\rangle = \{a + bs : a, b \in \mathbb{Q}, s^2 = -1 + s\}$.

10 $E = \mathbb{R}[x]/\langle x^2 + 1\rangle = \{a + br : a, b \in \mathbb{R}, r^2 = -1\}$. The map, $\theta: \mathbb{C} \to E$, given by $(a + bi)\theta = a + br$, is an isomorphism.   12 $F[x]/\langle p(x)\rangle \cong F$.

13 $\overline{g(x)} = \langle g(x)\rangle = N$, the zero of the factor ring. If $g(x) = f(x) \cdot h(x)$, where $\deg f \geq 1$, $\deg h \geq 1$, then $\overline{f(x)} \cdot \overline{h(x)} = N$, but $\overline{f(x)} \neq N$, and $\overline{h(x)} \neq N$.

16 $GF(2^2) = \mathbb{Z}_2/\langle x^2 + x + \bar{1}\rangle = \{a + br : a, b \in \mathbb{Z}_2, r^2 = \bar{1} + r\} = \{\bar{0}, \bar{1}, r, \bar{1} + r\}$.
   $(GF(2^2), +) \cong K_4$, and $(GF(2^2)^*, \cdot) \cong \mathbb{Z}_3$.

17 (a) $GF(5^2) = \mathbb{Z}_5[x]/\langle x^2 + x + \bar{1}\rangle = \{a + br : a, b \in \mathbb{Z}_5, r^2 = -\bar{1} - r\}$.
   (c) $GF(2^5) = \mathbb{Z}_2[x]/\langle x^5 + x + \bar{1}\rangle = \{a_0 + a_1r + a_2r^2 + a_3r^3 + a_4r^4 : a_i \in \mathbb{Z}_2, r^5 = \bar{1} + r\}$.

18 (a) $\mathbb{Z}_{243}, \mathbb{Z}_3 \times \mathbb{Z}_{81}, \mathbb{Z}_9 \times \mathbb{Z}_{27}, \mathbb{Z}_3 \times \mathbb{Z}_3 \times \mathbb{Z}_{27}, \mathbb{Z}_3 \times \mathbb{Z}_9 \times \mathbb{Z}_9, \mathbb{Z}_3 \times \mathbb{Z}_3 \times \mathbb{Z}_3 \times \mathbb{Z}_9$,
   $\mathbb{Z}_3 \times \mathbb{Z}_3 \times \mathbb{Z}_3 \times \mathbb{Z}_3 \times \mathbb{Z}_3$.   (b) 3.   (c) $\mathbb{Z}_3 \times \mathbb{Z}_3 \times \mathbb{Z}_3 \times \mathbb{Z}_3 \times \mathbb{Z}_3$.
   (d) $GF(p^n) \cong \{(a_1, a_2, \ldots, a_n) : a_i \in \mathbb{Z}_p\} = \mathbb{Z}_p \times \mathbb{Z}_p \times \cdots \times \mathbb{Z}_p$.

SECTION 6.7   (page 282)

1
	Root Fields	Splitting Field E	$A(E/\mathbb{Q})$
(a)	$\mathbb{R}, \mathbb{C}$	$\mathbb{Q}(\sqrt{7})$	S_2
(b)	$\mathbb{R}, \mathbb{C}$	$\mathbb{Q}(\sqrt{7})$	S_2
(c)	$\mathbb{Q}(\sqrt{n}), n = 3, -2$	$\mathbb{Q}$	S_1
(d)	$\mathbb{Q}(\sqrt{3}, \sqrt{-5}), \mathbb{R}$	$\mathbb{Q}(\sqrt{3})$	S_2

2 $E = \mathbb{Z}_7(r) = \{a_0 + a_1r + a_2r^2 : a_i \in \mathbb{Z}_7, r^3 = \bar{4}\}$. $|E| = 7^3 = 343$. $x^3 + \bar{3} = (x - r)(x - \bar{2}r)(x - \bar{4}r)$. Each automorphism is determined by the image of r. $\mathbb{Z}_3 \cong A(E/\mathbb{Z}_7) = \{\varepsilon, \alpha, \beta\}$, where $r\varepsilon = r$, $r\alpha = 2r$, and $r\beta = 4r$.

3 $E = \mathbb{Q}(\sqrt{3}, \sqrt{-3}) = \{a_1 + a_2\sqrt{3} + a_3\sqrt{-3} + a_4\sqrt{-1} : a_i \in \mathbb{Q}\}$, obtained as follows:
   $f(x) = (x^2 - 3)(x^2 + 3)$, and $E_1 = \mathbb{Q}[x]/\langle x^2 - 3\rangle = \mathbb{Q}(\sqrt{3}) = \{a + b\sqrt{3} : a, b \in \mathbb{Q}\}$. Then $E = E_1[x]/\langle x^2 + 3\rangle = E_1(\sqrt{-3}) = \{c + d\sqrt{-3} : c, d \in E_1\}$. The automorphisms of E that fix $\mathbb{Q}$ are the following:

   $\varepsilon: \sqrt{3} \to \sqrt{3}$ $\alpha: \sqrt{3} \to \sqrt{3}$ $\beta: \sqrt{3} \to -\sqrt{3}$ $\mu: \sqrt{3} \to -\sqrt{3}$
   $\sqrt{-3} \to \sqrt{-3}$ $\sqrt{-3} \to -\sqrt{-3}$ $\sqrt{-3} \to \sqrt{-3}$ $\sqrt{-3} \to -\sqrt{-3}$

   Galois group is the Klein four-group, K_4.

4 $x^4 + 4x^2 + 2 = (x - r)(x + r)(x - 3r - r^3)(x + 3r + r^3)$. $E = \mathbb{Q}(r)$. Automorphisms of E:

   $\varepsilon: r \to r$, $\mu: r \to -r$, $\rho: r \to 3r + r^3$, $\nu: r \to -3r + r^3$.

   $\rho^2 = \mu, \rho^3 = \nu, \rho^4 = \varepsilon$. So $A(E/\mathbb{Q})$ is isomorphic to $\mathbb{Z}_4$.

5 $p(x) = x^3 + 3x^2 + 4$,
   $E_1 = \mathbb{Q}[x]/\langle p(x)\rangle = \{a_0 + a_1r + a_2r^2 : a_i \in \mathbb{Q}, r^3 = -4 - 3r^2\}$,
   $p(x) = (x - r)(x^2 + (3 + r)x + (3r + r^2)) = (x - r) \cdot q(x)$,
   $E = E_1[x]/\langle q(x)\rangle = \{a + bs : a, b \in E_1, s^2 = -(3r + r^2) - (3 + r)s\} = \{a_0 + a_1r + a_2r^2 + b_0s + b_1rs + b_2r^2s : a_i, b_i \in \mathbb{Q}\}$.
   $p(x) = (x - r)(x - s)(x - (-3 - r - s))$. The roots of $p(x)$ are: $r, s,$ and $-3 - r - s$. Each one of the 3! permutations of the 3 roots determines an automorphism of E. So $A(E/\mathbb{Q}) \cong S_3$.

Answers to Selected Problems   357

6 Roots of $x^4 - 3$ are: $\sqrt[4]{3}, -\sqrt[4]{3}, \sqrt[4]{3}\cdot i, -\sqrt[4]{3}\cdot i$. The automorphisms are determined as follows:

$\varepsilon: \quad \sqrt[4]{3} \to \sqrt[4]{3} \qquad \theta_1: \quad \sqrt[4]{3} \to \sqrt[4]{3} \qquad \theta_2: \quad \sqrt[4]{3} \to -\sqrt[4]{3} \qquad \theta_3: \quad \sqrt[4]{3} \to -\sqrt[4]{3}$
$ i\sqrt[4]{3} \to i\sqrt[4]{3} i\sqrt[4]{3} \to -i\sqrt[4]{3} i\sqrt[4]{3} \to i\sqrt[4]{3} i\sqrt[4]{3} \to -i\sqrt[4]{3}$

$\theta_4: \quad \sqrt[4]{3} \to i\sqrt[4]{3} \qquad \theta_5: \quad \sqrt[4]{3} \to i\sqrt[4]{3} \qquad \theta_6: \quad \sqrt[4]{3} \to -i\sqrt[4]{3} \qquad \theta_7: \quad \sqrt[4]{3} \to -i\sqrt[4]{3}$
$ i\sqrt[4]{3} \to \sqrt[4]{3} i\sqrt[4]{3} \to -\sqrt[4]{3} i\sqrt[4]{3} \to \sqrt[4]{3} i\sqrt[4]{3} \to -\sqrt[4]{3}$

Then $A(E/\mathbf{Q}) = \{\varepsilon, \theta_1, \theta_2, \ldots, \theta_7\}$. Since $|\theta_5| = |\theta_6| = 4$, and $|\theta_i| = 2$ for all other i, $A(E/\mathbf{Q})$ is isomorphic to D_4.

11 $E_0 = \mathbf{Q}, 7 \in E_0$.
$E_1 = E_0(\sqrt{s_0}), s_0 = 7, 2 - \sqrt{7} \in E_1$.
$E_2 = E_1(\sqrt[3]{s_1}), s_1 = 2 - \sqrt{7}, \sqrt[3]{2} - \sqrt{7} \in E_2$.
$E_3 = E_2(\sqrt{s_2}), s_2 = 5, 3\sqrt{5} + \sqrt[3]{2} - \sqrt{7} \in E_3$.
$E = E_4 = E_3(\sqrt[4]{s_3}), s_3 = 6, 3\sqrt{5} + \sqrt[3]{2} - \sqrt{7} - \sqrt[4]{6} \in E_4$.

SECTION 6.8   (page 290)

1 $v_1 = \sqrt[3]{4}, v_2 = \omega\sqrt[3]{4}, v_3 = \omega^2\sqrt[3]{4}, r_i = v_i - \dfrac{2}{v_i} - 1$, for $i = 1, 2, 3$.
$r_1 = \sqrt[3]{4} - 2/\sqrt[3]{4} - 1 = \sqrt[3]{4} - \sqrt[3]{2} - 1 \approx -0.67252$.
Imaginary roots: $(\tfrac{1}{2})(-\sqrt[3]{4} + \sqrt[3]{2} - 2) \pm (\sqrt{3}/2)(\sqrt[3]{4} + \sqrt[3]{2})i \approx -1.1637 \pm 2.4659i$.

2 $t = (\tfrac{1}{3})\sqrt[3]{10 + 6\sqrt{3}}, r_1 = t - \tfrac{2}{9t} + \tfrac{1}{3}$.

$r_1 = 1 \Leftrightarrow t - \dfrac{2}{9t} + \tfrac{1}{3} = 1 \Leftrightarrow 9t^2 - 6t - 2 = 0 \Leftrightarrow t = \dfrac{1 + \sqrt{3}}{3}$.

$\sqrt[3]{10 + 6\sqrt{3}} = 1 + \sqrt{3} \Leftrightarrow 10 + 6\sqrt{3} = (1 + \sqrt{3})^3$.

4 $\mathbf{Q} < E_1 < E$,
$E_1 = \mathbf{Q}(\sqrt{2}) = \mathbf{Q}(r) = \{a + br : a, b \in \mathbf{Q}, r^2 = 2\}$,
$E = E_1(\sqrt[3]{s}) = E_1(t) = \{c + dt + et^2 : c, d, e \in E_1, t^3 = s = -3 + 2r\}$.
So $\mathbf{Q}(r, t) = \{a_1 + a_2 r + a_3 t + a_4 rt + a_5 t^2 + a_6 rt^2 : a_i \in \mathbf{Q}\}$.

List of Symbols

$p \Rightarrow q$	2, 294	$\mathbf{C}, \mathbf{C}*$	9
iff	2	$S \times T$	11
$p \Leftrightarrow q$	2, 295	$a \rho b$	12
$a \in S$	4	$a \sim b$	12
$b \notin S$	4	$\binom{n}{k}$	18
$\{a, b, \ldots\}$	4		
$\{a_1, a_2, \ldots, a_n\}$	4	$\bar{a}$	19
$|A|$	4	$\mu: S \to T$	22
$\emptyset$	4	$\mu: s \to t$	22
$\{x: p(x)\}$	5	$s\mu$	22
$\square$	5	$S\mu$	23
$S \subseteq T$	5	$\lambda \circ \mu$	28
$T \supseteq S$	5	$a_1 \otimes a_2 \otimes \cdots \otimes a_n$	35
$S \subset T$	6	$a^k, k \in Z^+$	35
$T \supset S$	6	$(S, \otimes)$	38
$P(S)$	6	$(S, \otimes, \star)$	38
$A \cup B$	6	e	39
$A \cap B$	6	a'	39
$A - B$	7	$(G, \otimes)$	46
$A + B$	7, 242	$H < G$	52
		$H \leq G$	52, 240
$\bigcup_{i=1}^{n} A_i$	8	N_a	52, 234
		C	53
$\bigcup_{i \in I} A_i$	8	Q_8	54
		K_4	54, 62
		S_A	55
$\bigcap_{i=1}^{n} A_i$	8	S_n	56
		$(a_1, a_2, \ldots, a_n)$	56, 100
$\bigcap_{i \in I} A_i$	8	D_n	62
(a, b)	8, 11, 100	A_n	60
$[a, b]$	8, 104	z	67
$[a, b)$	8	$-a$	67
$\mathbf{Z}, \mathbf{Z}^+$	9	u	67
$\mathbf{Q}, \mathbf{Q}^+, \mathbf{Q}*$	9	$M_2(S)$	68
$\mathbf{R}, \mathbf{R}^+, \mathbf{R}*$	9	a^{-1}	69, 183

List of Symbols

$(F, \oplus, \otimes)$	69
$(F^*, \otimes)$	69
$\|a\|$	81, 185
$a \mid c$	83
$a \nmid b$	83
$\mathbf{Z}(\sqrt{n}\,)$	84
$(a + b\sqrt{n}\,)^*$	84
$N(\lambda)$	84
UFD	86
$S \cong T$	90
g.c.d.	100
l.c.m.	104
$[a_1, a_2, \ldots, a_n]$	105
$a \equiv b \pmod{n}$	117
$\mathbf{Z}_n$	118
$\mathbf{Z}/<n>$	118
$\varnothing(n)$	125
U_n	127
$S[x]$	130
$F[x]$	135
$\dfrac{F[x]}{<m(x)>}$	143
$(a + bi)^*$	150
$F(r)$	161
$\mathbf{Q}[\sqrt{n}\,]$	169
a^0	183
$<a>$	188, 242
$<a_1, a_2, \ldots, a_s>$	192, 242
C_{4m}	194
gH	195
$g \otimes H$	195
Hg	196
$H \otimes g$	196
$a \equiv b \pmod{H}$	197
$[G:H]$	199
AB	202
$H \triangleleft G$	203
G/H	204
A_G	212
$i_g : G \to G$	214
I_G	214
$G_1 \times G_2 \times \cdots \times G_m$	217
C_a	234
PIR	242
PID	242
$(S/N, +, \cdot)$	246
$A(E/F)$	256, 278
$F[x]/<p(x)>$	269
$GF(p^n)$	274
$\sim p$	293
$p \wedge q$	293
$p \vee q$	293
$p \to q$	293
$p \leftrightarrow q$	293
$\forall x$	297
$\exists x$	298

Index

Abel, Neils Henrick, 50, 276, 284, 322
   theorem of, 285
Abelian group, 46
Absolute value, 81
Abstract algebra, 38
Abstract group, 221
Additive group, 69
Additive inverse, 67
Algebra, 38
   fundamental theorem of, 150
Algebraic extension field, 162, 270
   degree of, 164
Algebraic integer, 164
Algebraic number, 158
   over a field, 158
Algebraic number field, 162
Algorithm, 102
Alternating group, 60
Antecedent, 22
Antisymmetric relation, 13
Ascending chain condition for ideals, 245
Associate, 83
Associative property, 39
   generalized, 112
Automorphism, 211
   inner, 214
   ring, 251
   group, 211
Automorphisms that fix a subfield, 256, 278

Basic morphisms of field theory, 145
Basis for a vector space, 167
Bell, Eric Temple, 38
Biconditional, 293
Bijective map, 24
Binary operation, 33

Binomial theorem, 113
Boole, George, 66, 73, 325
Boolean algebra, 66
Boolean ring, 77

Cancellation laws, 44, 74
Cardan, Girolamo, 1, 283, 284, 291, 317
Cardan's cubic formula, 284
Cartesian product, 11, 217
Cases, proof by, 297
Cauchy, Augustin-Louis, 273, 276, 321
Cauchy's theorem for abelian groups, 234
Cayley, Arthur, 46, 228, 291, 325
   theorem of, 225
Cayley table, 46
Center of a group, 53
Chain of subgroups, 47
Characteristic of a ring, 257
Class, 4
Class equation of a group, 235
Classical algebra, 1
Closed interval, 8
Closed subset, 34
Codomain, 22
Coefficient field, 276
Coefficients of a polynomial, 129
Collection, 4
Combinations, 18
Common divisor, 100
Common multiple, 104, 105
Commutative
   diagram, 29
   group, 46
   property, 39
   ring, 67
   ring with unity, 67
Complex field, 69, 265

Complex numbers, 9, 266
Composite
  elements, 83
  maps, 28
  polynomials, 135
  statements, 293
Composition
  factors of a group, 206
  of maps, 37, 55
  series of a group, 206
Conditional, 293
Conditional proof, 297
Congruence modulo n, 117
  linear, 119
  quadratic, 119
Congruence modulo a
  polynomial, 143
  subgroup, 197
Conjugacy classes of a group, 234
Conjugate
  complex numbers, 84, 150
  elements of a group, 234
  numbers over a field, 161
  subgroups, 216, 238
Conjunction, 293
Connectives, 293
Constructible lengths, 287
Contradiction, proof by, 296
Contrapositive, 3, 295
Converse, 3, 295
Cosets of a subgroup, 195
  product of, 202
Cubic
  extension of a field, 162
  field, 162
Cycle, 56
Cyclic subgroup, 188

D'Alembert, Jean Le Rond, 240
Davenport, H., 179
Dedekind, Richard, 244, 245, 327
Dedekind cuts, 265
Degree of
  algebraic number field, 164
  element algebraic over a field, 159
  polynomial, 130
De Moivre's theorem, 112
Detachment, rule of, 296
Determinant function, 89

Dicyclic group, 194
Dihedral group of degree n, 62
Diophantine equation, 101
  linear, 107
Diophantus, 123, 316
Direct inference, 295
Direct product, 11
  of groups, 217
Disjoint
  collection of sets, 17
  cycles, 57
  sets, 6
Disjunction, 293
Distributive laws, 66
  generalized, 108
Division
  in a field, 76
  ring, 78, 266
Division theorem
  for polynomials, 137
  for rational integers, 97
Divisors, 83
Domain, 22
Doubling a cube, 288

Eisenstein, Ferdinand, 158, 326
  irreducibility criterion of, 156
Element, 4
Elimination of cases, proof by, 297
Embedding
  in a group, 217
  in a ring, 261
Empty set, 4
Endomorphism, 211
  ring, 251
  group, 211
Equality, 5
Equivalence class, 19
Equivalence relation, 12
Equivalent statements, 2
Euclid, 108, 316
Euclidean algorithm
  for polynomials, 142
  for rational integers, 103
Euclidean quadratic domain, 176
Euler, Leonhard, 202, 318
  theorem of, 201
Euler's phi-function, 125
Even permutation, 59

Existential quantifier, 298
Exponents, 183
  laws of, 183
Extension
  group, 217
  field, 135
  ring, 261
Extension by radicals, 282
Extracting a root, 283

Factorable polynomial, 135
Factor
  group, 125, 204
  field, 248
  ring, 125, 247
  set, 20
  set of **Z** modulo n, 118
  theorem, 146
Fermat, Pierre de, 202, 317
  theorem of, 201
  last theorem of, 318
Ferrari, 284
Field, 69
  algebraic number, 162
  coefficient, 276
  complex, 69
  cubic, 162
  cubic extension of, 162
  extension, 135
  quadratic, 162
  quadratic extension of, 162
  rational, 69
  real, 69
  root, 277
  splitting, 277
Finite group, 46
Finite set, 4
Finitely generated
  group, 192
  ideal, 242
Fontenelle, Bernard de, 1
Function, 22
  left-hand notation, 22
  right-hand notation, 22
Function ring, 73
Fundamental partition theorem, 20
Fundamental theorem
  of algebra, 150
  of arithmetic, 115

  on group morphisms, 232
  on ring morphisms, 253

Galois, Évariste, 1, 46, 50, 276, 323
Galois
  group, 279
  field, 274
  theory, 276
Gauss, Carl Friedrich, 158, 181, 320
  lemma of, 155
Gaussian
  integers, 181
  primes, 181
Generalized
  associative property, 112
  distributive property, 108
Generators
  of groups, 188, 192
  of ideals, 242
Goldstein, Larry Joel, 129
Graph of a relation on **R**, 11
Greatest common divisor
  of polynomials, 140
  of rational integers, 100
Group, 46
  abelian, 46
  abstract, 221
  alternating, 60
  center of, 53
  class equation of, 235
  commutative, 46
  composition factors of, 206
  composition series of, 206
  conjugacy classes of, 234
  conjugate elements of, 234
  cyclic, 188
  dihedral, 62
  embedded, 217
  extension of, 217
  finite, 46
  finitely generated, 192
  infinite, 46
  Klein four-, 54, 62
  left regular representation of, 228
  nonabelian, 46
  noncyclic, 192
  octic, 63
  quaternion, 54
  representation of, 221

Group (continued)
   right regular representation of, 226
   simple, 206
   solvable, 206
   special, 221
   symmetric, 56
Group morphism, 89, 209
   kernel of, 228
Groupoid, 46
Group postulates, 46, 54

Hamilton, William Rowan, 266, 268, 323
Hardy, Godfrey H., 291, 292, 329
Homomorphism, 88
Huntington, E. V., 55

Ideal, 241
   finitely generated, 242
   maximal, 247
   principal, 242
   proper, 243
   trivial, 243
Identity element, 39
   left, 42
   right, 42
Identity
   map, 26
   permutation, 26
Image, 22
Imaginary
   cubic field, 164
   quadratic domain, 172
   quadratic field, 169
Implication, 2, 293, 295
Inclusion diagram, 14
Indexing set, 8
Index of a subgroup, 199
Induction on n, 109
Induction principle, 109, 111, 305
Inequality, 80
Infinite group, 46
Infinite set, 4
Injective map, 24
Inner automorphism, 214
Integers, 3
   rational, 9
   irrational, 164
Integral domain, 69
   ordered, 78

Integral power of a group element, 183
Intersection of sets, 6, 8
Interval, 8
   open, 8
   closed, 8
Inverse element, 39
   additive, 67
   left, 43
   right, 43
Irrational integers, 164
Irreducible polynomial, 135, 136
Isomorphic map, 89
Isomorphism, 89
   group, 90, 210
   ring, 251

Jourdain, Philip, 291

Kasner, Edward, 291
Kernel
   of a group morphism, 228
   of a ring morphism, 252
Klein, Felix, 2, 66, 328
Klein four-group, 54, 62
Kronecker, Leopold, 95, 269, 276, 326
   theorem of, 273

Lagrange, Joseph-Louis, 202, 319
   theorem of, 198
Latin square, 48
Leading coefficient, 140
Least common multiple, 104, 105
Left
   cancellation property, 44
   distributive law, 66
   identity, 42
   inverse, 43
Lemma, 102
Linear
   algebra, 69
   combination, 100
   congruence, 119
Logic, 293
   mathematical, 299
   symbolic, 299
   two-valued, 2
Logical equivalence, 2, 295
Logical implication, 2, 294
Loop, 47

Map, 22
  bijective, 24
  injective, 24
  one-to-one, 24
  onto, 23
  surjective, 23
Mapping, 22
Mathematical induction, 109
Mathematical logic, 299
Matrices, two-by-two, 68
  addition of, 68
  equality of, 68
  multiplication of, 68
Maximal ideal, 247
Maximal normal subgroup, 206
Minimal polynomial, 159
Modern algebra, 1
Monic polynomial, 140
Monoid, 47
Morphic map, 88
Morphism, 88
  group, 209
  ring, 251
Multiple root, 149
Multiples, 83
Multiplication table, 46
Multiplicative group, 69
Multiplicative inverse, 69

Natural map, 229, 253
Natural numbers, 9, 305
Necessary conditions, 2
Negated implication, 295
Negative elements, 78
Newman, James R., 183, 291
Noether, Amalie Emmy, 1, 245, 246, 330
Noetherian ring, 245
Nonabelian group, 46
Normalizer of a group element, 53
Normal subgroup, 203
Norm of a complex number, 84, 89
Norm of an algebraic number, 167
Null set, 4
Numbers, theory of, 95

Octic group, 63
Odd permutation, 59
One-to-one correspondence, 24, 198
One-to-one map, 24

Onto map, 23
Open interval, 8
Operation, 33
  binary, 33
  ternary, 37
Operation-preserving map, 88
Operation table, 34
Ordered
  field, 79
  integral domain, 78
  pair, 11
Order of a group element, 185
Order of a set, 4

Pairwise disjoint sets, 17
Parity, 59
  opposite, 59
  same, 59
Partially ordered set, 14
Partial order relation, 14
Partition, 17
Peano, Guiseppe, 113, 328
Peano's postulates, 69, 113, 305, 328
Permutation, 26
  distance-preserving, 61
  even, 59
  odd, 59
Polynomial
  domain, 132
  ring, 130
  solvable by radicals, 283
Polynomial over a ring, 129
  coefficients of, 129
  degree of, 130
Polynomials over a field, 135
  composite, 135
  division theorem for, 137
  Euclidean algorithm for, 142
  factorable, 135
  greatest common divisor of, 140
  irreducible, 135, 136
  minimal, 159
  monic, 140
  prime, 135
  reducible, 135, 136
  roots of, 144
  zeros of, 144
Positive elements, 78
Power set, 6

Premise, 2
Prime
  elements, 83
  polynomials, 135
  statements, 293
Primitive polynomial, 155
Principal ideal, 242
  domain, 242
  ring, 242
Projections of a direct product group, 221
Proper
  divisor, 83
  ideal, 243
  subgroup, 47
  subset, 6

Quadratic
  congruence, 119
  extension of a field, 162
  field, 162
  unique factorization domains, 179
Quantified statements, 297
Quasigroup, 47
Quaternion, 78
  group, 54
  ring, 78, 268
Quotient field, 263
Quotient ring, 247

Range, 23
Rational
  field, 69
  integers, 9
  numbers, 9
  points, 286
Real
  cubic field, 164
  field, 69
  numbers, 9
  quadratic domain, 173
  quadratic field, 169
Reducible polynomial, 135, 136
Reflexive property, 12
Regular polygon, 62
Relation, 11
  antisymmetric, 13
  equivalence, 12
  graph of, 11
  partial order, 14

Relative complement, 7
Relatively prime integers, 100
Representation of a group, 221
Representation theorem for
  groups, 226
  rings, 256
Residue class ring, 247
Right
  cancellation property, 44
  distributive law, 66
  identity, 42
  inverse, 43
Ring, 66
  commutative, 67
  commutative, with unity, 67
  division, 78, 266
  factor, 247
  function, 73
  nontrivial, 69
  principal ideal, 242
  quaternion, 78
  quotient, 247
  residue class, 247
  trivial, 69
  with unity, 67
Ring morphism, 89, 251
  kernel of, 252
Ring postulates, 66
Root field, 277
Root of a polynomial, 126, 144
  multiple, 149
Roots of unity, 195

Scalar multiplication, 167
Scalars, 167
Semigroup, 47
Sequences, 265
Set, 4
  empty, 4
  finite, 4
  infinite, 4
  null, 4
  order of, 4
  partially ordered, 14
  power, 6
  universal, 5
Set multiplication, 202
Simple group, 206
Solvable group, 206

Special group, 221
Splitting field, 277
Square-free rational integer, 78, 84
Standard form of a positive integer, 115
Stark, Harold M., 179, 182, 332
Statement, 293
  composite, 293
  prime, 293
  quantified, 297
  valid, 294
Subdomain, 240
Subfield, 135, 240
Subgroup, 46
  conjugates of, 216
  cyclic, 188
  left coset of, 195
  maximal normal, 206
  normal, 203
  proper, 47
  right coset of, 196
Subring, 240
  proper, 240
Subset, 5
  closed, 34
  proper, 6
Substitution principle, 5
Subtraction in a ring, 74
Sufficient conditions, 2
Surjective map, 23
Sylow p-subgroup, 237
Sylow theorem, 237, 238
Symbolic logic, 299
Symmetric
  difference, 7
  group of degree n, 56
  property, 12
Symmetries of a polygon, 61
Synthetic division, 138, 139

Tartaglia, 317
Tautology, 294
Ternary operation, 37
Transcendental number, 158
  over a field, 158
Transformation, 22
Transitive property, 12, 80
Transposition, 58
Trisecting an angle, 288
Trivial ideals, 243
Trivial ring, 69
Truth table, 293
Truth value, 293
Two-valued logic, 2

Union of sets, 6, 8
Unique factorization domain, 86
Unit, 75
Unity, 67
Universal quantifier, 297
Universal set, 5

Valid statement, 294
Vectors, 167
Vector space, 167
Venn, John, 6
Venn diagram, 6

Well-defined operation, 123
Well-ordered set, 96
Well-ordering principle, 96
Whitehead, Alfred North, 291
Wilson, John, 128, 320
Wilson's theorem, 129

Zero, 67
  of a polynomial, 144
  divisor, 69